AF824044

SOTHIS PRESS

Sothis Press aims to completely reimagine the public domain book with an emphasis on universal access to captivatingly designed print editions. We believe that technology will enable every single book ever published to be perpetually accessible in beautiful print form. Sothis is beauty in ubiquity.

www.sothis.press
1 Waterhouse Square, London, ECIN 2ST

Water Supply of South Africa and Facilities for the Storage of It

John Croumbie Brown

SOTHIS PRESS

ISBN 978-1-0365-1665-9

Water Supply of South Africa and Facilities for the Storage of It is in the public domain.

This edition is published by Sothis Press © 2023. The company is registered in England and Wales. The Company Number is 14900389.

WATER SUPPLY OF SOUTH AFRICA.

Water Supply of South Africa

AND

FACILITIES FOR THE STORAGE OF IT.

COMPILED BY

JOHN CROUMBIE BROWN, LL.D.,

Formerly Government Botanist at the Cape of Good Hope, and Professor of Botany in the South African College, Capetown, Fellow of the Royal Geographical Society, Fellow of the Linnean Society, and Honorary Vice-President of the African Institute of Paris, etc.

EDINBURGH:
OLIVER & BOYD, TWEEDDALE COURT.
LONDON: SIMPKIN, MARSHALL, & CO.

1877.

Kirkcaldy: CRAWFORD & WALKER, *Printers, High Street.*

PREFACE.

Appended to the Report of the Colonial Botanist at the Cape of Good Hope for 1866 was an abstract of a Memoir prepared on the Hydrology of South Africa, which has since been embodied in a volume which has been published on that subject, and an abstract of a Memoir prepared on Irrigation and its application to agricultural operations in South Africa, which embraced a Report on the Water Supply of the Colony—its sources, its quantity, the modes of irrigation required in different circumstances, the facilities for the adoption of these in different districts, and the difficulties, physical and other, in the way of works of extensive irrigation being carried out there, and the means of accomplishing these which are at command.

In the following volume is embodied that portion of the Memoir which related to the Water Supply, and the existing facilities for the storage of this, with reports relative to this which were subsequently received, and similar information in regard to lands beyond the Colony of the Cape of Good Hope, which it has been sought to connect with the Colony by federation, or otherwise; and the information relative to irrigation has been transferred to a Report on the Rivers of the Colony, and the means of controlling floods, of preventing inundations, of regulating the flow of rivers, and utilising the water by irrigation or otherwise.

In the series of volumes to which this belongs its place is immediately after that on the "Hydrology of South Africa," which contains details of the former hydrographic condition of the Cape of Good Hope, and of causes of its present aridity, with suggestions of appropriate remedies for this aridity. And it has been prepared to show that, not in a vague and general use of the terms, but in strict accordance with the statement, the severe, protracted, and extensive droughts, and destructive floods and inundations, recorded in the former volume, find their counterpart in constantly alternating droughts and deluges in every district of the Colony,—and that, in every so-called division of it, notwithstanding the deluges, there were protracted sufferings from drought, and, notwithstanding the aridity, there was a supply of water at command, with existing facilities for the storage of the superabundant supply which at present proves productive of more evil than good.

Occasion has been taken to show that the statements are advanced in full view of all that relates to the case; occasion has been taken to show that crime, and more especially sheep-stealing, increases with drought, and decreases with an abundant or moderate water supply; and when occasion presented itself for supplying information in regard to the local character of the coloured population this has been done, in the knowledge that this is an important element in the estimate to be formed of the expediency or inexpediency of adopting expensive measures to store up the water supply in some important districts.

HADDINGTON, *1st October, 1877.*

CONTENTS AND ARGUMENT.

INDEX TO AUTHORITIES CITED.

WATER SUPPLY OF SOUTH AFRICA.

INTRODUCTION.

In a volume entitled "Hydrology of South Africa; or, Details of the former hydrographic condition of the Cape of Good Hope, and of causes of its present aridity, with suggestions of appropriate remedies for this aridity,"* I have, in treating of the former hydrographic condition of South Africa, cited testimony in regard to this, supplied by the physical geography of South Africa, and testimony in regard to the former geographical condition of South Africa supplied by geological observations, and indications of the former hydrographic condition of the country supplied by arborescent productions found in the interior of South Africa. And in regard to the hydrographic condition of the country within the historic period, I have cited observations made by Moffat, by Livingston, and by Chapman, and conclusions drawn by Fox Wilson in regard to the progress of desiccation; and shown that, in regard to the former hydrographic condition of the country, we have indications of a progressive desiccation having gone on continuously from the last emergence of the land from beneath the ocean up to the present day.

In regard to the cause or occasion of the desiccation of South Africa, I have shown, I consider, that the primary and principal cause of the desiccation and consequent aridity of South Africa has been the upheaval of the Continent and the consequent flowing off of the water and of the subsequent rainfall, and that the secondary cause of the desiccation and consequent aridity of South Africa has been evaporation, latterly promoted by the destruction of forests, herbage, and grass, removing what for centuries had in some measure checked the evaporation which was going on.

* Henry S. King & Co., London. 1875.

And in view of this it is stated that appropriate remedies are the erection of dams to prevent the escape of a portion of the rainfall to the sea; the abandonment or restriction of the burning of the veldt; the conservation and extension of existing forests; and the adoption of measures similar to the *reboisement* and *gazonnement* carried out in France, with a view to prevent the formation of torrents and the destruction of property occasioned by them.

It is with the first of these alone that we have here to do. The aridity is great, but it is not absolute. In the volume cited are given several accounts of deluges, alternating with droughts, which have occurred since 1863, with details of deluges and floods committing devastations at Port Elizabeth and other parts of the Colony in 1867, at Natal in 1868, at Capetown and the adjacent districts in 1869, at Beaufort in 1870, at Victoria West in 1871, at Capetown in 1872, and in various parts of the Colony at the breaking up of the drought in 1874.

That there is water to be had, if it be worth while to incur the expense of constructions to retain and utilise it, is beyond question. And in thinking of the expense which might thus be incurred, there may be taken into account the expense which is entailed by allowing the rainfall to rush unchecked to the sea. The damage to roads and to house property in Port Elizabeth alone, by the torrent of 1867, was estimated at the time at from £25,000 to £30,000. The damage done to public works alone at Natal, by the torrent of the year following (1868), was estimated at £50,000; while the loss to private persons was estimated variously from £50,000 to £100,000. In the following year, a torrent in the Western Province occasioned the fall of a railway bridge, which issued in loss of life, and in loss of property, and in personal injuries, —for one case alone of which the railway proprietors were prosecuted for damages amounting to £5000. In Beaufort West, a deluge of rain washed down the dam, and next year (1870) the town was flooded by the waters of the Gamka. In the year following (1871), Victoria West was visited with a similar disaster, and according to the report given in one of the Colonial newspapers, the damage done by the deluge of 1874 could not be estimated at much less than £300,000; according to the report given by another, the damages done to public works alone was estimated at £350,000.

When the damages occasioned year after year by torrents rushing to the sea are estimated on the spot at sums like these, this may with propriety be taken into account in estimating what would be the expense of constructions to arrest, and retain, and utilise the water thus lost.

Herr Wex, a distinguished Hydraulic Engineer in Austria, Councillor of State and Director-in-Chief of Works being executed to secure the regulation of the flow of the Danube, in writing* of important operations connected with the conservation and extension of forests, as a means of counteracting evils which have followed the extensive destruction of these, supplies some statements in regard to works of hydraulic engineering of which I gladly avail myself here :—

" Although, having found out the cause of these calamities, it will not be difficult to discover what measures and preventions should be adopted to prevent the recurrence of them in the future, yet we foresee that this can only be effected by overcoming many difficulties, and at an outlay of considerable capital, continued through a protracted series of years ; and withal, the measures necessary can only be carried out successfully when, besides the Government, the more intelligent portion of the population at least acknowledges the urgent necessity for the adoption of these measures, and the Government presses on the execution of these measures with the greatest energy.

" Seeing that it seems to be natural to man, before entering on any great undertaking, to look to the world and its previous history in the hope that perhaps something of the same kind has already been done in some similar case, and that thus there may be obtained some earnest of success and some suggestions whereby may be utilized and turned to account the experience gained in the execution of some previous enterprise, I shall, before advancing further, direct attention to an example well-deserving of consideration—one almost unique in the history of mankind—showing how the oldest civilised nation in the world (the Chinese) have delivered their entire country from deluge-like inundations, and transformed the devastated land into a ' Flowery Land,' as they delight to call it.

" In the middle of last century appeared a work, in forty volumes, entitled " *Shui-Hing-Kin-Kien*," or " The Hydrology of China," from which interesting extracts were given in the " *Allgem-Zeitung* "

* *Ueber die Wasser abnahme in den Quellen Flüssen und Ströme bei gleichzeitiger Steirung der Hoch wasser in den Culturländer, Von Gustave Wex, K.K. Ministerialrath und Oberbauleiter der Donauregulirung bei Wien. Wien, 1873. Druck und Verlag von R. v. Waldheim.*

(Vienna, 1858), from which I shall cite only those which have a bearing upon my present subject, and I shall do so as succinctly as possible.

"The Empire of China, at present covering an area of 250,000 German square miles, with a population of about 360 millions of men, lies on the eastern slope of the High-lands of Asia, and so, at the same time, of the highest mountain range of the world, by which two chains of mountains running from west to east toward the sea divide China into three parts. On the crests of this mountain range, covered with perpetual ice and snow, there arise very numerous great and mighty rivers, the floods of which are frequent, and attain a great height, and in former times the inundations spread over the whole land, like the Flood of old.

"About the year 2,300 before Christ,* there poured a new flood over China, and the Emperor of that day, Yao, said to his ministers: 'The enormous waters of the flood have spread themselves out and overflowed all; the mountains are buried in its bosom, the hills are sunk in it, and its upheaven billows threaten the skies; the people are calling for help. Who can give it?'

"After the Minister Kuan, who was commissioned to undertake the work, had, by labours continued for nine years, shown himself unequal to the work, the Emperor called a very intelligent man, Yue Schuen, the son of a simple peasant, to be co-Regent and successor to the throne, and handed over to him the direction of the works of defence against the inundation, which he carried out with extraordinary wisdom, and with great energy; and yet, after thirty years' continuance of hard work and erections, the waters still covered a part of the rich lands.

"When Schuen ascended the throne, he called the distinguished man Yue, first to be his Minister and after that to be his co-Regent, who during his labours, continued over sixty-two years, produced wonderful results.

"Yue, by canals and otherwise, regulated the flow of most of the greater rivers, changed the course of some, constructed strong dams, dug large reservoirs, created artificial lakes to draw off the waters of the great periodic floods, that they might afterwards be utilized in

* The date assigned to the flood spoken of in Genesis—varying, through the discrepancy in the chronology of the Patriarchs, in the existing Hebrew, the Samaritan, and the Septuagint versions—ranges from 2,348 to 2,400 before Christ.—*J. C. B.*

rendering productive lands which were sterile; he devised and executed a great net-work of canals and drains with sluices, with a view to moderate the very great rapidity of currents in rivers, to facilitate inland navigation, and to supply water to land in the quantities required for irrigation. And one of the boldest, and grandest, and most successfully-executed undertakings of Yue, beyond question, was the changing of the course of the Hoang-Ho, or Yellow River, in Upper Schen-Ssi.

"The mountains of Long-Men formerly turned off this mighty river to the east, and thereby the capital of the Empire, Ki-tscheu, was constantly exposed to the danger of inundation. In order to protect the city against this peril, and to give, at the same time, a better direction to the river, Yue caused a bed for it to be dug through the Mountains of Long-Men; he himself superintended the blasting of the rocks, and had the fragments and dèbris transported to a great distance from the spot. When this colossal work was completed, and the river took possession of its new bed, and made its majestic appearance on the other side of the Long-Men range of mountains, and with majestic flow rolled away to the south, Yue, satisfied with his work, caused to be cut in the rock a great inscription, which after four thousand years, although partly eaten into by the teeth of time, in its remains still testifies to the enterprising spirit and the high intelligence of the Prime Minister of Schuen.

"From these brief notices it is manifest that the two extraordinary men, Schuen and Yue, through their distinguished talents and knowledge of hydraulic engineering, may be regarded as the saviours and establishers of the Chinese Empire, inasmuch as they delivered the land from a deluge-like inundation, and taught, at the same time, their people how to remedy the evils occasioned by floods, and what measures to adopt in time to come; they showed to the Chinese that, along with the dangers to which they were exposed, and side by side with these, there were fountains of wealth, and that it was beyond all question necessary to maintain and to perfect the first works, and at the same time to battle without ceasing against the streaming waters.

"Although the Chinese have followed these instructions throughout the whole period of their subsequent existence as a nation, and maintained their hydraulic works at an extraordinary expenditure of power and money, from time to time there have occurred extraordinary inundations, which have devastated entire provinces with numerous villages and towns, and have engulfed hundreds of thousands of men; but the untiring industry and perseverance of this people have always been able completely to repair the catastrophe and to recover from its effects—stronger and ever stronger defensive works continually being produced. More than two hundred Emperors in succession have, through their arrangements and counsels and projects, and especially by munificent subsidies, contributed towards the completing and perfecting of the aforementioned works and artificial structures, which have continually required a large expenditure. Some Emperors did not consider it beneath them to prepare with their own hands plans and specifications for new works of hydraulic engineering; and even the Princes of the Mongolian Tartars, who had conquered China in the year 1280, found themselves necessitated to adopt the resolution to maintain the existing works, and to set about their extension, in order that they might make good their position on the throne of China.

"We shall first state, as succinctly as possible, what hydraulic works and structures, executed by the Chinese during the protracted period of 4000 years—throughout which they have bestowed upon them their best energy and unfailing diligence in the conquest, irrigation and utilizing of their land, threatened with tremendous inundation—have been brought into existence; and then, what results have followed.

"In the first place, all the rivers and streams throughout this extensive Empire, have had their flow completely regulated. Of these, very many by new beds have been turned into other directions from those in which they originally flowed; canals have been cut, and confined by huge embankments; overflow has been prevented. Further, many very great reservoirs and artificial lakes, covering areas of from 38 to 77 German square miles, have been excavated and confined by strong dams, and so prepared to receive the immense bodies of water which ever and anon come down from the high mountain ranges, and retain these, ready to be gradually drawn off and utilized in remote districts, to make these productive and to irrigate them.

"Throughout this extensive country, more than *four thousand* canals, many thousands of miles in length, have been dug as a network, uniting together the whole of the rivers and reservoirs, formed with the threefold object: to carry off the superabundant waters of individual rivers in flood; to diffuse these over districts which are naturally comparatively arid; and, lastly, to maintain a lively inland navigation, and keep as inexpensive as possible throughout the whole land the transport by water of agricultural and industrial products.

"The greatest and most interesting product of such engineering operations, however, which the Chinese have produced is Yue-Leang-Ho; or, as it is generally called, Yue-Ho—'The Imperial Stream,' or 'The Imperial Canal.'

"This canal traverses China from north to south without a break, and is according to some 250, according to others so much as 450, German miles in length. Further, according as occasion requires, it has a breadth of from 150 to 1000 feet; and it is throughout at least 8 feet in depth. This canal penetrates the mountains which interfere with its course, passes over valleys, crosses all rivers and streams flowing from west to east which lie in its way, and in this way the two greatest rivers of the empire—the Hoang-Ho and the Kiang—which it thus unites. This wonderful Imperial Canal—on all hands, and deservedly, extolled—fulfils also the aforementioned threefold purpose so efficiently that it has contributed in no measured or unimportant way to the advancement and promotion of the agricultural enterprise and of the commerce of China.

"Besides the aforementioned four thousand greater canals, which form a complete and skilfully combined system of inland navigation over the whole land, there are in China an incalculable number of smaller canals for drawing off water where it is in excess, and of canals of irrigation, and others, used to drain and fit for cultivation lands which were formerly boggy depressions. For the irrigation of lands under cultivation, and more especially extensive garden grounds, what are called Artesian Wells have been in use from the earliest times; indeed, in one district on the borders of Thibet somewhere about 10,000 such wells have been sunk, and the spring-wells in Sse-Tschuen reach a depth of 1000 metres, or 3333 feet.

"The results of these four thousand-years' long uninterrupted labours of this diligent, persevering Chinese people, are indeed stupendous, and fitted to excite our wonder.

"The tremendous bodies of water which to the present day still pour down from the high mountains of Asia and descend upon the plains of China, and which erstwhile covered the land as did the Deluge, are now confined in the great lakes and canals, and turned to account in the fructifying of lands but poorly supplied with water.

"Districts which aforetime were covered with water, or were extensive bogs and morasses soaked with stagnant water, having been completely drained, are cultivated with mulberries, cotton trees sugar, rice, and all kinds of cultivated vegetables, and belong to the most highly-cultivated and productive lands, in which neither weed nor wild beast any longer exist. It is to this excellent system of culture, and to the rich fertility of its soil, and to the industry which shows itself in every work alike, and lastly, to the inland navigation and commerce being so greatly facilitated by the regulated flow of its rivers, and by its numerous canals, that China is indebted for being in a position to be able to support its numerous population —a population so dense that not unfrequently they have to erect their dwellings upon the rivers and canals, in order that the land may not be withdrawn from cultivation, or deprived of its opportunities of yielding product, and thus has given rise to not a few large and fine cities, which, like unto Venice, stand in the midst of a labyrinth of canals.

"From the foregoing brief notices of the wonderfully-excellent results of hydraulic engineering in China, may we Europeans infer that for us also it is possible to arrest the prevailing inundations of our rivers and streams, which are but insignificant in comparison with the former deluge-like inundations of China, and be able to stop them altogether, or at least to render them innocuous, and to utilize their dangerous waters; but like the Chinese we must give ourselves to the execution of such works with intelligence, energy, and the determined perseverance, necessary to carry them to completion."

In South Africa the general complaint is of the aridity of the climate and the soil. In the volume I have published on the Hydrology of South Africa, to which reference has been made, I have given details in illustration of the aridity which prevails beyond the colonised portion of South Africa, and of the aridity which prevails within the Colony of the Cape of Good Hope; but I have with these given details illustrative of the water supply within and beyond the Colony named.* Some of the latter details are in keeping with some of the inundations from which China has suffered, but upon a much smaller scale than those against which the hydraulic engineers of China have had to contend, and which they have mastered and utilized—shall I say, tamed and domesticated. What they did in the days of Noah, and have been doing effectually throughout the *four thousand* years which have since elapsed, no British Colony, with the boasted appliances of modern science, need be afraid to

* "Hydrology of South Africa," pp. 216–258.

undertake upon a scale comparatively insignificant. What is required, including all that is required, cannot be accomplished without difficulty; but the work, once taken up in earnest, may prove less formidable than many imagine.

Solomon has told us: "The slothful man saith, There is a lion in the way; a lion in the streets." And Bunyan, in his "Pilgrim's Progress," has told us how his hero was frightened by seeing two lions barring his advance; but advancing, sword in hand, he found, when he came up to them, that they were chained!

The design of this volume is to make apparent the practicability of much being done by works of hydraulic engineering towards mitigating the inconveniences, connected with the prosecution of agricultural operations, resulting from the aridity which prevails in many parts of South Africa.

It may be less the engineering difficulties than the expense of executing undertakings commensurate with the requirements of the case, which prevents aught being done. The question of the amount of money required, and the value of the returns, direct *and indirect*, to be expected, I leave with others; this only do I consider that it comes within the sphere of my professional functions to state,—Every cupful of water kept back from the sea, retained and evaporated and again absorbed by the soil, promotes vegetation; and, from the first moment, the capital invested in retaining it begins thus to bring in a return.

PART I.

Meteorological Observations.

Chap. I.—The Humidity of the Air and Rainfall.

Pluviometrical observations, or measurements of the quantity of the rainfall, are being made at numerous stations in different parts of South Africa, and recorded, collated, and tabulated; and some of the results will be given in the sequel.

It is the rainfall on which we depend for the supply of all the water which can be utilized by hydraulic works, and to this alone attention may be given, if the construction of dams and water-leadings were alone to be considered; but while attention may be given here primarily and mainly to these, it is to these as means to an end, that end being the promotion of vegetation and the development of the agricultural capabilities of the land. And thus it comes to pass that along with the rainfall there are several connected phenomena which call for attention.

I do not attach to rainfall the importance that is done by many, and that chiefly because it does not supply to me information which I desire in studying the effect of moisture on vegetation, which is affected in every case by a portion only—and in many cases by a comparatively small portion—of what falls in rain, and in some cases by moisture obtained directly from the atmosphere by an attraction exercised upon it by the soil, without precipitation in the form of either dew, or rain, or snow.

But further, while I know that I may rely with perfect confidence on observations made, I find that observations made a few miles apart give very different results; and the mean daily, weekly, monthly, quarterly, or yearly rainfall—which is of more importance than either the maximum or the minimum—may be in a different ratio than either the *maxima* or *minima* of the rainfall in the localities at which the observations are made. Two inches of rainfall in autumn may produce disastrous consequences to vegetation, while one inch of rainfall at one stage of advancement or another of that vegetation might have been most beneficial.

I do not mean to intimate that the quantity of the rainfall is unimportant, very far from that; but I reckon this to be of secondary, not of primary importance; and the distribution of it, both in time and in space, varies greatly. Glasgow is proverbial for constant moist, showery weather. At the Observatory there, the rainfall in 1815, 1816, 1817,—the only years of which I happen to have memoranda at command,—was 22·344, 23·799, and 22·420 inches; that of the western climate of the Cape of Good Hope has been reported as 25·804 inches: but the distribution, how different! While the rainfall at the Observatory in Glasgow was what has been stated, during the same years the rainfall at Corbeth, twelve miles north-west of Glasgow, near the Campsie hills, at the height of 466½ feet above the level of the Clyde at Glasgow, according to a rain-gauge in every respect similar to that used there, and made by the same maker, was as follows: 1815, 41·393; 1816, 39·589; 1817, 44·965; a difference as great as the difference in the rainfall in the highest and the lowest of the South African climates noted below; and a difference as great has been noted in the rainfall over extensive districts at the Cape of Good Hope.

I have elsewhere had occasion to report, in regard to the rainfall at St. Helena, that at four stations, comprehended within a circle of little more than a mile radius, the quantities collected during a period of nine months, in 1841, varied thus:—

At 2644	feet of	elevation,	22·63	inches.
At 1991	,,	,,	27·11	,,
At 1782	,,	,,	43·42	,,
At 414	,,	,,	7·63	,,

In the Mauritius, according to a paper which appeared in "Nature," there was, in 1872, a variation of from 33 inches at one place to 146 inches at another.

Mr Marsh states:—"Recent observations go far to show that local conditions of very limited range have a much greater influence on the measurable rain-fall than has been generally supposed hitherto. Thus, by the Report of Dr B. A. Gould, Director of the Observatory at Cordova, in the Argentine Confederation, it appears that at the house of the secretary, which is one mile distant from the

Observatory, and 36·6 metres lower, during a period not given but apparently of some months, a pluviometer, three and a half metres above the ground, showed a precipitation of 496·5 millimetres. The measured fall at the Observatory during the same period was 459·1 millimetres, showing a difference of eight per cent. in favour of the lower station.

"In a period of some weeks, a rain-guage on the roof of a house near the Observatory, 4·02 metres above the standard instrument at the Observatory, registered 137·04 millimetres, the standard pluviometer 150 millimetres, a difference of eight per cent. in favour of the lower station.

"It 1859, at Charleston, S. C., there fell at the U. S. Arsenal, in a single rain-storm of an hour and a half, two inches of water; at the Registrar's Office, in the same city, at the distance of two miles from the Arsenal, the fall was but one third of an inch.

"In the same year, observations at three stations in the city of San Francisco gave a total rain-fall of 16″ 34; 25″ 41; and 21″ 39 respectively; and during the whole period from 1853 to 1860, the meteorological records at different stations in the same city show similar discrepancies."

The mean annual rainfall in the western part of the Colony, as determined by observations at Capetown, at Sea Point, at the Royal Observatory, at Simonstown, and at Somerset West, is reported as 25·804 inches. But the mean annual rainfall in an extensive district to the east, as determined by observations at Worcester, Swellendam, and Mossel Bay, is only 13·818 inches. And in an extensive district further to the east, as determined by observations at Graaffreinet and Nels Poort, though somewhat more, it is only 14·412 inches.

Similar differences within less distances have been observed in the vicinity of Capetown. Rain-gauges are kept and observations registered at Bishop's Court, near the southern face of Table Mountain; at Wynberg Hill, about two miles from the mountain, also to the south; and at the Observatory to the north-west of the mountain, at a like distance. During the first six months of 1866, and during the first six months of 1870, at these places it was as follows:—

	Bishop's Court.	Wynberg Hill.	Observatory.
1866,.........	22·544	17·603	11·243
1870,.........	23·968	23·437	13·100

I give these returns, and only these, because they, and they only, happen to be at my hand.

If my memory serve me right, all returns from these places have shown similar results. Those for 1866 were reported by Sir Thomas Maclear, Astronomer Royal; those for 1870, by Mr Blore, Secretary to the Meteorological Society, who gives the following as the average rainfall at the several places :

19·769 16·279 13·100

And a much greater difference is found to exist when a wider range of country is embraced in the sphere of observations. While at the stations mentioned, during the first half of the year 1866, the rainfall as reported by Sir Thomas Maclear had been as follows :—

	Bishop's Court. inches.	Wynberg. inches.	Observatory. inches.
January,.........	0·257	0·372	0·032
February,........	5·530	4·200	3·104
March,............	0·392	0·471	0·185
April,.............	2·180	2·017	1·501
May,..............	1·945	1·011	0·763
June,...... ...,...	12·240	9·532	5·658
	22·544	17·603	11·243

and in successive years the fall of rain on Wynberg Hill from 1st January to 30th June had been :—

1857,	... ...	21,390 inches.
1858,	... ...	16·200 „
1859,	... ...	27·525 „
1860,	... ...	14·533 „
1861,	... ...	16·823 „
1862,	... ...	26·991 „
1864,	... ...	14·544 „
1865,	... ...	10·292 „
1866,	... ...	17·603 „
Average,	...	18·433 „

and in 1866 (from 1st January till 4th July, 9 A.M.), 19.402 inches, —there was reported as the rainfall at Concordia, Namaqualand (A. von Schlicht, Esq., observer) :—

April 1866,	...	0·178	inches.
May 1866,	...	0·225	„
Total,	...	0·403	„

at Worcester (J. D. Hugo, Esq., observer) :—

January 1866,	...	0·000	
February 1866,	...	1·150	inches.
27th March 1866,	...	0·010	„
Total, ...	...	1·160	„

and at Bredasdorp (J. Bell, Esq., observer) :—

May 1866, ... 0·280 inches.

So greatly does the quantity of the rain-fall vary in the same district at different places.

The Astronomer-Royal, in reporting the first series of observations, had remarked, it would be interesting to compare with the above the falls at Green Point and in Capetown. Mr Blore, Fellow of the Scottish Meteorological Society, and Secretary of the Meteorological Society at Capetown, in reporting the latter observation, published also the following comparative statement of the rainfall at these places during the first six months of the years 1862–1866, from records kept at " Rockland's Villa " :—

	1862	1863	1864	1865	1866
January......	0·325	0·250	0·335	0·350	0·040
February,...	0·255	0·545	0·065	0·160	3·960
March,.......	0·305	1·933	0·220	0·185	0·247
April, ...	0·443	2·607	0·613	1·907	1·193
May,..........	0·790	5·066	2·030	3·100	0·672
June,.........	9·730	3·154	4·456	0·698	5·720
	11·848	13·555	7·719	6·400	11·832

And the rainfall in Capetown during the first six months of the years 1862–1866, from records kept in the Town-House :—

	1862	1863	1864	1865	1866
January,	——	——	0·530	——	——
February,	——	0·640	0·002	——	4·050
March,	0·750	2·640	0·300	0·500	0·300
April,	0·550	2·980	0·750	2·000	1·400
May,	0·950	5·760	2·450	4·100	0·850
June,	10·340	3·390	4·950	0·600	9·150
	12·590	15·410	8·982	7·200	15·750

He adds a note to the effect, that from the 1st to the 4th of July in that year (1866) 1·350 inches of rain fell.

Some two years before, Mr J. J. Steytler sent to one of the local newspapers, at the request of the editor, the following comparative statement of the rainfall as registered at his residence, Rocklands Villa, Sea Point, up to the date of his communication:—

	1862	1863	1864
January,	0·325	0·250	0·335
February,	0·255	0·545	0·065
March,	0·305	1·933	0·220
April,	0·443	2·607	0·613
May,	0·790	5·066	2·030
June,	9·730	3·154	4·456
July,	5·280	2·799	2·430
August,	3·870	2·601	1·975
September,	1·785	1·443	1·660
October,	3·991	2·010	——
November,	0·744	0·805	——
December,	0·033	0·232	——
	27·551	23·445	13·784

And in 1870 Mr Blore gave the following report of the rainfall at several stations for the first half of the year 1870, with the average for each station:—

	1870	Average.
Royal Observatory,	11·904	13·100
Bishop's Court,	23·968	19·769
Wynberg Hill,	20·437	16·279
Malmesbury,	9·940	8·947
Worcester,	5·775	5·314

And added: "The rainfall at Worcester from the 1st to 12th July was 4·605 inches, or nearly equal to that for the half year. The fall on Wynberg Hill during the same period was 11·256, making the fall from 1st January 31·693, against 35·017 for 1869. The Worcester rainfall sufficiently proves that mountain influence alone will not ensure an adequate rainfall. The mountains should be clothed with forest."

I have, in a separate volume, given my opinion in regard to the effect of forests on the rainfall. It is to the difference in the quantity and distribution of the rainfall at different places to which alone I at present direct attention.

In treating of forests and moisture, or effects of forests on humidity of climate, I have had occasion to cite the opinion of the late M. Cézanne, that some of the so-called caprices of rain, and the variations which they present in one quarter and in another, of even the same city, may often be attributable to irregularities in the contour of a country, in accordance with a law that the pluvial rainfall is greater in proportion as the atmospheric current, arrested by an obstacle, is compelled *to rise more rapidly,* * and not the comparative altitudes of places in which a difference in the quantity of the rainfall has been observed.

By Mr Marsh it is stated:—"It is but recently that attention has been drawn to the great influence of slight differences in station upon the result of observations of temperature and precipitation. Two thermometers, being but a few hundred yards from each other, differ not unfrequently five, sometimes ten, degrees in their readings; and when we are told that the annual fall of rain on the roof of the observatory at Paris is two inches less than on the ground by the side of it, we may see that the height of the rain-gauge above the earth is a point of much consequence in making estimates from its measurements."

He adds in a foot-note:—"Careful observations by the late lamented Dallas Bache appeared to show that there is no such difference in the quantity of precipitation falling at slightly different levels as has been generally supposed. The apparent difference was ascribed by Professor Bache to the irregular distribution of the drops of rain and flakes of snow, exposed as they are to local disturbances by the currents of air around the corners of buildings, or other

* "Forests and Moisture," p. 162, p. 275.

accidents of the surface." Indications of these may be seen in the forms of snow-wreaths, and this more especially in the varying curves of the surface of the snow half filling a road at right angles to the wind, traced some feet or yards both to windward and to leeward of the road; and something similar may sometimes be found in wreaths of drifting sand, or heaps of withered leaves swirled and deposited by the wind, or even of dust so borne by the breeze.

Mr Bache's conclusion, says Mr Marsh, seems not to be accepted by late experimenters in England; and he refers in proof to statements in the "Quarterly Journal of Science" for 1871, p. 12; and he states that the consideration of what is alleged much increases the importance of great care in the selection of positions for rain-gauges. Having made these remarks, Mr Marsh goes on to say in the text:—"The data from which results have been deduced with respect to the hygrometrical and thermometrical conditions, to the climate, in short, of different countries, have very often been derived from observations at single points in cities or districts separated by considerable distances. The tendency of errors and accidents to balance each other authorise us, indeed, to entertain greater confidence than we could otherwise feel in the conclusions drawn from such tables; but it is in the highest degree probable that they would be much modified by more numerous series of observations at different stations within narrow limits."

It is also stated by him—and I know of no one who has made so extensive an induction of facts bearing upon the subject, collected from so wide a range of observations, as he has done:—"There is one branch of research which is of the utmost importance in reference to these questions, but which, from the great difficulty of direct observations upon it, has been less successfully studied than almost any other problem of physical science. I refer to the proportions between precipitation, superficial drainage, absorption, and evaporation. Precise actual measurement of these quantities upon even a single acre of ground is impossible; and in all cabinet experiments on the subject the conditions on the surface observed are so different from these which occur in nature, that we cannot safely reason from one case to the other. In nature the inclination of the ground, the degree of freedom or obstruction of the flow of water over the surface, the composition and density of the soil, the presence or absence of perforations by worms and small burrowing quadrupeds—upon which the permeability of the ground by water, and the power of absorbing and retaining or transmitting moisture depend,—its temperature, and the dryness or saturation of the subsoil, vary at comparatively short distances; and though the precipitation upon very small geographical basins, and the superficial flow from them, may be estimated with an approach to precision, yet even here we have no present means of knowing how much of the water is absorbed by the earth is restored to the atmosphere by evaporation, and how much carried off by infiltration or other modes of underground discharge. When, therefore, we attempt to use the phenomena observed on a few square or cubic yards of earth, as a basis of reasoning upon the meteorology of a province, it is evident that our data must be insufficient to warrant positive general conclusions. In discussing the climatology of whole countries, or even of comparatively small local divisions, we may safely say that none can tell what percentage of the water they received from the atmosphere is evaporated, what absorbed by the ground and conveyed off by subterranean conduits, what carried down to the sea by superficial channels, what drawn from the earth or the air by a given extent of forest, of short pasture vegetation, or of tall meadow grass, what given out again by surfaces so covered, or by bare ground of various textures and composition, under different conditions of atmospheric temperature, pressure, or humidity; or what is the amount of evaporation from water, ice, or snow, under the varying exposures to which in actual nature they are constantly exposed."

And again: "As every astronomical observer has his *personal*, so every meteorological station has its *local*, equation, and the determination of these equations ought to be a cardinal object in every system of pluviometrical or thermometrical observations. Records of observations at the same hours and by the same methods, with the same or another carefully compared instrument, for a series of years, may authorise conclusions as to the essential or accidental climatal conditions of that precise locality; but in the present state of our knowledge, such records alone warrant no inference as to the meteorology of any other point, even within the distance of a mile, unless it may be in the case of observations with instruments absolutely insulated on great plains, or otherwise placed in exactly corresponding positions. Hence, until the equations we speak of are ascertained, comparisons of observations made at different periods and at different stations can have no scientific value whatever.

With these, and numerous similar facts before me, I attach much more importance to humidity of air than to copiousness of rain. This also has been made the subject of consideration. But even in regard to this there is not the precision in regard to the time and circumstances in which the observations are made, which I would require to satisfy me in a scientific determination of the supply of moisture thence obtainable for the promotion of vegetation.

In explanation of what is thus referred to, I may state that when moisture suspended in the atmosphere, dissolved as it were in the air, so that it this transparent and invisible, as is sugar or salt dissolved in water, has, by a fall of temperature, been deposited in the form of cloud, or dew, or rain, the atmosphere, relieved of the quantity of moisture beyond that which it could, at the pressure to which it was then subject, sustain in solution at the temperature to which it was reduced, may again become transparent, and mist and fog and cloud alike disappear.

But the air never is by any such deposit deprived of all the moisture it was holding in solution, and the quantity retained in, or the quantity existing in it, at any time can be ascertained.

Observations testing the quantity of moisture suspended in the atmosphere when clear and transparent have heen made at the Royal Observatory, Capetown, since 1842. The quantity, as might be expected, has varied in different days, in different months, and in different years. The following two tables show what was the mean or average humidity of the air in each month from 1842 to 1855, inclusive, and what was in each of these months the average or greatest difference between observations showing the greatest humidity and observations showing the least. They are published by Sir Thomas Maclear, the Astronomer-Royal at the Cape.

Humidity of the Atmosphere, from January 1st, 1842, to December 31st, 1855.

MONTHS.	1842.	1843.	1844.	1845	1846.	1847	1848	1849.	1850.	1851.	1852.	1853.	1854.	1855.	Mean for each Month
January,..................	77	70	71	72	71	65	68	65	72	75	65	66	62	63	68·7
February,.................	70	74	72	74	71	69	73	68	74	74	66	66	70	67	70·6
March,.....................	74	72	76	69	69	69	75	68	76	75	69	69	69	72	71·6
April,.......................	73	76	75	75	73	72	83	75	81	82	75	75	76	74	76·1
May,.........................	81	91	72	79	81	76	84	84	87	81	77	80	77	79	80·6
June,........................	82	86	81	76	82	80	85	85	88	83	79	79	76	83	81·8
July, ...	81	81	79	76	79	84	84	90	84	84	83	74	82	81	81·6
August,.....................	82	77	78	80	80	81	85	85	84	75	81	76	78	81	80·2
September,...............	81	76	73	81	73	78	81	80	84	77	77	72	71	80	77·4
October,....................	73	74	72	73	72	75	74	74	82	77	70	72	71	74	73·8
November,................	74	75	71	67	71	71	73	69	81	74	64	65	69	67	70·8
December,.................	70	71	71	66	67	68	71	70	74	68	68	65	65	65	68·5
Yearly Means,......	76·5	76·9	74·25	74	74·1	74	78	76·1	80·6	77·1	72·8	71·6	72·2	73·8	(75·14)

Extreme Range of Humidity of the Atmosphere, from January 1st, 1842, to December 31st, 1855.

MONTHS.	1842.	1843.	1844.	1845.	1846.	1847.	1848.	1849.	1850.	1851.	1852.	1853.	1854.	1855.
January,..................	46	57	55	66	64	55	51	55	55	61	56	58	59	71
February,.................	61	56	67	55	61	62	47	59	53	52	68	71	69	56
March,....................	57	63	68	62	61	60	51	58	45	75	65	76	64	55
April,.....................	63	72	56	65	72	59	41	65	49	63	60	77	72	70
May,......................	54	37	59	65	64	62	54	36	34	71	59	63	67	65
June,.....................	54	42	62	79	55	65	52	58	39	47	66	68	63	55
July,......................	61	48	52	59	56	50	41	29	75	47	59	73	60	54
August,...................	56	48	61	58	53	47	43	34	47	60	54	60	58	46
September, ...	42	47	74	47	55	53	38	42	31	47	61	56	58	56
October,..................	62	60	67	61	60	58	59	46	46	55	58	69	54	51
November, ...	52	69	56	60	57	64	52	49	47	49	64	61	57	71
December, ...	62	50	63	70	59	47	51	43	44	75	57	69	50	47

In regard to these tables it is stated: "If two thermometers be placed in the open air, but sheltered from the sun and reflected heat, at the distance of six or eight inches from each other, and a few feet above the ground, and the bulb of one of them be covered with thin muslin wetted with water (the purer the better), the wetted bulb will give the temperature of evaporation, while the other gives the corresponding temperature of the air; and by calculation from the indications of both,—the *tension* of the aqueous vapour suspended in the atmosphere at the time of observation,—the *dew* point, and the *humidity* of the atmosphere: properly speaking, of the stratum of the atmosphere in which the instruments are placed.

"By *humidity*, is here meant the ratio of the pressure of the aqueous vapour suspended in the atmosphere to the pressure which would hold if the atmosphere was saturated with vapour, as when a small increment would cause condensation or rain to fall.

"The table gives the humidity as being rather more expressive for the present purpose than the 'dew point.'

"The number 100 implies complete saturation, and the numbers in the table the per centage of saturation. Thus, the mean humidity of the atmosphere for the month of January, 1842, is 77-hundredths, or 77 per cent. of complete saturation."

From the first of the tables we learn that if full saturation, such as immediately precedes deposition, were represented by 100, the average throughout these years, and throughout every month of these years, would be represented by 75·14. In other words, the air held in solution above three-fourths of the whole quantity of moisture which it possibly could at the pressure and temperature at the time retain in solution without diminution of its transparency. And we learn further, that in no one month of these fourteen years did the mean or average of the observations made show that there was less than 62 per cent., or upwards of three-fifths of the full quantity of moisture which the air could sustain in solution present; and only in one month, January 1854, was the quantity so low as this; while from this to 90 per cent., or nine-tenths of the full quantity in July 1849, the range of the average in each month was pretty equable, giving as the mean quantity for the whole period the quantity stated, 75·14 per cent.

The actual range was greater. The range of the averages was 28. The range of the observations in no one month was less than 31. This the second table shows it to have been the case in September 1850, while in April 1853 it was 77.

These observations give us the humidity of the atmosphere at the Royal Observatory; but there is in this as in the rainfall a considerable variation not only in different days, and in different hours of the same day, but in different districts of the country. On this subject Sir Thomas Maclear remarks :—

"The varied features of the country, owing to its high mountains, rich valleys, and blotches of desert, are attended by corresponding variations in the temperature and humidity of the stratum of air near the surface. Hence the meteorological circumstances of one locality are not strictly applicable to other localities, though they may differ little in latitude. The climate of George presents a strong instance of this, when compared with the climate of the sweep from Cape Point to Clanwilliam. Even if the chief cause (the height and direction of the mountain masses) were absent, the humidity of the currents over the windward districts during the summer months would be altered before reaching the leeward shore. Yet it is probable that the *general type* of the area of the Colony (excluding the great Karroo and Bushman Flat), with respect to mean temperature, atmospheric pressure, and humidity, is approximately represented by the results obtained at this Observatory."

It is the rainfall alone of which the hydraulic engineer can avail himself for storage and irrigation, but not the local rainfall alone or chiefly. The river supply is in many districts most important. And the rainfall itself is itself but a precipitation of a portion of the moisture in the atmosphere which is available for the promotion of vegetation; and in estimating the importance—the great importance of the storage of water and of irrigation in an arid land—it is of some importance that this fact should not be lost sight of; and there are also phenomena connected with this which deserve consideration.

The two main elements which determine the precipitation of rain are the temperature and the humidity of the atmosphere. A fall of temperature below the point of saturation occasions a deposit of moisture in some one of the forms known as cloud, mist, dew, hoarfrost, rain, snow, and hail; a rise of temperature above the point of saturation occasions absorption. But there are other elements which require to be taken into account in the scientific study of these—amongst others, the law regulating gaseous diffusion, the elastic form of vapour, and the barometric pressure of the atmosphere at the time of observation.

According to the law regulating the diffusion of one gas—or kind of air, such as aqueous vapour, carbonic acid, oxygen, or nitrogen —through another, it seems, to express it in general terms, as if every gas were as a vacuum in so far as the diffusion of another gas for which it has no special chemical affinity. By Dalton it has been proved that under all circumstances gases penetrate each other, and become thoroughly mutually interpenetrated and co-mingled.

This fact may be proved by a very simple apparatus, and an illustration of the operation of the law may thus be obtained.

Provide two glass phials, each of the capacity of about an ounce measure, and also a tube open at both ends 10 inches long and $\frac{1}{20}$ inch bore, which tube is to be passed at each end through a perforated cork adapted to the necks of the phials. Fill one of the bottles with hydrogen gas and the other with oxygen gas; place the latter on the table with its mouth upwards, and into this insert the tube secured by its cork. Then, holding the phial filled with hydrogen with its mouth downwards, put it upon the cork at the top of the tube. Leave the bottles thus connected to remain for a time in this perpendicular position. When they have stood so two or three hours separate the phials and apply a lighted match to the mouth of each; and this will almost certainly occasion an explosion in both. Had a light been so applied before they were thus connected, the hydrogen would have burned with a blue lambent flame, and the oxygen would not have burned, but it would have caused the match to burn more brightly; now, being thoroughly co-mingled in both bottles in the proportions in which they both exist as water, they rapidly combine, occasioning an explosion and forming water. But the hydrogen, which was the lighter of the two gases, was contained in the higher bottle, and the oxygen, which was the heavier, was put into the lower bottle; and yet, after this lapse of time, both bottles contained mixtures of the two gases in the same proportion. It follows that, though separated by a tube 10 inches long and only $\frac{1}{20}$ of an inch in bore, the hydrogen must have been diffused equally through both bottles, and the oxygen must have been the same; hydrogen must have descended through the tube from the upper into the lower phial, and, contrary to what might have been expected from its great weight, oxygen (the weight of which is sixteen times that of hydrogen) must have ascended through the tube into the upper phial, the diffusion and interpenetration being in both phials alike.

Experiments of this kind, it has been shown by Dalton, may be extended to all the other gases, and they establish the conclusion thus enunciated: That elastic fluids which have a sufficiently free communication with each other tend to mutual diffusion, and that this tendency is even sufficient to overcome the obstacle of specific gravity—the heavier rising into the lighter, and the reverse.

It is of elastic fluids that this is alleged. Such are all gases; and by this are they distinguished from other forms of matter: in solids, the atoms of which they are composed are so held together by mutual attraction that a fracture is effected with difficulty; in liquids they are so held together by an attraction modified by a repulsion which leaves them free to move about in regard to each other, and to flow in any direction; while in gaseous bodies the atoms appear usually as if in a state of mutual repulsion, repelling each other to as great a distance as possible; and upon this depends what is known as their elastic force.

With the knowledge of this, the phenomena of gaseous diffusion becomes susceptible of easy explanation on the supposition, which is a legitimate one, that the particles of gas are thus so separated from each other, like the stars in the firmament which are separated by spaces in comparison with which their own bulk, though stupendous, is as nothing, and thus the atoms of the one may pass freely between the atoms of the other. But, looked at thus, it will be seen that the freedom of transmission is not absolute. Give more space to a given bulk of gas and it will expand, and expand, it is conjectured legitimately, simply by the wider separation of the atoms of which it is composed. In such a state of expansion it will afford wider room for the free passage of atoms of another gas. But compress it, and you reduce the inter-atomic spaces, and this may have the effect of impeding the diffusion of another gas.

Such compression may be effected by the attraction of gravitation acting on the stratum of air under observation, and on superincumbent strata pressing upon this; or it may result from the withdrawal of heat. But here there meet us observations which prevent us from reckoning this in accordance with a simple rule of arithmetical progression.

At and below the temperature at which water freezes, the solvent power of atmospheric air—or, as it has otherwise been designated, its chemical affinity for water—or, as I at present, in connection with the consideration of laws regulating gaseous diffusion, prefer considering it, the facility for the passage of gaseous vapour, is very great; but it rapidly increases, with a rise of temperature and consequent expansion of the air by the wider separation of its constituent atoms!

From experiments made by Mr Dalton, and calculations founded on these, it appears that atmospheric air at 32° Fahr. contains $\frac{1}{200}$ of its bulk of water; at 54°, $\frac{1}{100}$; 78°, $\frac{1}{50}$; 98°, $\frac{1}{25}$. And at a temperature above 167° or 170° the aqueous vapour and the air seem disposed to separate, as if the one would crush out the other altogether.

This varying affinity of air and moisture, or solvent power of the atmosphere at varying temperatures, is attributable, it is alleged by some, to atmospheric air, or some one or other of its constituents, and water expanding in different degrees with equal increments of heat.

In Prony's *Architecture Hydraulique* there is given a table of the elasticity or expansive force of pure vapour, determined by M. Betancourt, by experiment at every fifth degree of Reaumur above 0° to 110°, and by calculation at every intermediate degree, of which the following is an abstract:—

1°	2°	3°	4°	5°	10°	20°	30°	40°	80° = 212° Fahr.
·01	·03	·05	·07	·11	·28	·90	1·93	3·64	28·

M. Schmidt, another careful observer, made a series of experiments upon the dilatability of air made as *dry* as possible by exposure to hot tartarin; the following is a corresponding abstract of results obtained:—

1°	·0044675	10°	·0446750
2°	·0089350	20°	·0793500
3°	·0134025	20°	·1340250
4°	·0178700	40°	·1787000
5°	·0223375	80°	·3574000

From these tables it appears that 1000 inches or measures of dry air, at the temperature of the freezing point, would become 1004·4675 at 1° Reaum., and assume a corresponding increase of bulk at every corresponding increase of temperature; but 1000 inches or measures of vapour would assume a continually increasing increase of bulk at every corresponding increase of temperature, and at 1° Reaumur become 1010 inches.

Schmidt prepared also a table founded on observation and calculation of the dilatation, or bulk, that 1000 parts of dry air would acquire at each higher degree. The following is an abstract corresponding to those of the tables already referred to:—

1° 1010·56	5° 1028·58
2° 1010·78	10° 1064·72
3° 1016·45	20° 1152·83
4° 1022·21	30° 1610·02

From which it appears that while 1000 parts of dry air at the freezing point are at 10° expanded to 1044·67, 1000 parts of air saturated with moisture are expanded to 1064·72; and at higher temperatures the differences of evaporation are much more considerable.

I dislike the use of the phrase "solvent power of the air," as an assumption which otherwise than as a hypothesis is not warranted; and it is now more than 70 years since Luke Howard, who, in discussing the diffusion of water in air, and the several modes of its precipitation on chemical principles, regarding moist air as a ternary compound of base of air, water, and caloric, formed and subsisting by chemical attractions and decomposed by the operation of the same.

"Having perused," says he "the excellent work of my friend John Dalton, entitled 'Experimental Essays on the Constitution of Mixed Gases, &c.,' and having seen and repeated some of the experiments therein detailed, I must acknowledge that, although I might plead the authority of Dr Kirwan in support of the chemical theory of evaporation, I am now disposed to give up the idea of a proper *solution* of water in air *in toto.*"

But I accept the explanation given in accordance with these observations and calculations; and I refer to the matter in illustration of the number of different things to which attention is demanded in meteorological observations on the humidity of the atmosphere. And I return to the consideration of the effect of barometric pressure in restricting the expansive effects of elastic force, and the phenomena of condensation and deposition of moistnre when this is reduced.

With a rise of barometer clouds are in part or in whole dissolved; and with a fall in the barometer the contrary takes place; and simultaneously with this they rise higher into the atmosphere in the former case, and fall in the latter. The rise of the clouds is attributable to the increased specific gravity of the inferior strata of air.

At the same time the expansion of air is always accompanied by a fall of temperature, as the compression of it is accompanied by a development of heat. In exhausting a receiver on an air-pump, there may often be seen the formation of a cloud, or the condensation of vapour, on the glass at every successive movement of the handle ; and by means of a condensing syringe tinder may be ignited. Clouds are produced by the ascent of aerial currents in the tropics, and from a conflagration, a battle-field, and a city in which furnaces are numerous, and in their climbing and surmounting a lofty mountain range. This is attributable to reduction of temperature; but this reduction of temperature may in some, if not in all of the cases mentioned, be attributable to expansion consequent on reduction of pressure.

The exhilarating influence of the air at the Cape, which I can liken only to the exhilarating influence of air in intense frost, and which contrasts greatly with the enervating influence of air of the same temperature in many other lands, I attribute to its aridity; and the operation of this I consider two-fold. In the West Indies the air is heavily charged with moisture, and it has been described as stifling; nor is the cause far to seek: 1000 cubic inches of air saturated with moisture, at 86° Fahr., contain nearly 76 inches of moisture, which is useless to respiration. But this is not all; the drier the air, the further from saturation, the more rapidly does evaporation go on from the skin; and evaporation of perspiration from the skin lowers the temperature in the very way that the evaporation of *eau de cologne* from the forehead produces a sensation of coolness.

In consequence of this a temperature of 86° at the Cape produces less sensation of heat than does a temperature of 76° in London or in a greenhouse.

In view of what has now been advanced, and of much besides of the same character, in forming a scientific judgment of a climate, and of the amount of moisture or aqueous vapour which may be available for the promotion of agriculture, there is much besides the exact humidity of the atmosphere which requires to be known. Amongst other things these: The temperature, the barometric pressure, the dew-point, or temperature at which that vapour would exactly saturate the air, and below which it would deposit a portion of it.

Subjoined is a tabulated statement of these, as they have been determined by observations at the different places named, which may supply means of comparison between different districts in South Africa, and between these and other places. They also supply data from which the scientific meteorologist may be able to deduce information bearing upon his scientific researches; but they, together with what is subsequently stated, are adduced here in some measure with a view to satisfy practical men among the colonists, who do not pretend to be scientific, that such observations, made with meteorological instruments, and made in accordance with the requirements of science, have not been ignored. The table is as follows:—

Comparison of Climates in the Southern Hemisphere, and other comparative Mean conditions in the same places.

		Barometer at 32° Fahr.	Mean Temperature.	Mean Humidity.	Rain.	Mean Temperature.	Temperature of Evaporation.	Temperature of Dew Point.	Elastic force of Vapour.	Water required in the circ'mstances to saturate a Cubic Foot.	Water in a Cubic Foot.	Water required for full Saturation.
Cape Colony:		inches.	degrees.	per cent.	inches.	degrees.	degrees.	degrees.	inches.	grains.	grains.	grains.
Western Climate.........	Capetown, Sea Point,* Royal Observatory, Simonstown, Somerset West,	29·977	64·53	72·50	25·804	64·53	59·5	55·3	·437	1·65	4·85	6·50
Middle Climate	Mossel Bay, Worcester, Swellendam,*	29·390	66·06	61·89	13·818	66·06	58·5	52·4	·399	2·65	4·35	7·00
Eastern Climate..........	Graaffreinet, Nel's Poort,*	27·478	67·44	57·49	14·412	67·24	58·6	51·4	·381	3·10	4·20	7·30
Pietermaritzburg*⁎*.......	Means of 6 years (Natal Almanac),	27·879	64·57	72·20	30·077	64·57	59·5	55·2	·436	1·78	4·82	6·60
Port Louis....................	Mean of 9 years Mauritius (C. S. Almanac),	30·024	76·81	72·60	43·550	76·81	70·1	66·7	·659	2·78	6·72	9·50
Adelaide, South Australia	Mean of 6 years, 1857-1862,†	29·907	63·70	60·20	19·875	63·70	55·9	49·4	·353	3·62	3·91	6·52
St Helena.....................	Mean of 5 years, 1841–1845, hourly observations,††	28·285	61·40	87·00	47·198	61·40	59·0	57·0	·468	0·89	5·26	6·05

* Used for rain alone. *⁎* 2095·7 feet above Custom-house at D'Urban. † At Observatory, 140 feet above the sea. †† At Observatory, 1765 feet above the sea.

N.B. The notices respecting Adelaide and St Helena were furnished by Mr Mann, Assistant Astronomer-Royal.

CHAPTER II.

Clouds.

The existence of moisture in the atmosphere manifests itself in other ways besides the precipitation of rain. Amongst these may be reckoned haze, fog, mist, cloud, drizzle, snow, hail, hoar frost, and dew, some of which are rare at the Cape in comparison with the extent to which they are seen in England and elsewhere, but none of them are altogether unknown there. All of them apparently owe their origin to a condensation of moisture which was previously suspended in the atmosphere—and this a condensation occasioned by a reduction of temperature; and an acquaintance with the general principles regulating the general phenomena will make it easy by a consideration of the meteorological conditions on which these depend, to account both for what is seen, and to account for the absence of what is not so seen.

As the theory of Dr Hutton in regard to the formation of rain is substantially, and with the exception of a few, a very few, and these not very important, corrections introduced by later students, is a theory now very generally received, it may be well to look at this as it came from his hand.

It is founded on observations made on the action and effect of heat and cold on the atmosphere, and it is based on the principle that the moisture of portions of air saturated with water, and at different temperatures, will uniformly produce the condensation of a portion of water.

"This rule of condensation," says he, "may be applied to the theory of rain, which is the distillation of water first dissolved in the atmosphere and then condensed from that state of solution." But he adds—"The cause of rain, though often exerted, will not always produce the full effect; a scanty condensation of aqueous vapour produces *mists* on the earth and *clouds* in the atmosphere above; and, taking the gradation from one extreme of transparent atmosphere to the other of the densest cloud, from the falling of the gentlest mist to the heaviest rain, hail, and snow, we have an indefinite variety of appearances, all flowing from one simple principle."

But the effects of heat and cold in relation to air and vapour are not uniform, and he formulates the expression of the law thus: "The dissolving power of air on water may vary in different proportions to the heat. The solution may vary as the heat, or in a greater or less ratio, *i.e.*, the increments of each may be constant, or those of the heat being constant, the increments of the solution may be accelerated or retarded." By a geometrical diagram he shows the curves representative of the three classes of ordinates spoken of, and proceeds:—

"Thus the actual curve of evaporation being known, the effect of any mixture of two portions at different temperatures may be ascertained. If the solution of water in air increases equally with the heat, there will be neither super- nor under-saturation in a mixture of portions at different temperatures.

"If the solution increases with the heat, but in a decreasing ratio, there will be under-saturation; if it increases with the heat, but in an increasing ratio, there will be super-saturation.

"The last case applies to the phenomena of breath and steam, rendered visible in mixing with air colder than themselves, and to various appearances that may occur in mixing together several portions of air differently saturated with humidity, and at different temperatures: for it is not every mixture of air at different temperatures that forms a visible condensation, this effect depending on the degree of saturation with humidity. But if two portions of the atmosphere, both saturated with humidity, should be mixed, let there be but a difference in their temperature, and a condensation proportionate to this difference will take place.

"At present, from the influence of the ascending sun, two opposite currents of air are formed in the summer hemisphere, one moving along the surface of the earth from the pole to the equator, the other flowing above. These opposite currents, while separate, might pass each other without producing rain; but when sufficient portions are mixed, not only clouds but showers will be produced, since the sudden formation of a mean degree of heat in the mixture of two portions of different temperatures must condense a quantity of vapour sufficient to form rain.

"Without this law of condensation of evaporation, neither rain nor dew would take place anywhere in the summer hemisphere, perhaps not even in tropical latitudes; there would be evaporation, and a general tendency to saturate the atmosphere with water, but the mixture of different portions of air would only temper the saturation, without producing a condensation of vapour in the mean degrees of heat. At night, from the influence of the cold, the atmosphere would become gradually clouded; this cloudiness would increase to a general distillation of condensed vapour, which would be uniformly continued until the returning summer should change the state of condensation to evaporation; and instead of the beautiful return of seasons, tempered with various degrees of heat and refreshing showers, six months rain and six months drought would follow each other in an invariable succession."

Some objections were offered to this theory by De Luc in *Idees sur la Meteorologie (tome* ii. p. 67); and shortly after it was published exception was taken to it by Mr Luke Howard, in a paper read before the Askesian Society in the Session 1801–2, and printed in the "Philosophical Magazine" towards the close of 1802.

The subject both then and subsequently engaged the attention of Dr Dalton, the discoverer of the law regulating gaseous diffusion, or the interpenetration of each other by elastic vapours; and in a paper entitled "Observations on the Barometer, Thermometer, and Rain, at Manchester, from 1794 to 1818 inclusive," which he read before the Literary Society, at Manchester, on November 12th and December 11th, 1818, he said in regard to the theory under consideration:—"The late Dr Hutton, of Edinburgh, was, I conceive, the first person who published a correct notion of the cause of rain. Without deciding whether vapour be simply expanded by heat, and diffused through the atmosphere, or chemically combined with it, he maintained from the phenomena that the quantity of vapour capable of entering into the air increases in a greater ratio than the temperature; and hence he fairly infers, that whenever two volumes of air of different temperatures are mixed together, each being previously saturated with vapour, a precipitation of a portion of vapour must ensue, in consequence of the *mean* temperature not being able to support the *mean* quantity of vapour.

"This explanation may be well illustrated by contemplating a curve, convex towards its axis, in which case the ordinates increase in a greater ratio than the abscissæ. The abscissæ represent temperature, and the ordinates the quantity of steam which the corresponding temperatures are capable of retaining.

"In 1793 I published my Meteorological Observations and Essays, a few years after this theory of rain had been made known; as far as I was then acquainted with it from one of the Reviews, it appeared the most plausible of any I had seen; but on looking at my remarks, it is evident I had not been made acquainted with its distinguishing feature, and that on which its excellence depends; namely, a higher *solvent* power (if it may be so called) in the air, than what is proportionate to the increase of temperature; and that the precipitation of vapour in the form of clouds and rain is occasioned not by mere cold, but a mixture of comparatively warm and cold air.

"At the time of my publication of the Essay on Rain, &c., I had a strong bias to the opinion that the steam or vapour in the atmosphere exists in a state of combination with heat, but without any chemical union with the elements of the atmosphere; only it is subject to be wafted along mechanically by the great body of the atmosphere in its ordinary currents. This opinion was founded and supported on the authority of the late M. Saussure in part; he having determined by direct experiment that a cubic foot of dry air of the temperature of 66° would imbibe 12 grains of water for its saturation. Now, from experiments on the boiling of water in vacuo, I was persuaded that this quantity of vapour was nearly what would fill a cubic foot of empty space, in the temperature of 66°; and, by analogy, I concluded that the quantity of steam necessary to saturate any given volume of air at any temperature was the same that would be requisite to fill an equal void space at the same temperature. This reasoning was of course hypothetical at that time, and unsupported by any direct experiment.

"In 1801 a series of essays of mine were read before the Society, and subsequently published in the fifth volume of Memoirs; one object of experimental inquiry was, whether steam of any kind was the same in quantity in air and in a vacuum, all other circumstances being the same. The result was decidedly for the affirmative.

"Another object was to ascertain the true force of steam in all atmospheric temperatures. This was clearly proved to be progressively increasing with the temperature, as Dr Hutton had rightly conjectured. Indeed, with a slight modification of the thermometrical scale, the temperature is an *arithmetical* progression, and the force of steam a *geometrical* one. Hence the curve showing the force of steam is what mathematicians call the *logarithmic*, one remarkably convex to its axis.

"The cause of rain, therefore, is now, I consider, no longer an object of doubt. If two masses of air of unequal temperatures, by the ordinary currents of the winds, are intermixed, when saturated with vapour, a precipitation ensues. If the masses are under saturation, then less precipitation takes place, or none at all, according to the degree. Also the warmer the air, the greater is the quantity of vapour precipitated in like circumstances, as is evident to any one, on inspecting the logarithmic curve, or on considering that the increments of a geometrical progression are in proportion to the terms. Hence the reason why rains are heavier in summer than winter, and in warm countries than in cold.

"We may now inquire into the cause why less rain falls in the first six months of the year than in the last six months. The whole quantity of water in the atmosphere in January is usually about three inches, as appears from the dew point, which is then about 32°. Now the force of vapour at that temperature is 0·2 of an inch of mercury, which is equal to 2·8 or three inches of water. The dew point in July is usually about 58° or 59°, corresponding to 0·5 of an inch of mercury, which is equal to seven iuches of water; the difference is four inches of water, which the atmosphere then contains more than in the former month. Hence, supposing the usual intermixture of currents of air in both the intervening periods to be the same, the rain ought to be four inches less in the former period of the year than the average, and four inches more in the latter period, making a difference of eight inches between the two periods, which nearly accords with the preceding observations."

Mr Howard, while controverting some of the positions assumed by Dr Hutton, in speaking of the application of the hypothesis by the author of it, to natural phenomena, speaks of him as "seeming to experience no great difficulty in solving by it every problem that occurs, as the solstitial, equinoctial, and other periodical rains, and the comparative state of different climates, and of land and sea, as to its frequency and quantity, etc. And, if we except a short remark on hail and thunder storms (which seems only to give occasion to a confession of our imperfect knowledge of atmospheric electricity), in no instance does the doctor avail himself of any other principle than the mixture of currents of air in traversing the irregular surface of the globe, and the consequent precipitation of water, to account for the whole train of phenomena constituting the subject of meteorology."

Kirwan, in a paper "On the Variations of the Atmosphere," published in Dublin in 1801, writes: "Vapours issuing from water or moisture warmer than the air to which they unite, are soon cooled by it, and thence in a great measure dismissed: hence the morning mists observed in summer, and the winter mists of the colder regions; evening mists, on the contrary, proceed from the super-saturation of air with vapour previously dissolved arising from the supervening decreased temperature. The inferior strata of the atmosphere are scarce ever super-saturated by vapours arising from water or moisture warmer than the air into which they ascend; for, before the point of saturation can be attained, their affinity to the portion of air to which they are united is weakened, and thence exceeded by the unincumbered affinity of the superior strata: and this happens successively, on to the higher regions; but with diminished activity, by reason of the diminished density of the higher strata, until their ulterior progress is checked by saturation; but as they are still continually recruited from below, their quantity is at last so far increased that they coalesce into clouds. Here the process recommences; for from the surface of these clouds a fresh evaporation often takes place, which after some progress is again checked in its turn, and clouds are formed at a superior height; these again give room to a further evaporation, and a new stage of clouds is formed, until the process is at last arrested by the intense cold of the superior regions."

The expression may startle any reader who loses sight of what was proved by Dr Hutton, that what he calls the dissolving power of air and water vary in different proportions to the heat. And in connection with what has been quoted, Kirwan goes on to say: "The mere cold of congelation is not sufficient to arrest the process of evaporation and the reproduction of clouds at a higher altitude; for Bouguer informs us that clouds are formed 2500 feet above the lower line of congelation; and that ice itself evaporates, though cooled several degrees below the freezing point, is well known. The distance of the particles, both of air and vapour, from each other, when so far rarified as they must be in the superior regions of the atmosphere, prevents their coalescence in any but extreme degrees of cold."

As intimated by Kirwan: "By Bouguer, clouds were seen three or four hundred toises above Chimborazo, and consequently at a height of 22,528 English feet, or 4·3 miles above the level of the sea, —a height at which the barometer at the temperature of 32° would stand at 12·7 inches." And in view of the expansion which must take place under such a reduction of barometric pressure, it is easy to conceive that, under a reduction of pressure far in excess of this, not only may particles of air and particles of vapour be so separated that coalescence may be out of the question, but the particles of water may be so separated that the coalescence of these, which is a condition of their condensation into vapour, may be impossible, or if effected, the effect may be as far removed from the visible effect of haze, as the effect of haze is from that of the densiest fog or mist; and this, according to Kirwan, is the only difference between such products.

Writing of the phenomena of "vapours," he says: "In their first degree of coalescence, when separated from air, they form aggregates of exceeding minute particles, separated from air by the diminution of affinity, and also from each other by electrical atmospheres: these aggregates are of equal and often lower specific gravity than the air in which they are formed; and yet they are visible by reason of their opacity; when near the earth they are called *fogs*, *mists*, or *haze*, which differ only in density, and when at greater heights, *clouds*."

In warm countries and seasons there may be seen occasionally, and at the Cape not unfrequently, various strata of clouds, one above another, sometimes stationary and sometimes borne in different directions. Muschenbrook has given an interesting account of various phenomena observed by him, in connection with three such strata observed by him in Holland in August 1748. And the phenomenon is satisfactorily accounted for above.

Dr Kirwan attributes this combination of "vapours," or moisture, with atmospheric air, partly to the agency of heat and partly to that of affinity and of electricity; and he says: "They separate from it sometimes from a diminution of that degree of heat which they possessed in their nascent state, sometimes from a diminution of affinity, and sometimes from an alteration of their electrical state." I do not deem it necessary to draw these distinctions; I look upon them as correlated effects, and an opportunity of considering some of the effects of electricity will afterwards present itself.

Saussure estimates the weight of clouds at about one-third or one-fourth of that of the cubic foot of air in which they rest. According to Watt, as stated by De Luc ("Meteorology," v. iii. p. 145), the specific gravity of pure vapour is to that of air as 4 to 9, comparing it probably with air at the usual density of 29·30, and at some particular temperature; for at high temperatures the difference must be much greater.

"Saussure, ('Hygrometer,' p. 204,)" writes Kirwan, in the "Philosophical Magazine" for 1802, "has given us the specific gravity, not indeed of pure vapour, but of vapour dissolved in air, with more precision; for he tells us (1) That a cubic foot of perfectly dry air has its volume augmented by $\frac{1}{64}$th when saturated with ten grains of moisture, at about 65° Fahr. of heat, and barometer 28·77 inches (English). (2) That a cubic foot of pure or perfectly dry air of that density, and at that temperature, weighs 751 grains (French); and after dissolving ten grains of moisture, by which it is diluted $\frac{1}{64}$th, this new volume weighs $751 + 10 = 761$ grains; but a cubic foot of *pure* air, augmented by an accession of $\frac{1}{4}$th of its bulk of pure air, would weigh $751 + \frac{1}{64}751 = 765$ grains, that is 14 grains more. Hence he infers that in this case the specific gravity of the dissolved moisture is to that of dry air as 10 to 14; for $\frac{1}{64}$th of the cubic foot in the one case weighs 10 grains, and in the other 14 grains nearly."

Kirwan expresses a strong suspicion that the original experiment upon which this calculation was founded involved an error seriously affecting the result; but the principle involved is unquestioned.

The following is a tabulated statement of the extent of cloudy sky, observed at the Royal Observatoay, near Capetown, during a series of 14 years.

In regard to this Table it is stated: "The mean annual amount of cloud is 38 per cent. of the visible hemisphere: it should be stated, however, that since the Table is not founded upon instrumeutal measurement, but upon mere estimates by the eye, and perhaps often in haste, the numbers ought to be regarded as approximations only."

EXTENT OF CLOUDY SKY.

Date.	Jan.	Feb.	March.	April.	May.	June.	July.	August.	Sept.	Oct.	Nov.	Dec.	Means.
1842	22	23	36	53	52	54	42	48	55	43	48	32	42
1843	20	35	25	43	54	52	54	41	37	45	55	25	41
1844	22	26	34	47	22	44	38	56	43	36	43	27	37
1845	35	34	26	51	47	33	41	50	57	42	46	40	42
1846	40	33	32	35	59	49	44	55	60	29	33	17	41
1847	10	16	24	38	39	37	42	51	50	34	30	28	33
1848	24	31	23	50	45	60	59	51	54	44	35	30	42
1849	20	20	26	27	48	45	36	38	33	22	29	24	31
1850	35	33	31	40	47	60	28	25	25	40	43	19	36
1851	21	16	19	49	53	53	44	28	38	44	42	34	37
1852	30	27	35	41	52	38	52	46	44	37	35	26	39
1853	28	29	37	42	49	45	33	45	29	40	18	32	36
1854	24	25	33	28	38	29	44	36	38	49	35	32	34
1855	29	27	30	40	47	53	43	56	50	43	42	27	41
Sums..........	360	375	411	584	652	652	600	626	616	548	534	393	(38)
Means..........	26	27	29	42	47	47	43	45	44	39	38	28	(38)

And in regard to the subject of the Table it is stated:—"Assuming the humidity, and atmospheric pressure (30·045) to remain unaltered, the temperature of the air must descend about 12°, before precipitation in the form of cloud, fog, or rain, can take place. This circumstance is intimately connected with an interesting phenomenon in the immediate neighbourhood:—the Table Mountain *Cap*.

"By observation it appears that the temperature of the air decreases nearly 3°.38 for each 1000 feet of height, or about 300 feet for one degree of Fahrenheit's Thermometer. If, therefore, the stratum of air in the condition above specified should be suddenly lifted 3600 feet, viz: to about the height of Table Mountain, its temperature would be lowered by 12°, and a portion of the humidity would be condensed in the form of cloud or fog, irrespective of each cubic foot becoming lighter by about 61 grains.

"The strong and occasionally violent southerly winds which prevail during these months effect the displacement: Table Mountain, like a huge wall, receives some four miles in breadth of the current, which bounds up with diminishing temperature, and deposits the celebrated 'table cloth or cap' on the top.

"The upper surface of this majestic white cap is smoothed off like a well-dressed peruke: its north border hangs over the precipice, drapery fashion; but during very strong wind it pours down like a cataract to about 1000 feet from the top, where, entering a warmer temperature, it dissolves and disappears.

"Not the least interesting of the phenomena occasionally attending it is the *revolving* mass of detachments which hovers over Cape Town, fed by a stream of shreds from the cap on one side, which are disbursed or thrown off on the other, and float towards Table Bay, where they disappear.

"The 'black south-easter' cap differs from the preceding by the *nimbus* tint of a canopy of cloud, which projects on the southern side of the mountain over Rondebosch and Claremont, and from which light rain occasionally falls.

"The strong southerly winds are ushered in by the *tops* of the higher elevations on each side of False becoming covered in rapid succession from the southward: but these elevations seldom remain covered throughout the gale.

"The mean temperature of the winter months—June, July, and August—is 55°, the *humidity* 81·2, and the temperature of the *dew point* 49°,—viz., 6° below the mean temperature of the air. The prevailing winds are from the north, north-west, west, and occasionally from the south-west; and they are generally accompanied by rain. Hail-storm squalls are usually from the south-west. As seen from the Observatory, the first indication of a 'north-wester' is the appearance of a mass of condensed vapour rolling over the 'Lion' hill, and enveloping the signal station, also the air feels damp, and a swell sets into Table Bay: the tops of the ridges bordering the shore in the direction of Hooghtes Bay become covered, and next, but not always, Table Mountain.

"Strong wind, with squalls and showers, more or less heavy, follow these harbingers, and the fogs which now cover the elevations are of the usual European cast. The phenomena which are characteristic of *warm dry* air being forced upwards by strong wind, and again descending, are here entirely wanting.

"The duration of a North-wester fluctuates between two days and a week, sometimes ten days.

"The average temperature, *humidity*, and *dew point*, of the spring and autumn months—March, April, May; September, October, and November—are respectively 62°·4, 76°·1, 54°·3, and 61°, 74°, and 52°·1.

"May and September are the most delightful months of the year: the mean temperature of each is nearly the same (nearly 58°), but September presents to the eye luxuriant green vegetation, intermingled with beautiful flowers, and in common with the fine weather between the north-westers of the winter months, a transparent atmosphere, such as few climates can boast of; and blue-tinted mountains, whose sharp outlines deceptively plant them on the foreground.

"*Low* fogs occasionally occur, in the winter and autumn in particular—viz., the fog of calm weather, above which the tops of mountains, high hills, and topmasts of ships, are visible, and which is dispersed by the heat of the sun."

"It is a strange thing," says Ruskin, "how little, in general, people know about the sky. It is the part in creation in which Nature has done more for the sake of pleasing man, more for the sole and evident purpose of talking to him and teaching him, than in any other of her works; and it is just the part in which we least attend to her. There are not many of her other works in which some more material or essential purpose than the mere pleasing of man is not answered by every point of their organisation; but every essential purpose of the

F

sky might, so far as we know, be answered, if once in three days, or thereabouts, a great, ugly, black rain-cloud were brought up over the blue, and everything well watered, and so all left blue again till the next time, with, perhaps, a film of morning and evening mist for dew. And, instead of this, there is not a moment of any day in our lives when nature is not producing scene after scene, picture after picture, glory after glory, and working still upon such exquisite and constant principles of the most perfect beauty, that it is quite certain it is all done for us, and intended for our perpetual pleasure. And every man, wherever placed, however far from other sources of interest or of beauty, has this doing for him constantly. The noblest scenes of the earth can be seen and known but by few; it is not intended that man should live always among them; he injures them by his presence, he ceases to feel them if he be always with them: but the sky is for all; bright as it is, it is not

> 'Too bright, nor good,
> For human nature's daily food.'

It is fitted in all its functions for the perpetual comfort and exalting of the human heart, for soothing it and purifying it of its dross and dust. Sometimes gentle, sometimes capricious, sometimes awful, never the same for two moments together; almost human in its passions, almost spiritual in its tenderness, almost divine in its infinity,—its appeal to what is immortal in us is as distinct as its ministry of chastisement or of blessing to what is mortal is essential. And yet we never attend to it, we never make it a subject of thought, but as it has to do with our animal sensations: we look upon all by which it speaks to us more clearly than to brutes, upon all which bears witness to the intention of the Supreme that we are to receive more from the covering vault than the light and the dew which we share with the weed and the worm, only as a succession of meaningless and monotonous accident, too common and too rare to be worthy of a moment of watchfulness, or a glance of admiration. If in our moments of utter idleness and insipidity we turn our eyes to the sky as a last resource, which of its phenomena do we speak of? One says it has been wet; and another, it has been windy; and another, it has been warm. Who, among the whole chattering crowd, can tell one of the forms and the precipices of the chain of tall mountains that girded the horizon at noon yesterday? Who saw the narrow sunbeam that came out of the south and smote upon their summits until they melted and mouldered away in a dust of blue rain? Who saw the dance of the dead clouds when the sunlight left them last night, and the west wind blew them before it like withered leaves? All has passed, unregreted as unseen; or if the apathy be ever shaken off, even for an instant, it is only by what is gross, or what is extraordinary; and yet, it is not in the broad and fierce manifestations of the elemental energies, not in the clash of the hail nor the drift of the whirlwind, that the highest characters of the sublime are developed.

"God is not in the earthquake, nor in the fire, but in the still small voice. They are but the blunt and the low faculties of our nature which can only be addressed through lamp-black, and lightning. It is in quiet and subdued passages of unobtrusive majesty—the deep, the calm, and the perpetual; that which must be sought ere it is seen, and loved ere it is understood; things which the angels work out for us daily, and yet vary eternally; which are never wanting, and never repeated; which are to be found always, yet each found but once;—it is through these that the lesson of devotion is chiefly taught, and the blessing of beauty is given."

Attention having been directed to the beauty as well as the grandeur of cloud formation, I deem it not unjustifiable to open up this field of study more largely, believing that thus may be provided a source of gratification to intelligent colonists and their families, who live far from cities and crowded abodes of men, but under a sky such as is seldom seen in Central Europe; and it is a study intimately connected with the water supply of the country. In doing this I shall avail myself freely of the word-painting of Ruskin, while borrowing or supplying scientific expositions of what is described, required by us in our present inquiry, though not required by him for the accomplishment of what he had in view. The passages cited and to be cited are taken from the second part of his treatise on "Modern Painters; or Truths, and the importance of obtaining truthful impressions and giving truthful representations of what we see," and from that section of it which relates to truth of skies, of the open sky, of clouds, and of effects of light.

In South Africa I have witnessed more of the grandeur and the magnificence of cloud formations than I have seen in any land which I have visited. I have nowhere else seen anything in the heavens to compare with the magnificence of the "Table-Cloth" laid upon Table Mountain by a south-easter, as seen from near St. Andrew's Church on rounding the curve of the road in coming from Sea Point to Cape town—with the sky all round of a deep blue, and the sun shining brightly upon the gigantic fleeces rolling over the level summit, and descending continuously the precipitous front of the mountain, threatening to overwhelm the town and all its interests, as we may picture the Deluge rushing over some mountain range, bringing death to the inhabitants of some city nestling under its shelter,—death, sudden, inevitable, and involving every living thing, not only man, woman, and child, but "man, and bird, and beast." The only sight which has ever produced similar effects upon my mind is that of the Falls of Niagara. Of these I was reminded on first seeing the laying of the Table-Cloth; and of this I was again reminded on revisiting the Falls.

A cloud formation, not less grand and more rousing, if not more overpowering, is seen when cloud masses, such as are seldom seen in peaceful Europe, advancing in different directions, come into collision —rushing into conflict, as we picture armies on the field of battle, and ships of war at sea pouring against each other broadsides designed to carry devastation, and destruction, and death: an awful sight, which speaks of passions hellish and devilish rather than humane, angelic, or divine. At other times the mimic war assumes an appearance not less suggestive but more pleasing, more beautiful if less sublime; such a case I have cited in treating of the effects of forests on the humidity of the climate, in "Forests and Moisture."

"On Thursday evening," writes a correspondent from Sea Point, "during the hard south-east gale, some very remarkable and beautiful cloud effects were witnessed. Dark masses of cloud came driving across the Lion's Back at a furious rate, obscuring the brilliant moonlight. But in a few seconds after passing the zenith, their speed became checked, and, at the same time, they began rapidly to dissolve. Some of the cloud masses hung for a few seconds apparently motionless between the opposing currents of wind. Others, again, yielded to the counter curreut, and rushed back furiously from the north, meeting the driving masses from the south, and, almost in the moment of apparent meeting, dissolving utterly away. It was like the unavailing charge of a gallant body of light cavalry against a masked battery of artillery,—swept away before the cruel storm. The sudden alternations of clear blue sky, with black thunderous-looking masses of light fleecy clouds, and the changing iridescent hues of the vapours as they drifted across the moon, were very striking, and produced a scenic effect which those who witnessed it will not soon forget."

Another cloud scene, which I have also had occasion to refer to in the same connection, is not unfrequently seen at the Cape. It is seen occasionally, but not frequently, in Britain. It is one which seems to be the counterpart of what was seen from Carmel by the servant of Elijah after a long-continued drought of three years and six months—little clouds appearing in the heaven, such as was described as a cloud like a man's hand, which so increase and multiply that ere long the heavens are black with cloud. There may or there may not be an abundance of rain, but after a time the whole dissolves away, and 'leaves not a rack behind.'

In Europe such changes occasioned by the co-mingling of winds of different temperatures are not so frequent, but they are not unknown. The following narrative in the "Philosophical Journal," No. 132, dating back to 1812, suggests at least such scenes. The narrator, Mr Forster, states, "that on the 18th of May, the day was close and warm, and that in the afternoon different modifications of clouds might be observed at different heights of the atmosphere. In some places *cirro-stratus* might be distinguished; the clouds in others shewed the *cirro-cumulative* aggregation; *cumuli* increased in density, and cirrose fibres crossed their summits transversely, forming *cumulo-stratus,* which appeared like mountains in the horizon. The process of *nimbification* was going on rapidly, and distant thunder was heard. Towards evening, the sky between the clouds below the sun appeared of an unusual brown lake colour; and as the close of the day approached, the clouds in the horizon assumed a deep blackish blue colour, bordered with bright gold. Flocks of *cumulus* floated with the wind, and refracted a dark lake-coloured light; till at last all the separate characters were lost in one dark and dense mass; which terminated in rain during the night.

"The morning of the 19th was also rainy; but it was fair in the afternoon, though the continuous sheet of dark cloud still remained, notwithstanding a strong wind blew from the north.

"Early on the morning of the 20th, the same uniform sheet of cloud obscured the sky. As the day advanced it broke, and this dense sheet of *nimbus,* which had been originally formed by the collapse of several distinct modifications, appeared to dissolve itself into them again. As the sheet broke, part of it seemed to mount up into a higher and comparatively calm region, and formed itself into *cirro-cumulus,* in some places disposed like wind-rows of hay, in others consisting of small roundish nubeculæ of various sizes; and into *cirro-stratus,* consisting either of flat sheets of thin vapour with dentated edges, or disposed in streaks: other parts of the once continuous sheet of *nimbus* descended into a lower region, and floated along in flocks of *cumulus*, with a strong wind, and the day became very fine. In the evening the distinct modifications again seemed lost in a general mistiness of the atmosphere, which, as it became darker, seemed very red-coloured; and this vapour was seen to thicken in particular places, which became dense *nimbi* again, and gave forth vivid flashes of lightning; and thunder storms continued through the night.

"From the evident decrease in the quantity of cloud during the fine part of the day, it is evident, that, while part of the sheet of *nimbus*, which obscured the sky in the morning, divided itself again into the several modifications, the collision of which originally formed it, great part must have been absorbed by the air; this is further probable from the great transparency and dense blue colour of the sky between the clouds."

In explanation of terms used, for which we are indebted to an endeavour made by Howard, in the Askesian Lectures of 1802, to embody the chief characters of clouds in a determinate nomenclature, which nomenclature has been generally adopted, I may state that he divides clouds into three primary modifications: Cumulus, Stratus, and Cirrus, with intermediate forms graduating into one another under the names Cumulo-stratus, Cirro-stratus, Cirro-cumulus, and lastly, a composite form, resulting from a blending or confusion of the others, under the name Cirro-cumulo-stratus or Nimbus.

The term *Cumulus*, is in Latin applied to a heap; *Stratus*, to a strewing or scattering, or spreading out flat; *Cirrus*, to a lock of hair. "When a body of vapour," says Sir John Herschel, in writing on this subject in his volume on "Meteorology," which appeared previously as an article in the "Encyclopædia Brittannica," is generated from any warm evaporating surface, it ascends by its relative levity, losing sensible heat, as well by its own expansion as by its bodily transfer into and intermixture with colder air. Should the supply of vapour, however, not be very copious, and should it find itself, in its ascent, always in a region hygrometrically dry, it by no means follows that it will reach the point of precipitation; but should the reverse of these conditions obtain, viz., a copious and continued supply from below, and a state of vaporous tension already existing aloft approaching saturation, a portion will be precipitated in visible cloud on arriving at a certain level. When this process takes place in a calm state of the air, and the evaporating surface is limited in extent, or irregularly distributed in patches (as over marshy ground, rivers, lakes, &c.), or if any other cause dispose the vapour to rise in columnar bodies of greater or less extent, the summits of these are marked by protuberant masses or piles of cloud, with generally rounded outlines, which appear to repose on flat bases, indicating the 'vapour plane,' or that level where hygrometric saturation commences. To such clouds, in Howard's *Nomenclature of Clouds*, the name of Cumulus is assigned. They abound in the calm latitudes of the equatorial seas, and form a distinguishing feature in the meteorology of that region.

"Mr Howard defines *cumulus* as *sursum crescens*, increasing upward from a horizontal base. This is in conformity with its origin. When sharply terminated by rounded spherical forms, which under sunshine appear as snow-heaps, the form clearly indicates the act of self-dissipation in invisible vapour into the upper relatively dry air.

"*Stratus* consists of horizontal sheets. Its situation is low in the atmosphere, and may be considered as intermediate between cloud and fog, being chiefly formed at night, and under the influence of radiation either from the surface of a ground fog (as already described), or from impurities floating in the air itself. The latter is remarkably the case in great cities in which coal is chiefly consumed as fuel, and gives rise to those dense, yellow, suffocating fogs which infest London in the winter months. A very peculiar phenomenon exhibited by the London atmosphere has been described to us by the Astronomer-Royal, and appears to be referable to the same cause. In calm evenings after sunset, as seen from the Royal Observatory on Greenwich Hill, the vast irregular mass of smoke hovering over London appears to subside. Its heaped and turbulent outline becomes flat, and sinks rapidly into a low level cloud-bank, with a very definite outline, and fair sky above. It would seem that each particle of soot, acting as an insulated radiant, collects dew on itself, and sinks down rapidly as a heavy body. *Stratus* is also sometimes formed very suddenly on a higher level, when in a clear, calm night the general temperature of the air sinks by radiation, or by diminution of atmospheric pressure, till at some definite altitude above the surface the dew-point is attained. Thus, on the night of April 19, 1827, the sky, up to 16h. 16m. Sid. T, being perfectly cloudless, and not a breath of wind stirring, stratus at a high level commenced in the eastern horizon, and in eight minutes had exteuded to the western, obscuring the whole sky, the calm remaining unbroken. In this case the velocity of propagation of the edge of the cloud from east to west (or following the sun) could not have been less than 300 miles per hour (*Mem. Ast. Soc.* iii. 51).

"*Cirrus* consists of fibrous, wispy, or feathery clouds occupying the highest region of the atmosphere. The name of mares' tails, by which *cirri* are commonly known, describes their aspect well. Of all forms of cloud these present the greatest variety. Their filamentous structure clearly indicates them as either in the act of originating from the union of aerial currents running parallel to each other, or as the residues of dissolving cloud drawn out into fibres by wind. Mr Howard is disposed to assign to them an electric origin, or at least to attribute their fibrous appearance to electricity, but in this view we cannot coincide. From the great elevation of cirrus it is more than probable that its particles are frozen, and of course crystalline, and that to this constitution it is owing that halos, coronæ, and other optical appearances referable to reflections and refractions in ice crystals, appear almost invariably in this cloud and in its derivative forms, especially in the *cirro-cumulus*. It is said to be often a precursor of windy weather.

"The *Cirro-cumulus*, most characteristically known as a 'mackerel sky,' consists often of small roundish masses disposed with more or less regularity and connection. They usually float at great elevations, and often appear as a loftier stratum through the intervals of lower clouds, contrasting strongly by their slow (and sometimes contrary) movement, with the scud and drift of the inferior masses. They are frequent in summer, and attendant on dry and warm weather.

"*Cirro-stratus* 'appears to result from the subsidence of the fibres of *cirrus* to a horizontal position, at the same time approaching laterally. The form and relative position, when seen in the distance, frequently give the idea of shoals of fish.' It often precedes wind and rain, and often forms the ground on which halos and parhelia are projected.

"*Cumulo-stratus* would seem to be the modifications of *cumulus*, when the columns of rising vapour which go to form it arrive in an upper atmosphere not sufficiently dry to round off its summits by rapid evaporation, allowing them to spread horizontally and form flat-topped, mushroom-shaped masses, the upper parts of which are often curled by the wind of an upper current into cirrous wisps, or cleanly cut off by a horizontal plane, forming an "anvil-shaped cloud" with a lateral projection, generally considered as a precursor of wind below, and probably arising from its first impression at a higher level due to the greater mobility of the upper strata of the air, as less retarded by friction. The tendency of *Cumulo-stratus* is to spread, overcast the sky, and settle down into the *Nimbus*, and finally to fall in rain. When two strata of clouds on different levels tend to unite, it is evident that the intermediate region must be nearly or quite in a state of hygrometric saturation. The cloud then forms confusedly and in irregular masses through the whole region, and finally resolves itself into heavy rain."

All that has been stated I have seen at the Cape; at the same time I have remarked a well-defined outline in clouds, contrasting with the appearance of the rain cloud, which will afterwards be described, which is more prevalent elsewhere. The difference is like the difference between the appearance presented by the belching forth of the steam from a high-pressure steam-engine, on a frosty day, and on a day of fog: in the one case melting away, but well defined so long as it exists; in the other, mixing with the fog, rather than melting away, and exhibiting an ill-defined outline. And in both cases the cause is the same.

Ruskin, in writing of clouds, commences with remarks on the open sky, in regard to which he writes :—"The blue of the open sky may be spoken of as the colour of the pure atmospheric air, apart from the aqueous vapour commingled with it.

"It is constantly modified by the varying quantity of the aqueous vapour which is suspended in it, whose colour in its most imperfect, and therefore most visible, state of solution is pure white (as in steam); which receives like any other white the warm hues of the rays of the sun, and according to its quantity, and imperfect solution, makes the sky paler, and at the same time, more or less grey, by mixing warm tones with its blue.

"This grey aqueous vapour, when very decided, becomes mist, and when local, cloud. Hence the sky is to be considered as a transparent blue liquid, in which, at various elevations, clouds are suspended, these clouds being themselves only particular visible spaces of a substance with which the whole mass of this liquid is more or less impregnated.

"Yet we often find that even by the old masters the sky is thought of as a clear, high, material dome, beneath which the clouds as separate bodies are suspended, and in consequence, however delicate and exquisitely removed in tone their skies may be, you always look *at* them, not *through* them. In the words of Wordsworth :—

> 'The chasm of sky above my head
> Is heaven's profoundest azure. No domain
> For fickle, short-lived clouds to occupy,
> Or to pass through ;—but rather an abyss
> In which the everlasting stars abide,
> And whose soft gleam, and boundless depth, might tempt
> The curious eye to look for them by day.'

And if you look intensely at the pure blue of a serene sky you will see that there is a variety and fulness in its very repose.

" It is not flat dead colour, but a deep, quivering, transparent body of penetrable air, in which you trace or imagine short falling spots of deceiving light, and dim shades, faint veiled vestiges of dark vapour.

" It is something into which you can see, through what is near to what is far off; something which has no surface, and through which we can plunge far, and farther, and without stay or end, into the profundity of space."

Again he writes :—" The visible sunbeams occasionally seen and their effects produce a series of phenomena connected with the open blue of the sky, to the explanation of which it is necessary that we thoroughly understand the circumstances under which such effects take place.

" Aqueous vapour, or mist, suspended in the atmosphere, becomes visible exactly as dust does in the air of a room. In the shadows you not only cannot see the dust itself, because unillumined, but you cannot see other objects through the dust without obscurity, the air being thus actually rendered more transparent by deprivation of light.

" When a sunbeam enters, every particle of dust becomes visible, and a palpable interruption to the sight, so that a transverse sunbeam is a real obstacle to the vision—you cannot see things clearly through it.

" In the same way, wherever vapour is illuminated by transverse rays, there it becomes visible, as a whiteness more or less affecting the purity of the blue, and destroying it exactly in proportion to the degree of illumination. But where vapour is in shade, it has very little effect on the sky, perhaps making it a little deeper and greyer than it otherwise could be, but not itself, unless very dense, distinguishable or felt as mist.

" The appearance of mist, or whiteness in the blue of the sky, is thus a circumstance which more or less accompanies sunshine, and which, supposing the quantity of vapour constant, is greatest in the brightest sunlight. When there are no clouds in the sky, the whiteness, as it affects the whole sky equally, is not particularly noticeable. But when there are clouds between us and the sun, the sun being low, these clouds cast shadows along and through the mass of suspended vapour.

"Within the space of these shadows, the vapour becomes transparent and invisible, and the sky appears of a pure blue. But where the sunbeams strike, the vapour becomes visible in the form of the beams, occasioning these radiating shafts of light which are one of the most constant accompaniments of a low sun.

"The denser the mist, the more distinct and sharp-edged will these rays be; when the air is very clear, they are more vague, flashing, graduated passages of light; when it is very thick, they are keen-edged and decisive in a high degree.

"We see then, *first*, that a quantity of mist dispersed through the whole space of the sky, is necessary to this phenomenon; and *secondly*, that what we usually think of as beams of greater brightness than the rest of the sky, are in reality only a part of that sky in its natural state of illumination, cut off and rendered brilliant by the shadows from the clouds,—that these shadows are in reality the source of the appearance of beams,—that, therefore, no part of the sky can present such an appearance except when there are broken clouds between it and the sun; and lastly, that the shadows cast from such clouds are not necessarily grey or dark, but very nearly of the natural, pure, blue of a sky destitute of vapour.

"Now, as it has been proved that the appearance of beams can only take place in a part of the sky which has clouds between it and the sun, it is evident that no appearance of beams can ever begin from the orb itself, except when there is a cloud or solid body of some kind between it and us; but that such appearances will almost invariably begin on the dark side of some of the clouds around it, the orb itself remaining the centre of a broad blaze of united light.

"Wordsworth has given us in two lines, the only circumstances under which rays can ever appear to have origin in the orb itself:—

'But rays of light,
Now *suddenly* diverging from the orb
Retired behind the mountain top, or veiled
By the dense air, shot upwards.'"

I accept as beautifully simple, and in so far as it is descriptive of what is seen, the rationale given by Ruskin; but, viewed in the light of science, there are seen projections which may, without detriment to it, be removed. The haze is occasioned by what I would designate *moisture*, in contradistinction to *aqueous vapour*—the designation employed by him, because, technically speaking, aqueous vapour is a name applied to the moisture yielded, upon analysis, by dry air, being diffused in it, like the carbonic acid, in an invisible condition and moisture, to what would produce a sensation of humidity, being a portion of the vapour suspended in the air in excess of what, at the temperature at which it then is, it can sustain diffused in an invisible condition. And I consider it would be more correct to say that the blue spoken of is the colour produced by light being transmitted through the atmosphere, such as it is, having aqueous vapour diffused in it in an invisible state, than to speak of it as the colour of pure azote, or of nitrogen and oxygen. I have been told that a glacier seen from beneath it by transmitted light is of a bluish colour. The ocean, when of a great depth, is blue, proverbially spoken of as "the dark blue sea"; when of less depth, with yellow sand underneath, it is green, a colour which is produced by the commixture of blue and yellow. That there is aqueous vapour diffused in the atmosphere from the surface of the earth, at a height of at least 15,000 feet, appears from the fact that throughout that depth such vapour is deposited from it, forming clouds; and it is open to question whether the blue colour of the sky be not attributable to the moisture which is in the atmosphere? And an affirmative response would not be irreconcilable with the explanation given of the phenomena described, if the term aqueous vapour be accepted as the synonym of moisture, as explained above: vapour held no longer as a constituent of the air, but still suspended in it,—as carbonate of lime may be produced from a solution of the bicarbonate, or carbonate of magnesia, from Epsom salts,—vapour which may remain suspended for a considerable time before being precipitated, or may be again dissolved without precipitation taking place.

It is in this light that I read the description given by Ruskin of the formation of the haze so often seen when the temperature is near to what is known as the dew-point, or point of saturation.

From the consideration of the open sky, Ruskin proceeds to treat of the truth of clouds:—"The first and most important character of clouds," says he, "is dependent on the different altitudes at which they are formed. The atmosphere may be conveniently considered as divided into three spaces, each inhabited by clouds of specific character, altogether different, though there is in reality no distinct limit fixed between them by nature; clouds being formed at *every* altitude, and partaking, according to their altitudes, more or less of the characters of the upper or lower regions. The scenery of the skies is thus formed of an infinitely graduated series of systematic forms of clouds,—each of which has its own region, in which alone it is found,—and each of which has specific characters, which can alone be properly determined by comparing them as they are found, clearly distinguished by intervals of considerable space.

"The sky has accordingly been divided into three regions: the upper region, or region of the *cirrus;* the central region, or region of the *stratus;* the lower region, or region of the rain-cloud.

"The clouds of the upper region—the *cirrus*—never touch even the highest mountains of Europe; they may, therefore, be looked upon as never formed below an elevation of 15,000 feet; it is they which compose the multitudinous lines of delicate vapour with which the blue of the sky is commonly streaked or speckled after several days of fine weather.

"Their chief characteristics are: their symmetry, sharpness of edge, multitude, purity of colour, and variety.

"First, of their symmetry:—They are nearly always arranged in some definite and evident order, commonly in long ranks, reaching sometimes from the zenith to the horizon, each rank being composed of an infinite number of transverse bars of about the same length, —each bar thickest in the middle, and terminating in a traceless vaporous point at each side; the ranks are in the directions of the wind, and the bars, of course, at right angles to it, and these latter are commonly slightly bent in the middle. Frequently, two systems of this kind, indicative of two currents of wind at different altitudes, intersect each other, forming a net-work.

"Another frequent arrangement is a group of excessively fine silky parallel fibres, commonly radiating, or having a tendency to radiate, from one of their extremities, and terminating in a plumy sweep at the other: these are vulgarly known as 'mares' tails.' The plumy and expanded extremity of these is often bent upwards, sometimes back and up again, giving an appearance of great flexibility and unity at the same time, as if the clouds were tough, and could hold together however bent. The narrow extremity is invariably turned to the wind, and the fibres are parallel with its direction.

"The upper clouds always fall into some modification of one or other of these arrangements. They thus differ from all other clouds, in having a plan and system; whereas other clouds, though there are certain laws which they cannot break, have yet a perfect freedom from any thing like a relative and general system of government.

"The upper clouds are to the lower, what soldiers on parade are to a mixed multitude; no men walk on their heads or on their hands, and so there are certain laws which no clouds violate; but there is nothing except in the upper clouds resembling symmetrical discipline.

"Secondly, sharpness of edge:—The edges of the bars of the upper clouds which are turned to the wind, are often the sharpest which the sky shows; no outline whatever of any other kind of cloud, however marked and energetic, ever approaches the delicate decision of these edges.

"The outline of a black thunder-cloud is striking, from the great energy of the colour or shade of the general mass; but as a line it is soft and indistinct compared with the edge of the *cirrus*, in a clear sky with a brisk breeze.

"On the other hand the edge of the bar turned away from the wind is always soft, often imperceptible, melting into the blue interstice between it and its next neighbour.

"Commonly, the sharper one edge is, the softer is the other, and the clouds look flat, and as if they slipped over each other like the scales of a fish.

"When both edges are soft, as is always the case when the sky is clear and windless, the cloud looks solid, round, and fleecy.

"Thirdly, multitude:—The delicacy of these vapours is sometimes carried into such an infinity of division, that no other sensation of number that the earth or heaven can give is so impressive. Number is always most felt when it is symmetrical,—and therefore, no sea waves nor fresh leaves make their number so evident, or so impressive as these vapours. Nor is Nature content with an infinity of bars—or lines alone—each bar is in its turn severed into a number of small undulatory masses, more or less connected according to the violence of the wind.

"When this division is merely affected by undulation, the cloud exactly resembles sea sand ribbed by the tide; but where the division amounts to real separation we have the mottled or 'mackerel' sky.

"Commonly, the greater the division of its bars, the broader and more shapeless is the rank or field, so that in the mottled sky it is lost altogether, and we have large irregular fields of equal size, masses like flocks of sheep; such clouds are 3 or 4000 feet below the legitimate *cirrus.*

"They have been observed to cast a shadow on Mont Blanc at sunset, so that they must descend to nearly within 15,000 feet of the earth.

"Fourthly, purity of colour:—The nearest of these clouds, those over the observer's head, being at least three miles above him, and nearly all entering the ordinary sphere of vision, further from him still, their dark sides are much greyer and cooler than those of other clouds, owing to their distance. They are composed of the purest aqueous vapour, free from all foulness of earthly gases, and of this in the lightest and most ethereal state in which it can be, to be visible. Further, they receive the light of the sun in a state of far greater intensity than lower objects, the beams being transmitted to them through atmospheric air far less dense, and wholly unaffected by mist, smoke, or any other impurity. Hence their colours are more pure and vivid, and their white less sullied than those of any other clouds.

"Lastly, variety:—Variety is never so conspicuous as when it is united with symmetry. The perpetual changes of form in other clouds is monotonous in its very dissimilarity, nor is difference striking when no connection is implied. But through a range of barred clouds crossing half the heaven, all governed by the same forces, and falling into one general form, there is yet a marked and evident dissimilarity between each member of the great mass—one more finely drawn, the next more delicately moulded, the next more gracefully bent—each broken into differently-moulded and variously-numbered groups The variety is doubly striking because contrasted with the perfect symmetry of which it forms a part. Hence the importance of the truth that nature never lets one of the members of even her most disciplined groups of cloud be like another; but though each is adapted for the same function, and in its great features resembles all the others, not one, out of the millions with which the sky is checkered is without a separate beauty and character, appearing to have had distinct thought occupied in its conception, and distinct forces in its production; and in addition to this perpetual invention, visible in each member of each system, we find systems of separate clouds intersecting each other, the sweeping lines mingled and interwoven with the rigid bars,—these in their turn melting into banks of sand-like ripple, and flakes of drifted and irregular foam; under all, perhaps the massy outline of some lower cloud moves heavily across the motionless buoyancy of the upper lines, and indicates at once their elevation and their repose. . . . .

"Like Turner amongst the painters, Wordsworth amongst the poets excels in the truthfulness of his pictures; and thus does he describe a scene such as I have indicated :—

'But rays of light,
Now suddenly diverging from the orb
Retired behind the mountain tops, or veiled
By the dense air, shot upwards to the crown
Of the blue firmament—aloft—and wide:
And multitudes of little floating clouds,
Ere we, who saw, of change were conscious, pierced
Through their ethereal texture, had become
Vivid as fire,—clouds separately poised,
Innumerable multitudes of forms
Scattered through half the circle of the sky;
And giving back, and shedding each on each,
With prodigal communion, the bright hues
Which, from the unapparent fount of glory,
They had imbibed, and ceased not to receive.
That which the heavens displayed the liquid deep
Repeated, but with unity sublime.'*

The truthfulness of Wordsworth in his descriptions, and of Turner in his paintings, are frequently noted by Ruskin with admiration. In reference to these he says, after citing this passage:—" There is but one master whose works we can think of while we read this, one alone has taken notice of the neglected upper sky; it is his peculiar and favourite field; he has watched its every modification, and given its every phase and feature; at all hours, in all seasons, he has followed its passions and its changes, and he has brought down and laid open to the world another apocalypse of heaven."

In reading descriptions of Turner's paintings by Ruskin, I have often felt as if I were gazing upon the scene described, and this not less in his descriptions of his upper skies than in those of his other achievements in verisimilitude of representation.

"There is scarcely a painting of Turner's," says he, "in which serenity of sky and intensity of light are aimed at together, in which these clouds are not used, though there are not two cases in which they are used altogether alike. Sometimes they are crowded together in the masses of mingling light, as in the Shylock, every part and atom sympathising in that continuous expression of slow movement which Shelley has so beautifully touched:—

> 'Underneath the young grey dawn
> A multitude of dense, white fleecy clouds,
> Were wandering in thick flocks along the mountains,
> *Shepherded by the slow, unwilling wind.*'

"At other times they are blended with the sky itself, felt only here and there by a ray of light calling them into existence out of its misty shade, as in the Mercury and Argus; sometimes, they appear in a few detached, equal rounded flakes, which seem to hang motionless, each like the shadow of the other, in the deep blue of the zenith, as in the Acro-Corinth; sometimes they are scattered in fiery flying fragments, each burning with separate energy, as in the Téméraire; sometimes woven together with fine threads of intermediate darkness melting into the blue, as in the Napoleon. But in all cases each atom of the multitude has its own character and expression.

"Though they be countless as leaves, each has its portion of light, its shadow, its reflex, its peculiar and separating form.

"Take, for instance," says he, "the illustrated edition of Roger's Poems, and open it at the 80th page, where is the vignette of Sunrise on the Sea, and observe how every attribute of the upper sky, to which reference has been made, is there rendered with the faithfulness of a mirror; the long lines of parallel bars, the delicate curvature from the wind, which the inclination of the sail shows you to be from the west; the excessive sharpness of every edge which is turned to the wind, the faintness of every opposite one, the breaking up of each bar into rounded masses, and, finally, the inconceivable variety with which individual form has been given to every member of the multitude, and not only individual form, but roundness and substance even where there is scarcely a hair's breadth of cloud to express it in. Observe, above everything, the varying indication of space and depth in the whole, so that you may look through and through, from one cloud to another, feeling, not merely how they retire to the horizon, but how they melt back into the recesses of the sky; every interval being filled with absolute air, and all its spaces so melting and fluctuating and fraught with change as with repose, that as you look you will fancy that the rays shoot higher and higher into the vault of light, and that the pale streak of horizontal vapour is melting away from the cloud that it crosses. . . . .

"But it has been shown that sunbeams, or the appearance of them, are always sharper in their edge in proportion as the air is more misty, as they are most defined in a room where there is most dust flying; and here we find that though there is a blaze of light, its beams are never edged; a tendency to rays is visible, but you cannot in any part find a single marked edge of a rising sunbeam—the sky is merely more flushed in one place than another. . . .

"In striking contrast to this, and yet not less true to nature, is his Alps at daybreak—in the same volume (p. 193). Here we have the *cirri* again, but now they have no sharp edges; they are all fleecy and mingling with each other, though every one of them has the most exquisite indication of individual form, and they melt back, not till they are lost in exceeding light, as in the other plate, but into a mysterious, fluctuating, shadowy sky, of which, though the light penetrates through it all, you perceive every part of it charged with vapour.

"Notice particularly the half-indicated forms even where it is most serene, behind the snowy mountains; and now, how are the sunbeams drawn? No longer indecisive, flushing, palpitating, every one is sharp and clear, and terminated by definite shadows; note especially the marked lines on the upper cloud; finally, observe the difference in the modes of indicating the figures, which are here misty and indistinguishable, telling only as shadows, though they are near and large; while those in the former vignette came clear upon the eye, though they were so far off as to appear mere points."

"But it is necessary to notice not only the purity and the vividness, but also the variety and the changeableness of colour in these clouds.

"If you watched for the next sunset, when there are a considerable number of these *cirri* in the sky, you will see, especially at the zenith, that the sky does not remain of the same colour for ten inches together; one cloud has a dark side of cold blue, and a fringe of milky white; another above it has a dark side of purple, and an edge of red; another, nearer the sun, has an under-side of orange and an edge of gold; these you will find mingled with, and passing into the blue of the sky, which in places you will not be able to distinguish from the cool grey of the darker clouds, and which will be itself full of gradation, now pure and deep, now faint and feeble; and all this is done, not in large pieces, nor on a large scale, but over and over again on every square yard, so that there is no single part or portion of the whole sky which has not in itself variety of colour enough for a separate picture, and yet no single part which is like another, or which has not some peculiar source of beauty, and some peculiar arrangement of colour of its own."

To the *cirri* the designation goats' hair is given by agricultural labourers in some parts of Scotland. *Cirri, mares' tails,* and *goats' hair* speak of the same similitude having been remarked by scientific and unscientific observers alike.

In regard to the production of the mackerel sky, I had occasion, in treating of "Forests and Moisture," to say, in illustration of an opinion expressed by Sir William Thomson, at the meeting of the British Association for the advancement of Science, held in Glasgow last year (1876):—"Schoolboys amuse themselves sometimes with skimming flat stones along the surface of smooth water. The same phenomenon is produced upon a greater scale when a cannon ball, aimed at a floating target, strikes the water and rebounds into the air again and again ere it disappears. I have often, at the Cape of Good Hope, seen indications of the southeast wind advancing in waves, raising dust where it seemed to descend and strike the ground, but leaving undisturbed alternate stripes over which it seemed to bound or *rebound* at a higher level. Thus may it be with all currents, whether of water or of air, descending and striking upon other matter, solid, liquid, or gaseous, at an acute angle of inclination—as is indicated by the undulations or waves raised by the breath on a cup of tea, by the wind on a pond or placid stream, and by a storm upon the sea. And, according to the supposition of Sir William Thomson, a current of air striking upon a lower current, advancing in the same direction but with less velocity, or advancing in another direction, or upon a stratum of air in a state of quiescence, may rebound as does the skimming stone or the rebounding cannon ball on striking the water. And if the temperature of this current of air were near to that of the dew-point, or point of saturation, when it rebounded to a higher elevation, the temperature being thereby reduced, a condensation of moisture would ensue; but when the rebounding force was exhausted, and it again descended to a lower level, it would acquire a higher temperature, and the condensed vapour would be again dissolved in the air, producing the phenomenon of alternating stripes of cloud and of sky."

Upon the same principle may be explained the appearance of a flowing tail; if from obstruction to the wind, offered by a cloud of irregular outline, the current, diverging beyond the obstruction, flows at different altitudes, what is raised to an altitude where the temperature is below the dew-point, will appear as streaks of cloud. While what flows in at a lower altitude of higher temperature, will not have the aqueous vapour with which it is charged deposited as visible moisture.

"We have next," writes Ruskin, "to investigate the character of the Central Cloud Region, including all clouds which are the usual characteristic of ordinary serene weather. In Switzerland, these are often found touching and enveloping the mountains; but they never affect those of Britain; they may, therefore, be considered as occupying a space of air 10,000 feet in height, extending from 5 to 15,000 feet above the sea.

"These clouds, according to their elevation, appear with great variety of form, often partaking of the streaked or mottled character of the higher region, and as often, when the precursors of storms, manifesting clouds closely connected with the lowest rain-clouds; but the species especially characteristic of the central region is a white, ragged, irregular, and scattered vapour, which has little form and less colour. When this vapour collects into masses, it is partially rounded, clumsy, and ponderous, as if it would tumble out of the sky; it is shaded with a dull grey, and is generally totally devoid of any appearance of energy or motion. . . . But although this kind of cloud is typical of the central region, it is not one of which Nature is fond; and she scarcely ever lets an hour pass without some manifestation of finer forms, sometimes approaching the upper *cirri*, sometimes the lower *cumulus*. . . .

"I shall, as before, glance rapidly at the great laws of specific form.

"Clouds, it is to be remembered, are not so much local vapour, as vapour rendered locally visible by a fall of temperature. Thus, a cloud whose parts are in constant motion will hover on a snowy mountain, pursuing constantly the same track upon its flanks, and yet remaining of the same size, the same form, and in the same place for half-a-day together. No matter how violent or how capricious the wind may be, the instant it approaches the spot where the chilly influence of the snow extends, the moisture it carries becomes visible, and then and there the cloud forms on the instant, apparently maintaining its form against the wind, though the careful and keen eye can see its parts in the most rapid motion across the mountain.

"The outlines of such a cloud are, of course, not determined by the irregular impulses of the wind, but by the fixed lines of radiant heat which regulate the temperature of the atmosphere of the mountain. It is terminated, therefore, not by changing curves, but by steady right lines of more or less decision, often exactly correspondent with the outline of the mountain on which it is formed, and falling, therefore, into grotesque peaks and precipices.

"The marked and angular outline of the Grandes Jorasses, at Chamonix, has been seen mimicked in its every jag by a line of clouds above it.

"Another resultant phenomenon is the formation of cloud in the calm air to leeward of a steep summit; clouds whose edges are in rapid motion, when they are affected by the current of the wind above, stream from the peak like the smoke of a volcano, yet always vanish at a certain distance from it, as steam issuing from a chimney.

"When wet weather of some duration is approaching, a small white spot of cloud will sometimes appear low on the hill flanks; it will not move, but will increase gradually for some little time, then diminish, still without moving; disappear altogether, reappear ten minutes afterwards exactly in the same spot: increase to a greater extent than before, again disappear, again return, and at last permanently; other similar spots of cloud forming simultaneously, with various fluctuations, each in its own spot, and at the same level on the hill side, until all expand, join together, and form an unbroken veil of threatening grey, which darkens gradually into storm.

"What in such cases takes place, palpably and remarakably, is more or less a law of formation in all clouds whatsoever; they being bounded rather by lines expressive of changes of temperature in the atmosphere, than by the impulses of the current of wind in which these changes take place. Even when in rapid and visible motion across the sky, the variations which take place in their outlines are not so much alteration of position and arrangement of parts as they are the alternate formation and disappearance of parts. There is, therefore, usually a parallelism and consistency in their great outlines, which gives system to the smaller curves of which they are composed; and if these great lines be taken, rejecting the *minutiae* of variation, the resultant form will almost always be angular, and full of character and decision. In the flock-like fields of equal masses, each individual mass has the effect, not of an ellipse or circle, but of a rhomboid; the sky is crossed and chequered, not honeycombed; in the lower *cumulo*, even though the most rounded of all clouds, the groups are not like balloons or bubbles, but like towns or mountains And the result of this arrangement in masses more or less angular, varied with, and chiefly constructed of, curves of the utmost freedom and beauty, is that appearance of exhaustless and fantastic energy which gives every cloud a marked character of its own, suggesting resemblance to the specific outlines of organic objects.

"These accidental resemblances make manifest the originality and vigour of separate conception in cloud forms, which give to the scenery of the sky a force and variety no less delightful than that of the changes of mountain outlines in a hill district of great elevation; and there is added to this a spirit-like feeling, a capricious mocking imagery of passion and life, totally different from any effects of inanimate form that the earth can show.

"The minor contours, out of which the larger outlines of these clouds are composed, are indeed beautifully curvilinear; but they are never monotonous in their curves. First comes a concave line, then a convex one, then an angular jag, breaking off into spray, then a downright straight line, then a deep gap, and a place where all is lost and melted away, and so on; displaying in every inch of the form renewed and ceaseless invention, setting off grace with rigidity, and relieving flexibility with force, in a manner scarcely less admirable, and far less changeful than even in the muscular forms of the human frame.

"Nay, such is the exquisite composition of all this, that you may take any single fragment of any cloud in the sky and you will find it put together as if there had been a year's thought over the plan of it, arranged with the most studied inequality, with the most delicate symmetry, with the most elaborate contrast, a picture in itself. You may try every other piece of cloud in the heaven, and you will find them every one as perfect, and yet not one in the least like another."

In many paintings, which are in other respects of great merit, we find substituted for this variety, individuality, and angular character, a mass of convex curves, each precisely like its neighbour in all respects, and unbroken from beginning to end. And it is but the truth which is affirmed by Ruskin, when he says "that the clouds which God sends upon His earth as the ministers of dew, and rain, and shade, and with which He adorns His heaven, setting them in its vault for the thrones of his spirits, have not in one instant, or atom of their existence, one feature in common with such conceptions and creations. They are often much more like half-crowns than clouds.

"But it is not the outline only which is thus systematically false. The describing of the solid form is often worse still, for it is to be remembered that although clouds of course arrange themselves more or less into broad masses, with a light side and a dark side, both their light and shade are invariably composed of a series of divided masses, each of which has in its outline as much variety and character as the great outline of the cloud; presenting, therefore, a thousand times repeated, all the characteristics of the general form."

Further, these multitudinous divisions—being dependant on, are illustrative of, what is usually in a great degree overlooked—the enormous intervening space of solid clouds.

"Between the illumined edge of a heaped cloud and that part of its body which turns into shadow, there will generally be a clear distance of several miles, more or less of course according to the general size of the cloud. But we are little apt in watching the changes of a mountainous range of cloud, to reflect that the masses of vapour which compose it, are larger and higher than any mountain range of the earth; and the distances between mass and mass are not yards of air traversed in an instant by the flying form, but valleys of changing atmosphere leagues over; that the slow motion of ascending curves, which we can scarcely trace, is a boiling energy of exulting vapour rushing into the heaven a thousand feet in a minute; and that the toppling angle whose sharp edge almost escapes notice in the multitudinous forms around it, is a nodding precipice of storms—3000 feet from base to summit. It is not until we have actually compared the forms of the sky with the hill-ranges of the earth, and seen the soaring alp overtopped and buried in one surge of the sky, that we begin to conceive or appreciate the colossal scale of the phenomena of the latter. But of this there can be no doubt in the mind of any one accustomed to trace the form of cloud among hill ranges—as it is then a demonstrable and evident fact that the space of vapour visibly extended over an ordinarily clouded sky is not less, from the point nearest to the observer, to the horizon, than 20 leagues; that the size of every mass of separate form, if it be at all largely divided, is to be expressed in terms of *miles*; and that every boiling heap of illuminated mist in the nearer sky, is an enormous mountain, 15 or 20,000 feet in height, six or seven miles over in illuminated surface, furrowed by a thousand colossal ravines, torn by local tempests into peaks and promontories, and changing its features with the majestic velocity of the volcano. . . .

"If the quick eye of a painter has failed to catch the real expression of the clouds, to be measured not by yards, but by miles, and broken up into multiplicity of forms necessary to, and characteristic of, their very nature, those forms being subject to a thousand local changes, having no association with each other, and rendered visible in a thousand places by their own transparency or their cavities,—and if masters have been led to depict them as comparatively small, round, puffed-up white bodies, irregularly associated with other round and puffed-up white bodies, each with a white light side and a grey dark side, and a soft reflected light floating a great way below a blue dome, as has been done more or less by all the old masters with scarcely an exception,—we need not wonder that people should have come generally to think only of these clouds, even of the most magnificent of them, as about as large as they look—forty yards over perhaps—perhaps as large as a hay-stack, or it may be as the battlement of a castle.

"The battlement of a castle! Let any one go out on a day when, either before or after rain, the clouds arrange themselves into vigorous masses, and after arriving at something like a conception of their distance and size, from the mode in which they retire over the horizon, let him trace and watch their varieties of form and outline, as mass rises over mass in their illuminated bodies. Let him climb from step to step over their craggy and broken slopes, let him plunge into the long vistas of immeasurable perspective, that guide back to the blue sky, and he will find his imagination lost in their immensity.

"It is also to be remarked, that in her most ponderous mass 'Nature never leaves us without some evidence of transmitted sunshine. She perpetually gives us passages in which the vapour seems to become visible only by the sunshine which it arrests and holds within itself; not caught on its surface, but entangled in its masses, —floating fleeces, precious with the gold of heaven; and this translucency is especially indicated on the dark sides even of her heaviest wreaths, which possess opalescent and delicate hues of partial illumination, far more dependent upon the beams which pass through them than in those which are reflected upon them.'

"But this translucency is only one of many of the almost inimitable beauties and magnificent grandeurs of the form of the cloud I am attempting to recall.

"The *cumulus* masses, otherwise opaque, which are thus illumined, form but the thousandth part of the variety of the effects produced on these by Nature: 'She builds up a pyramid of their boiling volumes, bars this across like a mountain with the grey curves, envelopes it with black, ragged, drifting vapour, covers the open part of the sky with mottled horizontal fields, breaks through these with sudden and long sunbeams, tears up their edges with local winds, scatters over the gaps of blue the infinity of multitude of the high *cirri*, and melts even the unoccupied azure into palpitating shades; and all this is done, over and over again, in every quarter mile. Poussin or Claude have three *similar masses*, Nature has fifty pictures, fifty aisles penetrating through angelic chapels to the Shechinah of the blue, fifty hollow ways among bewildering hills—each with its own nodding rocks and cloven precipices and radiant summits, and robing vapours, but all unlike each other except in beauty, all bearing witness to the unwearied, inexhaustless operation of the Infinite Mind."

Would you have an artist's representation of such a sky, take Turner's Babylon, engraved in Finden's Bible Illustrations.

Thus is it described by Ruskin:—"Ten miles away, down the Euphrates, where it gleams last along the plain, he gives us a drift of dark elongated vapour, melting beneath into a dim haze which embraces the hills on the horizon. It is exhausted with its own motion, and broken up by the wind in its own body into numberless groups of billowy and tossing fragments, which, beaten by the weight of the storm down to the earth, are just lifting themselves on wearied wings, and perishing in the effort. Above these, and far beyond them, the eye goes back to a broad sea of white illuminated mist, or rather cloud melted into rain, and absorbed again before that rain has fallen, but penetrated throughout, whether it be vapour or whether it be dew, with soft sunshine, turning it as white as snow. Gradually, as it rises, the rainy fusion ceases. You cannot tell where the film of blue begins, but it is deepening, deepening still; and the cloud, with its edge first invisible, then all but imaginary, then just felt when the eye is *not* fixed upon it, and lost when it is, at last rises, keen from excessive distance, but soft and mantling in its body, as a swan's bosom fretted by faint wind; heaving fitfully against the delicate deep blue, with white waves, whose forms are traced by the pale lines of opalescent shadow, shade only because the light is within it, and not upon it, and which break with their own swiftness into a driven line of level spray, winnowed into threads by the wind, and flung before the following vapour like those swift shafts of arrowy water which a great cataract shoots into the air beside it, trying to find the earth.

"Beyond these, again, rises a colossal mountain of grey *cumulus*, through whose shadowed sides the sunbeams penetrate in dim, sloping, rain-like shafts, and over which they fall in a broad burst of streaming light, sinking to the earth, and showing through their own visible radiance, the three successive ranges of hills which connect its desolate plain with space. Above, the edgy summit of the *cumulus*, broken into fragments, recedes into the sky, which is peopled in its serenity with quiet multitudes of the white, soft, silent *cirrus*; and these, again, drift near the zenith under disturbed and impatient shadows of a darker spirit, seeking rest and finding none."

Many of the appearances described present themselves at the Cape. The atmosphere being generally perfectly transparent, in consequence of the temperature being far above the dew-point, and the aqueous vapour completely dissolved, if that phrase may be allowed me, hills and mountain ranges, 40 miles, 60 miles, or even 100 miles distant, are seen presenting a well-defined outline; and often are they seen crested, crowned or surmounted by a cloud formation such as has been described. And as with these cloud forms so is it also with several of the others.

The clouds which are characteristic of the lower rainy region differ not so much in their real nature from those of the central and uppermost regions, as in appearance. The differences in appearance, though dependent almost entirely on their proximity, are marked and important.

"In the first place, the clouds of the central region have, as has been stated, pure and aerial greys for their dark sides, owing to their necessary distance from the observer; and as the distance permits, a multitude of local phenomena, capable of influencing colour, such as accidental sunbeams, refractions, transparencies, or local mists or showers, to be collected into a space apparently small, the colours of these clouds are always changeful and palpitating; and whatever degree of grey or of gloom may be mixed with them, is invariably pure and aerial.

"But the nearness of the rain-cloud, rendering it impossible for a number of phenomena to be at once visible, makes its hue of grey monotonous, and (by losing the blue of distance) warm and brown compared to that of the upper clouds.

"This is especially remarkable on any part of it which may happen to be illumined, such part being of a brown, bricky, ochreous tone, never bright, always coming in dark outlines on the lights of the central clouds. But it is seldom that this takes place, and when it does, never over large spaces, little being seen of the rain-cloud but its under and dark side. This, when the cloud above is dense, becomes of an inky and cold grey, and sulphurous and lurid if there be thunder in the air.

"With these striking differences in colour, the rain-cloud presents no fewer, nor less important differences in form, chiefly from losing almost all definiteness of character and outline.

"It is sometimes nothing more than a thin mist, whose outline cannot be traced, rendering the landscape locally indistinct or dark; if its outline be visible, it is ragged and torn, rather a spray of cloud, taken off its edge and sifted by the wind, than an edge of the cloud itself. In fact, it rather partakes of the nature, and assumes the appearance, of real water in the state of spray, than of elastic vapour.

"This appearance is enhanced by the usual presence of formed rain, carried along with it in a columnar form, ordinarily, of course, reaching the ground like a veil, but very often suspended with the cloud, and hanging from it like a jagged fringe, or over it, in light, rain being always lighter than the cloud it falls from.

"These columns, or fringes of rain, are often waved and bent by the wind, or twisted, sometimes even swept upwards from the cloud.

"The velocity of these vapours, though not necessarily in reality greater than that of the central clouds, appears greater, owing to their proximity, and of course also to the usual presence of a violent wind. They are also apparently much more in the power of the wind, having less elastic force in themselves; but they are precisely subject to the same great laws of form which regulate the upper clouds. They are not solid bodies borne about with the wind, but they carry the wind with them.

"Every one knows, who has ever been in a storm, that the time when it rains heaviest is precisely the time when he cannot hold up his umbrella; and that the wind lulls when the cloud has passed. And every one who has ever seen rain in a hill country knows that a rain-cloud, like any other, may have all its parts in rapid motion, and yet as a whole remain in one spot."

Ruskin mentions then that on one occasion, in crossing the Tête Noire, having turned up the valley towards Trient, he noticed a rain-cloud forming on the Glacier de Trient. "With a west wind it proceeded towards the Col de Balme, being followed by a prolonged wreathe of vapour, always forming exactly at the same spot over the glacier. This long, serpent-like line of cloud went on at a great rate till it reached the valley leading down the Col de Balme, under the slate rocks of the Croix de Fer. There it turned sharp round and came down this valley at right angles to its former progress, and finally directly contrary to it, till it came down within 500 yards of the village, when it disappeared; the line behind always advancing, and always disappearing at the same spot.

"This continued for half an hour, the long line describing the curve of a horse-shoe; always coming into existence, and always vanishing at exactly the same places; traversing the space between with enormous swiftness.

"This cloud ten miles off would have looked like a perfectly, motionless wreath, in the form of a horse-shoe, hanging over the hills."

One effect peculiar to the rain-cloud is that its openness exhibits the purest blue which the sky ever shows. For as "the aqueous vapour always turns the sky more or less grey, it follows that we never can see the azure so intense as when the greater part of this vapour has just fallen in rain. Then, and then only, pure blue sky becomes visible in the first openings, distinguished especially by the manner in which the clouds melt into it; their edge passing off in faint white threads and fringes, through which the blue shines more and more intensely, till the last trace of vapour is lost in its perfect colour. It is only the upper white clouds, however, which do this, or the last fragments of rain-clouds, becoming white as they disappear, so that the blue is never *corrupted* by the cloud, but only paled and broken with pure white, the purest white which the sky ever shows.

"Thus we have a melting and palpitating colour, never the same for two inches together, deepening and broadening here and there into intensity of perfect azure, then drifted, and dying away through every tone of pure pale sky, into the snow white of the filmy cloud. Over this roll the determined edges of the rain-clouds, throwing it all far back, as a retired scene, into the upper sky.

Of this we have a representation in Turner's "Coventry." And in this fine specimen of drawing we find several illustrations of the cloud to which I am at present directing attention.

"The great mass of cloud which traverses the whole picture is characterised throughout by severe right lines, nearly parallel with each other, into which every one of its wreaths has a tendency to range itself; but no one of these right lines is actually and entirely parallel to any other, though all have a certain tendency, more or less defined in each, which impresses the mind with the most distinct *idea* of parallelism.

"Neither are any of the lines actually straight and unbroken; on the contrary, they are all made up of the most exquisite and varied curves, and it is the imagined line which joins the apices of these, a tangent to them all, which is in reality straight. They are suggested, not represented right lines: but the whole volume of cloud is visibly and totally bounded by them; and, in consequence, its whole body is felt to be dragged out and elongated by the force of the tempest which it carries with it, and every one of its wreaths to be not so much something borne *before* or *by* the wind, as the visible form and presence of the wind itself. At the same time, there is not one curve that repeats another, nor one curve in itself monotonous, nor without character, and yet every part and portion of the cloud is regularly subjected to the same forward, fierce, inevitable, influence of storm. . . .

"But, not only are the lines of the rolling clouds thus regular in their parallelism, but those of the falling rain are equally varied in their direction, indicating the gusty changefulness of the wind, and yet kept so straight and stern in their individual descent that we are not suffered to forget its strength. This impression is still further enhanced by the drawing of the smoke, which blows every way at once, yet turning perpetually in each of its swirls back in the direction of the wind, but so suddenly and violently as almost to assume the angular lines of lightning. Farther to complete the impression, all the cattle upon the near and distant hillside are represented as having left off grazing, and are standing stock-still and stiff, with their heads down and their backs to the wind; and finally, that we may be told not only what the storm is but what it has been, the gutter at the side of the road is gushing in a complete torrent, and particular attention is directed to it by the full burst of light in the sky being brought just above it, so that all its waves are bright with the reflection. . . .

"Impetuous clouds, twisted rain, flickering sunshine, fleeting shadow, gushing water, and oppressed cattle, all speak the same story of tumult, fitfulness, power, and velocity. Only one thing is wanted, a passage of repose to contrast with all, and it is given: High and far above the dark volumes of the swift rain-cloud, are seen on the left, through their opening, the quiet, horizontal, silent flakes of the highest *cirrus*, resting in the repose of the deep sky.

"Of all else that I have noticed in this drawing some faint idea can be formed from the engraving; but of the delicate and soft forms of these passing vapours, not the slightest, and still less of the exquisite depth and palpitating tenderness of the blue with which they are islanded."

Speaking of the rain-clouds, Ruskin says, as I have stated, they differ not so much in their real nature from those of the central and uppermost region as in appearance, owing to their greater nearness. "For the central clouds, and perhaps even the high *cirri*, deposit moisture, if not distinctly rain, as is sufficiently proved by the existence of snow on the highest peaks of the Himalaya; and when on any such mountains we are brought into close contact with the central clouds, we find them differing little from the ordinary rain-cloud of the plains, except by being slightly less dense and dark."

And he adds in a foot-note :—"I am unable to say to what height the real rain-cloud may extend; perhaps there are no mountains which rise altogether above storm.

"I have never been in a violent storm at a greater height than between 8000 and 9000 feet above the level of the sea. There the rain-cloud is exceeding light [in colour] compared with the ponderous darkness of the lower air."

He speaks apparently of the greatest altitude of these clouds.

The height at which the lowest clouds are formed varies with the latitude and the season, being generally greater in the warmer and less in the colder region. Schuckburgh, it is mentioned in the "Philosophical Transactions" for 1777 (p. 528), frequently observed clouds resting below the summit of Saleve, the height of which is 2831 feet.

In Dr Alton's "Meteorological Observations" (p. 41), it is stated that in Cumberland, lat. 34°, Mr Crosshwaite, in the course of several years' observations, observed none lower than 2700, and none higher than 3150 feet.

Lambert, in Berlin, lat. 52° 32′, found, it appears from the "Mem. Berlin," 1773 (p. 44), in the month of July, 1773, their height 7792 feet.

Gentil, at Ponticherry, lat. 12°, observed some at the height of 10,240 feet.

In the course of several years' observations in Cumberland by Mr Crosshwaite, the records of which are embodied in Dalton's "Meteorological Observations" (p. 41), none were observed lower than 2700 feet, and none higher than 3150. But it is suggested that this country, being mountainous, clouds are probably lower than in other districts under the same latitude.

At the Cape, when the clouds are at a higher level than the

summits of the Hottentot Holland range of mountains, Table Mountain, and the Lion's Head, it is considered that there is no probability of rain; but when there is a cap on the Lion's Head, it is considered that there is. The height of Table Mountain is 3502 feet, that of the Lion's Head is somewhat less.

"The clouds which commonly crown the summits even of low mountains, and often announce rain, are caused," writes Kirwan, "by the near approach to saturation at those elevations, and its actual attainment through the evaporation from those summits."

"To the region of the rain-cloud," says Ruskin, "belong also all those phenomena of drifted smoke, heat haze, local mists in the morning or evening, in valleys or over water, mirage, white steaming vapour rising in evaporation from moist and open surfaces, and every thing which visibly affects the condition of the atmosphere without actually assuming the form of cloud." . . .

Sometimes when a slight shower falls in a hot day of bright sunshine in summer, the streets may be seen to smoke with steam. I have seen the same thing in Russia during intense frost: the Neva smoking as if the water were about to boil; and yet it was a temperature below the freezing point. And the same thing have I seen at the Cape after a fall of rain. The rationale is simple: the earth and the river were of a much higher temperature than the air; evaporation proceeded from them as usual; had the temperature of the air been the same, or higher, the aqueous vapour would have remained suspended invisible, but this being in excess of what the superincumbent air could at its temperature sustain, it was like the steam which issues from the locomotive engine, invisible, but immediately condensed, to be like that steam again absorbed,—a process with which meteorologists are familiar, many clouds being often fed from below while evaporation is going on continuously from their upper surface.

Ruskin, speaking of this phenomenon, says: "I know of no effect more strikingly characteristic of the departure of a storm than the *smoking* of the mountain torrents. The exhausted air is so thirsty of moisture that every jet of spray is seized upon by it, and converted into vapour as it springs; and this vapour rises so densely from the surface of the stream as to give it the appearance of boiling water. I have seen the whole course of the Arve, at Chamounix, one line of dense cloud, dissipating as soon as it had risen ten or twelve feet from the surface, but entirely concealing the water from an observer placed above it."

And again: "When rain falls on a mountain composed chiefly of barren rocks, their surfaces, being violently heated by the sun, whose most intense warmth always precedes rain, occasion sudden and violent evaporation, actually converting the first shower into steam.

"Consequently, upon all such hills, on the commencement of rain, white volumes of vapour are instantaneously and universally formed, which rise, are absorbed by the atmosphere, and again descend in rain to rise in fresh volumes until the surfaces of the hills are cooled. Where there is grass or vegetation, this effect is diminished; where there is foliage it scarcely takes place at all.

"This has been beautifully rendered in Turner's 'Loch Coriskin,' in the illustrations of Scott, in which he had not only to relieve its jagged forms with veiling vapour, but to tell the tale which no pencilling could, the story of its utter barrenness of unlichened, dead, desolated rock:—

"'The wildest glen, but this, can show
Some touch of Nature's genial glow,
On high Benmore green mosses grow,
And heathbells bud in deep Glencoe,
And copse on Cruachan Ben;
But here, above, around, below,
On mountain, or in glen,
Nor tree, nor plant, nor shrub, nor flower,
Nor aught of vegetation power,
The wearied eye may ken;
But all is rocks at random thrown,
Black waves, bare crags, and banks of stone.'"

Lord of the Isles—Canto iii.

While reading this I seemed to see Bain's Kloof, not so utterly devoid of vegetation as Loch Coriskin thus depicted, but abounding in "bare crags and banks of stone," thus smoking after a shower.

Another atmospheric phenomena, which I have often witnessed with peculiar delight from the vicinity of Capetown, is the appearance of broad and ever-broadening rays proceeding from the sun when that glorious orb was low in the heavens, and concealed behind broken clouds. The rationale of this given by Ruskin has already been cited (ante. p. 51).

CHAPTER III.

Winds.

It is a prevalent idea that winds bring rain, and the popular writer who should attempt to throw ridicule upon the idea would run some risk of being laughed at for his pains. If it were alleged by him that all winds do not bring rain, this would only call forth the rejoinder that though rain does sometimes fall, and fall persistently, when there is no wind, yet in general when rain comes winds bring it. If he alleged that perhaps it would be more correct to say, in some cases, that they bring the aqueous vapours from which and of which the rain is formed, than that they bring the rain, this perhaps would be admitted; and if, taking heart of grace, it were further alleged that in some other cases it would be more correct to say that the wind cools down the air, and thus occasions a deposit of moisture which was suspended in it, than that it brings rain, perhaps this would not be denied.

A popular idea, and one widely prevalent, is that the rain comes from the clouds, and the clouds come from the sea, borne by the winds. But is this the case?

Often at the Cape I have seen the heavens without a cloud. At length there was seen a cloud, which might be described, like that seen by the servant of Elijah, as the size of a man's hand,—but rising not from the sea,—appearing, it may be, in the zenith, and soon the heavens were covered with blackness, and there was every promise, say rather a threatening of rain; but, instead of a deluge, the clouds after a time disappeared, and there was none: it was not that the clouds were dispersed or blown away,—they seemed to dissolve away, like the puff of steam from the locomotive on the rails.

As is the case with the cloud on Table Mountain, in its ceaseless change, they were not blown thither and again blown away, but there they were formed, and there they were again dissolved. How, and of what? A change of wind occasioned their formation, and a second change of wind their disappearance. And that of which they were formed was the moisture in the air, deposited in the form of cloud by a lowering of temperature, consequent on the first change of wind, and again dissolved through an elevation of temperature, consequent on the second change. And the quantity of the water thus suspended in that region of the heavens where this occurred can only be conjectured from the mass and apparent density of the clouds, for it was present there, and there it remains after the clouds have disappeared. These clouds were not borne thither from the sea by the wind, though the moisture of which they were formed may have come from the sea, raised by evaporation, in accordance with the laws of gaseous diffusion; and air charged with that moisture may at some time or another, it may be long before, have been borne thither as wind, but it cannot be said it was borne thither as cloud; and the whole process is of easy explanation.

I have had occasion, in writing of "Forests and Moisture: or the effects of Forests on Humidity of Climate," to refer to the constant ascent of air to a great elevation in the region of the equator, in consequence of the greater rarefaction of the air by the greater heat; and to the constant flow of air from the polar regions, and along the earth's surface, to supply the place of the air thus elevated, which air, bringing with it the velocity of rotation imparted to it by the earth's surface, from, and over which, it has come, which being less than that of the earth in the region to which it flows, this in its rotation leaves it behind, and imparts to it the combined character of an easterly and polar wind. Such are the north-east and the south-east trade winds.

In connection with this, I have had occasion to refer to the constant overflowing to north and south of this ever-ascending and expanding girdle of air in the region of the equator, and its flowing away northward and southward in currents above those proceeding towards the equator—but sometimes descending into and through these, bearing with it the velocity of rotation from west to east which it had acquired before its ascent, which being greater than that of the earth's surface in the latitudes in which it makes its descent, impresses on it the character of a westerly, a north-west, or a south-west wind. These upper and lower currents, the one set proceeding from the neighbourhood of the equator, and the other proceeding from the poles, are of course of different temperatures, each having more or less of the temperature of the latitude whence it last came; and to this I have again occasion to refer, but that briefly.

Within the tropics, the air which is raised to a great elevation is there cooled down, and, being unable to hold in solution as much moisture as it had in its composition immediately before its elevation, this is there condensed into cloud. Even when thus relieved somewhat of its load of moisture, when it comes in contact in its polar course with the colder currents below, or it descends to fill vacuums left either by these on their departure, or by the more rapid progress over one spot than another of the currents flowing towards the equator, more moisture is condensed, and a cloud is formed. As in the case of the cloud on Table Mountain, and the clouds in the tropics, the heat of the lower strata of air may be sufficient to prevent this descending below a certain altitude, but the formation of one layer of cloud above another layer or stratum of cloud, may so cool down the lower stratum that gravitation may drag down the water there condensed, and that as mist, as drizzling rain, or as rain of a character still more pronounced. And again by eddies of these currents, air of different temperatures become commingled, and the body of air of the resultant temperature may be unable to maintain diffused as vapour as much as was previously so maintained by the separate portions at their respective temperatures; or by a whirlwind or cyclone air may be elevated to an altitude at which the temperature of the air becomes so reduced that a deposit of a portion of the aqueous vapour in it must ensue; a warm wind charged with moisture, blowing over a cold space, may be so cooled down as to precipitate a portion of its vapour; or a cold wind blowing over a district in which air of a higher temperature is heavily charged with moisture, may so reduce its temperature as to cause it to drop a portion of the vapour with which it was charged: it is in such ways as these, and not by blowing from the sea clouds already formed there, that winds in general bring rain.

In the introduction to my treatise on "Reboisement in France," I have given a translation of M. Marschand's account of the *fœhn*, by which are produced the torrential rains of the Alps. And in my treatise on "Forests and Moisture," I have given illustrations from Cézanne and Raulin, of how the deposition of rain is affected by the contour, and by the relative position of a country over which a wind passes. Amongst other things it is stated by M. Cézanne:—"On the two sides of the Scandinavian Alps, the west wind and the east wind give inversely fine weather and rain, the one to Sweden and the other to Norway.

"When it rains at Narbonne the sun shines at Montauban.

"The rain comes from the west in Switzerland, and from the east in Lombardy. It is in consequence of its being sheltered from the south and southeast winds, which drench the basin of the Rhone, that the valley of the Durance presents that exceptionally dry climate which M. Surell points out to be favourable to the development of torrents. It rains upon an average at Marseilles, 57 days in the year; at Arles, 45 days; at Aix, 40; but in the region of the Durance it rains only 38 days in the year.

"Under the tropics, where prevail the trade-winds blowing from the east, the lands which incline towards the east are inundated with torrential rains: thus is it on the coast of Mozambique, and in the basin of the Amazon. On the contrary, it almost never rains on the western slope of the Andes. It is said that thunder has not been heard at Lima three times in as many centuries.

"In the Indian peninsula the eastern coast, or that of Coromandel, is watered by the North-east Monsoon; and the west coast, or that of Malabar, by the South-west Monsoon.

"When, after having climbed the slope of a mountain, the atmospheric current, greatly relieved of its load, comes upon a plateau, it freely expands; but its lowest layer, being in contact with the earth, is alone reduced to a lower temperature. From which it comes to pass that, from the same wind, a plateau, though more elevated, receives less rain than the ascending slope, but receives more than does the descending slope beyond."

At the Cape of Good Hope the same thing is seen. In the western districts of the Colony the rain falls chiefly in winter, when in the eastern districts of the Colony the sky is clear; and there the rain falls chiefly in summer, when in the western districts rain is rare; and the line on the two sides of which this difference is remarked is well defined. It is the wind, however, rather than the lay of the land, to which this difference is attributable. From the position of the Cape in relation to the ecliptic, south-easterly winds prevail in summer, and north-easterly winds in winter; these bring rain to the western districts; those do so to the eastern.

In view of this some information may be desired in regard to the prevalence of different winds at the Cape. The following is a tabulated statement of observations made and recorded at the Royal Observatory, near Capetown, in the years 1842 and 1855, the first and last years of a series of fourteen, in regard to which a report has been supplied to me; and of corresponding observations made in 1868, similarly arranged; and a tabulated statement of the mean force of the wind during each month of the period embraced by the observations, made in 1842 to 1855.

Mean Force of the Wind during each month of 14 years, in avoirdupois pounds weight pressure on the square foot.

Months.	1842.	1843.	1844.	1845.	1846.	1847.	1848.	1849.	1850.	1851.	1852.	1853.	1854.	1855.	Means.
	lbs.	lbs.	lbs.	lbs.	lbs.	lbs.	lbs.	lbs.	lbs.	lbs.	lbs.	lbs.	lbs.	lbs.	lbs.
January.....	2·30	2·16	2·54	2·71	2·79	2·17	2·39	1·39	1·36	1·27	1·07	1·20	1·61	1·24	1·87
February ...	2·30	1·10	1·86	2·08	2·97	2·61	1·02	1·54	1·21	1·87	1·55	1·06	1·38	1·47	1·72
March	1·70	2·07	1·55	1·43	2·25	2·27	1·21	0·74	0·79	1·05	1·25	1·05	1·06	1·13	1·40
April.........	0·95	1·00	1·13	0·98	1·75	1·19	0·42	0·82	0·63	0·40	0·92	0·93	0·94	0·83	0·92
May..........	0·90	1·10	1·22	0·91	1·04	0·71	0·60	0·52	0·70	0·78	0·92	0·61	0·90	0·56	0·82
June.........	0·89	1·24	0·88	0·70	1·11	1·02	0·71	0·85	1·57	0·56	0·78	0·75	0·90	0·93	0·92
July	1·25	1·10	0·79	0·93	1·35	0·45	0·85	0·83	1·30	0·81	1 09	0·77	0·87	0·71	0·94
August......	1·53	1·20	1·47	1·33	1·83	1·02	0·77	0·62	0·89	0·74	1·19	0·87	0·80	1·08	1·10
September..	1·80	1·43	1·11	1·33	2·28	1·57	1·44	0·74	1·16	1·06	1·12	1·14	1·21	1·34	1·34
October......	2·01	1·49	1·48	1·90	2·62	2·28	1·12	1·43	1·09	1·20	0·84	0·90	1·15	1·38	1·49
November...	1·42	1·30	1·72	1·44	2·76	1·59	1·65	1·38	1·23	0·62	1·18	1·61	1·22	1·45	1·47
December...	1·80	2·43	2·45	1·65	2·62	1·10	1·32	1·45	0·88	0·88	1·41	1·05	1·44	2·11	1·61
Means......	15·7	1·47	1·52	1·45	2·11	1·50	1·13	1·03	1·07	0·94	1·11	1·00	1·12	1·19	1·30

Number of Observations of the Wind at the several points of the Compass during each month of the year 1842 and the year 1855.

Points.	Jan.	Feb.	Mar.	Apr.	May.	June.	July	Aug.	Sept.	Oct.	Nov.	Dec.	Total	Jan.	Feb.	Mar.	Apr.	May.	June.	July.	Aug.	Sept.	Oct.	Nov.	Dec.	Total
S.	62	37	30	12	7	4	22	15	24	17	20	18	268	40	33	30	9	8	11	18	6	14	28	41	27	265
S.byW.	6	11	8	8	8	5	6	4	2	8	1	10	92	22	31	32	9	17	9	15	11	12	22	25	45	250
S.S.W.	-	12	6	6	8	4	8	1	8	6	1	20	80	14	11	13	11	8	4	16	9	8	15	18	15	152
S.W.byS.	-	-	-	1	1	3	1	3	-	4	3	10	26	5	4	6	4	5	2	4	2	4	4	-	2	42
S.W.	1	4	1	-	2	1	1	6	-	7	1	12	36	4	1	-	4	6	4	4	3	1	2	2	-	31
S.W.byW.	-	-	-	-	1	-	-	-	-	2	1	1	5	1	1	2	3	3	-	1	-	1	1	2	2	17
W.S.W.	-	-	-	1	1	1	-	1	1	4	2	3	14	1	3	-	4	3	4	-	4	2	1	-	-	22
W.byS.	-	1	-	1	-	1	-	1	-	1	1	1	7	3	1	2	2	-	1	3	5	-	3	2	1	23
W.	1	1	3	3	2	1	3	2	5	1	9	1	32	-	-	-	1	5	3	-	4	3	2	2	4	24
W.byN.	3	-	3	-	3	7	-	2	1	-	2	-	21	-	2	1	2	5	6	3	1	3	4	-	4	31
W.N.W.	-	1	-	10	1	6	-	-	3	3	2	2	28	3	4	6	4	6	7	7	6	6	9	4	6	68
N.W.byW.	-	-	1	3	7	4	-	3	1	2	-	-	21	6	2	2	1	6	7	3	7	6	4	3	3	50
N.W.	11	12	25	8	4	16	4	17	18	2	7	6	130	8	11	12	21	15	12	8	16	13	8	15	1	140
N.W.byN.	-	1	-	1	7	5	2	6	-	2	5	2	31	11	1	5	10	7	14	6	12	7	7	3	7	90
N.N.W.	3	-	2	9	10	9	3	7	3	2	6	3	57	3	3	5	14	8	14	11	16	14	5	3	1	97
N.byW.	2	-	7	4	4	6	5	8	4	1	3	1	45	-	-	1	4	20	15	16	12	5	3	2	-	78
N.	1	-	-	1	5	6	14	10	8	2	2	1	50	-	-	1	1	2	4	2	5	1	1	2	-	19
N.byE.	1	-	-	-	2	2	1	4	2	-	1	2	15	-	-	-	1	-	-	1	4	3	-	-	-	9
N.N.E.	-	-	1	2	3	-	3	2	2	4	-	-	17	-	-	-	2	-	1	-	-	2	-	-	-	5
N.E.byN.	-	-	-	-	-	-	1	-	1	-	-	-	2	-	-	-	-	-	-	-	1	1	-	-	-	2
N.E.	-	-	-	-	-	-	1	1	1	-	-	-	3	-	1	-	1	-	3	-	-	-	-	-	-	5
N.E.byE.	-	-	-	-	-	-	2	-	-	1	-	-	8	-	-	-	-	-	-	-	-	-	-	-	-	-
E.N.E.	-	-	-	-	1	-	-	-	-	-	2	-	3	-	1	-	-	-	-	1	-	2	-	-	-	4
E.byN.	-	-	-	-	1	-	-	1	-	-	-	-	2	-	-	-	-	-	-	-	-	-	-	-	-	-
E.	-	-	1	-	-	-	-	-	-	-	-	-	1	-	-	-	-	2	-	-	-	-	-	-	-	2
E.byS.	-	-	-	-	-	1	-	-	-	1	-	-	2	2	-	-	-	-	-	-	-	-	-	-	-	2
E.S.E.	-	-	-	-	4	1	1	-	-	-	-	-	6	-	-	1	-	-	-	1	-	-	-	-	-	2
S.E.byE.	-	-	-	1	6	1	1	-	-	1	-	-	10	-	-	-	-	-	-	-	-	-	-	-	-	-
S.E.	1	2	3	2	1	4	8	9	4	5	2	3	44	1	-	-	3	-	-	-	2	1	-	1	-	8
S.E.byS.	-	1	1	6	2	5	1	-	1	2	2	1	22	3	-	-	1	-	2	1	1	2	1	-	-	10
S.S.E.	3	8	8	4	7	7	2	-	9	9	2	3	62	2	3	9	3	3	4	4	4	3	3	1	2	41
S.byE.	5	4	3	16	7	3	6	2	4	14	3	6	73	6	7	7	11	6	3	5	4	1	12	4	10	76

Number of Observations of the Wind at the several Points of the Compass during each Month of the Year 1868, made at the Royal Observatory Station.

Points.	Jan.	Feb.	Mar.	Apr.	May.	June.	July.	Aug.	Sep.	Oct.	Nov.	Dec.	Totals.
1868.													
S.	73	46	80	53	44	37	44	50	36	49	43	33	588
S.byW.	11	15	10	15	4	1	12	10	8	4	8	11	109
S.S.W.	20	11	13	11	1	.	4	3	5	9	9	7	92
S.W.byS.	1	2	1	.	1	.	.	.	.	2	.	2	9
S.W.	3	2	2	4	2	2	1	1	5	.	6	4	32
S.W.byW.	.	.	.	.	.	.	.	.	.	.	.	.	.
W.S.W.	.	.	.	1	3	.	2	.	3	.	2	.	11
W.byS.	.	.	.	2	1	.	1	1	.	.	.	1	6
W.	1	8	3	14	1	7	5	7	7	12	9	15	89
W.byN.	1	1	1	4	5	5	3	6	4	3	5	5	43
W.N.W.	5	15	9	10	21	16	18	18	17	9	6	13	157
N.W.byW.	2	1	.	2	4	1	2	3	2	1	1	.	19
N.W.	14	12	17	17	41	32	34	18	47	18	12	25	287
N.W.byN.	1	.	.	.	1	3	2	1	.	.	.	3	11
N.N.W.	2	.	2	2	6	15	6	4	2	.	2	1	42
N.byW.	.	.	.	1	.	3	2	1	4	.	3	2	16
N.	1	.	.	.	4	6	3	5	2	4	4	.	30
N.byE.	.	.	.	.	1	.	.	.	.	.	.	.	1
N.N.E.	.	.	.	.	.	.	.	.	.	.	.	.	.
N.E.byN.	.	.	.	.	.	.	.	.	.	.	.	.	.
N.E.	.	.	1	.	.	.	.	.	1	.	.	.	2
N.E.byE.	.	.	.	.	.	.	.	.	.	.	.	.	.
E.N.E.	.	.	.	.	.	.	.	.	.	.	.	.	.
E.byN.	.	.	.	.	.	.	.	.	.	.	.	.	.
E.	.	1	.	.	.	.	.	.	.	.	.	.	1
E.byS.	.	.	.	.	.	.	.	.	.	.	.	.	.
E.S.E.	.	.	.	.	.	.	.	.	.	.	.	.	.
S.E.byE.	.	2	.	.	.	.	.	2	.	.	.	.	4
S.E.	.	2	.	4	5	7	1	13	2	7	4	6	51
S.E.byS.	1	1	.	2	.	.	.	1	.	2	1	2	10
S.S.E.	5	17	8	4	7	6	6	8	3	20	11	16	111
S.by.E.	14	9	8	4	3	9	9	3	2	15	24	9	109

In regard to these observations it is stated :—

"The south-westerly are in squalls, and generally accompanied, or followed, by rain or hail.

"The north-westerly winds are in general strong, sometimes violent, and almost invariably followed by rain. It is this wind which is dangerous to the shipping in Table Bay.

"One Table gives the mean force of the wind during each month of 14 years, in avoirdupois pounds weight pressure on the square foot.

"The construction of this Table has been a matter of some difficulty, owing to the earlier diagrams described by Osler's Anemometer not being available, and the slight pressure of light airs affecting the plate. But Mr Morton undertook, and carried out successfully, an elaborate comparison of the registered estimate of Beaufort's numbers, with the simultaneous indications of the Anemometer, and derived a numerical force value for the former, which is probably close to the truth.

"To the frequent winds may be attributed, in a great degree, the noted salubrity of the Cape climate,—particularly the absence of those fevers of the bilious remitting type, which in many other countries are produced by marsh miasma, or by continued calm hot weather. . . . .

"The putrid exudations from the ground are scoured off,—the unhealthiness which springs from inattention to domestic cleanliness is greatly diminished,—and the depressing effect of summer heat is considerably reduced."

One of the most approved instruments for measuring the force of the wind is Robinson's Hemispherical Cup Anemometer. It is thus spoken of in Admiral Fitz Roy's *Weather Book*:—"Wind-gauges have been tried; that by Lind—also a modification of it by Sir W. Snow Harris—and the well-known pressure plate; but these seem to yield only partial, if not equivocal results. The beautiful cup and dial anemometer (due to Robinson chiefly, if in some degree suggested by Beaufort or Edgeworth,) is more approved, after having been tried experimentally through several years of exposure. . . .

"Dr Robinson showed (in the Transactions of the Royal Irish Academy) that a current of air is opposed by a concave hemisphere, one-fourth more than by a convex one of the same size. Thence experimental trials and mathematical reasoning induced him to adopt the arrangement now general, namely, four hemispherical cups, on horizontal arms, revolving on a friction rollered axis, at a known proportional rate, one-third slower than the passage of air or wind current. Hence velocity, and, from it, pressure, are readily cal culated."

In illustration of its working, the following statement is given in Symons's Monthly Meteorological Magazine, for July 1867:—"Two modes of indicating the number of revolutions made by the cups have prevailed. The first consisted of an endless screw on the axis working into the right hand of a train of 5 dials, each of which revolved ten times for one revolution of the next to the left, the dial next the endless screw showing 10 revolutions, the next showed 100, then 1,000, 10,000, and lastly 100,000. The reading was, therefore, taken as in a gas meter, from left to right, the figures being noted down as read off, it being remembered that if the index points *between* two figures, the lesser of the two is to be taken. To take an imaginary case:—At 9 a.m., June 17th, the hands (counting from left to right) pointed to 176,420, and at the same hour on the 18th they read 297,340; then 297,340—176,420=120,920 has been the number of revolutions in the 24 hours. The dimensions of these instruments are understood to be such that the circle described by the centre of each cup is $\frac{1}{1500}$ of a mile—the amount of wind requisite to produce this being $\frac{1}{500}$ of a mile; 500 revolutions equal one mile. Therefore ($\frac{120920}{500}$ = 241·84 miles) in the case supposed, about 242 miles of wind had passed in the course of the day, that is to say, just 10 miles an hour. It is almost needless to point out that if the three left hand dials alone are read, (for example as above, 176 and 297, then 297—176=121,) and the result doubled, it gives at once the miles passed 121 × 2=242 miles.

"To Mr Casella, we believe, is due the improved mode of indication, whereby one dial is substituted for five, and the record is read off in miles at sight. Two further improvements have since been adopted, so that the record is now read off in miles and tenths, instead of miles and furlongs as shown; and secondly, by the adding another dial, it is made available for examination at long intervals. Another modification was introduced by Messrs Negretti and Zambra, in which the registering wheels can be disconnected from the cups, and wear and tear thus avoided. Of course, by this modification continuous records are not obtained. The price of these instruments ranges from £3 to £5."

With regard to the position desirable for anemometers it is stated:—"The only instructions we can find are, 'It should be fixed in an exposed situation, as high above the ground as may be convenient for reading.'* 'When in use, it may be screwed on a shaft or ordinary piece of gas pipe, which accompanies it, and elevated to any desirable altitude.'† In the instructions of the Scottish Meteorological Society, it is stated that 'A wind vane ought to be elevated 12 ft. at least above surrounding objects;' but though 'the Council would strongly recommend that every observatory be furnished with a Hemispherical Cup Anemometer,' they do not say where it should be fixed.

* Negretti's Treatise on Meteorological Instruments, (p. 121).
† Castella's Catalogue, (p. 29).

"In this recommendation of the cup anemometer for *general* use. we fully concur, but some preliminaries must first be attended to. Of these the most important are—(1) testing the anemometers, and (2) taking observations for a few months with several instruments differently placed, but near together. The desirability (not to say necessity) of testing the instruments is so obvious that it is un necessary to dwell upon it.

"We think the second preliminary indispensable, for the confusion and irregularity in the location of anemometers can only be realized by those who have visited a considerable number of observers. Let us jot down a few as examples. A has his anemometer on a pole 6 ft. above ground, while B has his on a pole 10 ft. above ground, and has a pair of steps to enable him to read it; C has it on a sentry box 12 ft. high, and with steps up the side; D goes higher still, having a stout pole 20 ft. high, with cross bars, up which he climbs like a bear; E being perhaps too stout for this performance, mounts his among the chimney pots on the top of his house, where the anemometer and himself get sadly begrimed; F has a long rod to bring the registration in doors, while G has an elaborate electrical arrangement whereby the motion of the cups on a distant hill is registered in his library. Surely the time has come when such differences should be swept away. What reliance can we place on records from instruments so variously circumstanced? Experiment alone can give us the relative motion recorded at different heights above the surface of the ground. We conceive that an open, level station should be selected, and anemometers placed with their cups 10, 20, and 30 ft. above the ground mounted on poles, or better still, on ladders steadied by guys—not that the observer need climb daily to the top (for he can easily read them with a telescope), but that access may be practicable for oiling, &c., when necessary. One should also be placed on the roof of the house, that we may learn the amount of motion due to such a position. Without forestalling the results of the investigation (if taken up), we may record our impression that such positions always register much in excess of the truth."

CHAPTER IV.

Thunder-Storms.

As it is believed by many in regard to wind, so is it believed by many in regard to thunder, that it brings rain. It might be well if, in the one case and the other, clear and accurate conceptions were formed in regard to the relation in which the rain stands to the meteorological disturbance to which it is attributed, though this may not be absolutely necessary to an intelligent appreciation of the effect of thunder-storms on the water supply of the country.

Currents of air in the atmosphere are often commingling.

Of the effects produced by the commingling of currents of different temperatures, frequent illustrations may be seen at the Cape. An eddy of the under-current flowing from the pole towards the equator dashing into the upper current flowing from the tropics towards the pole will sometimes cover a cloudless sky in half-an-hour with a sable pall of cloud, and in half-an-hour thereafter, though not a drop of water has fallen, not a cloud is to be seen; but often it is otherwise.

Descriptions of such scenes, produced by the commingling of currents, have been given in more than one of the preceding chapters.

Corresponding phenomena may sometimes be observed in connection with the thunder-storms, on which, in many parts of the Colony, the inhabitants are dependent for rain. Sometimes the clouds are formed in what was shortly before a cloudless sky. And it may be alleged that this is the case even when they appear to come rolling on massively from the horizon, or gathering from different quarters of the heaven like armies coming up against each other for battle; but it is enough here to direct attention to the circumstance that they come often from quarters other than that in the direction of the nearest sea, and that there is no reason to conclude that they have come in the form of cloud from oceans in the direction in which they have come. They, too, have been formed of vapour suspended transparent and invisible in the clear sky; but hurled, or hurling themselves, against each other, thunder follows or accompanies their collision: and in the mimic war of the elements, the lightnings flash, the thunder roars, and the rain pours down in torrents.

From the occurrence of this it appears that there is, even in the driest atmosphere of South Africa, sufficient moisture to produce such phenomena. Dr Moffat, writing of the effects in Namaqualand of winds from the north, says,—"These winds, I have learned from inquiry, come from within the tropics, where rain has fallen, and the cool air thereby produced rushes southward over the plains, filling up the space caused by the refraction of the air, owing to the approach of the sun to the tropics of Capricorn. The more boisterous these winds are, the more reason we have to expect rain. They cannot extend to any great height, as the thunder-storms which follow, and which often commence with a small cloud in the opposite direction, increasing into mountains of snow, with a tinge of yellow, pursue an opposite course. These are preceded by a dead stillness, which continues till the tornado bursts upon us with an awful violence, and the clouds have discharged their watery treasures. In such a case there are almost always two strata of clouds, frequently moving in opposite directions. The higher mountain-like masses, with their edges exactly defined, going one way, while the feelers, or loose misty vapour beneath, convulsed, and rolling in fearful velocity, are going another; while the peals of thunder are such as to make the very earth tremble."

At this point I might close my quotation, but what follows has an important bearing on the subject:—"The lightning is of three descriptions: one kind passing from cloud to cloud; this is seldom accompanied with any rain. Another kind is the forked, which may be seen passing through a cloud and striking the earth; this is considered the most dangerous. The most common, not always accompanied by rain, is what we are in the habit of calling stream or chain lightning. This appears to rise from the earth in figures of various shapes, crooked, zigzag, and oblique; and sometimes like a water-spout at sea: it continues several seconds, while the observer can distinctly see it dissolve in pieces like a broken chain. The perpetual roar of awful thunder on these occasions may be conceived, when twenty or more of these flashes may be counted in one minute. The lightning may also be seen passing upwards through the dense mass of vapour, and branching out like the limbs of a naked tree in the blue sky above. In such storms the rain frequently falls in torrents, and runs off very rapidly, not moistening the earth, except in sandy plains, more than six inches deep."

The whole phenomena, from the formation of the cloud at so great a distance from the sea, isolated and unconnected with clouds extending from the ocean thither, to the precipitation of the deluge of rain, appear to have been occasioned by a temporary disturbance and intermixture of the two aerial currents passing overhead between the equator and the poles.

In explanation of the thunder-shower it may be stated that it is more than probable that every change of temperature is accompanied by a disturbance of electric equilibrium. It is frequently, if not invariably, the case that a disturbance of electric equibrium induces a disturbance of equilibrium, the counterpart to itself, in matter adjacent to or contigious with that in which it originates. It has been found that a thunder-cloud passing over a electric-wire has given indications of its having a centre of either negative or positive electicity, surrounded by alternate concentric rings of the opposite and of the same character; and that it has induced a corresponding centre and system of concentric rings of electricity in these two polar conditions, on the earth below it, the counterpart, but not the same: the centre in the cloud, if positive, indicating a centre of negative electricity in the earth, and *vice versa*, with an arrangement of the concentric rings in accordance with this.

A pithball, or other matter charged with positive electricity, passes rapidly to another body in the opposite condition; whether this be produced by repulsion or by attraction we need not at present to inquire. Something similar occurs with water similarly charged. If a metallic jug, the bottom of which has been pierced with a number of small holes like those in the lid of a dredge-box or the bottom of a drainer, but more minute, be filled with water and suspended from the prime conductor of an electrical machine, the water will drop slowly, as it accumulates in the lower side of the bottom; but if the cylinder of the electrifying machine be put in motion, the water will stream in continuous flow from each of the holes.

So is it with the thuuder-shower: the fall of the rain being more rapid far in these than in the general rain. And thus are we prepared to account for the violence of the thunder-showers at the Cape.

The intermixture of large bodies of air of different temperatures occasions a large condensation of vapour in the form of cloud, but it often occasions also a great disturbance of electric equilibrium; and in consequence of this, that moisture descends, not in the form of mist or of gentle rain, but in the torrents characteristic of the thunder-shower.

The Colony is so situated that when the sun is to the south of the equator it is within a region which is swept over by the southeast currents flowing towards the heated regions there; and when the sun is to the north of the equator, it is in a region wherein the returning current flowing towards the south pole frequently strikes down upon the earth, and does so with violence, constituting the northwest gales so dreaded in Table Bay.

When, as sometimes happens, these atmospheric currents of different temperatures cross each other's paths, or meet in conflict, dashing into each other like two railway trains coming into collision from different directions at the maximum of their speed, the deluges of rain may be tremendous, and this, though no thunder be heard or lightning seen, but in general they are accompanied by these phenomena; and such are the thunder-showers occasionally flooding the villages of the Colony, and filling from bank to bank the old water-courses—called, by strange misnomer, dry rivers—and sweeping all before them to the sea.

The violence and the frequency of thunder-storms, and also the seasons of the year at which they prevail, vary greatly in different parts of South Africa. The following is a tabulated statement of the number of days on which lightning was observed between January 1842 and January 1856 at the Royal Observatory near Capetown:—

Date.	Jan.	Feb.	March.	April.	May.	June.	July.	August.	Sep.	Oct.	Nov.	Dec.	Annual total.
1842............	0	1	0	4	5	2	0	0	2	1	0	1	16
1843............	3	3	0	2	0	1	0	1	0	1	0	0	11
1844............	1	0	3	2	0	1	0	1	0	0	1	0	9
1845............	0	0	3	2	0	0	0	0	2	1	0	0	8
1846............	0	0	2	0	2	0	0	0	2	1	0	0	7
1847............	1	4	1	2	1	0	0	1	0	1	0	0	11
1848............	3	0	1	0	4	0	0	1	3	0	0	0	12
1849............	1	1	1	4	2	0	2	1	0	1	1	0	14
1850............	0	0	5	2	1	0	0	1	1	0	0	0	10
1851............	0	0	2	2	1	2	1	0	0	1	0	4	13
1852............	0	3	2	1	1	0	1	1	0	0	1	0	10
1853............	0	3	0	2	1	1	0	1	2	2	0	1	13
1854............	1	2	5	0	0	1	0	0	1	0	3	3	16
1855............	0	1	1	1	1	1	0	1	1	0	0	0	7
Monthly Total.	10	18	26	24	19	9	4	9	14	9	6	9	(13)

And the following is a general monthly abstract of meteorological observations during fourteen years at the Royal Observatory, Cape of Good Hope :—

Months.	Barometer at Temperature 32°.	Temperature.					Humidity.				Lightning.
		Mean Temperature.	Mean daily Range.	Mean of greatest range any one day of each month	Mean of least Range on any one day of each month.	Mean of greatest Range of each month	Mean for each month.	Mean daily range.	Rain.	Clouded sky.	Recorded No. of Days.
	From Table I.	From Table IV.	From Table IX.	From Table VII.	From Table VII	From Table VI.	From Table X.	From Table XIII.	From Table XIV.	From Table XVI.	
	inches.	degrees.	degrees.	degrees.	degrees.	degrees.	per cent.	per cent.	inches.	per cent. of visible hemisphere	total.
January	29·931	68·77	11·13	25·6	8·0	30·2	68·7	20·7	0·880	26	10
February.........	29·931	68·99	11·76	26·2	7·8	30·8	70·6	20·9	0·653	27	18
March	29·968	66·29	12·28	28·5	7·0	33·1	71·6	21·5	0·846	29	26
April...............	30·003	62·95	11·26	26·7	4·6	35·1	76·1	20·6	1·846	42	24
May.................	30·069	58·01	9·17	25·4	4·5	30 7	80·6	15·7	3·576	47	19
June................	30·129	55·35	8·66	24·2	4·2	29·1	81·8	12·8	4·311	47	9
July	30·160	54·57	9·07	22·7	4·8	29·0	81·6	13·4	2·921	43	4
August............	30·147	55·21	8·89	24·2	4·6	29·0	80·2	13·6	3·323	45	9
September	30·098	57·43	9·41	25·9	4·7	30 8	77·4	15·9	2·332	44	14
October...........	30·051	61·06	10·36	26·0	6·3	32·6	73·8	19·1	1 014	39	9
November.........	29 985	64·28	10·34	25·0	7·8	30·5	70·8	19·4	1·090	38	6
December.........	29·953	67·61	11·46	25·3	8·9	27·9	68·5	21·1	0·516	28	9
Yearly Means,	30·036	61 71	10·32	25·5	6·1	30·7	75·14	17·9	23·309	38	13

CHAPTER V.

Tabulated Abstracts of extensive and long-continued Meteorological Observations.

The Meteorological Commission at the Cape of Good Hope reported, for the year 1875, the following, as the materials at present existing for a study of the Meteorology of the Cape :—

"The earliest regular meteorological observations at the Cape of which we have any record, are those made by Lacaille in Strand Street in 1751 and 1752. The results are published in the Memoirs of the Royal Academy of Paris, 1855.

"In the year 1827 the Royal Observatory, Cape of Good Hope, was established, and since that time meteorological observations on a more or less complete plan have been made at the Observatory.

"During the years 1834 to 1838 meteorological observations were made by Sir J. Herschel at Mr Jones' house near Wynberg, but these observations have never been published.

"In the year 1841 a Meteorological and Magnetical Observatory was established at the Cape, and placed under the direction of Capt. Frederick Eardly Wilmot, R.A., who had under him a small detachment of the Royal Artillery. The Obseratory was erected (by permission of the Lords Commissioners of the Admiralty) in the Observatory grounds, and observations carried on from 1841 to 1846. The observations from October, 1841, until July, 1846, were made at every hour. The Observatory thus established was part of a general scheme of meteorological and magnetic observation carried out by the British Government at selected stations over the whole world. The results of the magnetic observations at the Cape were discussed and published by Lieut.-Colonel Edward Sabine, R.A., in the year 1851. The meteorological observations were discussed some years ago, but no steps for their publication and general distribution were taken until the year 1875, when Mr Stone, during a visit to England, made an arrangement with Mr Scott, the Director of the Meteorological Office, England, for the printing of a short preface and the distribution of the results.

"These hourly observations from 1841 to 1846 afford the only data at present available for a determination of the diurnal and semi-diurnal changes of the meteorological elements of the Cape. The corrections thus deduced will be found in the volume of Meteorological Results 1841 to 1870 published by Mr Stone in 1871. In the year 1846 the detachment of Artillery was recalled from the Cape and the Meteorological Observatory placed under the direction of the Astronomer Royal at the Cape. The meteorological observations at the Observatory have been continued since that period, but the hourly readings of the instruments had to be abandoned from the want of a sufficient number of assistants. The readings were at first taken at 1h., 5h., 9h., 17h., and 21h., and are now taken at noon, every six hours thereafter, and at 2 o'clock P.M.,—the latter for comparison with the simultaneous observations organized by the United States Government.

"The following are, so far as the committee are aware, the published results of meteorological observations made at the Cape previously to the present report:—

"1. Lacaille's Observations—1751, 1752—Memoirs of the Royal Academy, Paris, 1855.

"2. Meteorological Observations, Cape Observatory—1841 to 1846—Sabine, 1875.

"3. Monthly Abstracts derived from observations, 1842 to 1861—Maclear.

"4. Monthly Abstracts for the year 1862—Maclear.

"5. Rainfall, 1842 to 1862—Maclear.

"6. Number of days in each month during which rain fell from 1842 to 1862—Maclear.

"7. Committee's Report, containing results for stations at different points of the Colony, 1861 to 1868—Meteorological Committee.

"8. Meteorological Results, 1841 to 1870—Stone, 1871.

"On account of the breakage of instruments and other causes the observations made at most of the colonial stations after 1868 have been very incomplete, and the present committee have therefore decided to recommence the publication of the results only after the re-organization of the service."

They reported at the same time the results of the meteorological observations made at the following stations from 1875, July 1 to December 31, by the gentlemen whose names are attached:—

Royal Observatory, ...	The Staff.
Simon's Town, ...	Telegraph Officer.

Port Elizabeth, ...	Lighthouse Keeper.
Somerset East, ...	Professor Kyd.
King William's Town, ...	Dr Eagan.
Worcester, ...	Rev. G. P. Jeffreys.

The following were the directions given for observations of clouds:—"In estimating the quantity of cloud, divide by vision the hemisphere into two parts; one extending from the horizon all round to 30° altitude (being one half); the other comprising the remainder.

"Record whether the clouds are in the lower or upper half.

"A description of the clouds is attached:—

"Stratus.—Low creeping mists = st.

"Cirrus.—Light feathery clouds = ci.

"Cirro-stratus.—Light clouds stratified; mackerel sky = ci. st.

"Cumulus.—Clouds in heaps, or volumes like the dense rolling smoke from cannon = cu.

"Cirro-Cumulus.—A system of round clouds connected = ci. cu.

"Cumulo-Stratus.—Thunder clouds, gigantic looking clouds = cu. st.

"Nimbus.—Watery, showery clouds = nimb.

"Before a storm there is occasionally a collection of small semicircular clouds with the edges turned inwards; these are called Cymoid cirro-stratus = cy. c.s."

Explicit directions were given also in regard to the barometer, the proper mode of fixing it and observing its indications, and in regard to thermometers and rain-gauges; and in regard to the calculation of the dew-point, tension of aqueous vapour, and humidity, instructions had previously been issued.

With the barometer and the thermometer, their appearance and their use, most of my readers are probably familiar.

Rain-gauges in their simplest form may be described as composed of a cylinder of definite diameter, and a float supporting a graduated rod, which will rise above the level of the cylinder as the rain falling into this increases in quantity. By the observation of the height to which the rod is raised in any definite time, is ascertained what measurement of rain has fallen within that time.

In scientific observatories, maintained at the public expense, or by munificent contributions of private individuals, much more complicated appliance are employed.

The following description is given by Mr Draper, Director of the New York Meteorological Observatory, in the Central Park, of that which is employed in that Institution.

"The rain-gauge referred to differs from that described in the annual report for 1869, which was constructed on the principle of the Tantalus Cup; that is, when the receiving vessel had collected a certain amount of water, it emptied itself into another by means of a glass syphon, and returned to the top of the register to receive a new supply. But sometimes it happened that small insects or pieces of leaves, &c., would choke the syphon, either retarding or stopping it from emptying the receiver. This led me to the construction of the new one, which acts on the principle of a gravity-bucket: that is, after a certain amount of water has been received in the bucket, it tilts so as to empty all the water out, and after doing this returns to its former position.

"Outside of the building, about two feet above the roof, is placed the usual circular rain-gauge receiver; it is about eight inches in diameter and is equal to fifty square inches of surface. It is funnel-shaped at the bottom, and leads into a block-tin pipe, three-quarters of an inch in diameter. This gauge is mounted on a square box described below. The pipe descends from the gauge to the receiving or gravity-bucket, which is made of brass and is of a triangular prismatic shape, balanced just above and forward of its centre of gravity in a square frame by two pivots. These allow it to tilt when the water has reached a certain height or quantity A leaden weight is [employed as] an adjustable counterbalance for the regulating of the tilting of the bucket. The square frame which carries this bucket is suspended to two tempered steel springs placed parallel to the register sheet, the object being, that when the bucket is emptying, the pencil rises from the register and makes no return mark. I might here mention that there has been a great deal of discussion in regard to the permanence of such steel springs; those that I use have not changed in any perceptible degree. They are made by winding steel wire on a mandril and are tempered on it. If they were made of steel wire first tempered and then wound on the mandril they would be worthless, for the act of winding destroys the molecular elasticity of the metal. The upper ends of these springs are fastened by suitable means to the top of the case. Between the springs and attached to the square frame is an upright rod, going through a steadying bracket, and carrying on its upper end an ink pencil, which delicately touches the sheet of paper. This pencil consists of a glass tube drawn down to a capillary point, so as to make a fine mark. Past this ink pencil, by a clock, is drawn a board on which is fastened by means of two clamps a sheet of suitably ruled paper to receive the register.

"The square box alluded to in the preceding paragraph not only affords a support for the gauge, but retains warm air admitted from the room below, by means of a wooden pipe, which delivers the heated air close to the bottom of the funnel, melting any snow or sleet that may collect.

"To complete this description it is necessary to give and explain a register taken by the instrument. On the upper and lower lines of the register are marked the hours of the day; these are joined by vertical lines. On the sides are marked the fractions of an inch by ·05 down to half an inch (or decimal ·50), these are joined together by horizontal lines, and thus the register to receive the record is divided into squares."

By the ink pencil are marked the curves produced by the fall of rain, showing when it commenced, the rate at which it fell, when it stopped, and the amount of water. And in illustration is given the register for September 17th, 1874. According to this "it commenced raining at five minutes past nine, and continued to rain quite hard till ten o'clock, when it slacked off, there being a fall of one-tenth of an inch of water. It continued to rain till ten minutes of eleven, when there occurred a hard shower for fifteen minutes, after that the fall was more moderate till three o'clock, when the storm set in violently. By a quarter to six there had fallen half an inch, the bucket then tipped and emptied all the water into the lower vessel. It then returned to the upper edge of the register to take the next half inch, which required an hour and three quarters, when the bucket again tipped and rose to the upper line. The greatest fall of rain ocurred at nine P.M. It amounted to half an inch in three quarters of an hour, this being the maximum of the storm. It now commenced to fall more slowly, and after one o'clock the next morning there was a very decided diminution. It stopped for nearly a quarter of an hour; commencing again, it continued to fall moderately with lulls till nine, when the register was removed. On counting the curve lines we find that there are eight half inches and sixteen hundredths of an inch. Therefore, in the twenty-four hours, from nine A.M. to nine A.M. of the next day, there were four inches and sixteen hundredths of rain. The greatest fall we have on record at this place from our gauges," says Mr Draper, "is one inch in twenty minutes; on that occasion the city culverts could not carry away the water, the streets became flooded and the cellars deluged."

By a self-acting register such as this, the rain falling at any hour, or at any minute, may be noted, or registered and recorded for subsequent observations. But for ordinary use, a simple rain-guage such as I have described, or such an one with some simple modification, will suffice to supply the information generally desired.

The means of determining the humidity of the atmosphere, by means of dry and wet bulb thermometers, have been described above [ante p. 22].

Of instruments employed to determine with some measure of accuracy the humidity of the atmosphere, the first were constructed of strings, animal fibres, and other substances which elongate or contract with damp. Such is the toy constructed in the form of a house, from which comes out a figure enveloped in a cloak when the humidity is great, and one in summer guise when the air is dry. And such is the philosophical instrument known as Saussure's Hygrometer. This, in the form best known, consisted of a single hair, freed from unctuosity, and stretched by a slight weight, one end being fixed and the other attached to the short arm of a lever, the long arm of which multiplied the effect of the expansion and contraction of the hair, produced by variation of humidity, and thus rendered them visible. The hair was attached to the instrument near the top by means of a screw, and the lower end of the hair was attached to the back of the index, attached to which also was a small weight to insure continued tension; while the long arm of the index moved on a curved plate, graduated from 0 to 100, and a clamp was employed to prevent movement when the instrument had to be carried from place to place.

It appears that the variation in the length of a healthy hair between the extremes of dryness and humidity is $\frac{1}{42}$ of its length, affording in careful hands a useful indication.

Another instrument of the same kind has an awn of barley employed as the indicator.

Previous to the use of the hygrometer of Saussure, there were in use hygrometers based on the principle that the drier the air is the more water it can absorb. But these, it was alleged by Saussure, could not be relied on, while some led to enormous errors. That devised by him superseded these, being strongly recommended by its portability; but it also was objectionable, in consequence of a loss of sensibility in the hair gradually supervening; and Saussure himself points out that though such instruments were useful, no one had examined the effect on any of them of variations in the pressure, density, movement, etc., of the air, nor had they even examined if the varying indications were proportional to the amount of vapour in the air.

A third class of instruments had also been in use, and in them we find the germ of dew-point thermometers now in use. For example, the Academicians of Cimento took a conical glass vessel, which they kept full of broken ice or snow; they suspended this vessel point downwards in the open air, the moisture was condensed on the surface of the glass, and distilled drop by drop from the point of the cone, the greater or less frequency of the drops indicated to them the degree of humidity of the air. L'Abbé Fontana rendered this more portable by taking a plate of glass, quite clean and well polished, and of known weight; this he cooled to a certain temperature, exposed to the air for a certain time and then reweighed, the increase showing the amount of condensation. Lastly, M. Le Roi adopted a still more simple method, merely standing a glass of water in the air and at the air temperature, then gradually cooling it by the addition of iced-water, until dew was deposited on the glass, when he noted the temperature of the water, and "judged the air more or less damp according to the degree of cold necessary to produce this deposition."

In the use of a wet and dry bulb thermometer, we find these appliances perfected; but Saussure saw it not, and on the contrary wrote thus:—"These ingenious processes do honour to those who have invented them, and may even sometimes be useful, but if we consider that we can seldom make use of the closed vessel, and never employ any of them when the air is below freezing point, nor when the air is very dry, and that, besides, the least particle of grease, and other causes difficult to avoid, may disturb the precipitation of this dew, and cause uncertain results, we can easily see that it will be very difficult for any such hygrometer to become universal. I have often tried M. Le Roi's process, with this difference, that instead of using ice, which is not easily carried about, I employed powdered sal ammoniac, which I throw little by little into the water, and which, when the air is not very dry, cools it sufficiently to produce condensation, but when I repeat it at short intervals, and when the air apparently remains unchanged, I do not always find the deposition commences at the same temperature."

"It is rather a noteworthy coincidence," says a writer in "Symons' Monthly Meteorological Magazine," "that two of the most forward steps in hygrometry are mainly due to the illness of those by whom they were taken. Although Guy Lussac used a dry and wet bulb thermometer in 1822, it is to Dr Mason that we are indebted for its introduction into this country, he being led, while residing in Madeira for the benefit of his health, to devise the arrangement so well known as Mason's hygrometer, which, slightly modified by Mr Glaisher, now leaves nothing to be desired in that branch of research. So again with Saussure in 1780. Surprised by illness in the midst of his '*Voyages dans les Alpes*,' even on the ridge of a lofty mountain, he was compelled to pospone his laborious researches, to return to Geneva, and, during his forced residence there, he perfected the hygrometer, the rough idea of which he had conceived five years previously."

The instructions issued by the Meteorological Commission at the Cape of Good Hope, in regard to the computations to be made, were as follows:—" Dr Apjohn's formula for computing the dew-point from simultaneous readings of the dry and wet bulb thermometers and barometer, in other words, from the temperature of the air of evaporation, and atmospheric pressure, is

$$f'' = f - \frac{d}{68} \times \frac{h}{30} = f' \qquad \frac{dh}{6240}$$

" Where f'' represents the tension or force of vapour in the temperature of the dew-point.

" Where f' represents the tension in the temperature of evaporation.

" Where d represents the difference between the temperatures.

" Where h represents the barometric pressure.

EXAMPLE.

Temperature of the air,	... ...	72°·4 Fahrenheit.
Temperature of evaporation,	...	60°·6 „
Difference,	... ...	11°·8 „
Barometer,	... ...	29·8 inches.

" $\frac{29·8+11·8}{2640} = 0·133$, and $0·529 - 0·133 = 0·396 = f''$. 0·396 inch in the table of elastic force of aqueous vapour corresponds to 52°·5—the temperature of the dew-point.

" Regnault's table of 'the Elastic Force of Aqueous Vapour in inches of Mercury," here employed, is No. III of Mr Glaisher's very convenient Hygrometric Tables.

" The percentage of 'humidity' is obtained by dividing the elastic force in the temperature of the dew-point by the elastic force in the temperature of the air, as shown by the dry-bulb, and the quotient multiplied by 100. Thus, for the temperatures in the preceding example, $\frac{0.396}{0.796} \times 100 = 50$ nearly (correctly 49·7), Dr Apjohn's formual is superseded by a table of numbers generally termed 'the Greenwich factors,' from which the dew-point is obtained with comparative facility.

"These factors were derived from a long series of comparisons between the simultaneous indications of the wet and dry bulb thermometers, and the temperature of the dew-point obtained by means of Daniell's hygrometer.

"On arranging the differences in the order of ascending temperatures of the dry-bulb, a series of factors was obtained, which established the relation between the simultaneons differences of dry-bulb minus wet-bulb, and dry-bulb minus dew-point, viz.: temperature of air minus temperature of evaporation, and temperature of air minus temperature of dew-point.

"Table II, page iv., of Mr Glaisher's Hygrometric Tables, third edition contains the factors thus obtained for each degree of the dry-bulb thermometer from 10 to 100 Fah. They have been tested by Regnault's hygrometer at the Cape Observatory, and the Commission has been assured that they are as suitable for the climate of the Cape as for that of England, the coincidence being as close as the direct observation of the dew-point admits of.

"To diminish the trouble of interpolation, also the chance of committing errors when dealing with negative quantities, tables founded on Glaisher's numbers were printed for the observers, and were employed by them for calculating the dew-point and humidity."

The thermometer, or measurer of heat, is founded on the observation that bodies almost universally expand with every rise of the temperature. In consequence of the smaller force of cohesion to be overcome in liquids, heat expands them more than it does solids, and liquid thermometers are more sensitive to the influence of heat than metallic rods or bars, and they are generally employed.

Self-registering thermometers are so constructed as to indicate what has been the highest and what has been the lowest temperature experienced within a given time. Different plans are adopted: one frequently met with is known as Rutherford's register thermometer. It consists of two thermometers—one of them mercurial, the other spirit. In the former, *above* the mercury, there is a piece of steel which is pushed on by that fluid as it expands with a rise of temperature, and the instrument being fixed horizontally it remains at the place to which it has been pushed, marking what has been the highest temperature experienced. After being observed it is brought back to the surface of the mercury by the application by a hand of a magnet outside the tube. In the spirit thermometer there is index of ivory or enamel, which sinks with the spirit as it contracts under a fall of temperature, but which does not rise with the surface of the fluid under an increase of temperature, being so formed that the spirit easily passes it, and it remains to indicate what has been the lowest temperature experienced; and when this has been observed it is brought back to the surface of the liquid by inverting the thermometer and slightly shaking it.

With the thermometer, as with the rain-guage, there are instruments of much greater precision in use in expensively equiped observatories.

In Draper's pencil thermometer, used in the Observatory of New York, the meter is composed of two curved bars, consisting each of "two strips of zinc, each $\frac{1}{32}$ of an inch thick by one inch broad and twenty inches long, rivetted on the same side to a similiar single piece of iron. On warming this bar it curves from the zinc side, that metal being more expansible than iron. If it be cooled, the reverse takes place. On the right hand bar there are small troughs, with cotton threads to convey moisture to it. This part of the apparatus represents a wet-bulb thermometer. Several troughs are used, so that when the cotton becomes frozen in cold weather it will not resist the bending of the bar, as would be the case if there was only a single long one." These bars are firmly fastened to an iron bar projecting out at the window, and connected with the case in the room; "on this is raised a pillar which acts as a fulcrum to slender brass levers, one end of each lever being fastened by connecting links to the loose ends of the compound bars. On the other ends of the levers are ink pencils, one containing red the other blue ink; these mark on a register drawn by the clock, towards the right hand, the movements of the metallic bars. . . .

"On the vertical sides of the chart aremarked the days and hours; on the top, the degrees for the dry-bulb scale; at the bottom, the degrees for the wet-bulb. The scales are arranged ten degrees apart, so that the pencils will not interfere with each other when they are marking nearly the same temperature. A red line marks the dry-bulb curve, a blue the wet-bulb." And thus is registered permanently the varying temperature day and night continuously.

Mr Draper first constructed a thermometer with its indications depending on the difference of expansion between brass and glass.

But this instrument would not stand the sudden fluctuations it was subjected to. He "was, therefore, led to try other substances, and after many experiments found that bars made of strips of hard rubber and brass rivetted together would curve uniformily from the melting of ice to 160° Fahrenheit. Above that temperature the rubber softened and ceased to act. The advantage of hard rubber is, that if it be roughly bent at moderate temperatures it will not take a set." He then constructed a pair of bars of zinc and iron, which have the advantage that the temperature may be carried up to the boiling of water. He also constructed a self-recording sun thermometer. It is constructed on a principle similar to that of the dry-bulb thermometer, above described, except that "the metallic bar is enclosed in a glass shade, four inches in diameter and twenty inches long, mounted above on the roof, and exposed to the direct rays of the sun. There is no vacuum in the glass shade, it only serves to keep the action of the wind from the bar, which is connected by suitable means to the lever in the room below, in which is placed the case with its clock, lever, register, etc. When the sun shines the bar bends and raises the pencil on the end of the lever; when a cloud passes over the sun the temperature lowers; therefore, the lever descends on the register. In this manner a record is made, whether the day was overcast or not." Mr Draper has found that the readings follow closely those of Cassela's Maximum Thermometer, with this advantage, that we have the time of maximum heat and other fluctuations.

The barometer measures the weight or pressure of the atmosphere. We owe it to a discovery made by Torricelli, a disciple of Galileo. Previous to his day it was a prevalent opinion that Nature abhored a vacuum, and the following were the occurences which led to the discovery of the error which had thus been long and widely promulgated.

The superintendent of the water works of the Grand Duke of Tuscany, wishing to raise water by means of a pump to a considerable height, was surprised to find that while it followed his suction rod foot after foot, when it had reached the height of a little more than 32 feet it refused to follow him further, and by no force could he raise it higher than 32 or 33 feet. What could be the cause?—Could it be ascribed to any defect in the apparatus?—It was examined, and found to be complete. Why then did the water refuse to rise?—and why did it stop where it did?

After he had ascertained that this could not be ascribed to any defect in the construction of the pump, he mentioned the circumstance to Galileo, and requested him to give him an explanation of the anomaly. Galileo, either not questioning the justness of the opinion which then prevailed, or being unable to assign any other that was more plausible, replied,—" That water was raised to the height of 32 feet, on account of the horror which nature has for a vacuum; but that the horror was limited in its effects, and ceased to operate above the height of 32 feet." It appears, however, that Galileo was by no means satisfied with the explanation which he had given, and that he immediately began to suspect the agency of some external cause; but his death, which happened soon after, prevented him from bringing his thoughts to maturity. " His disciple, Torricelli, to whom it is supposed he had mentioned his ideas on the subject, was more successful. He suspected that the weight of the water was one of the elements which ought to be taken into consideration in investigating the cause of the ascent of that fluid in pumps, and that this was probably counterbalanced by the weight of something external pressing upon the surface below. To put this conjecture to the test of experiment, he took a glass tube about four feet long hermetically sealed at one end and open at the other. Having filled it with mercury, he shut the open end with his finger; he then inverted the tube, and introduced the open end of it under the surface of a small quantity of mercury in a basin. Lastly he placed the tube in a vertical position, and on withdrawing his finger, he observed that a part of the mercury descended in the tube, and that the rest of it was supported at the height of 27½ inches above the level of the mercury in the basin. By varying the experiment he found that in all cases, the mercury was supported at a perpendicular height above its surface in the basin, equal to about the 14th part of the height of the water in the pump. He therefore inferred, that the mercury in the tube, and the water in the pump, exerted equal pressures on the same base, their altitudes being inversely as their specific gravities, and that the weight of the column in either case, was counteracted by some fixed and determinate force. This force he supposed was the weight of the air."

The discovery thus made by Torricelli, was confirmed beyond all controversy by Pascal, and thus it became certain that air is a material, as well as is earth, and a material possessing the quality of weight,—or what is called weight, and not that of levity or lightness, as the speculative philosophers had supposed.

The explanation given by Torricelli of the phenomenon he had observed, however natural and obvious it may now seem, was by no means so readily admitted as might have been expected from its extreme plausibility. "The principle of the horror of a vacuum was too firmly believed to yield at once to the simplicity of truth. Attempts were accordingly made to reconcile the experiments of the pumps and the tube of Torricelli, with that absurd opinion. It was maintained, that a subtle fluid or aërial spirit was evaporated from the surface of the water and the mercury, which filled the upper part of the tube, and left only as much activity to the horror of a vacuum as was sufficient to sustain the column of those fluids.

"When Pascal, who was then at Rouen, was informed of the experiment of the Italian philosophers, he was anxious to repeat them; and soon after he obtained the same results. It does not appear, however, that he was aware of the conclusions which Torricelli had drawn; but, by reflecting on the nature of the experiment, he was soon convinced that the principle of a horror of a vacuum was altogether gratuitous and improbable, and that the suspension of the mercury was owing to some other cause. To place the matter beyond all dispute, he employed tubes of glass forty feet long, and having filled one of them with water, and another with wine, he inverted them respectively in basins of these fluids, after the manner of the experiment of Torricelli. The water remained suspended at the height of 31 feet 1 inch and 4 lines; and the wine at the height of 33 feet 3 inches. These experiments were performed at Rouen in 1646, in presence of several men of science, all of whom were attached to the old opinions. The conviction which they produced on their minds was complete, and they immediately embraced the new doctrine. Pascal published an account of the experiments the following year, in a work entitled '*Experiences nouvelles touchant le vide.*' This work was severely attacked, particularly by P. Noel, a Jesuit, who was then rector of the college of Paris. All the prejudices of a bad philosophy, and all the virulence of error, were summoned to the attack; and Pascal had the mortification to find, that many were still disposed to question the conclusions which he had drawn from his experiments.

"At length an experiment occurred to him which he saw would for ever silence the objections of his opponents, and establish his opinion beyond the possibility of controversy. If the mercury in the Torricellian tube, said he, is supported by the pressure of the air, it ought to stand higher or lower according to the length of the column of the atmosphere at the place of observation; on the contrary, if the weight of the air has no connection with the height of the mercury, the mercury ought to stand at the same elevation, at all heights in the atmosphere. He therefore prepared to make the experiment on a large scale; and in order that the difference between the heights of the mercury at the places of observation might be an appreciable quantity, he pitched upon the mountain Puy-de-dome, in the neighbourhood of Clermont, as well adapted to his purpose. Being at that time in Paris, he wrote to his brother-in-law Perrier, a man of distinguished talents, who was then going to Clermont, requesting him to perform the experiment on his arrival. Various circumstances preventing the experiment being tried till the 19th of Sept., 1648, when it was performed with equal accuracy and skill. The result coincided with the expectations of Pascal; as they ascended the side of the mountain, the mercury gradually subsided in the tube; and when they reached the summit it stood 3 inches 1½ lines lower than at the bottom. The experiment was repeated, on different sides of the mountain, and always with a similar result.

"Pascal was no sooner informed of the details of these experiments than he repeated them on a small scale at the top and bottom of the steeple of St. Jacques-la-Boucherie: and he observed a corresponding difference between the heights of the mercurial columns. There now remained no longer any pretext for ascribing the elevation of the mercury in the tube to the horror of a vacuum; for, it would have been absurd to pretend that nature had a greater abhorrence of a vacuum in a valley, than on the top of a mountain; and accordingly all those who were sincerely desirous of discovering the real cause of the phenomenon in question, admitted the conclusions of Pascal concerning the weight of the air, and applauded the simple and decisive method which he had taken to demonstrate its influence. On the whole, the history of this research affords a signal instance of the slow and gradual progress of human knowledge: Galileo proved that the air was possessed of weight; Torricelli conjectured that this fluid caused the ascent of water in pumps, as well as the suspension of the mercury in the tube which bears his name; and Pascal converted this conjecture into a demonstration."

This discovery of Pascal, that the mercury in the barometer fell as he ascended the mountain, suggested the application of the barometer, as an instrument wherewith to measure the heights of mountains, and other elevations; and Sir Henry Englefield constructed a barometer expressly for such investigations. Such instruments are now generally used, and they are called mountain barometers. They differ from the common barometer only, in being made of a portable form, and in having a much larger scale, and arrangements for observing very minute variations in the height of the column of mercury. This, at the surface of the earth, or the level of the sea, is in general about 30 inches in height; but on ascending

1000 feet the column falls to	28·91 inches
2000	27·86
3000	26·85
4000	25·87
5000	24·93
1 mile	24·67
2 miles	20·29

and if the barometer were carried to an elevation of 3 miles, the mercury would fall to 16·66 inches; to an elevation of

4 miles,	to 13·72 inches
5	1·28
10	4·24
15	1·60
20	·95

But it has also been ascertained that the temperature of the atmosphere diminishes 1° for every 350 feet of ascent. This is referable partly to the increased capacity of the air for heat in proportion as its density diminishes, and partly to the circumstance that the atmosphere is chiefly heated by the earth. It is therefore necessary in calculating heights by the barometer, to make allowance for the expansion and contraction both of the mercury and of the tube in which it is contained. There is therefore a thermometer attached to mountain barometers, with the bulb in the mercury cup; and there is appended to it a tabular statement of the addition or substraction necessary for the correction of the height indicated by the instrument.

The Aneroid Barometer is found more convenient as a portable instrument. It is generally seen in the form of a watch. The principle upon which it acts is the pressure of the atmosphere upon a metallic chamber, partially exhausted of air, and so constructed that by a system of levers a motion is given to an index hand which moves upon a dial. In this also a certain thermometrical correction is required.

In the observations embodied in the following tables carefully constructed mercurial barometers were employed.

ABSTRACT OF THE MEAN ANNUAL RESULTS FROM THE FOLLOWING METEOROLOGICAL STATIONS.

NAMES OF STATIONS.	NAMES OF OBSERVERS.	Mean height of the Barometer, reduced to temperature 32° Fahr. and to the Observatory Standard.	TEMPERATURES, Fahr. Mean annual	Mean daily range.	Mean of greatest range on any one day.	Mean of least range on any one day.	Mean of greatest range of each month.	Mean annual temperature of evaporation of 9 A.M., 1 and 5 P.M.	Mean annual humidity at 9 A.M., 1 and 5 P.M.	Mean annual falls of Rain.	Mean annual clouded sky; per cent. of visible hemisphere.	Mean annual number of lightning or thunderstorms	Approximate heights of the Stations above the sea level.	Years of observations from which the results are derived.
		inches.	deg.	deg.	deg.	deg.	deg.	deg.	per ct.	inches.	per ct.	No.	Feet.	
Graaff-Reinet ...	Professor Guthrie ... J. Atkinson, Esq. ...	27·508	64·41	24·52	37·88	11·58	48·74	58·17	55·98	13·196	36	23	2,517	1863–64–65
Worcester ...	J. D. Hugo, Esq.	29·265	62·88	24·91	41·10	9·98	48·97	58·17	54·99	11·619	31	14	776	1862–63–64–65 –66–67–68
Mossel Bay ...	H. W. Laws, Esq.	29·608	63·90	17·78	27·42	9·00	38·25	59·20	69·47	11·550	41	2	423	1862
Somerset West ..	Alexr. Graf, M.D.	29 931	61·89	20·88	35·08	8·12	42·87	59·93	68·83	26·671	33	9	124	1862-63
Port Elizabeth ..	C. Hammond, ...	29·878	63·21	14·16	25·63	6·43	33·77	59·96	71·24	26·685	...	...	190	1867–68
Simon's Town ...	Dr Churchill, ...	30·023	65·14	15·00	23·98	6·50	30·91	62·27	79·17	29·489	52	4	50	1862–63–64–65
Amalienstein ...	Rev. A. Schmidt	...	65·40	22·30	38·60	6·60	46·40	..	...	31·190	...	13	3,200	1867–68
Royal Observatory	Personal Establishment ..	30·031	62·38	14·11	26·72	6·41	34·50	56·62	72·00	22·476	42	15	37	1862–63–64–65 –66–67–68
Graham's Town...	*Royal Engineers	30·026	62·65	18·59	...	...	43·94	58·90	70·30	32·594	...	...	1,750	1855 to 1859
Aliwal North ...	R. Dowling, Esq.	...	59·30	23·50	38·30	12·30	45·10	...	...	18·910	...	21	4,500	1867
Maritzburg, ...	†J. R. Mann, M.D., F.R.A.S.	27·896	64·83	...	...	...	41·86	...	70·80	30·230	51	63	2,096	1856 to 1865
Adelaide	Cha. Todd, Esq., F.R.A.S. Australia	29·897	63·94	20·98	33·97	11·58	45·91	55·82	58·25	20·505	48	...	140	1858 to 1863
Adelaide	‡	29·876	64·60	20·36	33·83	11·57	45·00	56·30	58·00	22·763	49	...	140	1862 and 1863
Hobart Town ...	F. Abbot, Esq.	29·894	56·20	18·48	32·50	9·00	40·13	50·86	72·17	25·710	58	11	37	1859 and 1864
St Helena	Royal Artillery, ...	28·285	61·40	...	...	...	...	59·00	87·00	47·198	...	...	1,765	
Antananarivo ...	§Mr Cameron	25·606	...	...	...	...	...	...	64·69	38·710	66	...	4,418	1864
New Zealand ...	‖	30·610	55·3	Mean of greatest range for year,			50·3	51·4	77·0	44 88	...	...	27	1864

*Graham's Town.—"Abstracts of Meteorological Observations taken at the Stations of the Royal Engineers," published in 1862. The barometer is reduced to the sea level. The several means are derived from 5 years' observations Lat. 33 deg. 18 min. 30 sec., long. 8 deg. east of the Cape Observatory.

†Maritzburg.—Lat. 29 deg. 36 sec. S., long. 11 deg. 55 min. east of Cape Observatory. Mean of 9 years' observations, ending 1865, inclusive.

‡Meteorological Observations by Charles Todd, Esq., F.R.A.S. Lat. 34 deg. 57 min. S., long. 120 deg. 9 min. east of Cape Observatory.

§Greatest fall of rain in one day, 3·18 inches. Number of rainy days, 57. Mean of greatest range of barometer, 0·285 inches.

‖Meteorological Observations by Charles Rous Marten, Esq. Mean of 9 Stations—Martendale, lat. 46 deg. 17 min. S., long. 150 deg. east of Cape Observatory; Mongonui, Auckland, Taranaki, Napier, Wellington, Nelson, Christchurch, Dunedin.

NOTE.—The last six Stations are inserted for the sake of comparison.

RAINFALL AT THE ROYAL OBSERVATORY, CAPE OF GOOD HOPE, DURING THE YEARS

Months.	1859.	1860.	1861.	1862.	1863.	1864.	1865.	1866.	1867.	1868.	1869.	Year.	Total fall.	Difference from the mean of 30 years.	Year.	Total fall.	Difference from the mean of 30 years.
	inches.	inches.	inches.	inches.	inches.	inches.	inches.	inches.	inches.	inches.	inches.		inches.	inches.		ins.	inches.
Jan......	1·826	0·850	0·797	0·227	0·202	0·537	0·298	0·032	0·393	0·710	0·174	1842	26·3	+2·4	1856	21·9	—2·0
Feb......	0·964	1·045	0·068	0·233	0·673	0·010	0·135	3·104	1·064	1·037	0·065	1843	24·8	+0·9	1857	22·7	—1·2
Mar......	1·051	0·635	0·867	0·373	2·914	0·290	0·388	0·185	1·058	0·467	0·554	1844	18·8	—5·1	1858	24·1	+0·2
April....	0·780	1·186	1·582	0·939	2·560	1·011	1·850	1·501	2·443	2·165	1·876	1845	20·9	—3·0	1859	36·7	+12·8
May.....	6·925	6·572	4·315	1·247	5·338	2·780	3·878	0·763	3·041	1·870	8·057	1846	22·5	—1·4	1860	29·1	+5·2
June.....	5·457	4·961	7·608	10·783	3·206	4·365	0·934	5·658	3·542	3·359	9·515	1847	22·4	—1·5	1861	25·4	+1·5
July.....	6·524	4·926	4·297	6·279	2·371	2·684	4·911	2·380	4·343	2·653	3·060	1848	23·2	—0·7	1862	32·0	+8·1
Aug......	4·786	0·923	1·915	4·059	2·678	2·178	1·676	2·266	1·358	0·692	2·142	1849	24·6	+0·7	1863	25·6	+1·7
Sep......	3·215	5·019	2·544	2·244	1·715	1·982	0·652	1·439	1·441	0·904	...	1850	33·5	+6·6	1864	18·9	—5·0
Oct......	2·400	2·084	0·109	4·040	2·707	1·963	3·109	1·003	3·526	2·715	...	1851	20·3	—3·6	1865	18·7	—5·2
Nov......	2·579	0·237	1·287	1·584	0·924	0·995	0·568	0·396	0·213	2·382	...	1852	23·2	—0·7	1866	19·2	—4·7
Dec......	0·213	0·678	0·053	0·000	0·315	0·120	0·269	0·480	0·543	0·999	...	1853	21·2	—2·7	1867	23·0	—0·9
												1854	20·0	—3·9	1868	20·0	—3·9
												1855	24·6	+0·7			
Annual Fall,	36·720	22·116	25·442	32·008	25·608	18·915	18·668	19·207	22·965	19·953	8 Mon. 25·443	Mean annual fall = 23·671 inches.					

Rejecting 1839, we have 23·877 inches for thirty Years.

	inches.			inches.	
1838	—24·444	+0·5	1840	—25·058	+1·2
(1839	—17·483)		1841	—23·208	—0·7

RAINFALL IN INCHES, AT SEVERAL STATIONS IN THE COLONY.

	Bishop's Court Protea.	Observatory.	Wynberg.	Capetown.	Simonstown.	Somerset West.	Stellenbosch.	Groote Post.	Worcester.	Mossel Bay.	Bredasdorp.	Amalienstein.
1862 ..	..	32·008	68·500	28·750	36·090	28·707	..	..	21·235	11·550	..	..
1863 ..	..	25·603	..	26·900	33·940	24·634	23·460	..	14·551	..	..	..
1864 ..	..	18·915	34·226	17·720	23·090	..	21·860	..	8·789	..	..	..
1865	..	18·668	36·573	21·000	24·820	..	24·225	..	5·330	..	..	..
1866 ..	..	19·207	33·573	23·900	..	..	..	22·469	8·820	..	..	..
1867 ..	42·919	22·965	29·325	29·250	..	..	..	16·011	11·300	..	16·320	41·190
1868 ..	45·903	19·953	37·247	26·350	..	..	..	17·957	11·310	19·164	11·810	21·196
Means ..	44·447	22·474	39·907	24·839	29·485	26·670	23·181	18·812	11·619	15·357	14·065	31·190
Jan. ..	1·062	0·343	0·414	0·219	0·697	0·356	0·437	1·248	0·095	0·192	0·445	1·285
Feb. ..	1·823	0·894	1·116	1·201	0·470	0·742	0·540	1·607	1·189	4·151	1·880	3·475
March.	2·081	0·811	0·653	0·742	1·403	1·332	0·610	0·608	0·398	1·279	1·975	2·255
April..	4·350	1·781	2·070	1·897	1·957	1·448	1·970	1·729	0·622	0·811	1·195	2·770
May ...	3·343	2·702	3·438	3·280	3·395	2·475	4·240	0·986	1·149	1·160	1·680	5·155
June ..	7·110	4·550	8·588	5·269	5·103	5·727	3·303	2·769	1·861	0·669	1·045	1·385
July ..	7·075	3·660	8·471	3·714	5·282	4·299	3·330	3·204	1·806	2·172	1·270	3·290
Aug. ...	3·235	2·129	5·659	2·596	2·830	3·395	2·513	1·938	1·431	1·618	1·695	4·350
Sept. ..	4·635	1·483	3·001	1·767	2·753	2·456	1·890	1·098	0·981	0·730	0·480	0·950
Oct.	5·945	2·724	4·523	2·941	4·002	2·999	2·673	2·315	1·320	1·625	1·255	1·945
Nov. ...	2·735	1·009	1·614	0·895	1·353	0·962	1·308	0·752	0·661	0·514	1·045	3·915
Dec. ...	1·052	0·390	0·360	0·318	0·240	0·479	0·367	0·558	0·106	0·436	0·110	0·415
Mean Annual Fall.	44·446	22·476	39·907	24·839	29·485	26·670	23·181	18·812	11·619	15·357	14·065	31·190

	Rhenoster Fontein.	Keerom, Namaqualand	Concordia, Namaqualand.	Lower Nel's Poort.	Goliad's Kraal, near Graaff-Reinet.	Graaff-Reinet.	Wheatlands, Graaff-Reinet.	Port Elizabeth.	Aliwal North.	Sea Point.	Pietermaritz-burg, Natal.
1862 ..	..	..	13·600	..	..	..	..	..	..	27·551	29·970
1863 ..	..	8·400	..	..	16·980	16·521	..	..	..	23·445	34·660
1864 ..	..	9·620	..	12·060	14·370	14·010	..	..	..	18·071	37·310
1865	8·300	5·730	6·040	10·610	6·180	9·055	..	..	..	16·201	31·080
1866 ..	..	8·660	9·658	9·060	..	..	..	..	..	20·263	..
1867 ..	..	..	..	..	17·060	18·205	9·420	33·590	18·910	20·397	..
1868 ..	..	..	6·485	6·130	..	..	..	19·780	..	18·695	..
Means ..	8·300	8·103	8·946	9·465	13·648	14·448	9·420	26·685	18·910	20·648	33·255
Jan. ..	0·590	0·050	0·015	0·995	0·868	1·187	0·210	0·365	1·550	0·331	3·92
Feb. ..	0·530	0·678	0·145	2·187	2·147	2·618	1·815	2·750	2·500	0·999	4·41
March.	0·190	0·190	0·488	1·565	1·447	1·193	0·820	1·630	2·310	0·574	3·29
April..	0·700	0·389	0·602	1·608	1·315	0·964	0·640	1·600	1·060	1·564	1·44
May ...	0·790	0·523	1·010	0·263	0·812	0·978	0·935	1·770	0·620	2·531	0·95
June ..	0·130	2·158	1·657	0·022	0·163	0·319	0·160	2·525	1·350	4·104	0·26
July ..	0·470	1·018	1·525	0·580	0·445	0·405	0·200	2·915	0·000	3·276	0·23
Aug. ...	0·330	1·083	2·010	0·463	1·205	1·074	1·365	2·165	0·650	2·071	0·14
Sept. ..	2·010	0·529	0·645	0·720	0·405	0·941	0·165	2·145	0·650	1·479	1·32
Oct.	1·610	1·178	0·349	0·255	1·468	1·549	0·950	1·965	1·350	2·467	3·60
Nov. ...	0·560	0·175	0·195	0·487	1·798	1·903	1·580	5·720	2·380	0·807	4·58
Dec. ...	0·380	0·132	0·305	0·320	1·575	1·318	0·580	1·135	4·490	0·445	5·04
Mean Annual Fall.	8·300	8·103	8·946 9months in 1867 equal to 6·275 in.	9·465	13·648	14·449	9·420	26·685	18·910	20·648	29·18 For 8 years.

N.B.—1866.—Graham's Town, from January 1 to November 9, 18·120 in. Aliwal North, from March 7 to December 31, 12·730 in. Bredasdorp, from May 1 to December 31, 8·770 in. Port Elizabeth, from May 1 to December 31, 7·840 in. 1867—Amalienstein from February 1 to December 31, 39·990 in.

Abstract derived from observations in the seven years, 1862, 1863, 1864, 1865, 1866' 1867, and 1868.

Months.	Mean height of Barometer at Temperature 32°.	Temperature of the Air.					Mean Temperature of Evaporation for each Month.	Mean Humidity for each month	Rain.	Clouded sky.	Lightning.
		Mean Temperature.	Mean daily range.	Mean of greatest range any one day of each month	Mean of least Range any one day of each month.	Mean of greatest Range throughout each month					Recorded No. of Days.
	inches.	degrees.	degrees.	degrees.	degrees.	degrees.	degrees.	per cent.	inches.	per cent of visible hemisphere	
January	29·927	70·94	17·04	30·76	9·29	37·45	62·62	62·85	0·343	26	5
February.........	29·915	69·65	15·66	29·14	8·37	34·09	62·18	65·80	0·894	33	8
March	29·960	67·47	16·08	31·04	7·80	39·67	60·49	67·57	0·811	31	18
April.....	30·010	62·59	14·56	28·60	6·73	36·03	57·23	72·83	1·781	44	11
May...............	30·070	59·16	13·54	25·61	4·70	34·13	55·06	77·83	2·702	48	10
June...............	30·134	55·78	11·73	22·96	4·28	30·56	52·46	80·15	4·550	51	8
July...............	30·150	55·21	12·16	21·60	3·46	28·81	51·88	79·02	3·660	47	9
August.......	30·118	56·32	11·52	23·60	4·26	32·61	52·56	77·83	2·129	48	5
September	30·104	57·90	12·62	25·06	4·67	32·69	53·34	74·45	1·483	47	14
October	30·026	60·81	13·40	27·29	6·02	35·22	52·68	72·59	2·724	51	8
November	30·015	63·94	14·99	25·26	7·80	34·33	57·65	67·99	1·009	40	5
December...	29·947	68·77	15·99	29·75	9·53	38·40	61·32	64·99	0·390	32	7
Yearly Means.	30·031	62·38	14·11	26·72	6·41	34·50	56·62	72·00	22·476	42	15

EXTREME LIMITS OF TEMPERATURE AND DATES THEREOF.

	HIGHEST.			LOWEST.		
	Year	Date	degrees.	Year	Date	degrees.
Graaff-Reinet,	1865,	December 20,	105·0	1864,	June 24,	28·0
Worcester,	1868,	January 10,	106·5	1862, 1863,	May 30, June 28,	29·5
Mossel Bay,	1862,	March 24,	97·0	1862,	July 17,	39·0
Somerset West,	1862,	December 24,	98·0	1863,	July 1,	34·4
Port Elizabeth,	1867,	January 31,	92·0	1868,	August 22,	41·0
Simon's Town,	1865,	March 11,	95·0	1862,	September 25,	42·0
Amalienstein,	1868,	January 11,	110·7	1867	June 17, 21, July 21,	27·5
Royal Observatory,	1864,	March 18,	99·5	1864,	June 8,	39·8
Graham's Town,	1858,	February 6,	106·5	1856,	August 8,	32·5
Aliwal North,	1867,	December 24,	93·0	1867,	June 21,	20·0
Maritzburg,	1865,		97·6	1864,	July	29·0
Adelaide,	1862,		115·0	1863,		33·0
Hobart Town,	1864,		88·0	1864,		32·0
New Zealand,	1864,		91·0	1864,		17·0

PART II.

Sources from which is derived the Supply of Water which is at present available for Agricultural Operations, etc., in the Colony of the Cape of Good Hope and Regions beyond.

Having given consideration to the meteorological observations which have been made on the moisture existing in the atmosphere, and to phenomena connected with the existence and the precipitation of this, there next demands attention the sources, primary and secondary, whence is obtained what is at present available for agricultural operations, the quantities of water there existing, and of these what is available for purposes for which it may be desired.

It may be thought that it would be an easy matter to ascertain the quantity of water available for agricultural operations or the promotion of vegetation, at the Cape of Good Hope or in any country, by ascertaining by rain-guages the quantity of the rainfall; but it is otherwise. Pluviometric observations are being made at numerous stations in South Africa, and recorded, collated, and tabulated; but the laws regulating the distribution of the rainfall, not only in South Africa, but every where else, have not yet been evolved in their entirety; and it is not the local rainfall which anywhere measures the local supply. There is, therefore, an interest attaching to all that relates to the subject, and more especially to what is known on the points referred to.

CHAPTER I.

SUPPLY OF MOISTURE IN THE ATMOSPHERE.

IN the preceding chapter attention has been directed to the difference between the humidity of the air, occasioned by vapour, or water suspended in the atmosphere, and the rainfall. The latter is only a portion of what was previously so suspended, dissolved in the air and invisible, it may be, which was dropped because it happened to be in excess of the quantity which the air could hold in solution at a lower temperature, to which, by some cause, it had been reduced.

To quote from a Memoir which I addressed to the Government of the Cape of Good Hope, On the conservation and extension of forests as a means of counteracting disastrous consequences following the destruction of bush and herbage by fire, which was appended to the Report of the Colonial Botanist for 1863, when this, by command of the Governor, was laid before Parliament :—

"It is well known that if a wet cloth be exposed in the air it soon becomes dry, or if a quantity of water be left exposed in a basin it soon disappears if the air be hot and dry; even the large reservoir of water, if not replenished, dries up in the course of time. In all of these cases the water passes off in a state of vapour. This vapour is transparent; it is not like the cloud or mist, which consists of water in minute drops suspended in the air. It is in a state of gas, and is as invisible as the air itself. In accordance with the law of gaseous diffusion, which admits of an unlimited diffusion of one gas through another, the atmosphere is permeated with this vapour, which may be found everywhere, on the summit of the highest hills and at the bottom of the deepest mines alike. It may have been observed that a glass of iced water is immediately covered on the outside with dew-drops of moisture. Whence come these? Not through the glass from the water within, but from the atmosphere without—a proof that the moisture was there, though unseen.

But while moisture may be found in the atmosphere everywhere, it is not always present in the same quantity in the same space. In almost any circumstances, but more especially where there is an unlimited supply of moisture, the quantity varies with the temperature, and varies with the temperature to a very great degree. According to Sir John Leslie, it doubles with every rise of twenty-seven degrees of temperature; and, according to subsequent observers, it varies to a still greater extent, the doubling taking place at different intervals, which increase slowly with the temperature—the mean being 23° 4′ from the freezing point to 100° Fahrenheit.

According to Leslie's calculation, the air at a temperature of 59° or 60° will hold twice as much as at 32°, or the freezing point; and at 86° it will hold twice as much again. And, according to later observers, at 121° it will hold twice as much again, or sixteen times as much as at 32°, eight times as much as at 53°, or four times as much as at 75°.

It is customary to speak of this as the solvent power of the atmosphere, as if the moisture were dissolved in the air as sugar may be dissolved in water; but it is more in accordance with fact to consider that as water requires a certain amount of heat to keep it from passing into a state of ice, so it requires a certain amount of heat to keep it in a state of vapour and prevent it from passing into a state of water.

In accordance with this, it happens that if air of a higher temperature, having as much moisture in it as can be sustained in a state of vapour, be cooled down some degrees, a quantity of moisture which can no longer be maintained in a state of vapour will assume the form of water in minute drops, and appear as fog, or cloud, or mist; but let the temperature be again raised to the same elevation as before, and this water will again assume the state of vapour, and the air become transparent, the fog or cloud having melted away. In illustration of this fact, I may refer to what is seen in the laying of the cloth on Table Mountain when the south-east wind blows. The temperature on the top of Table Mountain is lower than that of the wind blowing over its summit, and the mountain cooling down the air, a quantity of the vapour intermixed with it is deposited in the form of a cloud, which is blown over the face of the mountain, and falls, threatening to bury the city below; but before it can reach Cape Town it must pass through an intermediate stratum of air of a higher temperature than that to which it has been reduced. By this it has its temperature again raised, and the cloud evaporates and disappears, followed by continuous masses of cloud, which on reaching the same level vanish into thin air, leaving no trace behind."

But the water is there, though it be no longer seen; and it is of this I speak.

Relatively to the other constituents of the atmosphere, the quantity of water thus suspended there may be small, but in itself it is great. Though invisible, its presence can in general easily be made apparent. There are districts in South Africa so arid that needles have been exposed to the atmosphere on the ground for months without rusting. But in most places, a glass of very cold water exposed to the air becomes bedewed outside with water deposited there in consequence of the reduction of temperature; and, without any marked change in the temperature of the air, common salt exposed to the air becomes moist, and potash so exposed deliquesces, or dissolves in its own brine, by the attraction of moisture from the atmosphere.

By analysis scientifically conducted it may be shown that 100 parts by measure of atmospheric air contains :—

	By measure.	By weight.
Nitrogen gas	77·50	77·55
Oxygen gas	21	25·32
Aqueous vapour	1·42	1·03
Carbonic acid	·08	10

The quantity of aqueous vapour varies with the temperature and facilities for saturation, but it does not vary much when compared with the quantities in which the two principal constituents exist in the air.

Of the greatness of the absolute quantity of the vapour suspended in the atmosphere, some idea may be suggested by the consideration that the most copious and long continued rain leaves the air, whence it has been precipitated, saturated. How much is required to saturate a cubic foot at any temperature under the conditions of barometric pressure, &c., has been ascertained, but how shall we ascertain the bulk of the air which remains so saturated? We might arrive at a remote approximation to an estimate of what remains by an estimate of what has fallen during the copious and protracted rainfall over the whole area over which it fell, and calculating what fractional proportion they must bear to what was, and what remains, suspended there. But where are the data for such a calculation?

From observations carefully made by others, the Rev. Dr Adamson compiled the data which furnished the tabulated results given in a preceding chapter, whereby a rough estimate may be formed of the quantity of water which must be thus suspended in the atmosphere over the land, and of how much of this there is, which may be considered a reserve supply, which may be employed in the promotion of vegetation. This depends on the quantity of water which would be required in the circumstances to saturate a cubic foot of air; and this varies with the temperature indicated by the thermometer, and the pressure indicated by the barometer. The measure in which it does so has been determined; and the results of an application of the formula, expressive of the ratio to the variation, to the temperature and pressure of the atmosphere at different places, are given in the Table of Comparative Climate [ante p. 30].

But this is only an approximate estimate founded on averages. In the last column but one is given the information referred to. From this it appears that from an average of all the observations in the Colony made, recorded, and examined, there was an average of nearly 4½ grains of water in every cubic foot of air in the stratum of the atmosphere in which the instruments were placed; and a corresponding quantity in every cubic foot above and below that stratum.

But in trying to estimate the elements required for the next calculation, the non-scientific student of the phenomena is again brought to a stand. What are the corresponding quantities; and what the cubic measurements of the different bodies of air to which the different corresponding quantities are confined? But though he is brought to a stand, enough may have been learned to show that though the quantity of aqueous vapour in the atmosphere be, in comparison to the quantities of nitrogen and of oxygen which are there, relatively small, it is in the mass absolutely great.

I can illustrate this numerically by a brief reference to the carbonic acid which exists in the atmosphere in a still smaller proportion. Experiments made by Saussure, with great exactitude in every season for a space of three years, show that the air contains on an average 0·000,415 of its own volume of carbonic acid, but say in accordance with the analysis given above $\frac{1}{1000}$ part of its weight; and only 27 per cent., or little more than one fourth of this, is carbon; yet this suffices for the production of the whole vegetation on the surface of the earth; and we can obtain data which enables us to calculate the quantity of carbon existing therein. This is done by Liebig thus:—" It is known that a column of air of 2216.66 lbs. weight, Hessian measure, rests upon every Hessian square foot of the surface of the earth; the diameter of the earth and its superficies are likewise known, so that the weight of the atmosphere can be calculated with the greatest exactness. The thousandth part of this is carbonic acid, which contains 27 per cent. carbon. By this calculation it can be shown, that the atmosphere contains three thousand billions of Hessian lbs. of carbon; a weight which amounts to more than the weight of all the plants and of all the strata of mineral and brown coal which exist upon the earth."

Let the same principle of calculation be applied to determine the quantity of aqueous vapour existing in the atmosphere, and the result will be found to be astounding.

The quantity is immense. It must be borne in mind that it is only the quantity which is in excess of what will saturate the air at any lower temperature to which it may be reduced, which will be precipitated as rain or dew, and become available in a tangible form. But while it is so, a portion of that immense supply beyond may be absorbed by the soil; and it is by moisture thus absorbed that vegetation is in a great manner promoted.

I might mislead my readers if I were to say that I attach more importance to this supply than I do to the rainfall; but I have facts at command which seem to show that by this, to a greater extent than by the immediate effects of what falls as rain, is vegetation promoted. My difficulty is, to state fully what I want to state, and yet not to convey an impression which goes beyond this.

In treating, in a separate volume, of the phenomena of vegetation on which the meteorological effects of forests affecting the humidity of climate depend, I have had occasion to show that the primary phenomena of vegetation depend on the absorption of moisture by endosmose,—to trace the liquid absorbed from the rootlets to the leaves, and thence by evaporation into the atmosphere,—and to cite facts illustrative of the quantity of moisture thus evaporated by different kinds of trees, such as the oak, the maple, and the birch.

The function of the moisture thus passing through the vegetable structure is mainly that of a solvent and a medium of transport. The quantity passing through any plant in a given time can be determined by appropriate appliances; and from experiments carefully made, both in England and on the continent of Europe, it appears that this is greatly in excess of what falls in the form of rain.

Calculations made on experiments in England show, that an acre of land planted with cabbages requires more than five millions of pounds of water for their growth; that an orchard stocked with dwarf apple trees requires an equal quantity; and that an acre planted with hops requires six or seven millions of pounds weight.

But observations made have shown that at most not more than 2,350,000 pounds weight of rain fall there in the course of the summer months; and it is conjectured that at least one-half of this, or 1,175,000 pounds of water, flow off or through the land to the sea, leaving an equal weight to be contributed in any form to the vegetation produced, and leaving more than four times that quantity to be supplied in some other form for the production of the cabbages and the apples, and nealy six times that quantity, or more than five millions of pounds weight of water to be so supplied for the growth of the hops.

If thus it is in England, thus must it also be at the Cape; and when we think over the matter, it appears that it must be chiefly by moisture suspended in the form of vapour in the atmosphere that vegetation in the Karoo, and other arid districts is promoted. There are there *Feiges* and other succulent plants in which moisture is retained, but this appears only to enable them to live. There are plants like the *Kann niet todt* which seem to grow, through the matter in the outer and lower leaves being employed in the nourishment of the inner and upper leaves of the plant, and sometimes in the production of a flower-stalk, of flowers, and of fruit; but how is the water requisite for this conveyed? Not by the torrent or the rain; for both haste away as fast as they can to the ocean whence they came, and what they lose by the way would likely, as in the case referred to, be found, if weighed, to be less than the moisture employed in the promotion of the vegetation following the passage of the thunder-storm; for we have to reckon, not only the water fixed in the resulting vegetable products, but also the water which has passed through the plant, entering by the roots, and escaping by the leaves after conveying thither the matter to be elaborated into nutriment, or material to be employed in the building up of the plant.

It is but an infinitesimally small portion of the water which is supplied to the plant which is retained in it, and a still smaller portion which is fixed in its structure, as may be shown by weighing the quantity of water applied to a plant from the period of its germination till the time at which you determine to proceed with the experiment, by then removing it from the ground and weighing it, and then drying it in an oven and weighing it again.

And accordingly vegetation is to be found where not a drop of rain has been known to fall from one end to the other of the year before, and it may be for more than one year before that.

CHAPTER II.

SUPPLY OF WATER AFFORDED BY THE RAINFALL.

WHILE it is the case that there are plants growing where no rain falls, and it can be explained how all the moisture necessary for their growth may be obtained from the atmosphere, irrespective both of rain and dew; and while the moisture in the atmosphere, relatively small in quantity when compared with the oxygen and the nitrogen which are there, is the source of the rainfall, which again relatively to it is only a fractional portion dropped, because in consequence of a fall of temperature, howsoever produced, it could no longer be retained; it is this fractional portion which is mainly by nature and by man's device rendered subservient to the promotion of vegetation; and we have come to look upon the rainfall as the most important source of supply of moisture available for agricultural and similar operations.

From the irregularity of its distribution it is difficult to determine the actual quantity of what moisture from this source of supply is at the command of man; perhaps it is as impossible to determine this with absolute precision, as it is to determine with absolute precision the quantity which is reserved in the heavens. Were the actual quantity of the rainfall determined, there are deductions necessary, and most if not all of these are of unknown amount: some falls on rocky places; some, on barren sand; much evaporates so soon, and so long, as the superincumbent air retains a temperature at which the moisture in it fails to saturate it; much percolates through the ground and sinks by fissures to subterranean depths; even the Chinese have failed to prevent the flow of much to the sea that falls as rain,—and though it may be theoretically possible it is actually impracticable to prevent some from so escaping. It is also subject to disturbing influences operating in an opposite direction: a great deal of what is precipitated in rain is evaporated and precipitated as rain again and again and again, and much is precipitated as dew, of which the pluviometer takes no cognisance.

How much of the rainfall is lost in the sea, and how much is apparently lost on the lower parts of the earth we do not know, nor do we know the quantity of the rainfall which is thus diminished. But we know that the quantity of the rainfall must be very great.

As has been intimated, the rainfall in different places of the Colony has been ascertained by careful measurements, with properly constructed instruments, day after day for several years in succession, so as to get an approximation to the average annual rainfall in the several localities in which the measurements have been made.

The difference in the quantity of the rainfall in different districts, and even in near-lying localities in the same district, we have seen to be great. It would therefore lead to an erroneous conclusion if we were to take even the average of these and multiply the number of square inches on any one farm, and give the product as a correct estimate of the quantity of rain falling annually upon that farm.

If it be desired to ascertain this, a nearer approximation to the facts of the case may be obtained if observations similar to those which have been referred to be made on the farm itself, and this on different parts of the farm at the same time, and these on parts of the farm differing as much as possible in altitude, in exposure, and in circumstance; the greater the number, and the greater the variety of conditions under which these may be done, the better; and as each of these will give indications of what quantity of rain falls upon that spot, by taking the average of the whole and multiplying this by the area of the farm a nearer approximation to a determination of the quantity of rain falling upon that farm may be obtained. It will only be a nearer approximation to this, but it may astonish some who have not previously thought upon the subject.

A common form of pluviometer is like a deep spittoon, with a rod attached to a float below, rising through the opening and passing freely through a hole in horizontal strips of tin soldered on across the mouth. This rod is graduated, and the height to which it is floated beyond the strip of metal corresponds to the depth of the water collected from the rain. And by a number of these, or even by one, may be determined, approximately, the quantity of the rainfall per square inch, or per square foot, and by calculation per morgen.

The quantity per morgen thus determined may be considered as equivalent to what is the supply of water from this source available for agriculture in that locality.

It does not supply an exact measurement of this, but only a rough and ready estimate; for, as has been intimated, as a matter of fact, much of what falls as rain runs off most lands; but every inch of rain is equivalent to 22,622·25 gallons, or upwards of 100 tons weight of water, 100·9935 tons per acre.

Much is carried off by evaporation, and much is carried off, to, or over the locality in streamlets from higher lying grounds, and in vapour from all the land around; and the quantities of these we have not, at this stage of our investigation, the means of ascertaining.

But if, notwithstanding this, the rainfall may, as is alleged, be reckoned as equivalent to the supply of water from this source available for agriculture in a limited locality, with much more confidence may this be reckoned equivalent to the measure of such a supply to the Colony, into which little is brought from higher-lying regions, and little or none need run off into the sea, if this it be determined to prevent and means are adopted accordingly—while the extent and the varied character of the surface may be supposed to equalise the loss of the supply from evaporation in different districts.

From observations made at the Royal Observatory during a period of fifteen years, extending from 1st January, 1842, to 31st December, 1856, the annual rainfall there varied from somewhat less than 19 inches upwards to 26¼ inches, and gave an average of 23¼ inches a year.

The observations have been continued, and, while in 1865 it was only 18·668 inches, in 1859 it was 36·720, and the average of the whole period of observations, embracing twenty-four years, was found to be 24·248; and dividing the twenty-four years into four equal parts, and taking the mean of each, we have for the first six years, 22·611 inches; second six years, 24·340; third six years, 25·066; fourth six years, 24·975.

Writing of this, Sir Thos. Maclear gives the following tabulated statement:—

Year	Inches	Inches (mean)
Year 1842	26·268	22·611
1843	24·821	
1844	18·783	
1845	20·913	
1846	22·503	
1847	22·378	
1848	23·246	24·340
1849	24·615	
1850	33·467	
1851	20·305	
1852	23·186	
1853	21·219	

1854,................20·048		Inches.
1855,................24·571		
1856,................21·927		25·066
1857,................22·670		
1858,................24·460		
1859,................36·720		
1860,................29·116		
1861,................25·442		
1862,................32·008		24·976
1863,................25·603		
1864,................18·915		
1865,................18·668		

Mean 24·248

And he says :—"Other combinations may be formed ; but the one just given is that which I regard as the most fit for dealing with the problem of declining moisture, and the result speaks for itself. If the first six years changed places with the last six years, a trace in favour of melancholic forebodings could have been produced, but, as before remarked, the analysis refers to only one locality of the Colony ; though I have no doubt it is in pretty close sympathy with the total area. The Cape is not the only part of the globe where occasional shortcomings of rain occur. I have before me a register of 48 years, kept at a place where the average was 48·842 inches. There in the year 1832 the fall was 18·450 inches (some 30 below the average) ; on the other hand, in the year 1827, it was inches 39·82 inches above the average."

A reference to the tabulated observations which have been given, will show that the mean rainfall in the western district is given as 25·804 inches ; that of middle district 13·878 inches ; and that of the eastern district, from which returns were collected, 14·412 inches, giving the mean of the rainfall at the ten stations at which the observations were made as 19·93, or nearly 20 inches.

If we ascertain the horizontal extent of the Colony, and calculate the quantity of water measured by that extent 20 inches deep, this will supply the means of forming a rough estimate of the quantity of water which is thus supplied.

A nearer approximation to a correct estimate of this will be obtained if an estimate be formed of the extent of country properly represented by the different localities at which the recorded observations were made ; and after making such a calculation for each of these, adding the products of the whole together. But even this will only supply a remote approximation to what we wish to determine; and it may be that an impression sufficiently near to the truth, for the purpose of these estimates, will be produced if it be stated, that a rainfall of 20 inches, the average or mean of the observations referred to, is equivalent to a supply of about four hundred and forty thousand gallons of water a year for every acre of land in the Colony.

This may give some idea of the immensity of the supply of water from this source which is at present available for agricultural operations in the country.

There are other elements to be taken into a calculation of what is actually available in an any one locality. Amongst these are the distribution of the rainfall in regard to the country, and in regard to the time at which it falls, and over which the fall of the different showers extend, and the character of the soil, and much besides, all of which would require separate consideration. And it will be found that on arable land the supply is not less but is greater, and much greater than what is thus indicated. But at present it is the quantity which is available throughout the Colony which is alone under consideration.

CHAPTER III.

Supply of Moisture as Affected by Rivers.

A prevalent idea is that rivers are the great fertilisers of a country, bringing supplies of moisture to what would otherwise be but a dry and thirsty land, where there is no water; and, as is the case with most prevalent ideas, there is a nucleus of truth in the supposition; but it is only a partial representation of the truth. If rivers bring supplies into a dry desert, they do so by running off with this supply from some other district, upon which it had descended as rain, or to which it had been brought as rain-water, drained off from a higher level. As rain is only water dropped from the atmosphere, being a quantity in excess of what it could retain in altered circumstances from those in which it had previously been able to hold all, so are rivers composed entirely of water which had fallen upon some higher-lying land, but which is in excess of what that land could retain and utilise; and, alike in low water and in flood, they tell of water running to waste,—running away from land which, like the heath in the desert, does not see when good cometh; laughing by the way; more than half-willing to be caught; but not less willing to lose itself in the sea, if no one have the good sense and the spirit of enterprise to lay hold on the playful imp and turn her powers to account.

I cannot paint; but one of my collaborateurs in this field of research has thus written of the rain and the rivers in the western provinces of the Cape Colony.

Having written of the interruption to travelling and to traffic occasioned by floods washing away bridges and allowing no one to cross till they have run their course, my correspondent writes in a more joyous mood:—

"But our friends may require some information as to the extraordinary power which has thus interfered with our communications.

"It is merely the steady brisk downpour of the rainfall in the vicinity of our mountain ranges, filling here a vlei and there a cavern on the mountain top; washing the rocky heights; pouring, trickling, roaring, or rippling over the ledges, ravines, or precipices of our time-worn defiles; and dashed forward or sideward into the depths of the chasm below by the stormy wind, or by the inclination of the slope over which this blessed boon of Providence flows. It is delightful to travel over Bain's Kloof on a thorough rainy day!

"There should be a twelve hours' downpour before starting from the Worcester side of the pass, and a continuation of the rainfall during the journey, so that the supply of water may be undiminished. An occasional lifting of the cloud, revealing just enough of the height aloft to make one think that the true height is something extraordinary, adds to the pleasure of the moment. The varied and thousand forms assumed by the countless waterfalls, sprays, streams, and cataracts of these glorious old hills; the dark and heavy storm-clouds whirling and driving around the inaccessible and lofty eminence; the soft and lightened vapour drifting away to the left over some hollowed curve in the summit; the sunlight glancing downwards between the two upon some beautiful cascade; the rushing, gushing, dashing, clashing, and splashing on every side; the surging, roaring, and restless torrent some thousand feet below; the changing cloud, the soughing wind, the tempest roar; the wild screech of the owl in the desert;—all tend to cause the traveller to exclaim with gladness: 'Let the floods clap their hands, and let the hills be joyful together.' Here is a bright fall of a fan-like form, proud of its beauty; there a mighty waterfall shooting forward and sporting in its strength; yonder a dancing, lightsome, gushing fountain, courting the sunlight as it steals upon its seclusion. In one part is to be seen the river below, foaming as though impatient of the restraint of the narrow channel, until it bursts forth upon a wider expanse, when it becomes more placid, and, with more even course, approaches some precipice, over which it flows in a graceful curve, which, in its evenness and contour, represents the Falls of Niagara in miniature. It is thus seen that a traveller may enjoy himself in a mountain pass on a rainy day."

I can enjoy such sights, and I do. But when I come to speak of lands suffering from desiccation and aridity, I can only think of rivers as of water lost; they speak to me, not of water supply, but of water wasted; and in now treating of the water supply as affected by rivers, I refer to this, as indicating, by the vastness of the quantity allowed to run to waste, how very vast must be the quantity at the command of the colonists to use, and to use for their own benefit if so disposed.

It is a fact that many, perhaps most, of the rivers of South Africa are known to many only as channels, dry water-courses, rivers without water; but all the rivers are not such. Journeying along the coast of the Colony of the Cape of Good Hope from west to east, we cross successively the Orange River (reckoned to be 1000 miles long), the Olifant's River, the Berg River, the Breede River, the Gauritz, the Kromme, the Gamtoos, the Sunday's River, the Kowie, the Great Fish River, the Keiskamma, the Buffalo, the Kei; and advancing further, the Bashee, the Umtata, the Umzimvooboo, the Umzimculu, Umkumanzi, and the Tugela—the boundary of Natal from Zululand beyond. And there are between one and two hundred more of lesser note: nearly a hundred small streams are found on the coast between the Kei and the Umzimculu alone.

All of the water poured by these rivers into the sea is a part of what had been precipitated as rain, but which is thus lost to the Colony, in so far as the promotion of vegetation and similar agricultural operations is concerned.

In estimating the quantity of the rainfall which is thus lost to the Colony, it is not enough that we look at these rivers when in their usual condition, we must look at them also when in flood. Herr Wex, to whose treatise I have referred in the Introduction, and from which I have quoted what is there said in regard to works of hydraulic engineering in China, has, as the title of that treatise shows, called attention to the fact that a continuous reduction of the water-level in the Rhine, the Elbe, the Oder, the Vistula, and the Danube at low water, and of the average flow of the river throughout the year, has been accompanied by a very great augmentation of floods in these rivers. He adduces evidence and testimony which shows that the change in the *régime* of these rivers has followed, and accompanied with corresponding progress the clearing away of forests. The fact and the coincidence have been confirmed by the Imperial Academies of Science of Vienna, and of St Petersburg, and have been formally accepted by several other Scientific Societies in Europe, and by numerous scientific and practical men in Europe and America. And several French students of Forest Science—viz., Surell, Cézanne, Marschand, Costa de Bastelica, and others, who had previously, by independent observations and research, arrived at corresponding conclusions—have shown how, in some cases, the effect had been produced by the destruction of forests, and how, in the same or other cases it had been remedied by the natural reproduction of forests, and have shown how it might be, and how it had been, remedied by artificial *réboisement*.

In accordance with what has thus been observed and advanced in regard to the *régime* of the rivers named, as affected by the destruction of forests, the rivers of South Africa draining, some of them, many thousands of square miles of treeless waste, present the contrast between the state of the river when at low water, and when in flood in a remarkable degree. I had almost said in an exaggerated form. In support of this statement I may refer to what most of the Colonists must have seen or heard of, the "coming down" of a river.

Captain Henry Hall, in his "Manual of South African Geography" (p. 93), makes the remark :—"There are several peculiar features in the rivers of South Africa. Those included in the coast-river system, with but few exceptions, as the Berg, Breede, and Olifant's Rivers, east and west, generally run in deep beds ploughed up by the torrents of waters which sometimes come rushing through the narrow gorges of the mountain ranges which run parallel to the coast, the effect of thunder-storms bursting over the Great Karroo desert, and the high mountain ranges bordering it, or over the elevated basins of the Division of Cradock, or Queenstown. Lichtenstein describes, in elegant terms, the frightful chasm of which the united streams of the Gamka, Dwyka, and other periodical streams pass through the Great Zwarteberg mountains, forming near the sea the Gauritz River. Along the coast of the George Division, where the mountain range approaches very close to the shore, the many streams there, although their course is but a few miles, have formed beds, many of them several hundred feet in depth, and running in deep valleys often many miles in length, rendering travelling along the coast very difficult. Such is the general nature of the rivers of the coast region as far east nearly as Natal."

And, in a passage which I have had occasion to cite in another volume, he says :—"The great majority of the rivers here are periodical, *i.e.* only flow after heavy rains. A few in the western district of the Cape Colony, the Breede, Berg, and Olifant's, for example, are permanent. In Great and Little Namaqualand, the Kalihari Desert, and the whole of the region situated on the southern slope of the Nieuweveld and Roggeveld mountains, whole years may elapse without the phenomena of a running stream, and yet the magnitude of the dry water-courses of the Buffalo, Hartebeest, and Oup, or Borradaile River, all tributaries of the Orange, show how immense must be the torrents that sometimes sweep along them. The writer of this has seen the bed of the Great Fish River perfectly dry, and within a few hours a torrent thirty feet deep and several hundred feet wide was roaring through it. In February, 1848, the Kat River suddenly rose upwards of fifty feet in the course of a few hours, sweeping seventeen feet above the road-way of a stone bridge at Fort Beaufort, supposed to have been built high enough to have a clear water-way to the highest flood ever before remembered. The Gamtoos, Gauritz, and all the other rivers draining the Karroo are also subject to very sudden rises, although generally but dry channels. The periodical rains falling on the mountains near its sources between September and March also swell the Orange River to a great extent, and large portions of land along its banks are then inundated."

Floods tearing along twenty feet deep, at places where at low water the river is passable on foot, are not unknown.

In narrow gorges floods have been seen similar to some of those spoken of by Captain Hall, forty, fifty, and sixty feet deep. In the conclusion of my volume on "Reboisement in France" I have given, as supplying a counterpart to the accounts given of the destruction of Toulouse by the flood of 1875, details of the flood of the Gamtoos in October, 1867, on which occasion at one place the river rose seventy feet (pp. 336–338). As a sequel to that I have given in the same volume (pp. 339–344) details of the inundation of the town of Port Elizabeth in the month of November in the same year, 1875, the damage occasioned by which to roads and to house property in that town alone was estimated at from £25,000 to £30,000.

In the last chapter of the volume on "Hydrology of South Africa" (pp. 228-252) I have given similar details of the same and of other floods at Natal, Capetown, Beaufort, Victoria, Bloomfontein, and other places in the Colonised parts of South Africa, which occurred within the years 1867–1874 inclusive, to which I have referred in the commencement of the Introduction to this volume; and in the Introduction to that volume I have stated what were my own experiences and observations in tours which I made in the Colony in 1847 and in 1863.

On my tour in 1863—"Everywhere I found that there had been copious rains, in some places torrents. For days I was confined to the house by drenching rains, which made the roads rivers. I met the Government Railway Engineer returning from the Eastern Province, at a farm house near Caledon at which we both outspanned, by whom I was told that I should find the roads in the Lang Kloof impassable for the vehicle in which I travelled, the clay in many places having been converted into mud three feet deep, the surface of which caked and hardened would in all probability, like rotten ice, again and again give way, and my horses would be unable to drag out my conveyance. My informant had himself only got through by the strength of his team, which was double that which I drove, and his vehicle lighter by one-half. I did get safely through, but I saw evidence of the correctness of my informant's observations.

At Riversdale I was detained forty-eight hours by the state of the river. The crossings of the Kroom River I found to be not unaccompanied with danger.

In approaching Van Staden's River I found one-half of the breadth of the road converted into a deep gully, leaving not breadth enough for my cart; and we passed it by letting one horse drag the cart, with one wheel on the road, the driver and I holding it level by bearing up the other. The Van Staden's River I found in flood, and I got through only by the help of a hired team.

At Grahamstown I was prevented day after day by rains from making excursions for botanical observations in the neighbourhood; and I learned that the Fish River was so full that to go to Kaffraria by Fort Peddie, as was my intention, was impracticable.

Leaving Grahamstown for Fort Beaufort I passed a streamlet where, during my sojourn of a few days at Grahamstown, a post-rider had been drowned in attempting to cross.

Further on my way, I found the Fish River had so flooded the road by its side that in the middle of the highway the water was some eighteen or twenty inches above the axle of my cart.

The Buffalo River I crossed with difficulty at a ford where a few hours before, and again a few hours later, it was impassable for man and beasts.

Leaving Queenstown for Burghersdorp, I travelled over ground high and dry, where within a few weeks before two farmers had been drowned.

At Burghersdorp I found the Civil Commissioner, after having been twice to the river, had given up all expectation of my being able to cross it.

On my return from beyond the Orange River, I re-crossed the Fish River, but within forty-eight hours after I had done so and was safely housed in Cradock, it was tearing along full from bank to bank impassable.

In proceeding to Bedford I had to cross the Tarka. But on the banks of the Tarka I was, along with many others, detained several days, the river, which the day before that on which I was to have crossed might almost have been crossed dryshod, having come down in the afternoon, and continuing day after day to rush and roar and tear along in a torrent, ranging in depth, as shown by the figures on the unfinished bridge, from twenty to four-and-twenty feet. Not deeming it expedient to remain longer, I was at length conveyed across in something like an old soap-box suspended from a strong rope attached to trees on the opposite banks, and to which were attached cords by which it could be drawn from one bank to the other, conveying the mails when the river was in such a condition.

My sensations, when suspended between heaven and earth above a roaring torrent, were something different from any I had previously experienced. But a gentleman well-known in the Eastern Province, who left Grahamstown for King William's Town about the same time as I did, but by another route, must have had more to tell of his sensations than I, if what was told me be true. I was informed that he had to cross the Buffalo in a way similar to that in which I had to cross the Tarka; but there the conveyance was a basket and not a box, and there were no cords attached to it whereby it could be drawn from the one bank to the other, but the passenger had by a hand-over-hand movement, on the suspending rope, to pull himself and the basket across; and this rope was not quite so tight as that by which I was suspended. Getting into the basket and being somewhat corpulent, his weight distended the rope into a beautiful curve, and the basket and he almost flew half way towards the further shore; but there it stopped! The slope of the rope made it equally difficult to go forward or to return; the weight of the aerial voyager, which so greatly facilitated his passage thither, seemed to say to his basket or to himself, what Canute is said to have said to the sea,—"Thus far but no further, and here shall thy proud course be stayed." At length, by strenuous effort, the further bank was reached in safety. My transit was comparatively a pleasant flight; but both speak of the abundance of rain which had fallen.

I proceeded to Adelaide and Bedford *en route* for Somerset and Graaff-Reinet. But at Bedford I was again detained through the Fish River being again impassable. It had come down and had continued to flow for a great many days, I think ten days or a fortnight, and that with a current so impetuous that the ferrymen had refused a fare of £10 to carry across a medical practitioner, whose services were required on the further side; and there being no appearance of subsidence, I had to abandon my purpose of crossing it, and to change entirely my route in returning to the Western Province.

Hearing everywhere of the drought, and seeing the deluges of water rushing in torrents to the sea, I was often ready to cry out, Wherefore is this waste? and the feeling prompting to this was the more intense that almost everywhere I saw facilities for preventing this, and indications that formerly the water was not so lost to the land. In elucidation of this remark, I must state I had previously travelled both in Finland and in the lake district of North America, and in many of the districts through which I passed in the Colony of the Cape I was reminded by what I then saw of what I had seen in these lands; in these lands were lakes of water—some of them like inland seas in extent,—in the country through which I was passing there were none; but there were what looked like basins of lakes, which it required no effort of the imagination to picture as filled with water, though then drained and dry.

The lakes of North America cover an area of 84,000 square miles: nearly a fourth part of the whole area of Canada; and two-fifths of the whole area of Finland are covered by lakes, a hundred of which may sometimes be seen in the course of a single day's journey. Lake Superior, in America, covers an area of 31,000 square miles, the others are of corresponding extent. In Finland—called by its inhabitants, The Land of a Thousand Lakes, and The Lost Daughter of the Sea—they are smaller, but of a magnitude not to be despised: one measures nearly 30 miles long by 30 miles broad, and Lake Ladoga, through which its waters flow in their progress to the sea, covers 6190 square miles, an area equal nearly to the whole principality of Wales. And such were the scenes recalled by what I saw of the general contour of the colony as I journed from place to place; but what was seen spoke of this as having been the case long, long ago. Now all was changed; the waters had all escaped and the dry basin alone remained, with, in some cases, gorges through which the waters had poured in seeking a lower level, and finally made their escape."

These floods are cited as supplying, along with the delivery of the permanent rivers of South Africa, some indication of the quantity of the rainfall which is at present lost for the promotion of vegetation and also for agricultural operations, but which might, in whole or in part, be retained and utilised, and if so retained would do much to remedy evils which are attributable to the aridity and the continuously progressive desiccation of the country.

While the rivers are drains, and they drain off only surplus water, that is water in excess of what can be retained by the ground of what falls upon it precipitated as rain; and while this is the aspect of them which chiefly concerns the inquirer into the best means of arresting and counteracting desiccation and aridity; the fact must not be ignored that they do promote vegetation along their course, in evidence of which effect a river course in South Africa can often be traced from a slight elevation by a line of arborescent vegetation which clothes its banks—by the percolation of its waters vegetation may be promoted at a greater distance, and by evaporation an addition may be made to the supply of moisture in the atmosphere.

From a statement which I shall afterwards have occasion to quote it appears that the flow of Zwartkops River, as it passes Uitenhage, is about 2,516,632 gallons in the twenty-four hours; and from another which I shall also have to quote, that the flow of the Van Staden's River, when measured in 1860, was 1,250,000 gallons per day; in 1862, 614,828 gallons; and in 1863, 431,723; and these numbers may give some idea of the quantities of water which are constantly flowing in such rivers as the Orange River, the Breede River, the Great Fish River, and the Buffalo, and by these rivers and others when they are in flood!

Looked at from this point of view, in considering how the supply of moisture in South Africa is affected by rivers, they can only be considered as mainly draining off to the sea the superfluous rainfall on the basin they have created and through which they flow.

There is another point of view, looked at from which they may be seen bringing the rainfall from regions where it was not retained into districts where by man's device it may be arrested and utilised, and to this view of the case attention will afterwards be given; but the other view is that to which alone attention can be given here.

CHAPTER IV.

Supply of Moisture as Affected by Fountains.

At first sight the water supplied by fountains and springs, some of them like the gushing forth of subterranean streams, may seem to be a clear gain and a great addition to the water supply of the country. This may be the case, and it doubtless is so to the districts in which they appear; but with them, as with rivers, this is only accomplished at loss to some other district. As rivers are the produce of a rainfall which the ground could not retain, but which flowed away over the surface or drained away through the superficial soil, so are these fountains the produce of another portion of the rainfall which subsides beyond the subsoil in unknown quantities to unknown depths.

What comes thus forth from the bowels of the earth may be considered as water recovered, but not as clear gain; and while it may legitimately be considered as water recovered, it suggests the thought that a quantity indefinitely greater may be lost.

I have spoken of the quantity of the rainfall which subsides beneath the subsoil as unknown.

Of the water which falls as rain a portion stagnates on the ground or flows away to the sea, a portion is absorbed into the composition of the soil and retained by what is known as hydroscopicity, a third portion is attracted into the interstices in the structure of the soil and retained there by what is known as capillary attraction, and yet another portion in many cases permeates the subsoil and strata beneath it and sinks to an unknown depth.

It is by this last-mentioned portion that fountains and warm springs, and subterranean rivers, and subterranean reservoirs are fed.

There are some strata which are impervious, or nearly so, to water, but there are others which are permeable in a high degree. And this depends on conditions altogether different from those whereby the phenomena of hydroscopicity and capillarity depend, or on conditions at least which can be studied apart from these.

Sand manifests little or no hydroscopicity, but it is extremely permeable; clay has great hydroscopicity, but it is in a high degree impermeable Travellers in Namaqualand and some other sand-covered districts of South Africa, when driven to extremity for water, have found a quantity of water, on hollowing out a basin in the sand, so deep as to reach a layer of other matter, collected there after a time more or less protracted. It had drained thither from and through the permeable sand, and there it was retained by impermeable matter. But great care was requisite in order that this stratum should not be broken through, as then the water was lost, draining off to a lower depth. The stratum has been found in some cases to be not thicker than a penny, formed probably by the washing down of minute quantities of clay, etc., from the superincumbent sand, but so impermeable was it that it had sufficed while covered, and it sufficed still, to prevent the escape of water to a lower level.

Something similar is met with in the sands of the Landes of the Gironde. At the depth of about two feet below the surface of the sand is an impermeable stratum of vegetable and calcareous matter about a foot in thickness, underneath which is sand of unknown depth, though the diggings have been made to a depth of upwards of sixty feet. In the winter season these Landes are covered with water, which has no fall, and cannot sink through this layer; but in summer neither pool nor moisture is to be seen, the water having been evaporated, and this layer preventing an ascent of moisture from below. But there are spots where this stratum is awanting, or has been broken, and on these grow bushes and trees; the ground neither being drenched in winter nor altogether devoid of moisture in summer.

Some limestones, again, are permeable in a high degree, through abounding cracks, and chinks, and vertical fissures, though they neither absorb water by hydroscopicity as does clay, nor retain by capillary attraction such quantities as does sand, nor part with what they do so retain so readily as does this.

In the case of the clay, its impermeability may be attributed to this—that its molecules, once moistened, though previously existing as an impalpable dust, cohere and present the appearance of a compact mass; and though the surface of a soil or subsoil of argillaceous earth absorbs, when moistened, a great quantity of water, when the surface has been saturated the transmission of water from particle to particle beyond goes on very slowly, and the permeability is almost nothing, in consequence of the capillarity resulting from the minute sub-divisions of the molecules, which are brought into the closest contact, and there is in the moist mass no such chinks as abound in the limestones of ooletic origin referred to; but with the sand the permeability is extreme.

Through the permeability of the earth in other lands, water precipitated as rain sinks in unknown quantities to unknown depths, and we have indications that in South Africa this is also the case. The evidence and indications of this occurring elsewhere are supplied by Artesian wells; the evidence and indications of this occurring in South Africa are supplied by the temperature and copious flow of the numerous hot springs within the Colony of the Cape of Good Hope and beyond it.

According to the latest theory—or, as I consider it, hypothesis—the old idea of the earth being formed of a solid crust enclosing a mass of matter in a state of fusion is founded on error, and much has been done by distinguished men to establish this; but their arguments have failed to convince me; and so long as I hold to the older view I feel free to avail myself of it for purposes of illustration. But even if it be the case that, in the process of cooling down, the earth has cooled to a temperature at which the whole has solidified, it may be admitted that it has cooled down to a greater degree on and near the surface than it has done at a greater depth; and if this be so, or be admitted as probable, my argument will hold equally good. That argument is, that the greater the depth to which water may percolate the higher must be the temperature which it is likely to attain; therefore, the higher the temperature of water in a spring it is probable that the greater is the depth from which it has come. I use this guarded expression because it has been alleged, with some measure of plausibility, that in mines of a great depth in which a rise of temperature has been observed at successive depths, this may be attributable to chemical action going on in the material exposed, or at the successive levels reached. I do not hold this view, but I have considered it proper not to ignore it; and in view of all the conditions referred to, I consider that the hot springs of the country are the produce of subterranean currents or reservoirs situated at a very great depth—a depth unknown; and unknown also is the quantity of the water which has sunk to these subterranean depths.

Besides fountains coming apparently from a great depth, according to the indications afforded by their high temperature, there are many more coming apparently from depths less profound, if temperature be accepted as an indication of depth.

They, too, are produced by the flow of water from subterranean reservoirs of unknown dimensions, which may have been tapped by a deep-reaching fissure, or it may be by their waters accumulating behind some impermeable dyke till they reach the surface and flow forth as they continue to accumulate. Nor are there awanting indications that this may be the case.

There are rents in the earth's crust with which both miners and geologists are familiar. In some cases a vertical displacement of one or other, or of both, of the two sides of the fracture has occurred, so that the corresponding strata on the opposite sides are not continuous; and while the order of succession of the different strata in both is the same, the altitude of these differs on the two sides of the rent by inches, or by feet, or by many, many feet. When this phenomenon is met within a mine it is called a *fault*. In other cases the two sides are parted by inches or feet, or many feet, and the spaces between have become filled up, sometimes with debris of the strata produced by friction accompanying the disturbance of continuity, in some cases by crystaline matter which must have been protruded from below in a state of fusion, or, in other cases, deposited from solution in water, finding its way thither, it may be, from above. To these the name of *dykes* is given.

Miniature specimens of what is referred to, less than an inch in thickness, may often be seen in rocks; indications of enormous specimens I have seen in South Africa in lines of vegetation, crossing mountains and penetrating valleys, different from the vegetation on either side, not in luxuriance but in generic character—vegetation of a kind which found only in the soil produced by the decomposition of the matter of which the dyke was formed a soil suited to its growth.

Such dykes, composed of material impermeable by water, may easily be conceived of as operating in the way described, preventing the further spread or flow of water and causing it to accumulate, whether in and above a cavern, or in a body of loose or porous material, in the interstices of which it could acccumulate to saturation, and whence what continued to come might issue, or cause that which had come before to issue, in a continuous flow.

In more than one district in the Colony I have traced something like a linear direction perceptible in the relative position of many fountains and springs; and if in this I was correct, the fact may be thus satisfactorily accounted for.

It is, moreover, not uncommon, but rather the reverse, to find subterranean water-courses in connection with stalactite caves, as might be expected by any who have made the formation of stalactites their study.

In many districts lying beyond the Colony the subterranean water courses have been exposed. Thus is it with one of the sources of the Mooi River in the Transvaal Republic.

I was informed by the Rev. Dr Andrew Murray, of Wellington, who had travelled extensively in the Transvaal territory, ministering to the immigrant population, then scattered more sparsely if not more extensively over the whole of that region, that he had been credibly informed that chaff thrown into a large cavern called Wonder Fontein, situated in the Gatzrand, has come out in the Mooi River, some eighteen miles distant, having been conveyed thither by a subterranean stream, which is by no means improbable.

Such caverns are by no means uncommon in certain limestone deposits. There are more than one within the Colony of the Cape of Good Hope. Elsewhere, there is not unfrequently found in connection with them a subterranean river, to which is attributable the creation of them; and so may it have been with all, even where no stream is now to be found. Limestone is, as has been stated, very permeable by water which flows through chinks and crevices more copiously than can be prevented by its hydroscopicity or its capillarity. Existing, as it often does, in the form of bi-carbonate of lime, which is soluble in water, it may be largely dissolved, forming what is known in domestic use as *hard water*. What is thus removed must leave a vacant space, and the aggregation of millions of these may form a cavern. By the continued dissolution of limestone above a cavern thus formed, and the percolation of the water thus charged with lime, and by the evaporation, in whole or in part, of the water on its reaching the vacant cavern, the lime becomes again deposited, and beautiful stalactites are produced, some of them in fantastic shapes like gigantic icicles, petrified water spouts, drums and pulpits, trumpets and horns, for the sight of which such caverns are visited by multitudes of the curious; but not less interesting to the students of nature are the caverns themselves in their hidden depths.

It may, to the Colonist who has not had the opportunity to visit other lands, make the nature of such caves more intelligible if there it introduced some account of the phenomenon of these as presented and studied elsewhere.

A famous Cave of the kind is Weyer's Cave, near Lexington,

(Virginia). It is occasionally lighted up by some 2000 or 3000 candles for the accommodation of visitors who are thereby attracted from great distances. On one of these occasions it was visited by Drs Reed and Mathieson, who were at the time in the vicinity visiting the American Churches as a deputation from the Congregational Union of England and Wales; and they have given of it the following description:—"We reached the spot about one o'clock. The party of visitors had been, and were just sitting down to dinner. But the lights were burning, and the guides were willing; and this was, to my taste, the very time to see it, free from the noise and confusion of 200 visitors. So, turning our back on the dinner, away we went.

"The cave is found in a ridge of limestone hills, running parallel with the Blue Bridge. In going to it, you pass by Madison's Cave, which was once an object of much interest, but is now neglected, from the greater attractions of the one we are about to explore. You ascend the side of the hill by a zig-zag path of about 150 yards long, and then find yourself opposite a wooden door, which is the entrance, and having a bench, on which you rest, to get cool, or to prepare otherwise for the ingress. Having passed the door, you find yourself in a small cave, which may be regarded as a lobby to the whole apartments. With raised expectations, you look about you, by the aid of the day-light, which is struggling to enter, for some openings more considerable. All that you see, however, is the mouth of what appears a dark recess about four feet square; and you are told this is your passage onward. You have no alternative, but to double yourself up into the smallest possible dimensions, and move along, after the lights of your guides, as well as you may, by the assistance of your hands.

"Having scrambled along for about twenty-five feet, you come into some larger openings, which allow you the free use of your person. You look upward and around you, and find yourself surrounded by the most grotesque figures, formed, through ages, by the percolation of the waters through the heavy arches of rockwork over your head; while the eye, glancing onward, catches the dim and distant glimmer of the lights—some in the deeps below, and some in the galleries above. On quitting these smaller rooms and galleries, you enter an ascending passage, of easy access; and on coming to its extremity, you see the opening of a large cavern spread before you, and the commencement of some steps, by which you are to descend. Your care is engaged in getting safely down; but when you have obtained your standing on the floor, you are delighted to find yourself in a large cavern, of irregular formation, and full of wild beauty. It is about thirty by fifty feet, and is called Solomon's Temple. The incrustations become finer here. At your right hand, they hang just like a sheet of water, that had been frozen as it fell. There they rise before you in a beautiful stalactitic pillar; and yonder they compose an elevated seat, surrounded by sparry pinnacles, which sparkle beautifully in the light. The one is called Solomon's Throne, and the other his Pillar.

"On leaving the Temple, you enter another room, more irregular, but more beautiful. Besides having ornaments in common, it spreads over you a roof of most admirable and singular formation. It is entirely covered with stalactites, which are suspended from it like inverted pinnacles. They are of the finest material, and are most beautifully shaped and embossed.

"You now make an ascent of several feet, and move along a passage, and through two or three lobbies, and come to what is called the Twin-room, and find your way is just on the verge of a dark cavern, which is yawning at your feet; and is named the Devil's Oven. A descent is now made of some difficulty, and from an elevation of about forty feet; and you enter a large room, which is called the Tan-yard. This, like some of the rest, is an absurd name; but it has been adopted from the force of association. There are in the rocky floor of this room large cavities, which may be thought to resemble the tan-pits; and from the ceiling are suspended large sheets of beautiful stalactites, which resemble the tanner's hides. You advance to an upper floor in this room, which has chiefly one ornament, and that is sufficient. There is, extending along the room, and from roof to floor, an immense sheet of the finest stalactite. When it is struck with the hand it emits deep and mellow sounds, like those of a muffled drum, and it is called the Drum-room.

"You now rise by some natural steps to a platform, which you have again to descend, and then find yourself in what is named the Ball-room. It is a handsome and large apartment, 100 feet long, 36 wide, and 26 high. Its floor is so level as to admit of dancing, and it has been used for this purpose. There is in the centre of it a large calcareous deposit, which has received the name of Paganini's Statue; the whole room is relieved by grotesque concretions; and the effects of the lights burning at every elevation, and leaving hidden more than they revealed, is exceedingly fine.

"From the ball-room you make an ascent of 40 feet. This is named Frenchman's Hill; from the circumstance that a visitor from France, with his guide, had their lights extinguished at this spot. Happily the guide had such an accurate knowledge of the locality, that, after much difficulty, they got safely back, a distance of more than 500 feet. You wind your way through passages, and make a descent of nearly 30 feet, by what is known as Jacob's Ladder, with pits and caverns opening about you, and come into the Senate Chamber, and afterwards to Congress Hall. The last is a fine room, very like the Ball-room, but with an uneven floor. As you leave it, an immense cavern spreads itself before you, with the dim lights gleaming over its mouth, so as to make its unfathomed darkness horrible. You gaze on it with amazement, and instinctively long to pass on, lest it should drink you up. It has received the name of 'Infernal Regions.' By another lobby, and another descent, you enter Washington Hall. This is the most wonderful opening of the whole. It is 250 feet long, and 33 feet high. There is a fine sheet of rock-work running up the centre of this room, and giving it the aspect of two separate and noble galleries, till you look above, where you observe the partition rises only 20 feet towards the roof, and leaves the fine arch expanding over your head untouched. There is a beautiful concretion here standing out in the room, which certainly has the form and drapery of a gigantic statue; it bears the name of the Nation's Hero, and the whole place is filled with those projections, appearances which excite the imagination by suggesting resemblances, and leaving them unfinished. The general effect, too, was perhaps indescribable. The fine perspective of this room, four times the length of an ordinary church; the numerous tapers, when near you, so encumbered by deep shadows as to give only a dim religious light; and when at a distance, appearing in their various attitudes like twinkling stars on a deep dark heaven; the amazing vaulted roof spread over you, with its carved and knotted surface, to which the streaming lights below in vain endeavoured to convey their radiance; together with the impression that you had made so deep an entrance, and were so entirely cut off from the living world and ordinary things; produces an effect which, perhaps, the mind can receive but once, and will retain for ever.

"On leaving these striking apartments, you pass through a passage in which is standing some grand formations, named Cleopatra's Needle and the Pyramids; and then enter a room called the Church. The appearances in this instance suggest the name. It has about the dimensions of a church, and has an elevation of about 50 feet. There is at one end an elevated recess, which has the air of a gallery. At the back of this gallery there are a great number of pendant stalactites, of an unusual size and beauty. They are as large as the pipes of a full-sized organ, and are ranged similarly. They emit, when struck, mellow sounds of various keys; and if a stick is run over them, as we run the finger over musical glasses, they make pleasant music. There is nothing forced in giving this instrument the appellation of organ; it is one of the best that nature ever made; and the most remarkable that I ever beheld. At the other extremity there rises from the ground (not stuck on a roof, as we frequently see) a beautiful spire of considerable height; and this is the steeple.

"You pass by the steeple, and come into an apartment, which has the name of the Dining-room. It has similar dimensions to the Church, and on its left side there is a continued elevation, resembling a table. You now enter an immense gallery, about 10 feet wide, and some 121 feet long, and from 80 to 100 feet high. You turn aside to visit a small apartment, but of exquisite beauty. Here the most singular sparry concretions hang pendant from the roof, while an equal number are growing up from the ground in several degrees of progress, many of them meeting in the centre and becoming one. Winding passages are left amongst them, which make a sort of labyrinth; and as they are semi-pellucid, the passing of the lights through the several alleys has a very singular effect. This has the name of the Garden of Eden.

"You return to the Dining-room, and pass by a dark opening at your feet, which is the mouth of a cavern, into which the foot of man has never been. It can only be explored by rope ladders; and it is supposed, though I think without sufficient reason, to be charged with mephitic gases, fatal to life. You may now make an ascent of some 50 feet, if your nerves allow, and your reward will be adequate to your pains. You must climb over the face of the rock, which has nearly a perpendicular pitch, and you will then find yourself on an elevated platform, and surrounded by loop-holes and striking figures. You may now look down from your eminence, which is the Giant's Causeway, into the large illuminated rooms you have left, and perhaps see a small party moving over the floors in misty shadow. Here stands out in relief before you, and on the very verge of the platform, a fine group of stalagmites, white as alabaster, and suggesting to the fancy the figures of a small party of horse moving over high and dangerous precipices. They are Bonaparte and his Guards. There is a fine arch expanding before you over the scene below; you may with caution ascend on its head, and by this means gain a more commanding view of the objects so far beneath you.

"But we must hasten on. When you have made your descent to the ordinary level, and move on your returning course, you pass by an enormous and most beautiful concretion. It is a tower, about 30 feet each way at the base, and rising in diminished squares to the height of 30 feet. It is a stalagmite, nearly as white and clear as alabaster, and dazzles you by its capacity to reflect lights.

"You pass, also, some fine springs, at which you may refresh yourself on the way. There is one I must distinguish before we leave. You ascend, in getting to it, a steep of 12 feet, by a ladder, and then, by a little hard climbing, attain to the end of the recess, and stand before what is named the Source of the Nile. It is a fine transparent spring, and is very remarkable for being covered with a thin pellicle of stalagmite. It is strong enough to bear you; and has a hole cut in the centre, which gives you access to the water.

"I hope you will not think you have been detained too long on this spectacle. My regret is, that I have only described one-half of what it unfolds, and that with haste and imperfection. It is, in my judgment, one of the great natural wonders of this New World; and, for its eminence in its own class, deserves to be ranked with the Natural Bridge and Niagara, while it is far less known than either. Its dimensions, by the most direct course, are more than 1600 feet; and by the more winding paths, twice that length; and its objects are remarkable for their variety, formation, and beauty. In both respects it will, I think, compare, without injury to itself, with the celebrated Grotto of Antiparos.

"For myself, I acknowledge the spectacle to have been most interesting; but, to be so, it must be illuminated, as on this occasion. I had thought that this circumstance might give to the whole a toyish effect; but the influence of 2000 or 3000 lights on these immense caverns is only such as to reveal the objects, without disturbing the solemn and sublime obscurity which sleeps on every thing. Scarcely any scenes can awaken so many passions at once, and so deeply. Curiosity, apprehension, terror, surprise, admiration, and delight, by turns and together, arrest and possess you. I have had before, from other objects, one simple impression made with greater power; but I never had so many impressions made, and with so much power, before. If the interesting and the awful are the elements of the sublime, here sublimity reigns, as in her own domain, in darkness, silence, and deeps profound."

Not unlike to this celebrated cave are some of the limestone caves at the Cape.

Of a visit to the cavern through which flows the subterranean affluent of the Mooi River the following account has been given:—"Footsteps announced the approach of a fellow-creature, and a young Boer appeared, provided with two candlesticks of primitive construction and a few lucifer matches. He was the guide come to take us to the cave of Wonder Fontein. We followed him over a barren deserted stretch of limestone rock, and suddenly found ourselves standing in a hollow of some hundred feet in diameter, shut in by steep walls. On the western side is a crevice scarcely big enough to admit a man, but we managed to creep through in a stooping posture; and a few steps along a very steep passage brought us into a round room of about eighty feet in diameter, the roof of which is formed by a perfectly horizontal slab of rock. Opposite the entrance is a wider arched passage, the unfathomable recesses of which die away in the darkness in a most impressive manner, whilst the scene in the foreground is fairy-like, the stalactites, white as alabaster and pure as undriven Arctic snow, hanging from the gothic-like roof, now in the form of columns, now in that of delicate draperies. Making our way down the arched passage, our lights were reflected in ghostly fashion from the white stonework, glistening apparently with thousands of gems, and on either hand opened gothic arches, revealing glimpses of a chaos of grand and delicate forms, beyond which was the weird and awful darkness, concealing we knew not what.

"On we went, our footsteps falling noiselessly on the black, india-rubber-like ground; presently we heard a soft rushing and murmuring sound, and we were standing on the brink of the subterranean waters of the Mooi River—a modern Styx, black as a pall, reflecting the light of our candles in a grudging sullen manner. What a melancholy, lonely, awfully silent spot! We wandered about the subterranean palace for a good half-hour, like shades seeking in vain the dread boatman Charon. Nothing could exceed the gloom of this new aspect of the Mooi River. Here no clouds were reflected in its waters, no breeze ruffled its surface, the note of birds, the shout of the human voice, were alike unknown; unnoticed and alone it sped on its course, mourning and darkness its only companions.

"We regained the entrance of the cave not a moment too soon, for our candles were nearly burnt out. I have spent many a stormy night at sea, when the morning light was indeed a welcome sight, but never in all my life was I more glad to greet the golden sunbeams than when I had left the dim recesses of the cave of Wonder Fontein behind me, for beautiful as they are, they are very terrible, and as solemn as if amongst them the best hopes of the human race had been petrified—converted into cold and lifeless, irresponsive stones."

The following account of a visit to the Cango caverns, near Oudshorn, is given by Mr Freeman, a Secretary of the London Missionary Society:—"Before quitting this part of the Colony, I was glad to embrace the opportunity of paying a visit to the famous Cango Caverns, in the district of George, of which accounts have been published by Thompson and other travellers. They are said, by those who have seen the Caves of Flora and Elephanta, to be far superior to them, both in extent and grandeur. They are wholly natural, and are among the stupendous wonders of creation.

"The road along which we had to proceed forms the 'Poort,' or 'Pass' of the river Grobbelaar; it is a gorge, or defile, between two ranges of lofty and precipitous mountains. The river winds most coquettishly along the bottom; now gentle, placid, and inviting, then abruptly dashing aside, frowning, threatening, and concealing its course amidst the dense umbrage of the jungle. In rainy seasons, when the torrents are immensely swollen, the stream becomes dangerous and impassable. At the time of my visit it was most obligingly shallow, seldom exceeding two feet in depth. With due caution we crossed it, without difficulty or accident. But from its extremely serpentine course, we had to cross it thirty times. The scenery is magnificent. The convulsions have been fearful. Many of the rocks are thrown up perpendicularly. Many are broken into wild and awful forms. They are for the most part covered with vegetation. Along the course we took, the principal trees are mimosas, with their long and sharp thorns; where an equestrian, with a closely-buttoned leather jacket, has far less danger of interruption and laceration than one clad in the flowing costume of an oriental. These mimosas are interspersed with thousands of beautiful crimson geraniums, large and splendid plants of *palma Christi*, and a kind of strong-scented lilac. After proceeding about five miles through the gorge, we came to an open space, and presently reached the residence of Mr Botha, a farmer who has lately purchased the property, including the Caverns. It was advanced in the afternoon when we arrived, and the farmer expressed a little reluctance at our going in so late; but as my time was precious, we pressed it, and having taken a cup of tea, which the hostess soon provided, we mounted our horses, and at a mile's distance came to the entrance of the Caverns, on the side of a lofty limestone mountain. The entrance is vast and imposing, exceedingly lofty and spacious,—a porch befitting these subterranean and 'crystal' palaces. A fire was kindled, and we were provided with long bamboo canes, spiked, not with daggers, but with candles. On entering the caverns, we descended into a dark and gloomy passage, implicitly following our guides, whose lighted candles, however, were all we could distinguish. We were now soon arrested, by reaching the brink of a yawning precipice, and 'darkness visible' beyond it. Down we descended, by means of a ladder placed there for the purpose, and which is drawn up again every time it has been used. With due precaution, holding the steps of the ladder with one hand and our friendly candle with the other, we safely reached the lower regions, say thirty feet from the top of the descent, and we then commenced our subterranean pilgrimage, and proceeded to inspect one spacious apartment after another, all the while filled with awe, wonder, and admiration. Many of the rooms are very lofty—thirty, forty, fifty, sixty feet high. Their extent cannot be seen at one view, nor indeed of any of them, unless, perhaps, an immense number of torches were placed in the room; even then, I suspect, only the lights would be seen in the distance, and not the objects themselves. Many rooms are filled with millions of stalactites, descending in all forms from the ceiling, and meeting their kindred *stalagma* on the ground. Many of them retain only the grandeur of their forms: they are vast, magnificent, and exceedingly diversified. Others retain beauty, as well as colossal dimensions, consisting of fluted columns, towering to amazing heights, and resembling, at some distance, immense cathedral-organs. In other instances, they stand like primeval trees, such as I have seen in the quarries of Portland. Some of the specimens were of a remarkably white and glittering character, and some perfectly transparent. Here were niches, columns, cornices, fretted-work roofs in all variety of form and of beauty, far beyond verbal description. The detail would be insufferably tedious, but the impression of it as a whole is most effective. You feel at once transported into a region where you stand amidst the silent work of untold ages, perhaps thousands of ages. You have nothing in nature above ground to compare with it, and nothing of the work of man that can compete with it. The process of crystallisation is still going forward, but not in all places. The crystal palace advances, though without the magic hand of Paxton. The stalactite is still gradually forming in innumerable places; in others it has ceased, and the slow process of decay and disintegration is going forward. It seems an established law that it must live and increase; or else, in becoming quiescent and stationary, it decays. The exterior becomes first moist and clammy, then the crystals are destroyed; the adhesion ceases, and they crumble to powder. One room, called the 'Sand-room,' is strewn with fine sand—I presume, the decayed crystals of decomposed stalactites. Part of this is beautifully white, and part beautifully red; the colour of the latter occasioned, no doubt, by the presence of iron, which may be found in the vicinity of the limestone rock, through which the water had oozed, carrying an oxide in solution with the lime."

There is preserved in the office of the Surveyor-General at Cape Town a series of seven sketches of different compartments of this cavern—"*Afgeteekent van J. W. Mollendorf*"—accompanied by a representation of a section of the hill, dated 14th October, 1816, and entitled—"Plan of the Grotto in the Cango."

From this sketch it appears that the hill in which it is rises by a slope of about 1589 feet to a perpendicular height of about 550 or 580 feet above a streamlet or small river running towards the south at its base. The cavern is situated about a fourth of these measurements from the river bed. The entrance is about 20 feet high, and extends, nearly horizontally, about 200 feet, when there is a descent of about 30 feet into what is called Van Zulg's Hall, which is about 100 feet wide and from 60 to 70 feet in height. It contracts to a continuation of it, about 35 feet wide and 25 feet high, called the Registry, from many visitors having there written their names upon the wall. Beyond this is Botha's Hall, which is about 110 feet wide and about 50 feet; and beyond this is a smaller chamber, called the South Chamber, about 25 feet long, 14 feet wide, and 22 feet high. The measurement from the descent, 200 feet from the entrance to the further end of this chamber is 530 feet.

The grotto is entered by a N.N.W. course, but here the compass pointed due south in the direction further pursued by the cavern.

In this direction there is a passage leading inwards, called Botha's Poort, but it is so narrow that a large person cannot pass. Beyond is Vander Westhuisen's Chamber, which is about 8 or 10 feet long. Another passage, called Nel's Poort, similar to that by which it is entered, leads on to what is designated Thom's Chamber, which measures 14 feet long, 8 feet broad, and 15 feet high.

These passages and chambers lead on 70 feet beyond the south chamber, or some 800 feet from the entrance, 600 from the first descent.

Beyond this is a cavern, the floor of which seemed to be about 14 feet or more below the level of those traversed, but which was then explored.

The floor of this unexplored cavern is represented as horizontal, that of the explored chambers as almost perfectly so, and that of the entrance passage at a higher level as nearly so.

There are diverging recesses referred to, but not represented in the plan—it representing only a section of the hill.

One of these is the Fleder-muis-hoek, or Bat Corner, "on the right as you descend the latter;" another is the Kombuis, to the right of the large entrance chamber called Van Zulg's Hall; another is the Pyp of Ysigel Chamber, on the right of the large inner chamber called Botha's Hall; and another is the Bath-house, on the same side of the same chamber.

It is intimated that the names given to the different departments were chiefly the names of the first discoverers, and that those beyond the south chamber were first discovered by G. Thom, and entered by Vander Westhens, Nel, and Botha, with him. And there are annotations bearing the initials "G. T.," in which is said: "The unexplored cavern is a dark, dismal place, having few of the beauties of the other parts of the grotto. We saw numbers of bats there. The whole length of the grotto is about 800 feet, and I suppose that I saw 100 feet more into the unexplored cavern, but as the descent into it is about 14 feet I could not enter; the end, therefore, of this wonder in nature is not yet known."

When I was in the Colony I was allowed to peruse a translation of an account of a visit made to similar grottoes near Cape Agulhas, drawn up by the late Mr P. B. Borcherd's, for a long time Civil Commissioner in the Cape Division, and a narrative of a visit by a lady to the Cango Caves.*

I have, oftener than once, in travelling in South Africa, heard what is known as a hollow sound, produced under my conveyance or my

* They, with a number of other papers, were found amongst the papers of Mr Borcherds, in the possession of his son, Dr Borcherds, along with a prospectus for a volume "On New Objects in Geography, Geology, and Botany," projected for publication in 1817, with a list of subscribers, which were shown to me in the knowledge that I would take an interest in finding that so far back in the history of the Cape as a British Colony these objects attracted attention.

horses feet, from which I infer that extensive caverns are by no means few in number there. Such caverns have been studied elsewhere; and from the observations made some idea may be gathered of what has been going on long, and probably is going on still, beneath the surface of the ground. This, if known, may enlarge our conceptions of the quantity of water flowing away in these subterranean currents; and it may modify considerably the views entertained in regard to the way in which the supply of moisture in South Africa is affected by fountains.

The accepted theory is, that the caverns in limestone rocks have been created by subterranean currents, the precursors of those which now flow through them to profounder depths. And to those to whom the consideration of the subject is new some information in regard to existing indications that this has been the case may be acceptable. To supply this I avail myself of ample illustrations supplied by Mr Boyd Dawkins, in a valuable work entitled "Cave-hunting researches on the evidence of Caves respecting the early inhabitants of Europe."

Of subterranean caverns there are some which have been hollowed out by the action of the sea. "The set of the current, the tremendous force of the breakers, and the grinding of the shingle," says the author cited, "inevitably discover the weak places in the cliff and leave caves as the result of their work, modified in each case by the local conditions of the rock. Caves formed in this manner have certain characters which are easily recognised. Their floors are very rarely much out of the horizontal, their outlook is over the sea, and they very seldom penetrate far into the cliff. A general parallelism is also to be observed in a group in the same district, and their entrances are all in the same horizontal plane, or in a succession of horizontal and parallel planes. In some cases they are elevated above the present reach of the waves, and mark the line at which the sea formerly stood. . . .

"In volcanic regions also there are caves formed by the passage of lava to the surface of the ground, or by the imprisoned steam and gases in the lava while it was in a molten state. . .

"Caves also occur sometimes in sand-stones, in which case they are the result of the erosion of the lines of the joints by the passage of sub-aërial water, and if the joints happen to traverse a stratum less compacted than the rest, the weak point is discovered, and a hollow is formed extending laterally from the original fissure.

"But it is chiefly in calcareous strata that caverns abound. There their natural history is easily traced. They are constantly being formed where the material is sufficiently hard and compact to form a roof; where it is not the roof falls in, and a ravine is the result, and this in course of time may be moulded into the form of a valley. By characteristics pertaining to such valleys they may in many cases be easily distinguished from those which are formed by superficial erosion alone, and we are thus enabled to understand how some of the most beautiful scenery in Europe has been formed, and to realise the mode by which all precipices and gorges have been carved out of the calcareous rock, and to see why it is that the combination of hill and valley, ravine and precipice, present the same general feature in all limestone districts—why, for instance, the ravines of Carniola are the same as those of Greece, and both are identical with those of Palestine."

To the characteristic appearances presented by limestone formations Ruskin calls attention, that he may show how truthfully they have been represented in paintings which he specifies, and how false to Nature have been the representations given of them in others.

He states:—"The massive limestones separate generally into irregular blocks, tending to the form of cubes or parallelopipeds, and terminated by tolerably smooth planes. The weather, acting on the edges of these blocks, rounds them off; but the frost, which, while it cannot penetrate nor split the body of the stone, acts energetically on the angles, splits off rounded fragments, and supplies sharp, fresh, and complicated edges. Hence the angles of such blocks are usually marked by a series of steps and fractures, in which the peculiar character of the rock is most distinctly seen; the effect being increased in many limestones by the interposition of two or three thinner beds between the large strata of which the block has been a part; these thin laminae breaking easily, and supplying a number of fissures and lines at the edge of the detached mass. Thus, as a general principle, if a rock have character any where, it will be on the angle; and however even and smooth its great planes may be, it will usually break into variety when it turns a corner. . . . It follows from this structure that the edges of all rock being partially truncated, first by large fractures, and then by the rounding of the fine edges of these by the weather, perpetually present convex transitions from light to the dark side, the planes of the rock almost always swelling a little *from* the angle."

Of the caverns in limestone, carefully examined by Mr Boyd Dawkins in the prosecution of ethnological researches, this author gives the following account:—"The caverns are hollowed in calcareous rock. They open for the most part on the abrupt sides of valleys and ravines at various levels, being arranged round the main axis of erosion just as branches are arranged round the trunk of a tree—as, for example, in Cheddar Pass. The transitions in some cases from the valley to the ravine, and from the ravine to the cave, is so gradual that it is impossible to deny that all three are due to the same cause. The caves themselves ramify in the same irregular fashion as the valleys, and are to be viewed merely as the capillaries in the general valley system, through which the rainfall passes to join the main channels. Very frequently, however, the drainage has found an outlet at a lower level, and its ancient passage is left dry; but in all cases unmistakable proof of the erosive action is to be seen in the sand, gravel, and clay which compose the floor, as well as in the worn surfaces of the sides and the bottom.

"In districts in which caves occur are funnel-shaped cavities of various sizes, . . . in which the rainfall is collected before it finally disappears in the subterranean passages.

"They are to be seen in all stages; sometimes being mere shallow funnels that only contain water after excessive rain, and at others profound vertical shafts into which the water is continually falling." And it is alleged "that the chemical action of the carbonic acid in the rain water, and the mechanical friction of the sand and gravel, set in motion by the water, by which Professor Phillips explained the origin of caves, will equally explain the 'pot-holes' and ravines by which they are invariably accompanied. . . .

"In some cases, in the Swiss and French Jura, they afford passage to powerful streams; and in others they are more or less filled with ice. Similar caverns are found in rocks composed of gypsum, or sulphate of lime; but in some cases in saliferous gypsum, caverns, not dissimilar to these, may have been formed by the removal, by the passage of water, of salt embedded in the gypsum; as in Cheshire the pumping of the salt brine from the saliferous and gypseous strata produces subterranean hollows which sometimes fall in, and eventually cause depressions in the surface, such as those which are now destroying the town of Northwich, and causing the neighbouring tidal estuary to extend over what was formally meadow land."

By the study of these phenomena, some light may be thrown upon the phenomena presented by the salt-pans, which are numerous in South Africa. There is one of these near Bethelsdorp, in the division of Uitenhage, which at certain seasons gets filled with salt water, and as the water evaporates the shallow borders of the lake are filled with crystals of salt, which with solid wooden rakes is pulled ashore like half-melted snow from a river; and there drained and dried it is fit for the market. And though this has been continued for half a century and more there is no apparent diminution of the supply. I think it not improbable that it is renewed by water percolating the ground at some higher level, flowing over a deposit of salt, and afterwards arising in the lake charged with salt.

Besides such, there are many salt-pans of much greater extent, generally dry, and covered with an efflorescence or crust of salt, the salt of which may be otherwise accounted for. One of these is the Commissioner's Pan in Bushmanland—a shallow basin, with a circuit of eighteen or twenty miles. Dr Livingstone estimates the length of a "pan," at Ntetwe, at a hundred miles, and the breadth at ten. A still larger "pan" was discovered by Mr Chapman, one into which several channels of periodical torrents appear to enter from the east, while the waters of the Zouga appear to reach it from the west. By a writer on the Geology of the Diamond Fields, in "Silver's Hand-book to South Africa," it is stated that the term "pan," as the termination of a compound name, always indicates the presence of a geological depression, collecting all and sundry of the substances which, under the influence of rains and winds, have drifted into them and been retained by rings of porphyry, or greenstone; but how those subsidences have been occasioned may be matter of conjecture. For the salt found in these, the commonly received theory may suffice to account. According to that theory, all bodies of water into which rivers flow, but from which there is no outlet, are salt; and it is inferred, from the position of these receptacles, skirted by hills abounding in alkaline mineral, that the combination of chloride of sodium with nitrate and sulphate of soda is sufficient to account for the saline properties with which they are impregnated. A more popular supposition is that the salt is the residuum of sea water, which had been retained there on the upheaval of the land, and been subsequently evaporated. Neither of these theories satisfactorily account for the phenomena at Bethelsdorp, as these were detailed to me on the spot, witnessing the collection of the salt; but I find in the supposition I have advanced a solution of the difficulty of accounting for these, which meets the case as a key fits into the wards of the lock to which it belongs.

I have stated that, according to information given on the spot to Dr A. Murray, chaff introduced into the current at Wonder Fontein has been carried into the Mooi River. Something similar has been observed near Wells, in Somerset. Mr John Beaumont brought before the notice of the Royal Society, in the year 1680, a cavern called Wookey Hole, which was amongst the first, if not the first, of these caverns examined with care in Great Britain. It was observed by him that the water used there in washing lead ore at the village of Priddy, about two miles off, on the Mendip Hills, found its way into the River Axe, which, like the Kuruman at its source, here issues in full current out of the cave, and poisoned cattle in the valley of Wookey. "And this observation," says Mr Boyd Dawkins, "has been verified, during the last few years, by throwing in colour and chipped straw. The stream at Priddy sinks into a swollen hole, and has its course directed by the southerly dip of the rock, by which the joints running north and south afford a more free passage to the water than those running east and west." . . .

In Ireland also may be seen a counterpart to the phenomena presented by the Wonder Fontein and the Mooi River, and by the fountain whence flows the Kuruman. Midway between Donegal and Ballyshannon, running through the grounds of Major Hamilton of Brownhall, is a small river, which bores its way underground, from cavern to cavern, for nearly two miles. The caverns are formed in the limestone, and the power of petrifaction possessed by the water is very astonishing. Sticks, straws, leaves, etc., are soon coated and crusted over, and by degrees hardened into stone; and so rapid is this process that we may see one half of the same leaf of moss green and growing, and the other half crusted over in the process of petrifaction. The caves are, some of them, very beautiful, and when lighted up by the aid of a fragment of magnesium wire, the effect is most brilliant and striking, the walls sparkling as though studded by countless diamonds. In one of them the river makes a fine waterfall. The roof of the cavern is broken, and as the spectator stands in the subdued green light, looking up through a mass of ferns and stems of trees, with the hoarse roar of the waterfall plunging into an unseen depth through a cloud of mist at his feet, the dim cave partially lighted up around, but dark and mysterious in its recesses—the effect is inexpressibly charming—it is a perfect chamber of romance, and one can fancy superstition formerly holding high court here. Almost equally striking in another way is the manner in which the river once more makes its appearance into the light, welling up from a dark unfathomable-looking hole under a wide rift of rock—stealthily emerging from the ground, it oozes forth in a still black stream—a veritable kelpie's hole.

Of Wookey Hole, that mentioned above, Mr Boyd Dawkins writes:—"The hamlet of Wookey Hole nestles in a valley through which passes the river Axe, and the valley passes insensibly, at its upper end, into a ravine, which is closed abruptly by a wall of rock about two hundred feet high, covered with long streamers and festoons of ivy, and affording scanty hold on its ledges and in its fissures to ferns, brambles, and ash sapplings. At its base the river Axe issues in full current out of the cave, the lower entrance of which it completely blocks up, since the water has been kept back by a weir, for the use of a paper mill a little distance away. A narrow path through the woods, on the north side of the ravine, leads to the only entrance now open. Thence a narrow passage leads downward into the rock, until suddenly you find yourself in a large chamber at the water level. There you pass over a ridge covered with delicate fret work of dripstone, with each tiny hollow full of water and ornamented with brilliant lime crystals. One shapeless mass of dripstone is known, in local tradition, as the Witch of Wookey, turned into stone by the prayers of a Glastonbury monk. Beyond this the chamber extends considerably, being some seventy or eighty feet high, and adorned with beautiful stalactites far out of the reach of visitors. The water which bars further entrance forms a deep pool, which Mr J. Parker managed to cross on a raft into another chamber, which was apparently easy of access before the construction of the weir. It was in this chamber that Dr Buckland found human remains and pottery. . . .

"The cave is merely a subterranean extension of the ravine in the same line, as far as the swallow-hole; and all those have been hollowed, as we shall see presently, by the action of the stream and of carbonic acid in the water."

Mr Boyd Dawkins looks chiefly at the evidence supplied by caves respecting the early inhabitants of Europe, the subject of his study; but this evidence could not be elicited without careful study of all the phenomena presented by the caves examined. There follow details of these presented by several, and the effect is to add confirmation to the view of their physical history and of their formation, which is now generally received, which is, that they are not the result of subterranean disturbance, but of the mechanical action of rain-water, and the chemical action of carbonic acid, both operating from above.

Carbonic acid exists in the atmosphere; and additions are constantly being made to this by combustion, and by the breath of every man and beast and bird, and flying insect and creeping thing, yet it experiences little or no increase. As water and carbonic acid have a great affinity for each other, it is absorbed by the moisture of the atmosphere, and the rain falls partially charged with it; and more is also absorbed from the humus of the soil.

Mechanically the rainfall acts on calcareous rocks as it does on other formations, but the carbonic acid with which it is charged has also its action on the limestone. "Universally," says Mr Boyd Dawkins, "the pot-holes, ravine, and caverns are so associated together that there can be but little doubt that they are due to the operation of the same cause.

"It requires but a cursory glance to see at once that running water was the main agent. The limestone is so traversed by joints and lines of shrinkage that the water rapidly sinks down into its mass and collects in small streams, which owe their direction to the dip of the strata and the position of the fissures. These channels are being continually deepened and widened by the mere mechanical action of the passage of stones and silt. But this is not the only way in which the rock is gradually eroded. The limestone is composed in great part of pure carbonate of lime, which is insoluble in water. It is, however, readily dissolved in any liquid containing carbonic acid, which, as an essential part of our atmosphere, is invariably present in the rain-water, and is given off by all organic bodies. By this invisible agent the hard crystalline rock is always being attacked in some form or another. The very snails that take refuge in its crannies leave an enduring mark of their presence in a surface fretted with their acid exhalations, which sometimes pass current among geologists for the borings of pholades, and are the innocent cause of much speculation as to the depression of the mountain-tops beneath the sea in comparatively modern times. The carbonic acid taken up by the rain is derived, in the main, from the decomposing vegetable matter which generally forms the surface soil of the limestone."

Amongst other caverns described by him is the Goat Church Cave, the largest cavern in the Mendip Hills, which opens on the eastern side of the lower of two ravines, which branch from the magnificent defile of Barrington Combe, about two miles from the village of Urington, at the height of about 120 feet from the bottom of the ravine. Entering and descending, passing through successive chambers, and still descending, Mr Boyd Dawkins and two friends, by whom he was accompained, found themselves in one which, says he, "was traversed by a subterranean stream, doubtless in part the same which disappears in the ravine at a point eighty feet above by aneroid measurement. The temperature of the water as compared with that of the stream outside (49° : 59°) renders it very probable that between the point of disappearance in the ravine and re-appearance in the cave it is joined by a stream of considerable subterranean length, since the water could not have lost ten degrees in the short interval it had to traverse, were it supplied only from the stream in the ravine. From the point of its disappearance in the cave, the water passes downwards to join the main current, flowing underneath Barrington Combe, that gushes forth in great volumes at Reckford. The lowest portion of the cave was eighteen or twenty feet below the stream, and two hundred and twenty feet below the entrance of the cavern.

"On examining the floors of the chambers and passages we discovered that they were composed of the same kind of sediment as that which is now being deposited by the water in Wookey Hole, and there could be no doubt that they had been originally traversed by water. For this to have taken place it is necessary to suppose that while the Goat Church was a water cave, the ravine on which it opened was not deeper than the entrance—in other words, that in the interval between the formation and excavation of the passages to the present time, the ravine has been excavated in the limestone to a depth of a hundred and twenty feet, and the water which originally passed through the entrance has found its way by a new series of passages, to the point where it appears at the bottom of the cave.

"We obtained evidence that the horizontal passage immediately below the first vertical descent had been inhabited at a very remote period. . . . It would follow from this, that the date of the formation of this part of the cave was before the time when the traces of elephants, bears, and of man were introduced.

Of the water caves of Yorkshire he writes :—"The caves in the mountain limestone of Yorkshire rival those of Carniola, or those of Greece, and they are to be seen in all stages of formation. In their gloomy recesses all the higher qualities of a mountaineer may be exercised, and there is sufficient danger to give a keen zest to their exploration. The mountain streams sometimes plunge into a yawning chasm, locally known as a pot, and at others emerge from the dark portals of a cave in full current. There is perhaps no place in the world where subterranean circulation of water may be studied with better advantage.

"Ingleborough forms a centre from which the rainfall on every side finds its way into the dales through a system of caves more or less complicated, which, during the last forty years, have been thoroughly explored by Mr Farrer, Mr Birkbeck, and Mr Metcalfe. On the south it collects in a ravine, and then leaps into a deep bottle-shaped hole, called 'Gaping Gill,' into which Mr Birkbeck unsuccessfully attempted to descend, the sharp ages of the rock cutting the rope and very nearly causing a serious accident. Its depth is about three hundred feet. The stream thence finds its way through a series of chambers and passages until it re-appears in the famous Ingleborough cave that was explored by Mr Farrer in 1837, and proved to pass into the rock between seven and eight hundred yards.

"The present entrance to the cave is dry, except after heavy rains, when the current reverts to its old channel."

Professor Phillip, in his volume entitled "Rivers, Mountains, and Sea Coast of Yorkshire," says of this cave:—"From Mr Farrer's plan and description, as given in the 'Proceedings of the Geological Society,' 14th June, 1848, and from information obligingly communicated to me, a clear notion of the history of this most instructive spar grotto may be formed. For about eighty yards from the entrance the cave has been known immemorially. At this point Josiah Harrison, a gardener in Mr Farrer's service, broke through a stalagmitical barrier, which the water had formed, and obtained access to a series of expanded cavities and contracted passages, stretched first to the north and then to the north-west, afterwards to the north and north-east, and finally to the east, till after two years spent in the interesting task of discovery, at a distance of seven hundred and two yards from the mouth, the explorers rested from their labours in a large and lofty irregular grotto, in which they heard the sound of waters in a still more advanced subterranean recess. It has been ascertained, at no inconsiderable personal risk, that this water falls into a deep pool, or linn at a lower level, beyond which further progress appears to be impracticable: in fact, Mr Farrer explored this dark lake by swimming—a candle in his cap and a rope round his body."

The cave is the property of Mr Farrer. By his care it is admirably preserved, and may be visited without any difficulty. To Professor Phillip's account of it I shall afterwards revert.

"The view from the ancient camp on the top of Ingleborough offers a striking example of the effect of rain-water in eroding the surface of the limestone. As you look down over the dark crags of millstone grit, great grey pavement-like masses of limestone strike the eye, standing above the heather, perfectly bare, and in the distance resembling clearings, and in rainy weather sheets of snow. On approaching them the surface of erosion becomes more and more apparent, and the shapes, due to the mere accident of varying hardness in the rock, or the varying quantity of water passing over it, present a most astonishing variety. There are, however, general principles underlying the confusion. The lines of joints in the strata being lines of weakness, searched out by the acid-laden water, have been widened into chasms, sometimes of considerable depth; and as they cross at right angles the whole surface is formed of rectangular masses, each insulated from its fellow, and some of them detached from the strata beneath so as to form rocking-stones." And speaking of effects seen in the limestones of Doveholes in Derbyshire, along with those seen here, he says:—"Each of the upper surfaces of the blocks is traversed by small depressions, which are valley systems in miniature, in which the tiny valleys converge into a main trunk leading into the nearest chasm. There are also tiny caves and hollows, that are sometimes mistaken for borings made by pholas. In these pavements every feature of limestone scenery is represented on a minute scale. There are the valley systems in the surface determined by the direction of the drainage; the long chasms represent the open valleys and ravines, and the caves and hollows for the most part run in the line of the joints."

"The carbonic acid," writes Mr Boyd Dawkins, "has left precisely the same kind of proof of its work within the caves as we find above ground; and it would naturally follow that to it, as well as to the mechanical powers of the waters flowing through them, their formation and enlargement must be due, as Professor Phillip has pointed out in his 'Rivers, Mountains, and Sea Coast of Yorkshire,' (pp. 303).

"From the preceding pages it will be seen that caves in calcareous rocks are merely passages that have been hollowed out by water, which has sought out the lines of weakness, or the joints formed by the shrinkage of the strata during their consolidation. The work of the carbonic acid is proved, not merely by the acid-worn surfaces of the interior of the cave, but also by the large quantity of the carbonate of lime which is carried away by the water in solution. That, on the other hand, of the mechanical friction of the stone and sand against the sides and bottom of the water caverns is sufficiently demonstrated by their ground, scratched, and polished surfaces, and by the sand, silt, and gravel carried along by the currents. The generally received hypothesis that they have been the result of a subterranean convulsion is disproved by the floor and roof being formed, in very nearly every case, of solid rock; for it would be unreasonable to hold that any subterranean force could act from below in such a manner as to hollow out the complicated and branching passages at different levels without affecting the whole mass of the rock. . . . The causes at present at work, operating through long periods of time, offer a reasonable explanation of their existence in every limestone district; and those which are no longer water-courses can generally be proved to have been formerly traversed by running water by the silt, sand, and rounded pebbles which they contain. In their case, either the drainage of the district has been changed by the upheaval or depression of the rock, or the streams have searched out for themselves a passage at a lower level.

"But if caves have been thus excavated, it is obvious that ravines and valleys in limestone districts are due to the operation of the same causes. . . . The ravine is merely a cave which has lost its roof, and the valley is merely the result of the withering of the sides of the ravine." And referring to diagrams of the Wookey Hole cave and ravine, and of the subterranean water-course of Dalebeck, he says:—"There can be no manner of doubt but that, in both these cases, the ravine is gradually encroaching on the cave, and the valley on the ravine; and if the strata be exposed to atmospheric influences or agencies long enough, the valley of the Axe will extend as far as Priddy, and that of Dalebeck to the water-shed above the Gatekirk cave. In the same manner the lofty precipice of Malham Cave, near Settle in Yorkshire, is slowly falling away and uncovering the subterranean course of the Aire. Eventually the ravine thus formed will extend as far as Malham Tarn, and the Aire flow exposed to the light of day from its source to the sea.

"This view is applicable to many if not to all ravines and valleys in calcareous rocks, such as the Pass at Cheddar, or the gorge of the Avon at Clifton, and those of Derbyshire, Yorkshire, and Wales. And since the agents by which the work is done are universal, and calcareous rock for the most part of the same chemical composition, the results are the same, and the calcareous scenery everywhere of the same type. In the lapse of past time, so enormous as to be incapable of being grasped by the human intellect, these agents are fully capable of producing the deepest ravines, the widest valleys, and the largest caves.

"This view of the relation of caves to ravines was so strongly held by M. Desmayers [who with M. Duport held a different theory in regard to the production of the caverns] that he terms the latter [the ravines] '*cavernes á ciel ouvert.*' I arrived independently at the same conclusion after the study of the scenery of limestone for many years.

"In many cases, however, in northern latitudes and in high altitudes, the ravine or valley so formed has been subsequently widened and deepened by glacial action. That, for instance, of Chapel-en-le-Dale bears unmistakeable evidence of the former flow of a glacier in the *roches moutonnées* and travelled blocks that it contains. To this is due the flowing contour and even slope of its lower portion.

"The pot-holes and 'cirques' in calcareous rocks with no outlet at the surface, may also be accounted for by the operation of the same causes as those which have produced caves. Each represents the weak point towards which the rain has converged, caused, very generally, by the intersection of the joints. This has gradually been widened out, because the upper portions of the rock would be the first to seize the atoms of carbonic acid, and thus be dissolved more quickly than the lower portions. Hence the funnel shape which they generally assume, and which can be studied equally in the compact limestone or in the soft upper chalk. They are to be seen also in all limestone 'pavements.' Sometimes, however, the first chance which the upper portions of the funnels have of being eroded by the acidulated water is more than counterbalanced by the increased quantity converging at the bottom, and the funnel ends in a vertical shaft. If the area in the rock thus excavated be sufficiently large to allow of the development of a current of water, the mechanical action of the fragments swept along its course will have an important share in the work, as we have seen to be the case in Helln Pot."

Of Helln Pot, a tremendous chasm near Selside, on the east of Simon's Hall—the name of which Helln Pot, Ællan Pot, a Mouth of Hell, speaks the feelings with which it was regarded by the angles of former times—Mr Boyd Dawkins, writing of the subterranean passages which group around this, says:—"Those which have been explored constitute the Long Churns cavern, which is comparatively easy of access, through a hole known as Diccan Pot. On descending into it the visitor finds himself in the bed of a stream that now roars in a waterfall, now gurgles over the large fallen blocks from the roof, and here and there has worn for itself deep pools, by the mechanical friction of sand and pebbles brought down by the current. If it be followed down after passing over a waterfall, the light of day is seen streaming upwards from beneath the feet from the point where the water leaps into the great chasm of Helln Pot. Above the entrance there is a complicated net-work of passages, some dry, and some containing streams which have not yet been fully explored.

"The two actions by which caves are hewn out of the calcareous rocks are seen here in operation side by side. Below the level of the stream the rock is seen smoothed and polished by the mechanical action of materials swept down by the current above the water level, the sides of the caves are honeycombed and eaten into the most fantastic and complex shapes, the resultant surface bearing small points and keen knife-edges of stone, that stand out in relief and mark the less soluble portions of the rock. This is due to the chemical effect of the carbonic acid in the water percolating through the strata.

"The Helln Pot, into which the stream flowing through the Long Churn falls, is a fissure, a hundred feet long by thirty feet wide, that ingulfs the waters of a little stream on the surface, which are dissipated in spray long before they reach the bottom. From the top you look down on a series of ledges, green with ferns and mosses, and from about a hundred feet from the surface an enormous fragment of stone forms a natural bridge across the chasm, from one ledge to another. A little above this is the debouchement of the stream flowing through the Long Churn Cave, through which Mr Birkbeck and Mr Metcalfe made the first perilous descent in 1847. The party, consisting of ten persons, ventured into this awful chasm with no other apparatus than ropes, planks, a turn-tree, and a fire-escape belt. On emerging from the Long Churn Cave, they stood on a ledge of rock about twelve feet wide, which gave them free access to the Bridge. This was a rock ten feet long which rested obliquely on the ledges. Having crossed over these they crept behind the waterfall which descended from the top, and fixed their pulley, five being let down, while the rest of the party remained behind to hoist them up again. In this way they reached the bottom of the Pot, which had never been trodden by the foot of man. Thence they followed the stream downwards as far as the first great waterfall, down which

Mr Metcalfe was venturesome enough to let himself with a rope, and to push onward until daylight failed. He was within a very little of arriving at the end of the cave into which the stream flows, but was obliged to turn back again to the daylight without having accomplished his purpose. The whole party eventually, after considerable danger and trouble, returned safely from this most bold adventure.

"A second descent was made in 1848 from the surface, and a third in the spring of 1870, in both of which Mr Birkbeck took the lead. The apparatus employed consisted of a windlass supported on two baulks of timber, and a bucket covered with a shield, sufficiently large to hold two people, and two guiding ropes, to prevent the revolution of the bucket in mid air. There was also a party of navvies to look after the mechanical contrivances, and two ladders, about eight feet long, to provide for contingencies at the bottom. Thirteen of us went down, including three ladies. As we descended, the fissure gradually narrowed, until, at the bottom, it was not more than ten feet wide. The actual vertical descent was a hundred and ninety-eight feet. After running the gauntlet of the water-fall we landed in the bed of the stream, which hurried downwards over large boulders of limestone and lost itself in the darkness of a large cave about seventy feet high. We traced it downwards, through pools and rapids, to the first waterfall of about twenty feet. This obstacle prevented most of the party from going further, for the ladders were too short to reach the bottom. By lashing them together, however, and letting them down, we were able to reach the first round with the aid of a rope, and to cross over the deep pool at the bottom. Thence we went on, downwards through smaller waterfalls and rapids, until we arrived at a descent into the chamber, where the roar of water was deafening. Down to this point the daylight glimmered feebly, but here our torches made but little impression on the darkness. One of the party volunteered to go down with a rope, and was suddenly immersed in a deep pool; the rest, profiting by his misadventure, managed to cling on to small points of rock, and eventually to reach the floor of the chamber. We stood at last on the lowest accessible point of the cave, about three hundred feet from the surface. It was, indeed, one of the most remarkable sights that could be imagined. Besides the waterfall down which we came, a powerful stream poured out of a cave too high up for our torches to penetrate the darkness, and fell into a deep pool in the middle of the floor, causing such a powerful current of air that all our torches were blown out except one. The two streams eventually united and disappeared in a small black circling pool, which completely barred further ingress.

"The floor of the pot and the cave was strewn by masses of limestone, rounded by the action of the streams; and the water-channels were smoothed and grooved and polished in a most extraordinary way by the silt and stones carried along by the current. Some of the layers of limestone were jet black, and others were of a light fawn colour, and as the strata were nearly horizontal the alternation of colours gave a peculiarly striking effect to the walls. Beneath each waterfall was a pool more or less deep, and here and there in the bed of the stream were holes drilled in the rock by stones whirled round by the force of the water. High up, out of the present reach of the water, were old channels, which had evidently been water-courses before the pot and cave had been cut down to their present level. In the sides of the pot there are two vertical grooves, reaching very nearly from the top to the bottom, which are unmistakeably the work of ancient waterfalls. There was no stalactite, but everywhere the water was wearing away the rock and enlarging the cave. We found our way back without any difficulty, a small passage on the right-hand side enabling us to execute the very unpleasant task of scrambling up two of the waterfalls. We arrived finally at the top, after about five hours' work in the cave, wet to the skin. . . .

"At the bottom of Helln Pot it was impossible not to realise that the enormous chasm had been formed by the same action as that by which it was being deepened before our eyes. It was merely a portion of the vast cave into which it led, which had been deprived of its roof and opened out to the light of heaven. The bridge was but a fragment of the roof which happened to fall upon the two ledges; the rounded masses of rock at the bottom are fragments that have fallen, probably, within comparatively modern times. The absence of stalactites and stalagmites proves that the destructive action is rapidly going on."

Now, as with these so is it with the limestone caverns of South Africa. The extent and magnitude of some of these tell of the quantity of water which must have flowed for ages through them; but they also tell of the quantity of the rainfall which must be carried off by subterranean currents where there is no limestone. Limestone affords special facilities for the percolation of water; but the probability is that most of the water by which the limestone has been dissolved from these caverns, has come, not from the superficial area of their covering, but from land beyond, flowing thither in channels formed by friction, till here its solvent powers came into operation, and thus supplied some indication, in the absence of other data, of the magnitude of the quantity of water which has thus been, and is thus being, conveyed away.

The subterranean river probably has its first origin at a great distance from the cavern, but into it there may percolate, throughout its course, from first to last, water from above, feeding and increasing its size. And, in the stalactites and stalagmites with which the caverns are adorned, we find a means of measuring the quantity percolating the ground covering the area of the cavern.

"If the velocity of the stream in a water cave be lessened," says Mr Boyd Dawkins, "the silt, sand, or pebbles it was hurrying along will be dropped, and may ultimately block up the entire watercourse. In bringing this to pass, however, the carbonate of lime in the water plays a most important part. If the excess of carbonic acid by which it is held in solution be lost by evaporation, it immediately re-assumes its crystalline form, and shoots over the surface of the pool like plates of ice, or is deposited in loose botryoidal masses at their sides and on their bottoms; and since the atmospheric water very generally percolates through the crevices in the rock, the sides and roof of the channel above the level of the water are adorned with a stony drapery of every conceivable shape." It is to this I refer as supplying indications of the quantity of water coming into the cavern from the ground overlying it.

Professor Phillips, in continuation of his account of the series of expanded cavities and contracted passages in the Ingleborough cave, says:—"In this long and winding gallery, fashioned by Nature in the marble heart of the mountain, floor, roof, and sides are everywhere intersected by fissures, which were formed in the consolidation of the stone. To these fissures and the water which has passed down them, we owe the formation of the cave and its rich furniture of stalactites. The direction of the most marked fissures is almost invariably northwest and south-east, and where certain of these (which in my geological work I have called master-fissures) occur, the roof of the cave is usually more elevated, the sides spread out right and left, and often ribs and pendants of brilliant stalactites, placed at regular distances, convert the rude fissure into a beautiful aisle of primæval architecture. Below most of the smaller fissures hang multitudes of delicate translucent tubules, each giving passage to drops of water.

Splitting the rock above, these fissures admit, or formerly admitted, dropping water; continued through the floor, the large rifts permit, or formerly permitted, water to enter or flow out of the cave. By this passage of water, continued for ages on ages, the original fissure was in the first instance enlarged, through the corrosive action of acidulated water; by the withdrawal of the streams to other fissures, a different process was called into operation. Then the fissure was bathed by drops instead of streams of water, and these drops, exposed to air, currents, and evaporation, yielded up the free carbonic acid to the air, and the salt of lime to the rock. Every line of drip became the axis of a stalactitical pipe from the roof; every surface bathed by their films of liquid became a sheet of sparry deposit. The floor grew up under droppings into fantastic heaps of stalagmite, which sometimes reaching the pipes united roof and floor by pillars of exquisite beauty."

Mr Boyd Dawkins adds, in regard to this cavern:—"At the time of its exploration the water stood at a considerably higher level inside than at the present time, and formed deep pools. The barrier of dripstone has been cut through, and the water level lowered, and a passage made for a considerable distance. Inside, the old water line which separated the sub-aërial from the sub-aqueous dripstone is very distinct; the former being deposited in thick bosses, crumpled curtains, drops, straws, pyramids, and other fantastic drip-structures; while the latter is honey-combed, and comprised of rounded, grape-like masses. Between them an ice-like coating of stalagmite forms a dividing line, now supported in mid-air, but which formerly shot across the surface of the pools that have been drained, or rested on the mud and stones which had been brought down by the stream in ancient times. In some places it still rests on the surface of the pools."

The sub-aërial formations, of an age subsequent to the draining off of the water, are thus easily distinguished from the sub-aqueous formations of an earlier day.

Some attempts have been made to determine the age of stalactites and stalagmites, or the time passed during their formation; but this must vary with many varying circumstances and conditions. In one case a rough estimate from somewhat uncertain data led to the conclusion that a large stalagmite, called the Jockey Cap, had increased at the rate of a quarter of an inch per annum, and it was calculated that at the rate of a quarter of an inch per annum twenty feet of stalagmite might be formed in 1000 years. But the estimate is inapplicable to a determination of the age of the stalagmite in the caverns at the Cape of Good Hope. Could we but determine their age by determining the quantity of water evaporated in the course of their formation, we might obtain an approximation to a determination of the annual percolation throughout the period of their formation.*

On the subject of subterranean rivers, Sir John Herschel, in his treatise on Physical Geography, writes:—"The welling forth of streams from perennials prings is of the most ordinary occurrence; but it is seldom more than a rivulet which rises in this manner. There are, however, some instances of considerable streams so originating. When this is the case they issue from caverns, and these occur usually either in ice or in limestone. In the former case, they are evidently only the drainage of melted snow, which pours out at the foot of a glacier by the contribution of sub-glacial streams converging to the lowest point. . . .

"Limestone formations are very apt to be hollowed into caverns by the solvent power of carbonic, and perhaps also of other kinds of acids derived from vegetable decomposition, held in solution in the percolating water. Such caverns often run to great distances

* It is with the water that we are more immediately concerned; but it may be allowed to us to take a passing look at the transformations of the carbonate of lime as it goes on beyond the little space of time during which it comes thus necessarily within our sphere of study. "The circulation of carbonate of lime in nature presents us," says Mr Boyd Dawkins, "with a never-ending cycle of change. It is conveyed into the sea to be built up into the tissues of the animal and vegetable inhabitants. It appears in the gorgeous corallines, nullipori, calcareous sea-weeds, sea-shells, and in the armour of crustaceans. In the tissues of coral-zoophytes it assumes the form of stony groves, of which each tree is a colony of animals, and in the wave-defying reef it reverts to its original state of limestone. Or, again, it is seized upon by tiny masses of protoplasm, and fashioned into chambers of endless variety and of infinite beauty, and accumulated at the bottom of the deeper seas, forming a deposit analogous to our chalk. In the revolution of ages the bottom of the sea becomes dry land, the calcareous *débris* of vegetable and animal life is more or less compacted together by pressure and by the infiltration of acid-laden rain-water, and appear as limestone of various hardness and constitution. Then the destruction begins again, and caves, pot-holes, and ravines are again carved out of the solid rock."

From a calculation given by Mr Prestwich, in a presidential address to the Geological Society in 1871, p. xvii, it appears that about 1000 tons of carbonate of lime, and 238 tons of sulphate of lime and gypsum, are carried to the sea by the Thames alone every twenty-four hours, and apportioning this, he concludes that from the chalk, upper green-sand and ooletic strata alone of the Thames' basin, there is thus removed 140 tons from every square mile daily, or 14,000 tons in a century.

Lime is a valuable constituent of arable soil; it may be worth a moment's thought how much is lost to the Colony by subterranean currents, and whether means might not be found to recover water thus charged with it.

underground, and frequently contain running streams, even considerable rivers, as is the case in the caverns of the Peak and Castleton, in Derbyshire, and in that of the Nicojack Cavern, Georgia, U.S., on the Tenessee River, where a waterfall occurs at a distance of three miles under ground (Ed. Ph. Jour., i., 426).

"Where such springs emerge to day, we have the phenomena in question—as in the cavern of the Gancheros, in the valley of Caripe, in Cumana—described by Humboldt; in the celebrated fountain of Vaucluse, which issues as a considerable stream from a cave at the foot of a perpendicular limestone cliff; and in a great number of caves in Carniola and Illyria, where 'almost every lake or river has a subterranean course, and often a subterranean exit. The Laibach River rises twice from the limestone rock, and is twice again swallowed up by the earth before it makes its final appearance' (Davy). The rivers Sarapa and Blanco, which flow from the lake of Yojoa, in Honduras, both enter subterranean channels, through which having passed, in the one case a mile, in the other a mile and a half underground, they reappear. . . .

"In some places in the Mediterranean springs of fresh water, and some of a magnitude to justify their being described as subterranean rivers, well up to the surface from considerable depths. The most remarkable of these," says Sir J. Herschel, "is at Anavola, in the Sinus Argolicus, between Kiveri and Astros, where a body of fresh water, fifty feet in diameter, rises with such a force, at a quarter of a mile from the shore, as to produce a visible convexity of surface, and to disturb the sea for several hundred feet around (Leak's Travels in Moria, ii., 480). In the Gulf of Xagua, southwest of the port of Batabano, on the south coast of Cuba, a similar cave occurs (Humboldt Aspects, p. 233), and others are said to exist in the Pacific, among the Sandwich Islands."

Subterranean currents similar to these which traverse such caverns, but varying in magnitude from such as connect Wonder Fontein with the Mooi River to a thread-like dimension, come to the surface as springs and fountains; and the quantity of water thus brought to the surface and made available for the use of man is very great.

To what is brought by rivers into different districts from higher lying lands may legitimately be added, in an attempt to estimate the water supply of these districts, what is brought from the same or similarly situated regions by subterranean water-courses, and issuing thence by fountains or springs, or otherwise giving indications of their proximity to the surface. But with regard to this supply, and with regard to the supply derived from rivers alike, it must be borne in mind that it is only a transfer of a portion of the rainfall on another district brought thither, and not an absolute addition to the water supply of the country. Brought into the district it may be considered a godsend to the district, to be received with thanksgiving. It has been brought from another district where it was not used or stored for future use; but brought and delivered there, it is a gift of God which man is free to use and to enjoy.

The quantity of water brought by this means, in addition to the supply received naturally from other sources in different districts of the Colony, I have not the data necessary to enable me to determine; but, to enable others to estimate some of the fountains thus bringing in an extra supply, I may state that the formula by which such quantities are calculated are these:—

(a) v=Velocity per second. d=Diameter of pipe.
h=Height of reservoir. l=Length of pipe.

$$v = 48\sqrt{\frac{hd}{l+54d}}$$

(b) a=Area of pipe=$\cdot 7854 \times d^2$.
$v \times (7854 \times d^2)$=Quantity per second.
1 cubic foot water = 6·23 gallons.

Ex. $h = 65\frac{3}{5}$ feet. $l = 9843$ feet. $d = 3\frac{3}{20}$ inches.
(a) $v = 48\sqrt{\frac{65\cdot 6 \times \cdot 2625}{9843 + 54 \times \cdot 2625}} = 2$ feet per second.

$2 \times \cdot 0541 = \cdot 1082$ cubic feet per second = ·674 gall. = $\frac{674}{1000}$ of a gallon per second.

In open canals:—
c=Wet contour.
s=Area of section of fluid.
$\frac{s}{c}$=Hydraulic mean depth.
g=Force of gravity.
a=Angle of inclination of surface of stream.
v=Mean velocity.

$$v = \sqrt{\left(\frac{50}{3}\right)^2 g.\frac{s}{c} \sin. a + \left(\frac{1}{100}\right)^2} - \frac{1}{100} = \text{velocity in feet.}$$

CHAPTER V.

Supply of Moisture from Subterranean Streams and Reservoirs.

The supply of water produced by fountains tells not only of water which has fallen as rain on higher-lying land, and which was lost and buried, but has been raised again from the grave, as living water restored to man for use—but it awakens a suspicion that much more has been interred to await a resurrection by other means,—or to remain for ages in the depths of the earth, until by some great convulsion the fountains of the great deep be broken up; and much, it may be, to escape by an underground course to the sea, which never says—"It is enough!"

Excepting when retained by some counteracting means, water, like much besides, always descends under the influence of gravitation to the lowest level it can reach. Subterranean currents, whether thread-like or more like a mill-lead, are in their flow acting under this influence; and if we would learn their resting-place we must follow their lead, or reason upon what we do know. We find such subterranean currents issuing in fountains and springs. This must result from their finding there an outlet at a lower level than that from which they have come; but their issue is equally explicable on the supposition of this being, like the outlet from Wonder Fontein on the Mooi River, the termination of the water course,—or of the water having so accumulated as to have risen above the level of the spring or fountain, which carries off only the overflow. That the latter is the case with some fountains and springs can be demonstrated, and that it is the case with many can scarcely be doubted.

That the accumulation extends in some cases to a very great depth, we have evidence in the great depth from which water has been drawn by Artesian wells. The immensity of the quantity conveyed to such depths is evidenced by the quantities which have been thus obtained. And the pressure to which it is subjected in some cases is indicated by the height to which it is ejected when a way of escape has been given to it; which pressure speaks again of the greatness of the quantity accumulated, as this is attributable either to its having reached a much higher level in the bowels of the earth than that to which it is projected into the air, or, what is more probable, such a level that, in accordance with the hydraulic paradox of a little water forced into a pipe of small dimension raising a considerable weight, or acting on a hydraulic press, it counterbalanced such a pressure of earth that the elasticity of this projected the water so high on its thus obtaining relief.

Nor is it improbable that the outlet to some of these subterranean currents is under the sea, to which these waters are conveyed by such underground water-courses, which thus deny to man all of the adventitious benefits which river drainage gives to him. The mention of this may call forth the exclamation, uttered in a somewhat indignant tone—Impossible. But second thoughts may show it is not impossible. If the subterranean reservoirs lie above the sea level it is not impossible; and if, to take an extreme case, the water must descend to a depth of five miles below the sea level before it find an exit under the ocean, even there will the ocean draw it forth: the low specific gravity of fresh water compared with that of sea water will of itself give rise to an upward flow, and this, operating as a drain, will tell upon the ground above the reservoir, on some elevated plateau, situated, it may be, a hundred miles from the sea. The case of Anavola is a proof.

In order to conceive of this, and of much besides connected with such phenomena, it is not necessary to conceive of a continuous water course extending through the whole mass of earth traversed, or to conceive of the water being contained in immense subterranean caverns; it is enough to conceive of the water accumulating in some loose-lying strata, such as is the sand of Damaraland, from which, as has been stated, water may be collected by simply with the hand shovelling out a hollow; nor is it necessary to conceive of a continuous bed of such loose matter—a patch here, a patch there, some at one level, some at another, connected with earth more or less permeable by water, is all that is required in order to an intelligible conception of what it is alleged may be the case.

That at the Cape of Good Hope there are subterranean reservoirs of water at a great depth seems to be indicated by the high temperature of many fountains, the warm baths at Namaqualand, the warm baths at Caledon, Brand Vley, and others. That there are subterranean waters in many places at a comparatively superficial depth is indicated by the fact of water having been obtained by sinking wells to depths ranging from ten to thirty feet. And that there are in numerous places water immediately under the soil or the subjacent subsoil, or at no great depth below, seems to be proved by the success with which in many places water has been found by those who knew how to seek for it.

To indicate the appropriate spots for striking and tapping such subterranean stores is the profession of the so-called "Water-finder." From the great success with which one of these, named Kohl, followed this calling in various districts and localities of the Colony, shortly before I left the Cape, it would almost seem as if the whole country were permeated by subterranean streams, or reservoirs of water. And there need be no mystery in the practice of the art.

Many of the indications of the proximity of water are such as may be determined by the most cursory observation. Habits of observation, quickened by occasional privation, have taught many of the travellers in South Africa to find water where the unexperienced would not think of looking for it.

I have had occasion, in writing of the effects of forests on humidity, to refer to the first discovery by Mr James Chapman, the South African traveller, of water in the trunk of the Kushé tree. It occurred on a day when he and his companions were suffering from thirst. Observing one of his native servants placing a forked bough against one of these trees, the first he had seen, and preparing to climb, he asked what he was going to do. "Look for water," was the reply; and the man having mounted called out—"Here it is!" The heart-wood of the tree had decayed, leaving a very deep hole which almost closed atop, apparently by the spreading of subsequently formed bark. The water filling the well thus formed, and which had, it was conjectured, fallen as rain, was thus protected against loss by evaporation. One of the companions of the man, cutting then a twig or shoot from the root, cut through the bark of this at two places, two feet or thirty inches apart, treated this as boys in Europe treat branches of elder of which they wish to make whistles, beat and pressed it, and drawing out the wood as a sword is drawn from its sheath handed up the bark to the man aloft, who, using this as a suction pipe, soon quenched his thirst, and made way for the others in succession to do the same. Frequently did Mr Chapman afterwards avail himself of the supply of water thus retained in the Kushé tree.

From the same traveller, and subsequently from others, I learned that in some parts of Damaraland, in the way I have described, water is sought for in the sand; however dry this may be a hollow is dug till a clay like stratum is met with, where, in a short time, water collects; but this stratum is thin and brittle, and every care is taken against its being broken, for, if this happens, the water drains off and is lost in sand below. And another of my correspondents, the gentleman who furnished me with the list of springs on the western border of the Karroo, which will afterwards be cited, informed me that on his first journey across that region he was at first surprised in searching for water to find that while he went to hollows his companion went to mounds, and when asked what he was doing there he answered—Looking for water! and he found that he was right. He has furnished me with a list of many places in which, on the summit of a little mound, a spring exists, while all around is dry. The mounds appeared to me, from the description of them given to me, to have been formed of deposits from the water rising there; and something similar to such formations I have seen around a hot spring at Balmoral, near Uitenhage.

Habits of observation thus developed may easily direct any one to places where superficial subterranean water-courses may be tapped.

The indications of the proximity of water are various.

Dr Livingstone writes:—"When passing through a dry district the native hunter knows where to expect water by the animals he sees. The presence of the gemsbuck, duiker or diver, springbucks, or elephants, is no proof that water is near; for these animals roam over vast tracts of country, and may be met scores of miles from it. Not so however the zebra, pallah, buffalo, and rhinoceros; their spoor gives assurance that water is not far off, as they never stray any distance from its neighbourhood. But when amidst the solemn stillness of the woods, the singing of joyous birds falls upon the ear, it is certain that water is close at hand."

Another traveller in the interior of South Africa, Dr Fritsch, informed me that the sight of a Doornboom, but still more that of a wild dog, and still more than that the sight of a quail, or flight of quails, was to him an indication of whereabout water was to be found.

My own experience in the Colony was limited, but in my successive tours I have observed in arid districts plants which grow only in stagnant water, and on marshy or undrained land. I have found very many of these giving a character to the vegetation of the suurfeldt; and I have observed the natural vegetation, below the level of a water-leading, or of a reservoir on a hillside, to be different from that growing above the same, which is attributable to the greater moisture existing there in consequence of percolation from the reservoir: an illustration of this may be seen on the banks of the reservoir above Sea-point. In the country the Kraalbosch and Wasserbosch give indications of the same thing—a comparatively moist soil.

I shall afterwards have occasion to cite numerous observations of indications of subterranean supplies of water made, in various parts of the eastern province of the Colony, by Mr Kohl, the professional water-finder; and I shall therefore give here some additional information in regard to indications by which the so-called water-finder may be guided.

There may be many appearances of vegetation, soil, moisture, etc., recognised and easily recognisable, which cannot be described. It is thus with the diagnosis of disease; it is thus also with the recognition of the figure, features, voice, or hand-writing of different persons; and so may it be with the appearances in question.

When I was in the Colony, some thirty years ago, a robbery was committed in the house of the Rev. Mr Judge of Simon's Town. A sharp youth, whose boots had been stolen, was required to identify them, and having testified that the boots were his he was asked by the Chief-Justice—at least so the story went—"You say these boots are yours; how do you know?" And the answer is said to have been—"You are Sir John Wylde—and I know that you are Sir John Wylde; but I cannot tell you how I know it. And I know that these are my boots—though I cannot tell you how I do so."

And thus may it be with the "water-finder." It would not shake my confidence in a "water-finder," who had given me proof of his skill, that he could not tell me how he knew where water was to be got. Formerly such men carried what was called a "divining-rod," which, suspended or balanced like a weighing-beam at the middle, dipped, or inclined toward the spot where water was indicated. By some this may have been used for the purpose of mystifying; in some cases the necessary inclination of it may have been given intentionally by the diviner; in other cases this may have been done by him unconsciously; but, whether it were consciously or unconsciously that the divining rod of old was made to dip by the water-finder who used it, it was by the physical conditions of the place, I doubt not, that Mr Kohl was directed in his decisions.

Captain Baird Smith, in his Report to the Directors of the East India Company on Italian Irrigation, states that the springs most in request for irrigation in Piedmont and Lombardy "are connected with permeable alluvial beds which stretch away towards the rising lands on the northward of the plain. In seeking for these springs there are a number of natural signs which are followed as guides; and though the faith in the magic virtues of the divining-rod, which once prevailed universally throughout these districts, may now have passed away, the fountain seeker—a much-employed member of the community—has still his traditional signs, which gather round the source he is in search of. Where in the spring-time the verdure of a meadow is of a deeper green than the general mass, or the soil of a field has a darker or damper appearance than the rest, there he considers it desirable to try his fortune. When in the summer the gnats are seen hovering in masses over a particular spot and resting very close to the soil he suspects that aqueous vapour is ascending from below, and that a spring must be near the surface. At all seasons of the year light vapours frequently rest over the hidden springs at night and morning; and to these the searcher gives attention with the sort of practical instinct, in recognising their indications, which long observation creates."

Indications must vary with circumstances and conditions. The operation adopted by a water-finder in Ireland was to bury a dry stone in a dry spot at night and examine it in the morning. If he found it moistened, he said there was a probability of water being found by digging; if it was dry, he said there was none. The method may have been with him empirical, used and applied without science, but the proceeding can be justified by science; and though not applicable everywhere, the mention of it may suggest that the indications of the existence of subterranean currents or reservoirs are numerous and developed, and by appropriate observations the discoveries which were made by Kohl may be multiplied.*

To the sources of supply whence water is available for the pro-

* Careful and continued observations have enabled seamen and shippers, and others who are much in the open air, to prognosticate the weather with remarkable correctness.

"The evening red and the morning gray,
That 's the sign of a bonnie day."

"When it is evening, ye say, it will be fair weather; for the sky is red: and in the morning, it will be foul weather to-day; for the sky is red and lowring."

In different lands different phenomena have been observed to precede changes of weather; and in our own language the designation *Weatherwise* has come into use as applicable, seriously or in jest, to the man who professes to be able to prognosticate the weather. In illustration of the number of the weather-signs which are in popular use I may cite the following lines, which were sent by Dr Jenner, the discoverer of vaccination, as embodying reasons why he did not accept an invitation from a friend

motion of vegetation and for the arrest and remedy of the desiccation of South Africa, which have been treated of in preceding chapters, may be added the subterranean water-currents and reservoirs of the land. By observations similar to those made by Kohl or others, appropriate to the locality, there may be discerned, as there have been, many unknown reservoirs; and in some cases springs may be taken as indications of these, which, if followed, may lead to the discovery of unknown treasures.

to make with him an excursion, and which may suggest that were equal attention given to indications of subterranean water stores it may be that many may be found :—

"The hollow winds begin to blow,
The clouds look black, the glass is low;
The soot falls down, the spaniels sleep,
And spiders from their cobwebs peep;
Last night the sun went pale to bed,
The moon in halves hid her head;
The boding shepherd heaves a sigh,
For see—a rainbow spans the sky;
The walls are damp, the ditches smell,
Closed is the pink-eyed pimpernel.
Hark, how the chairs and tables crack,
Old Betty's joints are on the rack:
Loud quack the ducks, the peacocks cry;
The distant hills are seeming nigh;
How restless are the snorting swine;
The busy flies disturb the kine;
Low o'er the grass the swallow wings;
The cricket, too, how sharp he sings;
Puss on the hearth, with velvet paws,
Sits wiping o'er her wiskered jaws;
Through the clear stream the fishes rise,
And nimbly catch the incautious flies;
The glow-worms, numerous and bright,
Illumed the dewy dell last night.
At dusk the squalid toad was seen,
Hopping and crawling o'er the green;
The whirling wind the dust obeys,
And in the rapid eddy plays;
The frog has changed his yellow vest,
And in a russet coat is dressed.
Though June, the air is cold and still,
The mellow black-bird's voice is shrill;
My dog, so altered in his taste,
Quits mutton bones on grass to feast;
And see yon rooks, how odd their flight,
They imitate the gliding kite,
And seem precipitate to fall,
As if they felt the piercing ball;—
'Twill surely rain, I see with sorrow,
Our jaunt must be put off to-morrow."

The spring or outlet is not always the termination of a subterranean water-course; and, perhaps, in inland districts, it seldom is such. The water may be in course of passage, by some crack, or crevice, or channel—which the flow of water has enlarged, if it have not originally created or formed it—to a lower level, and the fountain or spring is but a leakage from this; and by an artificial enlargement of this, or by clearing away hindrances to a freer escape of the water existing there, the supply there obtained may be greatly increased.

Before leaving the Cape I found it stated in the journals of the day, that, by simply opening up a fountain on his property, Mr C. Guest, near Oudshorn, had obtained such a supply of water as had raised the value of his property a thousand pounds.

The method to be adopted in opening up a fountain must vary with circumstances. The following is the account given by Captain Baird Smith of the procedure followed in Piedmont and Lombardy:—"Supposing that by these or other similar signs the existence of springs is sufficiently indicated, excavations are made through the surface-soil to those beds of sand or gravel which form the subsoil of the whole irrigated plain; and as new threads of water, rising to the surface, continue to show themselves, so does the excavation extend, until a supply adequate to the demand has been obtained, or until the area of the head of the spring reaches the limit beyond which it cannot profitably be carried, owing to the value of the land. The small jets of water forming the springs rise, of course, in variable volume, and with variable force; and in the large spring-heads (locally termed *teste dei fontane*), it is only the more important of these that receive special attention.

"In the irrigated plains of Piedmont and Lombardy, the depth to which the excavation for springs has to be carried very rarely exceeds 8 or 9 feet; and, owing to the natural slope of the country from north to south, this depth is soon worked out, and the channel brought to the level of the soil. The shape of the spring-head is determined entirely by the manner in which the jets (termed, on the spot, the *ochi*, or eyes of the spring) are distributed. Among the immense number I saw, there were none that affected any regularity of form, though the banks were trimmed and cared for with more or less attention, according to the tastes or habits of the proprietors. I observed generally that those in the Milanese were kept in better condition than either in Piedmont or the other provinces of Lombardy.

"Passing through sand strata, the jets are liable to obstruction from the matter the water carries with it, or from collections of earthy or vegetable deposits in the head. It is therefore necessary to protect them from these, and this is done by enclosing each jet within a wooden tube of variable depth, but generally ranging between 6 and 10 feet. This tube, of about 3 or 4 feet in diameter, slightly conical in form, is made of alder-wood or oak, about 1 or 1½ inches in thickness, bound with three or four strong iron hoops. It is, in short, very like a cask without a bottom, and it is forced into the soil until its upper rim is a few inches above the surface of the water in the spring-head. On one side of the rim a small cut is made, through which the discharge of the water takes place, and occasionally, though rarely, the top is covered over.

"The excavation of the flooring of the spring-head is carried to a depth of about 1 foot or 18 inches below the level at which the jets are distinctly visible. The sides are either formed in long slopes and sown with grass, or, as in many instances I saw near Milan, neatly rivetted with small piles and planks, and having trees planted all round, so that the spring-heads are generally rather picturesque spots. Occasionally it happens that the entire basin becomes filled with water, in such a manner as to make it difficult to discover the localities of the jets. My kind friend, Mr Charles Noe, drew my attention to an example of this kind in the Vercellese, and informed me that the method usually employed to discover the jets was to leave the water undisturbed for some time, when the surface became covered with water-cresses, which were invariably much denser over the sites of the jets than anywhere else. I saw this very distinctly in the case in question, and the indication is considered on the spot an almost infallible one.

"When the dimensions of the head and consequent volume of water have been tolerably well ascertained, the next process is to excavate the channel of irrigation (locally termed the *asta de fontana*). Regarding this there is nothing special to remark, as it is merely an ordinary channel, having dimensions and slopes varying according to local circumstances. When the soil is very porous, wooden or metal pipes are occasionally in use to economise the water, though it is but rarely that such expedients have to be employed."

He states further:—"The vicinity of Milan is remarkable for its abundance of irrigating springs, or *fontanili*—some of very considerable dimensions. I was taken by M. Mazzeri, one of the most eminent of the Milanese engineer-advocates, whose acquaintance I had the pleasure of making, to see one situated about two miles from Milan, which, as an example of the class of works, may deserve a brief notice. Its head was formed by an excavation about 200 feet long, 100 wide, and about 8 deep. At this depth the water-bearing stratum was reached; and over the surface thus laid open no less than forty-two separate springs, each enclosed within its wooden case, were to be seen throwing their small supplies into the main reservoir. The united discharge of the whole amounted to the considerable quantity of nearly 12 cubic feet per second, which, at the ordinary value of water on the spot, was worth in all nearly £4000. Some of the wooden tubes, or *tinelli*, as they were termed, were of unusually large dimensions, being as much as 8 feet in diameter. But this is in practice found to be too much, the expense of sinking them being excessive, as only one man can work at a time with an instrument which is like a magnified hoe, and is, in fact, precisely the same as the Indian *jham*, by which the undersinking of masonary wells, under circumstances analogous to those of the Milanese *tinelli*, is effected. M. Mazzeri informed me, that to sink one of these huge tubes to the necessary depth of about 12 feet, through the hard gravelly soil, required about five months' daily labour. The sides of the excavation for the fountain-head were neatly finished, turfed, and, where necessary, rivetted with planks. Trees were planted round the edges; and the whole place had a pleasant, fresh, and rather picturesque look, with its clear, pure stream, drooping foliage, and constant gurgling of the spring waters."

The supplies of water procured by sinking wells are obtained by tapping such subterranean reservoirs, from which the water flows into the vacant receptacle often in small thread-like currents—in number, numberless—upon the same principle that water collects in a hollowed basin in the sand; by what are known as Abyssinian wells such supplies may be obtained from greater depths than those to which wells are usually sunk; and by Artesian wells, if it were thought expedient to sink, such supplies might possibly be obtained from a still greater depth.

I believe it will be many years before the Colony will derive from Artesian wells the benefits they may be likely to yield. But still these subterranean treasures may be reckoned among the available supplies of water to which the colonists have access for employment in the promotion of vegetation and the modification of the existing aridity of the country, and the arrest of the continuously progressive advancing desiccation from which they suffer.

But it holds equally true of what may thus be obtained, as of what is naturally supplied by fountains and springs and of the supply which may be obtained from rivers, that it is only a supply brought from a higher-lying region—a portion of the rainfall recovered after having been lost to the land, and a gain to the district in which it has been recovered; but not an absolute addition to the water supply of the country.

CHAPTER VI.

SUPPLY OF MOISTURE IN THE SEA.

THE sea is generally considered the primary source of all the waters which are in the heavens above, and in the earth beneath, and of the waters which are under the earth; and I am not prepared to deny that it is so, or to question the truth of the allegation. Rivers are fed by the rainfall; the rain comes from the clouds; and clouds come from the sea. But whither go the rivers? Back to the sea. Whence they came, thither they go. But there arises yet another question—Whence came the sea?

A circle has neither beginning nor end, and the water seems to circulate in a circle. But it would apparently be more correct to speak of the sea as the reservoir rather than the source; and traced chronologically the atmosphere rather than the sea should be considered the source, for there the water existed long before it found a resting place in the ocean basin. On this subject the very ancient cosmogomy embodied in the Hebrew Scriptures, like a fossil relic of some more ancient time, and the discoveries of modern science are at one. According to that very ancient cosmogomy there was a time when "the earth was without form and void, and darkness was upon the face of the deep. And" mighty winds like unto the breath of God, or, as it stands in our version, "the spirit of God, moved upon the face of the waters. . . . And God said, Let there be a firmament in the midst of the waters, and let it divide the waters from the waters. And God made the firmament, and divided the waters which were under the firmament from the waters which were above the firmament. And God called the firmament Heaven. And God said, Let the waters under the firmament be gathered together into one place, and let the dry land appear, and it was so. And God called the dry land Earth, and the gathering of the waters called he Seas."

In illustration of the accordance of this old, old statement with the latest discoveries of modern science, I cite these as embodied in two papers, headed "When the Sea was Young," which appeared last year (1876) in the numbers of the *Cornhill Magazine* for August and October.

From these it may be gathered that, according to deductions drawn from observations of phenomena which were for a time considered anomalous, Saturn and Jupiter are in a condition similar to the description here given of the earth, while as yet a far-reaching atmosphere of condensed vapour, rent as if by mighty winds, enveloped the earth, and from which the condensed vapour was only being, by an enormously protracted process oft interrupted, but again renewed, precipitated upon the earth, leaving what is called a firmament, forming a marked division between the waters which were beneath it from the waters which were above it. And the solid earth was invisible, while the atmosphere which constituted its outer envelope was tossed and torn by winds, and it was without form and void.

Of these planets it is alleged, as intimated, that we see not, and cannot see, except at intervals upon which we cannot calculate, the matter of the orb itself whether solid or in a state of fusion, but only the reflection of the sun's light from the atmosphere of clouds surrounding it—an atmosphere measuring not miles but thousands of miles in thickness around the nucleus enclosed, and this rent at times as if blown aside by some mighty rushing wind, such as is now never is experienced or witnessed on earth.

From the papers cited we learn that it is recorded by Admiral Smyth, one of the most careful and skilful of modern astronomers :—" On Thursday, June 26, 1828, the moon being nearly full and the evening extremely fine, I was watching the second satellite of Jupiter, as it gradually approached to transit its [the planet's] disc. My instrument was an excellent refracter of 3¼ inches aperture, and five feet focal length, with a power of one hundred. It appeared in contact at about half-past ten by inference, and for some minutes remained on the edge of the limb" (that is on the outside of the disc), " presenting an appearance not unlike that of the lunar mountains coming into view in the first quarter of the moon, until it finally disappeared on the body of the planet. At least twelve or fifteen minutes must have elapsed, when, accidentally turning to Jupiter again, I perceived the satellite *outside the disc.* It was in the same position," as to level, " where it remained distinctly visible for at least four minutes, and then suddenly vanished." And he adds :—" As I had observed the phenomena of Jupiter and his satellites for many years, without any remarkable irregularites, I could not but imagine that some optical or other error prevailed, especially as the satellite was on this" (*i.e.*, the hither) " side of the planet. . . . A few days afterwards I received a letter from Mr Maclear, Biggleswade," afterwards Sir

Thos. Maclear, of the Royal of Observatory, Capetown, "informing me that he also observed the same, but that he had considered it a 'Kichener's wonder'" (Kichener, the telescopist, having been apt to recount every optical illusion by which he was perplexed as a real phenomenon). "And, about the same time, Dr Pearson, having favoured me with a visit, asked me whether I had noticed anything remarkable on the 26th; for that he had, in accidentally looking at Jupiter, seen the second satellite re-appear! Here, then, were three observers, at distant stations, with telescopes of different apertures, all positive as to the extraordinary deviation from rule. It may be borne in view that Biggleswade is twelve miles from Bedford," the place of Smyth's observatory; and South Kilworth, Dr Pearson's residence is thirty-five. Mr Maclear's telescope was rather smaller than Admiral Smyth's; while Dr Pearson's was a much more powerful instrument, twelve feet long and nearly seven inches in aperture.

The facts of the phenomenon were satisfactorily attested, and they were accepted, but they were inexplicable. Different hypotheses were tried, but they were found inapplicable or unsatisfactory. "There remained," says the writer I quote, "only one possible interpretation—viz., that the outline of Jupiter's disc had changed in position." And of this interpretation he now, after a lapse of half a century, avails himself to explain what may have happened on our earth "when the sea was young." In doing this he assumes, assigning satisfactory reasons for the assumption, that it was not the solid nucleus of the planet but the atmosphere of clouds surrounding it, the outline of which, on the occasion in question, had been disturbed. And he proceeds to show what the depth of that atmosphere must have been.

In writing of the extent of change which would be supposed satisfactorily to account for the phenomena, he says:—"Smyth may have been mistaken as to the true intervals he mentions in his account, since he does not seem to have taken them from the clock. The interval which he supposed to have lasted twelve or fifteen minutes, may in reality not have lasted more than five or six; and the time during which, after re-appearing, the satellite continued visible, may not have lasted more than two minutes instead of four, as roughly estimated. But taking only eight minutes as the total interval between the first and second disappearance, . . . in eight minutes the satellite would travel about 4,000 miles, and the outline of Jupiter must have changed by that amount, seeing that at the first disappearance the visual line of the satellite just touched the planet's apparent edge, while at the second disappearance the visual line to the second portion of the satellite, 4,000 miles from the first, touched the planet's edge in its now changed position. Probably the difference was even greater; Smyth's over-estimate of the time would make it at least 8,000 miles; but 4,000 miles will be enough to deal with." And he says further:—"It is not necessary to suppose that the planet's apparent outline *as ordinarily seen* shrank inwards by the whole of the amount, the amount required to account for the time during which the satellite was seen on its re-appearance, during which time the satellite must have travelled onwards some 4,000 miles. . . More probably the outline bulged beyond its normal position at the time of the first disappearance, and presently shrank below its normal position, bringing the satellite again into view, and remaining thus depressed until the second disappearance had taken place. . . .

"Two thousand miles being less than the fortieth part of the diameter of Jupiter, we can readily understand why even so enormous an apparent expansion or contraction should not have noticably affected the symmetry of the planet's apparent figure."

In a previous paper* he had shown that "*caeteris paribus*, the atmosphere of Jupiter, would be much shallower, layer for layer, than that of our earth, simply because the planet's mighty attractive power would more strongly compress it. That it is manifestly not thus compressed indicates the intensity of the heat pervading its whole extent. But that it should range to a height of thousands of miles above the true surface of the planet does certainly seem at first amazing. Yet be it remembered that not only is such an inference demonstrably correct, but it also follows necessarily from the comparison instituted between Jupiter and the earth in respect of mass and density. If we assign to the solid globe of Jupiter the same mean density as the earth has—or rather, if we imagine the totality of material, whence, millions of years hence, his solid globe is to be formed, gathered into a globe having the same mean density as the earth—we find for this globe a diameter of 53,000 miles, less than his apparent diameter by nearly 32,000 miles; so that the level of his surface in that condition would be 16,000 miles below his present surface, the space between the two surfaces, or the total shrinkage of Jupiter's volume, amounting to about 930 times the volume of this earth on which we live. As we have every reason to believe that (in a general sense) all the planets are constructed of the same materials,

* *Cornhill Magazine* for May, 1872.

not very differently proportioned, we are compelled to admit this vast expansion of Jupiter's present dimensions, and can therefore very well understand even such mighty changes of apparent surface-level, as the observation of Admiral Smyth, Sir T. Maclear, and Dr. Pearson certainly shows to have taken place."

Numerous observations of Jupiter, Saturn, and others of the planets, are adduced to prove that what we see of these larger planets are clouds in which they are enveloped to an immense thickness; and it is alleged that so must it erstwhile have been with the earth. "One fact in the past history of our earth," says he, "stands out with unmistakable distinctness. The whole frame of the globe on which we live, and move, and have our being, was once glowing with intense heat. Whether we consider the earth's frame with the geologist, or study with the astronomer the nature of the planets' movements and the evidence so afforded respecting prior conditions of the solar system, we are alike forced to this conclusion. At a very remote period the whole substance of the earth must have been molten with intensity of heat; at a still more remote period the whole of that substance must have been gaseous with a heat still more intense; and these stages of the earth's history, remote though they were, and continuing so long that, according to our modes of measuring time, they were practically everlasting, were yet but two among a series of eras whose real number, no doubt, was to all intents and purposes *infinite.*

"Now when we go back to even the nearer of those two eras we find that we must conceive of our ocean during that era as utterly unlike the seas which now encompass the earth. Its substance was the same, or nearly so, but its condition must have been altogether different. No water could for a moment rest upon the intensely hot surface of a globe raging with heat exceeding that of a smelting furnace. There could not have been during that era oceans of liquid water, though all the water of our present oceans surrounded the earth then as now. The water must at that time have existed in the form of mixed vapour and cloud; that is, it must have been spread through the air partly as pure aqueous vapour, and partly in those aggregations of minute liquid globules and vesicles of water forming visible cloud-masses. There must also at that time, as now, have been various kinds of cloud-forms—an outside layer consisting of the light feathery cirrus clouds, below that a layer of the cumulus or 'woolpack' clouds, and below that again a deep layer of the densest nimbus or rain-clouds, from which perfect sheets of rain must at all times have been falling; not, however, to reach the glowing surface of the earth, but to be vaporised in their fall, and in the form of vapour to pass upwards again. We say that all this *must* have been; because, in point of fact, however doubtful we may feel as to many details of the earth's condition in the remote era we are considering, there can be no doubt whatever as to the general facts indicated above. We have only to inquire what would happen at the present day if the earth's whole frame were to be gradually heated until at last the surface glowed with a heat equal to that of white-hot iron, to perceive that, whatever changes might take place, the ocean certainly would be entirely evaporated—boiled off, so to speak. But the water thus added to the earth's atmospheric envelope in the form of vapour could not possibly remain *wholly* in that form. At a great distance from the glowing earth the aqueous vapour would find a cooler region, and higher still would be exposed to the actual cold of space. Hence there would follow inevitably the formation of clouds of the various orders, *cirrus, cumulus,* and *nimbus,* not probably in absolutely distinct layers, but the *cirrus* commingled with the *cumulus,* the *cumulus* with the *nimbus,* and the whole series of cloud-layers affected by the most violent disturbances, partly from the continual rushing upwards of freshly-formed vapour, partly from the continual rarefactions and condensations of the air under the varying conditions to which it would be subjected through the continual changes of the watery envelope. For at every change from the form of pure aqueous vapour to the cloud-form, an enormous amount of heat would be developed, while corresponding quantities of heat would be withdrawn in vaporising other masses of watery matter. The depth of the atmospheric region throughout which these stupendous processes were continually in progress must far have exceeded the depth of the cloud-regions of our own atmosphere. For the same heat which prevented the water from resting on the earth's surface must have prevented the heavier rain-clouds from approaching within many miles of that surface without being turned into pure aqueous vapour. Again, not only would the layer of rain-clouds, thus raised many miles above the earth's surface, be also many miles in depth, but the heat prevailing throughout the layer would in turn prevent a layer of cumulus clouds from being formed, except at a great height above the rain-cloud layer. In like manner the cirrus or snow-cloud layer would be raised high above the layer of the cumulus clouds. And each of these layers, besides being separated from the next below by a deep intermediate space of commingled cloud-forms, would also be of great thickness. Hence we may fairly assume that the extreme range of the lightest and highest clouds in that era of the earth's history must have been many miles from the earth's surface, even if the atmosphere then contained no greater amount of matter (other than its watery constituents) than at present. But we have reason for believing that, besides the oxygen and nitrogen now present in the air, there must have been at that remote era enormous quantities of carbonic, chloric, and sulphurous gases besides an excess of oxygen; and all these, with the aqueous vapour (alone far exceeding the entire present atmosphere of the earth), expanded by a tremendous heat. This heavily-loaded atmosphere must therefore have extended much farther, we may even say *many times* farther, from the earth than her present aërial envelope. It is not at all unlikely that the outermost part of the cloud-envelope was then several hundred miles from the earth's surface, itself raised, through the expansive effects of heat, many miles above the level it was to assume when cooled. In attempting, indeed, to conceive the effects produced by that tremendous heat with which, most certainly, the whole frame of our earth was once instinct, we are far more likely to fall short of the reality than to exceed it, partly because the physical processes concerned are so far beyond our ordinary experience, but much more because they operated on so inconceivably vast a scale."

The writer, drawing inferences from phenomena seen in Jupiter and Saturn, and others of the planets, justifies the postulate that, though, according to the nebular hypothesis of Laplace, the larger planets must be of greater age than the earth, they may still be in the condition in which the earth was in its young days. And after citing and comparing numbers and calculations, of which he says,—"The definiteness of such statements as these makes them more attractive to many than mere general statements, but they cannot be relied on;" he adds:—"All that can be safely alleged—and manifestly so much *can* be safely alleged—is that planets like Jupiter and Saturn, exceeding the earth enormously in quantity of matter, must have required far longer periods of time for the various stages of planetary development, and must consequently be as yet far less advanced towards planetary maturity. It follows, equally, of course, that bodies like Mars, Mercury, and the Moon, as well as the moons of Jupiter, Saturn, and Uranus, being so much less than the earth in mass, must require much less time for the various stages of their development, and may be regarded as having probably long since passed the era corresponding to that through which the earth is now passing."

In accordance with these views, it is the Moon we should study if we would discover something of what may be the condition of our earth when there shall literally "be no more sea;" but, as remarked by the writer, "It would be, therefore, to Saturn and Jupiter that the telescopist would turn for indications of the existence of ocean waters in the state wherein our own ocean must once have existed." And he goes on to say:—"Without entering here at length into the evidence relating to the age of the planets Jupiter and Saturn, or rather to their present stage of development, we shall consider how their appearance corresponds with that which the earth must be supposed to have presented when the waters now forming her oceans enveloped her in the form of commingled, vaporous, and cloudy masses.

"We have seen that at that remote epoch the earth must not only have been completely cloud-enwrapped, but that the innermost of her cloud-layers must have been raised hundreds of miles from her real surface. . . .

"It is next to be noticed that certain very striking phenomena would result from the great depth of the earth's vapour-laden and cloud-laden atmosphere, disturbed not only by tremendous hurricanes moving horizontally, but also by vertical movements of great energy and velocity. Conceive the descent of vast sheets of water towards some intensely-heated portion of the earth's surface, and the effect of their rapid conversion into vapour. The mass of vapour thus formed, being much lighter than the surrounding atmosphere, would rise just as heated air from a chimney rises in the surrounding cooler and therefore heavier air; only with much greater rapidity, because the vapour of water is far lighter than heated air, and the atmosphere of the remote period we are considering was far denser than our present air. The mass of vapour would rush upwards to an enormous height in a very short time, and, coming from a region relatively near the centre of the earth to a region farther away, it would be affected by the difference in the rate of rotational movement at these different levels. For instance, at the present surface of the equator the movement due to rotation has a velocity of rather more than a thousand miles an hour, while at a height of a hundred miles above the surface the air is carried round with a velocity twenty-five miles greater per hour. If, then, a body or a mass of vapour were shot upwards from the equator to a height of a hundred miles, it would, while at that height, lag behind the surrounding parts of the air, and, in fact, would travel backwards at the rate of twenty-five miles an hour.

"If the matter propelled upwards were vaporous, and when at the higher level became condensed into cloud, a trail of clouds would be formed along a latitude-parallel, and, as observed from some other planet, the earth would appear to be girt round by a whitish band parallel to the equator. The deeper the envelope of mixed vapour and cloud, the more readily would such bands form; and remembering the tremendous energy of the causes at work, the whole frame of the earth glowing with intensest heat, and keeping the whole mass of water now forming our oceans in the form of mixed cloud and vapour, we cannot doubt that well-marked belts must almost at all times have existed in the earth's cloud-envelope. The earth, then, would have appeared as *a belted planet*, resembling the planet Jupiter (or Saturn without his rings), but on a miniature scale. It is, indeed, common enough to find the belted aspect of Jupiter and Saturn compared with the probable present aspect of the earth, because of the existence of a zone of calms near the equator, bounded on the north and south by the trade-wind zones, and these in their turn by the zones of the counter-trades. But there is not the slightest reason for supposing that these so-called zones could be recognised by an observer viewing the earth from without. Still less reason is there for supposing that they would, even if recognisable, resemble in the remotest degree the well-defined bands surrounding the globes of Saturn and Jupiter. Such as they are, too, they would be found obeying the influence of the sun as the ruler of the day and also of the seasons; they would be also limited to sea-covered regions; and, in fine, they would correspond much more nearly with the appearances presented by the planet Mars (where occasionally for a few hours portions of bands, not complete zones, are seen across the Martian seas) than with anything shown on the discs of Jupiter and Saturn. What we see on these giant planets corresponds closely, however, with what we should expect to find in the case of planets whose vapour-laden and cloud-laden atmospheres are so deep as to form a considerable portion of the disc seen and measured by astronomers. For the belts of these giant planets show no dependence whatever upon the progress of day and night, or of the long years of Saturn and Jupiter, but behave in all respects as if generated by forces residing in the planets themselves; their well-defined shapes also corresponding exactly with what we should expect from the mode of formation indicated above.

"But, returning to the earth, it is manifest that cloud-belts formed in the way we have described would not be permanent. Sometimes they might continue for several weeks, sometimes perhaps even for months; but frequently they would be formed in a few hours, and last but for a few days, or not even, perhaps, for an entire day. So that the belts of the planet earth, viewed in those times from some remote world, would present changes of appearances, sometimes occurring slowly, sometimes rapidly. Now this precisely corresponds with what is observed in the case of the belted planets Jupiter and Saturn. Sometimes the belts remain, though undergoing constant changes of form, for weeks or months together, while sometimes they vanish very soon after their formation.

"Again, it is clear that other changes than the formation or dissipation of cloud-belts would affect the deep cloud-laden atmosphere of the planet. Hurricanes and tornadoes would rage from time to time, and sometimes for long periods together, in an atmosphere where processes of evaporation and condensation, with all the rapid variations of temperature occasioned by them, were continually taking place on a scale compared with which that of the most tremendous tropical storm on the earth in our time is utterly insignificant. The effects of such hurricanes and whirling storms would be visible from without through the displacement of the great cloud-masses forming the belts. Sometimes cylonic storms would produce great circular openings in the cloud-belts, through which the darker depths below would be brought into view. These openings would be visible from without as dark spots on the lighter background of the belts. At other times the uprush of columns of heated vapour, condensing as soon as it reached the higher regions of the planet's atmosphere, would cause the appearance (to an observer outside the earth) of rounded masses of cloud, which, because of their strong reflective power, would seem like spots of white upon the background even of a light belt, and show still more markedly if they appeared above one of the dusky bands corresponding to lower cloud-levels. And besides changes due to great disturbances and rapid movements in the cloud-envelopes, the changes resulting from evaporation and condensation proceeding quietly over extensive portions of these cloud-regions, would be discernible from without. The observer would see dark spaces rapidly forming, where some higher cloud-mass which had been reflecting the sun's light brightly, evaporated, and so allowed part of a lower cloud-layer to be seen. Where the reverse process took place, large masses of transparent aqueous vapour rapidly condensing into cloud, the formation of bright spots would be observed. How closely all this corresponds with what now takes place in the deep vapour-laden atmosphere of Jupiter will appear from the following account by South of the appearance and rapid disappearance of an enormous dark spot on one of the belts of Jupiter: 'On June 3, 1839, I saw with my large achromatic, immediately below the lowest [edge] of the principal belt of Jupiter, a spot larger than I had seen before; it was of a dark colour, but certainly not absolutely black. I estimated it at a fourth of the planet's "longer" diameter. I showed it to some gentlemen who were present; its enormous extent was such that, on my wishing to have a portrait of it, one of the gentlemen, who was a good draughtsman, kindly undertook to draw me one; whilst I, on the other hand, extremely desirous that its actual magnitude should not rest on estimation, proposed, on account of the scandalous unsteadiness of the large instrument, to measure it with a telescope five feet in length.

'Having obtained for my companion the necessary drawing instruments, I went to work, he preparing himself to commence his. On my looking, however, into the telescope of five feet, I was astonished to find that the large dark spot, except at its eastern and western extremities, had become much whiter than any of the other parts of the planet, and' in thirty-four minutes from the first observation, 'these miserable scraps' (that is, the two extremities of the original spot) 'were the only remains of a spot which, but a few minutes before, had extended over at least 22,000 miles.' . . .

"Next, we may note yet another remarkable feature which the earth must have presented to observers on other worlds during the first stage of our ocean's history. With an atmosphere so deep as she then had, in which many layers of clouds were floating at various depths, it could not but happen that from time to time such changes would take place, either by the rapid appearance or by the rapid disappearance of extensive cloud-masses at high levels, that her shape would seem to be distorted. Indeed, this is only supposing that from time to time high cloud-layers formed or vanished in a part of the earth's atmosphere chancing at the moment to form a portion of the *outline* of her visible disc, instead of forming part of a belt in the mid portions of the disc. Accordingly, to an observer viewing the earth from without, her shape would not always appear perfectly circular, or rather of that figure almost circular, but very slightly elliptical, which in those remote times, as now, must have corresponded to the proportions of her real globe. Cloud-layers floating very high in the earth's extensive atmosphere would cause her disc to bulge out slightly but perceptibly, if they chanced to be so placed as to form the outline of that disc, while regions where for a while the higher layers were wanting would (under the same circumstances) appear slightly depressed below the mean outline of the disc. It might very well happen that these irregularities would usually be too minute to be detected; that effect called eradiation, which slightly expands the apparent outline of every bright object seen on a dark back-ground, would go far to hide such peculiarities. Yet sometimes they would be too marked, probably, to escape notice, supposing only the observer's station were well placed for the observation of the earth; as, for instance, if at that remote time there were creatures living on the moon, and able to examine the earth from that convenient distance. Especially when it chanced that raised portions of the earth's outline lay between two depressed portions, or a depressed portion between two raised portions, the observer would have a good opportunity of recognising the irregularity so resulting. He would perceive in one case that the outline had two somewhat flattened parts with a sort of corner between them, while in the second case there would be flattening between two corners. Of course, in neither case would the corners or the flattened parts be well marked; they would, in fact, only be just discernible by the most scrutinising observation. It might, however, have happened at times that whole zones of cloud-layers would lie higher than usual, while adjacent to them were zones where only the lower cloud-layers were formed for the time being. During such periods the whole disc would appear out of shape, at least to very keen vision.

"Now, precisely such peculiarities have been recognised in the case of Jupiter and Saturn, the two planets which, as already seen, we should expect from *à priori* considerations to be in the cloud enveloped condition."

And again he says, in conclusion :—"It appears to us, in fine, that all the evidence, both *à priori* and *à posteriori*, correspond with the theory, which we have brought before the reader, that a planet during its extreme youth has its oceans floating in the form of cloud-masses and cloud-layers in a very deep atmosphere. We have seen reason, first, for believing that the intense heat of a planet, for many ages after its first formation, would keep the ocean in this cloud-like condition. Then, looking around for planets such as we might suppose to be much younger than the earth, we have seen that Jupiter and Saturn, the giant planets of the solar system, are probably the youngest (in this sense), always excepting the Sun, which is in an earlier stage than any member of his family. And, considering what appearances a planet with a very deep cloud-laden atmosphere might be expected to present, we have found that just such appearances are presented by the planets Jupiter and Saturn, the phenomenon described not being seen at all times but only occasionally, and in varying degree, precisely as we should expect from the varying causes producing them. We have also seen that the small density of the giant planets cannot readily be otherwise explained than by the theory that we do not see their real surface, but the outer-surface of cloud-layers enveloping them. Moreover, while not a single fact known about the great planets is opposed to this theory, there are some facts, as we have seen, which cannot be *possibly* explained on any other theory. But when so much as this can be said of any theory, the theory may be regarded as established.

"When the earth and the sea were young, then the earth's whole frame was intensely heated. Her real surface was doubtless partly solid and partly liquid then as now; but the solid portion glowed with ruddy, and, in some places, with white heat, while the liquid portions, instead of being water as now, were formed of molten rock. Above this surface, with its 'tracts of fluent heat,' was the fiery atmosphere of that primeval time, enormously deep, complex in constitution, bearing enormous masses of aqueous vapour, and every form of cloud and cloud-layer swept by mighty hurricanes whose breath was flame, drenched with showers so heavy that they might rather be called floods, and tortured by the uprush of the vaporous masses formed as these floods fell hissing on the earth's fiery surface.

"After myriads of centuries came the time when the surface so far cooled as no longer to glow with ruddy light, and no longer to reject by vaporising the waters which fell upon it. Then a fearful darkness prevailed beneath the still mighty canopy of cloud; for only little by little, by very slow degrees, would the water descend upon the earth's surface."

And in making these remarks he calls attention to the accordance of the alleged phenomena with the ancient statement,—"The earth was without form and void, and darkness was upon the face of the earth;" and light following the separation of waters under the firmament from waters above the firmament, the waters under the firmament being next gathered together into one place—the gathering together of the waters into one place being called Seas.

Such appears to have been the genesis of the sea, which thus appears to be but the reservoir of moisture from the atmosphere in excess of what at the present temperature of the earth and its atmosphere that atmosphere can retain diffused,—ever replenishing this as atmospheric disturbances occasion precipitation, but receiving again, by rivers innumerable, the drainage of the lands drenched by the rain, and of the mountains covered with the snow as that snow melts again under the solar rays.

About three fourths of the whole surface of the earth is taken up as a reservoir for ocean waters. "It appears that, till within the last few years, a continuous series of soundings in deep water was a difficult operation, every sounding costing the ship a fresh line. This difficulty has been conquered by the Americans. The weight used in the process, on touching the bottom, becomes detached, and the line is then drawn back with ease. The depths of the Pacific are as yet imperfectly explored; those of the Atlantic, however, have been better ascertained; and from Maury's map we may derive a good general idea of the shoals and abysses of this great sea bed. The deepest depression of the Atlantic basin seems to lie between 30° and 40° N. lat., where the plummet (though we must allow for errors occasioned by the possible deflection of the line) has been lowered to the depth of 30,000 and even 40,000 feet. At short distances from Madeira, the Cape de Verd Islands, and the Bermudas, the sea deepens to 12,000 and 15,000 feet; so that, seen from the ocean ground, yon isle-clusters would appear as the summits of mighty mountain-lands, grand and imposing as the Andes. Between Ireland and Cumberland, the depth seems nowhere to exceed 11,000 feet. The enclosed European seas are comparatively very shallow. The depth of the Baltic seldom exceeds 250 feet, and there is only one spot where the sounding line finds a depression of 840 feet. Between the Orkneys and Norway, the North Sea has its maximum depth of 800 feet. The Mediterranean, in some parts, attains a depth of more than 6,000 feet; the Black Sea, with the same reservation, a depth of more than 3,000; while the waters of the Adriatic everywhere roll over a shallow bed."

In accordance with these facts are the observations made by the scientific expedition in the Challenger. The distinguished naturalists sent out on that expedition were furnished with every appliance they desired for the prosecution of their enterprise; and with these they were sent out to investigate the physical and biological condition of the great ocean basin; but under these general headings were certain more minute instructions. They were instructed to traverse the Atlantic and Pacific Oceans and the Southern Sea so far as that was possible. They were instructed to select certain stations and to determine certain points. They were to determine in the first place the exact position of the station with the appliances at their disposal. They were to determine the precise depth. They were to bring up, by means of the apparatus, a certain amount of the material at the bottom, for examination and for chemical analysis. They were to bring up a specimen of the water from the bottom for analysis and examination, to determine the bottom temperature with accuracy, and the temperature of the sea at different levels from the surface and the bottom; to get specimens, if possible, of the sea water from various depths; and lastly, by using the dredge or any other instrument answering the purpose, to get, as they went along, a certain amount of material from the bottom, and of the creatures which inhabited the bottom at that particular locality, and to endeavour to get a general idea of fauna inhabiting the sea. The Challenger left Sheerness on 7th December, 1872. Crossing the Atlantic four times during the year 1873, and running a course of nearly 20,000 miles, they established 150 observing stations, at each of which, without exception, all the required observations were made. In 1874 they went southward to the Cape of Good Hope, and spent a month there. Going beyond the Antarctic Circle as far as possible, the Challenger then traversed the seas of Australia and New Zealand, and made some stay in the Malay Archipelago, they finally arrived at Hong-Kong, after having run a course of 17,000 miles, along which 66 stations had been established. In 1875 the Challenger traversed the Pacific in a course of 20,000 miles with 100 stations. In the early part of 1876 she crossed the Atlantic for the fifth time, filling up here and there blanks in the earlier observations suggested by increased experience, and reached England on 24th May in that year. The cruise on the whole proved a singularly fruitful one. It was only occasionally under very exceptional circumstances that they were prevented by the weather from doing their work. The health of the party was generally good, and the loss by death was exceptionally small. As to the contour of the sea bottom, it appeared that the average depth of the ocean was somewhere about 2000 fathoms —probably 2500 fathoms. A very large portion of the ocean had a depth somewhat less than this, but the depth of 2000 fathoms appeared to be extremely general. Where they had 2500 and 3000 fathoms they appeared to be getting into submarine valleys, except in the Pacific, where they had an enormous extension of water of a very great depth, in many cases going beyond 3000 fathoms. In the Atlantic they had a great portion of the northern part at a depth of somewhere about 2000 fathoms. Passing southward, they had a ridge running down the middle of the Atlantic, where the depth was less; but beyond that ridge they found a depth of 3000 fathoms.

It follows that we have in the sea a reservoir of water covering in round numbers three fourths of the entire surface of the earth to an average depth of about two miles, but in some places well nigh four miles in depth.

The sea is not only a body of water, but that water is highly charged with brine, containing salt (chloride of sodium), sulphuric acid, magnesia, soda, lime, and other substances in abundance. Brine forms on the average about 3½ per cent. of sea-water. "Hence," says the writer I have quoted in this chapter so extensively, "if we take the average depth of the ocean at two miles, or roundly 10,000 feet, it follows that if all the water were evaporated there would be left a deposit of salt averaging 350 feet in depth all over the present floor of the sea. This would correspond in quantity to salt covering all the present land surface of the earth to a depth of a 1,000 feet—or to a deposit of *two hundred feet deep over the entire surface of the globe.*" And the question—Whence has come this immense body of salt? may have suggested itself to some of my readers.

There are agencies at work by which some of the constituents of the brine are being constantly withdrawn from the sea. This is being done both by molluscs and by fishes, and by other means. But what is thus withdrawn is probably replaced more or less completely by equivalents dissolved out of the ground and borne thither in river water. Such salts existing in minute quantities may be detected in the waters of many streams; and such salts apparently accumulate in inland seas and ponds and lakes, from which there is no outlet. But the question returns—Whence have come the *immense quantities* of these which exist in the ocean? "The theory that these substances have been washed from the earth's surface by causes such as are now in progress, would not, we think," says the writer I am quoting, "be seriously entertained if the vast amount of matter thus present in the waters of the sea were remembered and considered." And he refers to a paper by Dr Sterry Hunt, an eminent Canadian chemist and geologist, as supplying a much more satisfactory explanation of the case. In this paper, after showing that carbonic acid, chlorine, and sulphurous acids, would be present in enormous quantities in the primeval atmosphere, besides, of course, still vaster quantities of the vapour of water, he proceeds :—" These gases, with nitrogen and an excess of oxygen, would form an atmosphere of great density. In such an atmosphere, condensation would only take place at a temperature far above the present boiling point; and the lower levels of the earth's slowly cooling crust would be drenched with a heated solution of hydrochloric acid [muriatic acid, or spirit of salt] whose decomposing action, aided by its temperature, would be exceedingly rapid. The primitive igneous rocks on which these showers fell probably resembled in composition certain furnace slags or volcanic glasses." The process of decomposition would continue " under the action of the heavy showers until the affinities of the hydrochloric acid were satisfied. Later, larger quantities of sulphuric acid would be formed, and drenched showers of heated solutions of this energetic dissolvent would fall upon the earth's heated crust. After the compounds of sulphur and chlorine had been separated from the air, carbonic acid would still continue to be an important constituent of the atmosphere. It would be gradually diminished in gravity," through chemical processes resulting in the formation of various clays, " while the separated lime, magnesia, and alkalies, changed into bi-carbonates, would be carried down to the sea in a state of solution."

Thus does it appear that the ocean is not only a reservoir of water but a receptacle of all the refuse washed down from the laboratory of Nature; and, in consequence of many of the constituents of the brine, the water of the sea is not available immediately for the promotion of agricultural operations. But as there may be, and apparently there are, fresh water springs in the basin of the sea, fed probably by percolation of the rainfall on dry land, so there may be, and probably there are, fresh water springs on dry land, fed by percolation of water from the sea. There are islands far from the mainland on which there are fresh water springs, and in which fresh water is found by sinking wells to some depth, and these may be thus produced.

It was found by Berzelius that a saline solution filtering through a long tube filled with sand runs out more or less completely deprived of salt.

Matteucci, who carefully investigated the effects of passing liquids through organic tissues, confirmed the discovery of Berzelius by using a tube about 26 feet long filled with sand, and he found that the density of the liquid introduced at the upper extremity of the tube was to that of the liquid escaping from the other end, as 1 : ·91* But he found that this difference of density was not always maintained. After a time the saline solution was found to be as dense at its exit from the tube as at its entrance—proving that the decomposition of the solution takes place in the first action of contact between it and the particles of sand.

We can thus account for the origin of many fresh-water springs on islands; and the rising of the water in the spring, or well, may be attributable to the superincumbent pressure on subterranean reservoirs of the sea basin subsiding under the pressure of an ever-increasing deposit of earthy matter brought down by rivers and dispersed by currents, or by pressure on air intervening between the source of the spring and remote reservoirs with which it has communication, or by pressure otherwise produced. But in the evaporation which takes place from the surface of the sea, it is mainly water which is evaporated, leaving the constituents of the brine as a residuum; and it is mainly by the evaporation from its surface, and the dispersion of the vapour far and wide by diffusion in the atmosphere, and the precipitation of it by the chemical affinity of the soil for moisture, and by disturbances of the temperature of the air by the winds, that the sea supplies moisture for the promotion of terrestrial vegetation.

* The results varied with the salt employed—*e.g.*, a solution of carbonate of soda, passing through a tube 10 feet long filled with sand, had its density at its exit as 1·005 : 1.

CHAPTER VII.

Supply of Water Available for Storage for the Use of Man which is at Command.

In view of the meteorological observations which have been brought forward in the first chapter of this part of the treatise, which relates to the sources of the supply of water which is at present available for agricultural operations, and for otherwise compensating for the aridity which prevails in the Colony of the Cape of Good Hope and in regions beyond,—and in view of the supply of moisture in the atmosphere, the supply afforded by the rainfall and by rivers, and by fountains and by subterranean reservoirs, and the supply of moisture in the sea,—it appears that we have only at command for storage for use such portion of what is precipitated from the atmosphere as we can arrest on its return journey to the sea.

Such is the conclusion to which we are brought by a common sense view of the case; and we get no further insight by a scientific consideration of the facts. "All the rivers run into the sea, yet the sea is not full." Thus wrote the Hebrew King, who has been proverbial for his wisdom for three thousand years. And why is the sea, though constantly receiving such supplies, not full? The King gives the reply. Because "unto the place from whence the rivers came thither they return again;" the waters only return to the place from which they went, meanwhile others have followed on the same circuit, and they only fill the place which these have vacated. But if we can arrest any portion on its way we are free to do so, and to cause it to do our bidding.

We have traced the genesis and some parts of the subsequent history of the waters. For what purpose the immense body of waters are stored up, reserved, and kept in store in the basin of the sea, we know not. The history of the future of the treasures of the deep might be more interesting to us than even the history of the past, but this only do we discover, that in the present—and we are of the present, the creatures of a day, and know nothing—it is only droppings of what is passing over us, with a little of what has fallen elsewhere which is at our command. The aqueous vapour in the atmosphere is only the remains, or the equivalent and representative of the remains, of what was precipitated of old and gathered together and called sea, and remains apparently infinitesimally small in quantity in comparison with what was precipitate. What is now precipitated as rain, together with all that is precipitated as dew and hail and snow, is only what happens to be in excess of what can be retained, diffused, or suspended in the atmosphere at the temperature to which a portion of the air has been reduced by the operation of adventitious causes; and what is thus precipitated is as nothing in comparison with what is reserved in the atmosphere. And the delivery of the streams and rivers, whether by gentle flow or by devastating torrents, is again but the portion of the rainfall which was in excess of what the thirsty earth could drink up and retain; and springs and subterranean reservoirs are only supplied by the percolation above what the earth could retain of what it had imbibed: so that fraction of a fraction is all that is devoted to the production of food for man and beast. Combined with lavish expenditure, commensurate with exhaustless supplies, there is an economy which casts all the economy of man into shade.

But while so long as this fraction of a fraction supplies all that is needed for the support of man and beast, all that is flowing past him to the sea is placed at man's command, if he will himself make the arrangements needed to arrest and utilise it, with a promise of far more abundant bonus than is ever flashed before the poor to induce them to begin the practice of saving and providing against a future more or less remote—a bonus consisting largely, but not exclusively, in an increased humidity of atmosphere, and the effect of this in the promotion of vegetation, over and above the direct and immediate benefits, for the sake of which the works of hydraulic engineering in connection with agricultural improvements may be undertaken; and together with this, the physical, intellectual, and moral development which generally accompanies sustained endeavours to surmount difficulties in our way, which, though great, are not insuperable; and the satisfaction which is experienced in serving our generation according to the will of God—like David the king of Israel, of whom this was testified, and like many besides who have had this testimony: "They pleased God."

PART III.

SUPPLY OF WATER AND FACILITIES FOR STORING IT EXISTING AT THE CAPE OF GOOD HOPE.

CHAP. I.—GENERAL SUPPLY OF WATER AND FACILITIES FOR STORAGE OF IT EXISTING IN THE COLONY OF THE CAPE OF GOOD HOPE.

IN the preceding chapters there have been noticed all the supplies of water, or sources of the supply of water, available for agricultural operations. And in view of these it appears that what seems first to demand attention is—What means can be employed to retain a portion, be it great or small, if we cannot retain the whole, of the water now being carried off in floods to the sea? What can be done in the condition of the country, in the circumstances in which the inhabitants who have a personal interest in the adoption of such measures are placed, and in view of the amount of capital, be it great or small, which can be commanded with a reasonable expectation of satisfactory returns—ultimate, if not immediate, indirect, if not direct—being obtained.

Not a little may be done by the conservation and extension of forests.

I have, in a preceding chapter, detailed experiments by which the quantity of water embodied in soils may be determined, and stated that if the soil, treated as is there proposed, be afterwards exposed for some time to what appears to be a dry atmosphere, either within or out of doors, it will regain a great measure if not the whole of its original weight, and a repetition of the original experiment will prove that this must have been by the absorption of moisture from the atmosphere, dry as it appeared to be. The measure of this affinity of soil for moisture is a point to which attention is generally given in analyses of soil, and it is found to vary with the constituents of the soil, and the proportions in which these are found. While an addition of sand to any soil will diminish, an increase of clay or of humus (the product of vegetable decay) will increase the affinity of the soil for moisture; and experiments of Professor Church, at the Royal Agricultural College, Cirencister, adduced in connection with the discussion of this subject, show that by deep and thorough tillage the affinity of the soil for moisture may be increased from what is represented by 19·2 per cent. of water in the soil to what is represented by 28·6, or from two gallons to three gallons in every hundred pounds weight of soil; and by appropriate appliances, as I have attempted to show in one and another of the Memoirs and Annexations to my Report, which I submitted as Colonial Botanist at the Cape, the humidity of the atmosphere, or quantity of water retained in it, can be greatly increased.

This may be reckoned an addition to the supply obtained from the atmosphere in the form of rain; but still more important than this is the effect of trees in arresting and preventing the flowing off of the rainfall to the sea.

More tangible and sensible results may be obtained by sinking wells of greater or less depth to cut off the flow of subterranean currents; and thus might water be obtained for domestic use and for the supply of cattle, and even for irrigation upon a limited scale.

But in sight of river floods, dashing headlong to the sea, all of these means of increasing the supply of water available for agricultural operations appear of secondary importance. And it seems to be natural that we should next inquire—What facilities exist for the storage and utilisation of the water supply in the several districts of the Colony of the Cape of Good Hope and the regions beyond?

In a preceding chapter I have cited my observations and experience in travelling through the Colony on my return to it in 1863—seeing rivers pouring their waters headlong to be lost in the sea, while on the right hand and on the left were what looked like dry basins of former lakes, in which much of the treasure might be stored. Similar were my feelings on my first tour through the Colony, some fifteen years before. In narrating my experience of drought on that journey, in the introduction to "The Hydrology of South Africa," I have stated:—

"Dry and arid as was the Karroo, I was told that within a few days after a thundershower it is clothed with verdure on every spot on which this may have fallen; and if the rain were copious, within perhaps three weeks thereafter it would be studded with flowers, many of them of exquisite beauty, delicate in structure, and brilliant in hue. So copious at times, I was told, are the showers which fall in connection with a thunderstorm, that they deluge the land. Shortly before it had happened, I was told, that the inhabitants of a village which was named to me, but the name of which I have forgotten, were roused during the night by the noise of a rush of waters threatening to carry all before them; and one man stepping to the door to see what it might be, found himself, on crossing the threshold, more than knee-deep in the stream, and scarcely able to maintain his footing against its flow. And I subsequently witnessed what satisfied me of the verisimilitude of what I had heard. I was on one of my journeys overtaken in a Kloof by a thunderstorm, in little more than five minutes the road was a river, the waters rushing along with a rapidity such as is seldom seen in a river-bed, and from six to ten inches in depth from bank to bank.

In accordance with this is the testimony of others. The late Dr Rubidge, of Port Elizabeth, in a paper on irrigation and tree planting, which appeared in a volume entitled *The Cape and its People*, published in 1869, says,—'In October 1866, I passed over the bare plains between the Milk River and Graaffreinnet just after a heavy shower of rain. The "by-road" was running knee-deep, in every hollow was a fine running stream, while the gullies were great torrents now, which in a few hours you could pass dry shod. Yet most of these channels presented spots where much of the water that was rushing uselessly and destructively to the sea might have been stored.

'It was sad to contrast the beautiful orchards, vineyards, and corn fields we had just left, with the dreary monotonous flats, and to reflect that all that was wanting to convert that wilderness into a smiling garden was speeding away to swell the rivers into dangerous torrents, and carrying along with it some of the most fertile soil of the country.'

Similar were my feelings in witnessing what I did on the journey to which I am now referring. While I witnessed what I have detailed, it appeared to me that it was quite practicable greatly to modify the condition of the Colony and of its inhabitants, by a proper storage of the waters which fell from the heavens during the rainy seasons in districts in which these annually occur, and the water which fell in thunder showers, and tropical torrents of rain in districts in which annual rainy seasons were unknown.

It seemed to me that the country offered considerable facilities for storing up such waters. There might be difficulties to be overcome, but I know of nothing great which has been accomplished by man without difficulty. It might be difficult to get labourers; it might be difficult to carry to them needed provisions for their support while engaged in the work; it might be difficult to find the money required; and it might be difficult to do a hundred other things;—but the practical questions resolved themselves into two—Was it practicable? and, Would it pay? On the latter point I was not then, nor am I now, in possession of the data necessary for a solution of the question, and therefore I could not speak. But in answer to the first I could say I see no physical hindrance which may not with reasonable effort be overcome; and on my return to Capetown I communicated to others the impressions I had received of the practicability of greatly modifying the effects produced by the aridity of the climate."

In other tours which I have since made in different districts of the Colony I have been struck with what has also struck others—the facilities afforded, in almost every locality, by the contour of the country, for the storage of water, and for the application of irrigation to the development of the agricultural capabilities of the land.

Inquiring of Mr Bourne, then the Colonial Railway Engineer, to what extent his observations accorded with my own, he wrote to me in reply, under date of 20th June, 1865:—"In reply to your note of to-day's date I beg to assure you that 'valleys and hollows,' suitable for dams for the purpose of irrigation, exist in numbers on every farm that I have seen in the Colony. Along our line of railway surveys they are too numerous to specify.

"I agree with you, that in an average of years there is no lack of water in the form of rain and dew; but it is allowed to run off the land into the sea almost as soon as it falls."

In accordance with this is the following evidence given by him before a Select Committee of the House of Assembly, appointed to consider and report on the Railways Bill:—

"*Mr Munnik.*] Are you sufficiently acquainted with the nature of the country along the line to Malmesbury, its supply of water, and so on?—I do not know any part of the country where water is not to be obtained by the construction of dams, or by boring.

"I mean the nature and character of the soil. You are aware that in this Colony the value of the ground depends upon the quantity of water you can obtain from it?—I do not know any farm in the country that I could not supply with water if it were mine.

"*Mr Molteno.*] Are you aware of the depth they went to at Wellington to get water, and that they were obliged to abandon the well after all, and get their water from the Berg River?—Yes, I am quite aware of it; but I should never dream of boring into such a rock to get water.

"*Mr Munnik.*] Do you know the depth of the well?—I should say I went down about fifty feet, perhaps to where they commenced boring. I do not know the depth of the boring. Mr Heaford can tell you.

"Talking of seeing no difficulty in supplying any part of the country with water, do you mean a sufficient quantity to irrigate ground, so that people could live by the produce of a garden, for instance, so irrigated?—Yes, sufficient water for all purposes can be obtained by the proper means. The Worcester Valley is sufficiently watered by streams from the mountains, but the farmers there told me that it was not worth while growing any quantity of produce as they could not get it to a market at paying rates. The cost of carriage almost equals the market value of the produce in some cases.

"Are you aware of any places along the Malmesbury line sufficiently supplied with water?—There are on the route between Capetown and Malmesbury large numbers of fine farms, producing large quantities of wheat and grain, and I take it for granted they must have water enough to produce it. These farms are not new ones; they have been in existence for a great many years, and they would long ago have been abandoned if they could not have obtained sufficient water.

"The farms alluded to produce chiefly grain, and very little else?—Yes.

"Are you aware that the grain is grown in the winter months, and is entirely dependent upon the rains?—The rains can be collected in dams; there is no occasion to let it run away as they do.

"Is not all cultivation throughout the whole of that part of the country entirely dependent upon rain?—Yes, and the fogs and morning dew, which is almost better than rain.

"Taking into consideration the character of the country along the line, do you think it of such a nature as to be capable of supporting a dense population?—I should say that it would keep a large population. If a number of Chinese were put down there they would be able to produce a large quantity of everything they wanted."

Some six months later I had occasion to write on the subject to Mr Bain, of the Road Department. His reply to me under date of 27th November, 1865, was as follows:—"I have received your letter inquiring for information about the geology of the Colony, etc. A few weeks ago I got a letter from Mr Bourne on the same subject, and I sent him most of my late father's papers on the subject, together with a copy of his map and sections of the Colony, which gives a complete geological view of the Cape of Good Hope as far as Port Natal. When Mr Bourne returns them to me I shall endeavour to let you have them.

"The construction of dams, wells, and opening up of likely places, are intimately connected with geology; and I feel assured you would find a great deal of information, in the documents alluded to, to complete your irrigation schemes. I do not know a part of the Colony better adapted (geologically) for dam making, and containing more latent springs, than the Karroo. Those greenstone dykes—subterraneous wells—which traverse the Karroo in almost every direction, with a small break in them, are very tempting to shut up, perhaps at a cost of a couple of hundred pounds, and thereby form a lake large enough in many instances for small vessels to float; and invariably, for obvious reasons, there is a spring just in the break of the dyke. If not visible, you may safely sink there for water. How I would like to travel about the Colony, with a couple of hundred convicts, to open up springs and make dams at all those inviting places!—and in a few years' time convert the barren Karroo into a fine agricultural country! I may say, with my intimate knowledge of the Colony, I know all the most likely places for such dams. The farmers are too poor to do it themselves. A company or the Government ought to take the matter in hands. Make the whole Colony a producing place first, and then have railways. The soil of the Karroo is unequalled, if there were only the means of irrigation."

Captain Hall, in his manual of South African Geography, states "that in many of the upland valleys near the sources of the rivers of the coast region, such as the Warm Bokkeveld, Twenty-four Rivers, Olifant's River (east and west), Zwagershoek, the valleys of the Sneeuwberg and Great Winterberg, the Kat River, and Division of Queenstown, irrigation is practicable to a great extent; and in many other places, by a scientific system of damming, the available arable land of the Cape Colony could be increased to a great extent. The heads of the Gamka and Dwyka Rivers, and their branches, and the sources of many other streams now draining the Great Karroo, could, by proper damming the narrow valleys in the Nieuweveld and Roggeveld Mountains in which they rise, be made to afford a sufficient supply of water to make the now sterile Gouph in reality what its

Hottentot name denotes—*i.e.*, a fat field. The same observations apply to nearly all the narrow kloofs and valleys of the mountain ranges of South Africa. The water-courses have too great a fall, and the country is pretty much in the condition of a slated roof with good draining;—a heavy shower falls, the pipes sometimes, perhaps, overflow, but the water eventually runs off and leaves the roof dry as before."

And he goes on to say:—"The great majority of the rivers here are periodical—*i.e.*, only flow after heavy rains. A few in the western districts of the Cape Colony, the Breede, Berg, and Olifant's, for example, are permanent. In Great and Little Namaqualand, the Kalihari Desert, and the whole of the region situated on the southern slope of the Nieuweveld and Roggeveld Mountains, whole years may elapse without the phenomenon of a running stream, and yet the magnitude of the dry water-couses of the Buffalo, Hartebeest, and Oup or Borradaile River, all tributaries of the Orange, show how immense must be the torrents that sometimes sweep along them. The writer of this has seen the bed of the Great Fish River perfectly dry, and within twenty-four hours a torrent thirty feet deep and several hundred feet wide was roaring through it. In February, 1848, the Kat River, suddenly rose upwards of fifty feet in the course of a few hours, sweeping seventeen feet above the roadway of a stone bridge at Fort Beaufort, supposed to have been built high enough to leave a clear waterway to the highest flood ever before remembered. The Gamptoos, Gauritz, and all the other rivers draining the Karroo, are also subject to very sudden rises, although generally but dry water channels. The periodical rains falling in the mountains near its sources, between September and March, also swell the Orange River to a great extent, and large portions of land along its banks are then inundated."

In accordance with this testimony is the testimony given by so many others that it seems to be impossible for any observing man, when once his attention has been directed to the subject, to travel in any district of the Colony without having his eye arrested, again and again, by features of the country indicative of the adaptation of the land to the extensive application of works for irrigation in one form or another.

But something more than general impressions are needed before actual operations would be warranted. In the hope of obtaining, not all that is desirable, but important contributions towards this,

I issued a circular in the following terms, under date of 1st June, 1865 :—" I send by this day's post to editors of newspapers a copy of the Report of the Colonial Botanist for 1864, with a request that it may be lent to any gentleman in the neighbourhood desiring it for perusal, and afterwards returned to me, unless it be required for future use. This request I have to make in consequence of the number of copies which can be placed at my disposal being insufficient to enable me to submit the Report to more than a very small portion of those whose views on one or more of the subjects embraced in it, I desire to elicit for the benefit of the Colony.

I shall be glad to hear of any of the statements or suggestions advanced receiving a fair, full, and free discussion. But there are three subjects referred to in the Appendix to the Report to which I solicit your attention, in the hope that you may be able to furnish, either to me or to others, information which may be useful. These are, irrigation, pasture herbs, and bushes.

In a letter to J. A. Merrington, Esq., London, on the sour, the sweet, and the mixed veldts, and the karoo, etc., I have made the following statement :—

'I consider it probable that in the sweet veldt and the karoo there are stored up resources which are inexhaustible. What is chiefly wanted is a supply of moisture. In not a few of the districts over which the sweet veldt and karoo extends, there falls at present enough of rain to render the whole land productive, and by a proper system of arboriculture this supply might be indefinitely increased. The land may be too undulating to admit of a system of universal irrigation in the usual and perhaps proper signification of the term ; but I am led to question whether the usual mode of irrigation be in any circumstances the best which can be employed, and more especially, whether it be in the circumstances of the Colony the most economical which, at this stage of our progress, can be adopted. Taking into account the irregularity of our supply of water, and the mode of operation whereby water promotes the germination and the growth of vegetable productions, I attach much more importance to the presence, in a district, of a large sheet of shallow water, and of a broad, winding, almost stagnant watercourse, from which a free evaporation may take place, than I do to irrigation, as usually practised ; and, paradoxical as it may seem, I should prefer for the bed, both of the reservoir and of the watercourse, a porous to a retentive soil,—that the water might percolate slowly, but extensively, through the ground of the surrounding districts, affecting it, it may be, many miles from the locality. The construction of a dam in such circumstances might not be remunerative to the holder of a very limited extent of land; but, to the holder of an extensive district, it might. And to the community at large, I doubt not, it would be advantageous if suitable land, in the districts of the sweet veldt still belonging to the community and at the command of their representative, the Government, were made reservoirs of such sheets of water, and so constructed as to impede the progress to the sea of all the water which falls on the district in the form of rain or which passes through it from regions beyond. Evaporation from these would lower the temperature and increase the humidity of the atmosphere; and percolation would produce similar effects on the soil.

'Numerous are the valleys and places in which there are indications that at a period not very remote such sheets of water were there, and that the reservoirs of such sheets of water again they might easily be made. Should this be attempted, arrangements might be made at the same time for land lying near to the reservoir being irrigated in the usual way, if this should be considered desirable; and also arrangements for lands lying nearer to the sea, deriving from the rainfall in the upper districts a greater benefit than they at present do from the river-beds proceeding therefrom, which are at one time dry and at another time filled from bank to bank with water rushing on impetuously to be lost in the great abyss. Let this be done, let manure be used as it is in Britain, let drainage be introduced in connection with irrigation, and the difficulty will not be to raise crops, but to get the crops which are raised in the karoo and sweet veldt conveyed to a profitable market, for consumption and exportation.'

And in the same letter, writing of information I had supplied at the desire of my correspondent, I said:—

'Such statements speak to the aridity of the soil and climate; but they also show upon a small scale that we have the means of counteracting this at command.

'The quotations are made from "Report and Proceedings of the Committee of the Legislative Council on Irrigation, published by order of Legislative Council, July, 1862."

'About the same time, there was printed and presented to both Houses of Parliament a "Copy of a circular addressed by the Colonial Secretary to the several Civil Commissioners, and their replies thereto, on the subject of Irrigation."

'In both of these official publications I find much which is in accordance with the views I have been led to form of the difficulty of damming up rivers with a view to irrigation, but of the comparative ease with which dams might be formed by embankments thrown up where two elevated ridges nearly meet, or where a natural hollow exists in a comparatively level plain, or even water be led off from swollen rivers into such plains,—though this would be more difficult.

'I have seen the Tarka, the Fish River, the Keiskamma, and the Buffalo in their might. I have crossed the bed of the former in a box suspended from a rope stretching from trees on the opposite banks, while the river torrent was tearing along below twenty-two feet deep, as ascertained by measurement, where forty-eight hours before the depth was only sixteen inches. I have been told, by a gentleman who had given special attention to the subject, of the mean rise in a number of rivers in the same district being twenty-eight feet; I have been told by the same gentleman of a maximum rise of sixty feet; and I have gone over the scene of devastation occasioned by the sudden rise of a river to a height of seventy feet above its usual level.

'No dam of ordinary structure could withstand such torrents as these. But what the dam in the river cannot effect, the dam on dry land may; it may even bridle the river's speed, by arresting its waters ere they reach its bed.

'The subject to which the attention of the Legislature and of the parties whose views were collected appears to have been principally given, was the practicability or impracticability of employing the water in the irrigation of cultivated fields by means of water-leadings; while the subject which forced itself upon my consideration, when making the tour of the Colony, was the practicability of securing similar results over extensive districts by climatal changes through the creation of artificial lakes and forests.

'This I have had in view in all the suggestions I have made, either in regard to arboriculture or irrigation; but the consummation is in a future so distant that I have left each suggestion to be tried by its adaptation to the wants of the present; so that, in the meantime, though no perceptible climatal change should follow the adoption of my suggestions, the labour and expense incurred in the execution of them will not be thrown away, as the trees planted will be useful as timber or as fuel, and the water stored up will be available for the watering of cattle and the irrigation of land.

'The irrigation I have suggested as applicable to the sweet veldt and the karoo I consider equally applicable to the mixed veldt nearer the coast. Throughout the districts so designated, whatever lack there may be of moisture in the soil and of humidity in the atmosphere, there is not that lack of water which is so characteristic of the sweet veldt or karoo. And what is required is only a more equable distribution of what is rolling onwards towards the sea. Even where there is no river, the rainfall in a series of years is abundant, but it is irregular and periodic, and it must be caught when it falls, and retained; and, in this matter, we have a great deal in our power.'

I shall feel greatly obliged if you can supply or procure for me explicit information in regard to the situation, extent, and depth of hollows or valleys such as are indicated in the first of these passages, and in regard to facilities which exist for carrying out such a scheme of irrigation as I have advocated, or any other, in the district in which you reside, or in any other district with which you are acquainted.

In other letters the subjects of pasture herbs, and bushes, and grasses are adverted to.

I am very desirous of specimens of the twig, leaf, flower, and fruit of bushes characteristic of different parts of the country, with memoranda of their local name, of their adaptation for the support of cattle, horses, mules, sheep, or goats, of the kind of soil on which they grow, and of anything else in regard to the plant deserving of notice.

I would value, also, similar information in regard to grasses.

If the communications addressed to me on any or all of these subjects be made on separate sheets of paper, this will greatly facilitate my arrangement of the information received.

This information may be supplied either in Dutch or English.

If marked 'On Her Majesty's Service,' and sent to me through the Colonial Office, it will come post-free. And specimens in small packets may be forwarded in a similar way.—I am, etc.

P.S.—I have endeavoured in vain to procure the addresses of Land-Surveyors, a class of gentlemen to whom I wished specially to address myself. You will oblige me greatly, if you will submit this communication to any such with whom you are acquainted, and respectfully request them, from me, to communicate to me any such information as I have desired of you."

A copy of this circular I sent to all parties from whom I considered it probable that I might receive the information desired. In the Report of the Colonial Botanist for 1865, it is stated that "the information elicited by this means, embracing information relative to geological formations, and the pre-adamic condition of that land, in so far as these tend to elucidate the past hydrographic history, and the present hydrographic condition of the country, together with the information specially desired is forthcoming when it may be desired."

The former has been embodied in the treatise on "Hydrology,"* of which this volume may be considered a continuation; the latter, combined with information which has been obtained by my own observation, and from documents which have been placed at my service by correspondents taking an interest in the subject, I shall embody in the following chapters. Amongst other documents placed at my command are the following "Reports of Surveys of the Olifant's River, conducted by Mr P. Fletcher, with the view of ascertaining its capabilities for purposes of Navigation and Irrigation," presented to both Houses of Parliament, by command of His Excellency the Lord-Governor, May, 1860. "Copy of a circular addressed by the Colonial Secretary to the several Civil Commissioners, and there replies thereto on the subject of Irrigation," presented to both Houses of Parliament, by command of His Excellency the Governor, July, 1862. "Report and Proceedings of the Committee of the Legislative Council on Irrigation," published by order of the Legislative Council, July, 1862. "Report of Hydraulic Engineer for 1876." And some Annual Reports to the Government, made by Civil Commissioners, on the state of the Divisions in which they are severally located.

From all of these I shall have occasion to quote largely; and I consider that the statements I am thus enabled to make prove alike that there is no lack of facilities for storing water, and no lack of water flowing off the land to store. Though far from even approximating what is required for warranting expensive operations, my report may not be without its use, as a preliminary survey of the field, to ascertain whether there be sufficient *prima facie* probability of remuneration to warrant the expenditure necessary to procure in regard to any district or locality a more precise report.

* "Hydrology of South Africa: or, Details of the former hydrographic condition of the Cape of Good Hope and of causes of its present aridity, with suggestions of appropriate remedies for this aridity." London: Henry S. King and Co. 1875.

CHAPTER II.

Supply of Water and Facilities for the Storage of it in Divisions along the Western Coast of the Colony of the Cape of Good Hope.

Section I.—*Division of the Cape.*

The Cape Division is bounded on the south by False Bay, on the west by Table Bay, on the north by the division of Malmesbury, and on the east by the divisions of the Paarl and Stellenbosch. It is sub-divided into three minor districts, or divisions: those of the Cape, of Wynberg, and of Simonstown.

The Cape Division is, in its greatest breadth from east to west, about thirty miles; in its extreme length, from Cape Point to the most northern extremity of the division, it measures about one hundred and ninety miles.

The district of this division, known as that of the Cape, includes the country in the immediate neighbourhood of Capetown and the slopes of Table Mountain. It has an area of only some ten square miles. That of Wynberg, in which is produced the celebrated wine of Constantia, is more extensive. Simonstown is situated on the west side of False Bay, and has a small sheltered anchorage, where are the Admiralty dockyards and magazines for the naval stations on the east coast of South Africa.

Commencing with the district of the Cape, I have to state that in the immediate vicinity of Table Mountain there is a copious supply of water, and no lack of facilities afforded by the ground for storing this and applying it to agricultural purposes.

Of some of these facilities for collecting and storing water advantage has been taken to secure a supply for the town and for shipping visiting the port. But there are others which have not been so appropriated; and there may be others which only require to be observed and developed to become important additions to these, if not their equals in importance.

The sources from which the water supply of Capetown was obtained in 1865 were—(1) A spring of water rising on the estate Orangezigt, then commonly called the Main spring, but previously named "Stads-fontein," yielding on an average per day 229,942 gallons. (2) "The Vineyard Spring," as it was called, rising on the same estate, and yielding on an average 62,271 gallons. (3) A spring rising on the estate Waterhof, and called "The Waterhof Spring," yielding on an average 80,489 gallons. (4) Several small springs on the estate Leeuwenhof, called "Kotze's Springs," yielding on an average *per diem* 30,000 gallons.

There were three reservoirs. The first, or distributing reservoir, was built, or rather completed, in 1819, costing £1997 14s., and contained 223,251 gallons. The second reservoir was constructed in 1850, costing £2710 3s. 1d., and contained 2,337,534 gallons. The third reservoir was constructed in 1860, costing £8131 6s. 5d., and contained 12,500,000 gallons.

The total average yield of the combined springs was 402,702 gallons per day.

The three reservoirs together could store 15,060,785 gallons.

Besides the springs above mentioned there were several wells of brackish water in different parts of the town; these were made available as fire-fountains in cases of fire, and also for cleaning the butchers' shambles. For the latter purpose, 12,000 gallons were daily appropriated from a well in Plein Street.

The annual yield of the Main spring had been about the same from the year 1813, when it was properly covered over and led out in iron pipes. But prior to that time, when the spring was more exposed, and wooden troughs were used to lead the water to the town, the spring yielded much less, and a large quantity was lost by leakage. (In 1812, it yielded 134,000 gallons per day.)

The Vineyard spring was first made available in 1853—that is, it was then led out in iron pipes to the reservoir. Previously the water was suffered to run to waste, and measured only 30,000 gallons per day.

The Waterhof spring was up to 1863 almost exclusively used for irrigation purposes, and for turning the Hope Mill; but about that time the town authorities acquired by purchase a title to the water for sixteen hours daily, and having made great improvements in the collection, and impounding of the same, they nearly doubled the yield.

Kotze's springs were purchased I think in 1864, and yielded, before they were utilised for the supply of the town, about 14,000 gallons on an average daily.

There are other supplies which it was considered might be made available if required.

One is the Platteklip stream. This stream it was considered could easily be impounded by the construction of a reservoir in the ravine immediately below the confluence of the Platteklip and the Silver stream. And an immense quantity of water might be collected in this locality.

The second is a spring rising in a ravine behind the property called Prospect Hill. This spring is in itself not of much value in summer, but the ravine seems destined by nature for a gigantic reservoir, which could be constructed, at a comparatively small outlay, by building a retaining wall across it.

On the estate of Orangezigt there were other springs not then utilised which could easily be led into the reservoir. Another spring, the "Lemmetje's Spring," was one about which there was then an action pending in the Supreme Court. It was considered very desirable to secure it for the town.

These springs were produced, in all probability, in a great measure by the percollation through the masses of Table Mountain of moisture deposited on the summit by the cloud produced on the mountain by the South-easters in the summer season.

The quantities must vary from time to time and from year to year. August, September, and October—and in some cases with these months, July, November, and December—are the months in which the supply is most copious.

In 1857 a good deal of evidence on the sanitary state of Capetown was collected by a Committee of the Legislative Assembly, which produced an impression that the supply of water was not sufficient for all the purposes of cleanliness and for health. And early in 1858 there were published, in one of the Capetown newspapers, "Tables of Quantities, etc., etc.," which showed that the average daily flow throughout 1857 had been—From the Main Spring, 204,295 gallons; from the Vineyard Spring, 90,000, of which the town received an average daily supply of 52,911 gallons, and the remainder was employed in the irrigation of gardens; from the Waterhof Spring, 51,569 gallons.

It is not improbable that the supply of Capetown might be greatly increased. The late Mr James Cameron, who had been engineer to the Municipality, and had had occasion in that capacity to consider the state of the water-works of the town, stated in evidence given before the Supreme Court in the case Prince & Co., *v.* Breda and others, on February 26th, 1858:—"I know the water-house on the place Orangezigt. There is a collection of springs in the water-house, flowing almost from every direction, chiefly from Table Mountain. I consider that it is built over the outlet of a number of springs. Before the alteration was made, in 1853, there was a scarcity of water in Capetown, so much so that the supply was stopped during the night, and this led the Commissioners to consider how to increase the supply. The increase was about 50,000 gallons. I am not aware what was the quantity coming out of the stream before that, but I believe that the quantity was less than that now flowing, independent of the increase then obtained, I should say by one-fifth, or one-fourth. When I saw the spot before the alteration was made, the water issuing out of the vineyard ran first into a stream, which runs parallel to the front of Mr Breda's house, about 50 or 60 feet from where that stream joins the Platteklip stream. That is what some call the mill-stream and some Breda's stream. It is a transverse cut. [Described the locality.] There are numerous small springs that soak out of the vineyard, and flow into the cut. That was the general condition of the place before 1853, but there was one place where the water flowed stronger, about the middle. The quantity running before must have been more than after it was led to the pipes. I should say about one-fourth. The then Superintendent of Water-works made his report to the Commissioners, I believe monthly. The return always stated the quantity running. When I was subpœned to appear in August last I visited the water-house, and again when your lordships were there, and I made a sketch. There are two apartments; at the end of the inner one, at each side, there are two strong springs rushing in, or, I should say, streams. They had the appearance of being from springs, but whether one or more I could not say. The excavation is 8 or 10 feet below the surface of the ground, so that the water rushes in from beneath some stones, and comes out in a body. The side of the water-house is the exposed rock. In addition to these two streams there are a series of smaller streams issuing in on the side of Mr Breda's house, distant 15 or 20 feet from the principal streams. The whole of the water, I should say, comes from a distance. Springs flow into the water-house from the direction of the vineyard, if not the diagonal house farther up. The water-house is placed in a position where the various kloofs of Table Mountain converge; and I can state further, what I have often stated to the Commissioners, that if they would dig deeper into the rock more water would be obtained than has ever yet been got."

This is in accordance with my own observations on examining the spot. These led me to conclude that there is a very great quantity of water percolating downwards through the sand composing the soil and subsoil, and covering the rock above that place to the level of the bay, which might in all probability be intercepted by transverse cuts at successive levels. There are also other supplies which are not yet utilised as they might be.

In a leading article in a number of the *South African Commercial Advertiser and Mail*, published early in 1858, it is stated:—"A large portion of the water which falls in the Wynberg district, on this side of the hill, escapes to the sea by the Liesbeck River. As much of it passes through cultivated ground and near villages, it must be rich in the most valuable sorts of manures, especially during the rainy season. The *surplus* during those seasons might be conveyed to the sandy downs on its right bank."

And in a leading article of that journal, published apparently about the same time, there occurs the following statement:—"It has been said that there is something like a marsh, or even a lake of water, on the summit of Table Mountain. It is certain that vast torrents of water rush down its ravines during many months of the year, and that at all times very copious streams and springs issue from its base. We need only to mention the springs in Table Valley, the streams or rivulets in Camp's Bay, or Hout Bay, and on the Wynberg and the Rondebosch sides of the vast mass of sandstone and granite. It is possible, however, that a large quantity of the moisture deposited on the level summit, forming this alleged marsh or lake, may evaporate, and might be previously secured and drained off for the supply of Capetown. If so, it could easily be conducted in pipes down the face of the mountain, which is not three thousand feet above the level of the springs that at present supply the town. Following the ravines, the pipes would probably not exceed one or two miles in length. A survey by a competent person would cost a mere trifle, and without troubling Government or the Municipality on the subject—who, as they never die, seem to think that nobody grows old—we would recommend a private subscription to raise the necessary funds, which may amount to some £40 or £50. There may be no such thing as that we speak of, but to ascertain its non-existence is worth the money, for then we should hear no more of it."

In reference to this it is stated in the number published, May 6 :—

"We suggested a survey of the higher parts of Table Mountain, with a view of ascertaining how far the watery deposits there might be made available when the lower springs were deficient. Without troubling Goverment or the Municipality on the subject, for obvious reasons, a few private persons, among whom Mr W. L. Blore was the most active, proposed at once to request a well-qualified surveyor to proceed to the mountain top and examine the localities, so far as to enable him to state generally whether any good was likely to arise from a survey of the whole, scientific and complete. This was done in the course of a few days, and the following is the Report made :—

"CAPETOWN, 27th April, 1858.

"W. L. BLORE, Esq.

"MY DEAR SIR,—Enclosed is an ideal sketch of 'Table Mountain,' which, along with following remarks, will give you a tolerable idea of the capabilities of the mountain for giving a permanent supply of water to Capetown.

"This I hope will, at all events, put you in possession of your required information,—*i.e.*, Is it worth being at the expense of making a systematic survey of 'Table Mountain' and its adjoining plateau, with a view of providing Capetown with sufficient water for domestic and other purposes?

"Since you spoke to me on the subject, I have made a cursory examination of the mountain and its back spurs and base, from Kasteel's Stream, on the Camp's Bay side, round to beyond 'Kersten Bosch,' and I have arrived at the following approximate conclusions :—

"1st. About three-fourths of the top of Table Mountain and its back plateau, has its superficial drainage in one stream, which runs in the direction of Hout's Bay. For future reference I will call it the Back Stream.

"2nd. In the space described (Table Valley excepted) there are only six waters worth noticing,—*i.e.*, the Kasteel, Back, and Stink-water Streams, and three small feeders at Kersten Bosch, the source of the Liesbeck River.

"I may also mention, however, a small gathering saved by private parties on the Newlands side, and conveyed over the debris by means of a 4½ inch leader pipe.

"3rd. The configuration of the valleys and krantzes in the course of the 'Back Stream' are alone suitable for impounding reservoirs in such a manner that the cost of embankment will be in proportion to to their capacity, and the latter, say from 6 to 14,000,000 gallons.

"The resource of the stream itself is greater than any other that I have observed. On the 9th and 15th inst., it yielded about 90 or 100,000 gallons per diem.

"This quantity can be considerably increased by geological and superficial drainage; though to what extent can only be ascertained by a minute investigation of the number, etc., of surface or land springs, and smaller streams, which lose themselves on the sides of mountain, but which can be turned to account.

"I therefore consider that 200,000 gallons per diem is a low estimate of what can be got from the 'Back Stream,' at the close of the present dry season. Its water shed will, at a rough guess, be about 600 acres, embracing an elevation of about 900 feet. Four-fifths of the soil, from the 9th to the 15th instant, was at the point of saturation.

"I believe, from what I have observed, that there is much more moisture deposited in the mountain in summer than is generally supposed, and that it is possible that I have undervalued this water resource.

"From 'Tremble Krantz' the water can be led in pipes either by Newlands or Camp's Bay side. Upon a second inspection I found that it can be brought by Kasteel Ravine without difficulty.

"At the commencement of Hout's Bay Ravine there is a suitable place for a reservoir on the same stream, which is at a lower level, and has the advantage of having an additional drainage of about 200 acres; a leading will however be required of about one mile further round than the above, and a portion of the route is rather rugged; both are represented on sketch.

"4th. I examined 'Stinkwater' on the 10th, 15th, and yesterday, the 26th. I consider that this stream of itself will not compensate for the expense of leading it into town, particularly in the absence of a suitable place for a reservoir.

"Its character and capabilities ought, nevertheless, to be embraced in your future investigation on the water question; you will observe by the sketch that this Stinkwater shares a considerable portion of the mountain drainage.

"It has no less than five of those precipitous ravines joining at the commencement of the mountain debris. The principal one of those is the middle one, and has its source at the landing of the common foot-path from Capetown to the top of the mountain. A few hundred yards west of the spot Stinkwater first appears in the shape of a weak spring; farther down the kloof it issues over the precipice in a small, but apparently a constant flow. This and the Waterfall ravine to the south are in the dry seasons the only visible contributions to the part of the stream known as the 'Stinkwater.'

"In its course, however, there is evidence of a certain amount of under-drainage, as the flow is stronger or weaker, according to the nature of the debris through which it passes; this under-drainage can be saved at any point below the junction of the ravines. On the 10th and 15th I estimated that Stinkwater would give at least 30,000 gallons per diem, at the junction of the ravines, or at the same level as the top of the 'Kloof.'

"I may further remark, that the stream in question is, of all the others, the most sensitive to rain. This, no doubt, is principally owing to the extent of its vertical surface drainage, which faces the prevailing rainy winds. Yesterday, the 26th, from the effects of the rain on the previous day (the first of the season), upwards of one million gallons must have rolled down its course.

"5th. Kasteel stream is about half-a-mile beyond Stinkwater, and is in nature different from the latter. Its superficial drainage is comparatively small, its flow more steady, and in the dry season is stronger, as it approaches the sea, than the Stinkwater. Its source can be traced for about 100 yards on the plateau above. The run of water which is here trifling disappears about the edge of the ravine for a few hundred yards, when it re-appears and gradually gains strength as the sandstone debris diminishes over the primary formation towards the sea. Its water source being to a great extent from geological drainage, rain has much less direct influence on it than on the Stinkwater. The rain of the 25th increased the latter one-third more than the former.

"The low level at which this stream must be arrested materially destroys its value as a water contributor to Capetown.

"6th. I would next draw your attention to the north-east corner of Table Mountain, where a hull-shaped reservoir can be formed by constructing a bank about 20 × 30 feet, and which would contain between 2 and 3 million gallons of water. The drainage of it on the 9th inst., did not appear to be more than 10 or 15,000 gallons *per diem;* it is, however, the most sensitive for the South Easter and other mists, and consequently is well worthy of further examination.

"The area of its water shed and its greatest capacity as a site for a reservoir ought to be ascertained. The soil, which is from 2 to 3 feet deep, being scooped out, any partings of the strata or fissures, may be secured. This can be easily effected, as the strata are nearly horizontal, smooth; and, owing to a hardened crust formed on the sandstone, a large portion of the rock is impermeable to water. Thus it may be possible to have a reservoir in the living rock containing several million gallons of water, at an elevation of nearly 3,000 feet, comparatively free of evaporation, receiving and containing water in the heat of summer, pure as the cloud itself. Although the discharge is towards Newlands, it is still perfectly practicable to fix a pipe round the Bluff.

"7th. In conclusion, I beg to remind you that the above remarks are made without that full local knowledge and examination which are so absolutely necessary, before giving a definite opinion upon such an important subject.

"One thing is, however, certain, that until such time as you get a systematic and verified survey of your existing water resource, with and without servitude, and lay the same before a commission of enquiry, you will labour under the disadvantage of conflicting opinions, and doubtful supplies.

"No one knows better than yourself the necessity of having a plan showing position of all summer springs and streams at a given height, say seven miles off Capetown, guaging or measuring the same at select spots, ascertaining and sectioning the most convenient sites for impounding reservoirs, showing their proximity to embanking material, etc., testing the extent of percolation which might be expected by forming reservoirs on the Table Mountain sandstone, ascertaining the probable amount of evaporation going on in the summer on the top of the mountain and back plateau, and also the quantity of moisture deposited from south-easters and summer mists.

"These and other details collected from individual testimony, and put into the hands of competent judges, will alone ensure for the public of Capetown the best and most judicious system of water supplies.

"I may mention that Table Valley is altogether excluded in the above remarks, and that the configuration of the ravine formed by the Devil's Hill and Table Mountain appears to be very favourable for concentrating a considerable portion of the drainage of both mountains; it would be well to examine about the base of the sandstone, as it is possible that there exists a water channel worth tapping."—I remain, etc., P. FLETCHER.

In a subsequent number of the same journal their appeared the following letter to the editor of the *S. A. C. Advertiser and Capetown Mail,* from Mr Blore, dated Capetown, 7th May, 1858:—"You did me the kindness to publish Mr Fletcher's report upon the drainage of Table Mountain towards the west, south, and east; and I beg now to acknowledge the assistance of several kind friends, who have enabled me to meet the expenditure incurred thereon.

"I also beg to state that as the present reservoirs are, from their low elevation, incapable of supplying the greater portions of district 12, of Scotch Kloof, and of Green Point with water, the contemplated supply will admit of the construction of a reservoir as high as the Kloof, to contain, say 200,000 gallons, or even 100,000 only, as the supply would be constant. The daily supply would be 200,000 gallons from the back stream, and, therefore, a reservoir of small extent would suffice, as it would not be required for the retention, but merely for the distribution of the water in various directions; therefore, perhaps a reservoir containing, say 50,000 gallons, might be sufficient.

"The two basins on the top of the mountain contain an area of 500 acres, which, in one day's average rain, will yield six millions of gallons; so that, independently of the Back Stream, which is constant, a natural reservoir might be made, 900 feet lower than the summit, capable of containing some twenty millions of gallons of water, at a comparatively small outlay, that is, as far as we at present know.

"A three-inch pipe, laid from Tremble Krantz, will deliver 246,450 gallons at the top of the gardens, with a velocity of 9·5318 feet in a second; but for all purposes it would be advisable to employ a pipe capable of delivering at least 500,000 *per diem,* as the town is rapidly increasing; all the high and remote parts are destitute of water from the want of sufficient pressure; and no system of drainage or purification can be carried out without it.

"I intend, therefore, to submit the plan and report for the consideration of the Capetown Municipality, who have been complaining for some time of the scarcity of water, and have never been able to supply the upper districts with that necessary article. The whole subject is, therefore, worthy a thorough scientific survey, and is respectfully recommended to them for further inquiry,—yours, etc."

Besides the supply secured for the use of Capetown and its suburbs, extending along the coast to Sea Point, there is also at Sea Point a reservoir from which water is supplied for irrigation of gardens, and for household use from that point to Three Anchor Bay. It is filled from a spring appearing a little beneath the Round-house. The flow is about 25,000 gallons a day in summer, and it varies from that quantity to nearly 70,000 gallons a-day in winter. The formation of the reservoir and the conveyance of the water into it from the spring cost about £2000; an expenditure of about £1000 more was incurred in laying pipes from the reservoir to Three Anchor Bay. It was alleged that by an additional outlay of about £500 the supply in summer might be doubled, by bringing into the reservoir a supply from the top of the Kloof. And by a further outlay supplies might be obtained from the Back stream flowing into Hout's Bay, from the Stinkwater flowing into Camp's Bay, and from Kasteel stream flowing into the sea a little beyond Camp's Bay, all the waters of which streams were at that time lost to the community, as was the water from the spring below the Round-house previous to its being led to Sea Point. And thus might be secured all the benefits spoken of as likely to follow the leading of water into Capetown from Tremble Krantz.

Mr Blore, ever alive to the interests of the Colony, and specially devoted to the study of meteorological phenomena, wrote again, as he did often, to one of the newspapers of the Colony:—"We have had a delightful fall of rain upon our hills on this side of Table Mountain, and the large cataract above Kerstenbosch is now, with a thundering noise, pouring its twenty million gallons of water into the Liesbeek. The rain commenced at five P.M. on the 2nd, and poured down continuously till five A.M. this morning. The measured fall during the sixteen hours has been 2·910 inches, which, distributed over an acre of ground, yielded about 65,000 gallons. It appears to me that very much might be done by the Municipality during the present year in impounding the torrents of rain-water which pour down the sides of the Devil's Hill during the winter, and which are unfettered by any servitude. There is a patch of pot-clay ground above the new prison where a large reservoir might be constructed to receive one or more of these mountain torrents. A structure, to contain five million gallons, might be commenced forthwith, and partially finished, and afterwards be built up as funds come in; and three times that quantity could be impounded. Of course, the excavation should be made to the full depth required for the entire quantity; but the mason-work could be gradually raised. It would require a surface or watershed of only thirty acres, with a rainfall equal to the Capetown average, to produce fifteen million gallons. If a farmer has a watershed sloping off at an angle of 45°, and exposing a horizontal surface of thirty acres, with a rainfall equal to that of Namaqualand for 1865, which was only six inches, he should secure, in a suitable locality and with a proper subsoil, at least three millions. This may be done with a small winter supply, upon a very limited surface with a good incline. But, I believe, that upon almost every farm in the Colony of 100 morgen and upwards, at least one million gallons may be securely and advantageously impounded for summer use."

The testimony of such facts as have thus been reported goes to prove that in the immediate vicinity of Table Mountain there is a copious supply of water, and no lack of facilities afforded by the contour of the ground for storing this for domestic, manufacturing, horticultural, or, if need be, for agricultural purposes.

A practical and professional gardener, Mr Colin Sydney Farquharson, who accompanied Mr Fletcher on his examination of Table Mountain, was so struck with the appearance and conditions, in regard to water supply, of some plots of ground on the table-land of the range, with sufficient depth of soil for cultivation, that he applied to Government for a grant of the table-land of Table Mountain and its adjoining range, so far as it is accessable towards Hout's Bay, for experimental gardening and farming purposes; requesting that it might be granted as a perpetual quit-rent farm, or for a long term of lease, as it would require time, labour, and expense to prove the results; stating that, owing to the laborious and expensive method he should be obliged to adopt to transport things up and off there, he could not afford to give a very high rent for the land in question, in regard to which he states,—"It is my opinion, from its great elevation, that the climate in the winter season is suitable for many kinds of fruit trees and vegetables that require cold to drive the sap down and give rest during the winter season, which cannot be sufficiently obtained on the lower lands around the Cape. I think hence arises the failure. The following fruits, etc., might be grown with success:—Cherries, plums, gooseberries, currants, raspberries, strawberries, rhubarb, seakale, and various other things, which at present can only be obtained in Capetown at high prices, or in a preserved or pickled state."

The expediency of complying with the proposal was questioned, and the subject was discussed in the journals of the day. One of the defenders of the proposal, in whose initials I recognise those of one who has given proof of his interest in the development of all the resources of the Colony, wrote thus in regard to it:—"Without knowing anything at all about Mr F. personally, I must maintain that the proposition he makes is not so preposterous and 'insane' as may at first sight appear.

"In all deliberative bodies at the Cape there exists an inclination to condemn and hold up to ridicule, without the trouble of inquiry, any scheme, the object of which may be either difficult of appreciation or not at once perfectly understood. To nine-tenths of the inhabitants of the Cape division the summit of Table Mountain is as much a '*terra incognita*' as the banks of the Zambezi, or the Mountains of the Moon. The untimely fate of more than one explorer has probably given rise to the idea that the locality in question is not only uninhabitable during the greater part of the year, but totally unfit for either grazing or agricultural pursuits.

"After having visited the place at different seasons, some four or five times, and after having examined particularly into its farming capabilities, I feel convinced that the back part of Table Mountain and the south-eastern slope of the Devil's Peak afford as good pasturage for breeding cattle, goats, and horses, as any farm in the Cape district. The atmosphere is moist and genial during eight months of the year, and peculiarly fitted for the promotion of vegetation. There is abundance of excellent grass at all times, and several beautiful and never-failing springs of water. The weather during a few months in winter may be somewhat cold and damp, but I am far from coinciding with Mr ——— in the idea that it would be impossible to sustain animal life there. I think if that gentleman had spent one or two winters in the Highlands of Scotland he would not have given his opinion so decidedly on this particular point. It is well known that some of the best grazing farms in this Colony, situated on the top of the Sneeuwberg and Nieuwveldt mountains, are 2,000 feet more above the level of the sea than the highest point of Table Mountain. The wise-acres of a by-gone generation maintained that the rearing of sheep or cattle in such localities was wholly impracticable; but enterprise, industry, and experience have proved the fallacy of such an argument. Many propositions which, at the outset, appear ridiculous and impracticable, turn out eventually not only feasible but of immense benefit to succeeding generations. When Stephenson stated, before a committee of the House of Commons, that he could construct a locomotive capable of drawing a load at the rate of twenty-five miles an hour, he was regarded by the self-sufficient sages of the day as little better than a madman. The Cape divisional council seem to look upon Mr Farquharson much in the same light; but why should he be prevented from carrying out his ideas when he merely requested to be supplied with land in a particular place, which, up to this time, has been wholly waste and unproductive, on usual terms of quit-rent? If he succeeds, a difficult question will have been solved; if he fails, the Colonial revenue will not have been injured, but to some extent benefited by the attempt. In my humble judgment, the vicissitudes of the climate might, by a little industry and management, be easily overcome. A substantial house, erected in a sheltered situation, with the requisite extent of sheds and out-buildings for affording refuge to the live-stock in bad weather, would supply sufficient protection. From the Constantia end of the range, a good bridle road might be cheaply constructed, by which produce could be conveyed to a market. Dairy operations might, I think, be successfully carried on, and the excellent pasturage turned to good account in that line.

"Not being acquainted with the culture of hemp or flax, I cannot pretend to offer an opinion on that part of Mr Farquharson's plan; but surely, if he undertakes to bear all the responsibility, and defray all charges, he might be allowed to make the experiment. If he fails to conform with the existing regulations for the disposal of Crown lands, of course there will be an end of the matter; but on the other hand, if he is willing to submit to those rules, I do hope that no unnecessary obstacles will be thrown in his way by the authorities."

The application and the defence of it are cited as indicative of the opinions entertained in regard to the sufficiency of the water supply for such horticultural operations on the very summit of the mountain.

More than ten years have elapsed since this statement was prepared; a lengthened *Résumé* of it was appended to the Report of the Colonial Botanist for 1866. And it is well-nigh twenty years since the *interim* report by Mr Fletcher, upon which it is mainly founded, was made (April, 1858). On the 25th November, 1875, an official professional report was made to the Secretary of the Municipality of Capetown by Mr Gamble, the Hydraulic Engineer of the Colony, which treats successively of the various resources whence a supply of water for use in Capetown is or may be obtained,—the existing storage and supply,—proposals which have been made for improving the water supply,—the comparative levels of the various springs, etc.,—and the recommendations which he had to make.

It is such a report as is greatly to be desired in regard to every part of the Colony. I attach the greatest importance to this report, and place implicit confidence in its statments in regard to existing supplies, and provision for the storage and delivery of these. And in doing so I have great satisfaction in stating that I find there is no necessity to modify the statements of Mr Fletcher and others embodied in that which I have prepared and have given above.

While the supply of water which may be obtained from Table Mountain is so great, many complaints are made of the insufficiency of the supply to meet the wishes of the inhabitants in the vicinity; and yet not a little is allowed to run to waste, a good deal of which, if properly husbanded, might be utilised; and some which is utilised is unnecessarily polluted and defiled. A few months after I left the Colony there was sent to me a copy of the *Cape Argus*, in which was a letter to the editor in regard to the Liesbeek River, in which the writer says:—"Those limpid streams and pellucid pools of which the Hottentots drank in days of yore are now sinks and drains for every impurity; and to these the ladies, the gentlemen, the tradesmen, and mechanics—a large population—send their linen; and there also the Malay, the Mozambique, and the Coolie, the diseased, and even the leprous, send theirs. In these drains, and fouler pools, the filthy garments of the whole population are washed together. Bad as this is it is not the worst. It is a well-known fact that of these fœtid waters, especially the waters of the Liesbeek, many householders still drink.

"Before the Colony was blessed with a Parliament, the Ordinance No. 6, 1862, was passed 'to prevent the commission of nuisances in the river Liesbeek.' The preamble states that 'a large and increasing number of families, resident in Rondebosch and its neighbourhood, are solely dependent for their supply of water, for drinking and domestic purposes, upon the water of the river Liesbeek; that owing to the commission of nuisances in the said river, and the absence of proper regulations for securing the purity thereof, the water of the said river is likely to be rendered, more especially in the summer season, so unwholesome as to be unfit for use.' Then follow certain regulations which have never been enforced. The people still drink the water, having first washed their linen in it.

"In the *Government Gazette* of this day there is a Bill to amend Ordinance No. 6, of 1862. Surely no better opportunity is likely to occur for putting an end, once and for ever, to the abominable practices referred to. It is necessary that a fund should be raised for the payment of an overseer, and for the purchase of wash-troughs for free use at public washing-places, and the use of which should be made compulsory, in terms of the Ordinance just cited.

"There is no time to be lost. Should the Mauritius fever, or any other epidemic, visit this Colony, nothing could be devised more calculated to promote its rapid spread than the washing of the linen of the whole population in the same waters."

I did not, while in the Colony, form an opinion adverse to the habits of the people in regard to cleanliness, taking into account the quantity of water which was utilised and which alone could be utilised unless some little trouble or expense was incurred; but I often saw done what the Chinese Government would never have allowed to be done, in view of what might be done and should be done, where the utilisation of water was of importance; and I often saw what British Officers of health would condemn, and which Sanitary Commissioners would not sanction. I found in some parts of the Colony that it was a proverb: "Don't throw away your dirty water till you have got clean water to take its place." And I saw a state of things which might well give occasion for the usage referred to in the proverb. I was often reminded of what in the early part of this century was alleged of Mrs M'Clarty in "The Cottagers of Glenburnie." In many cases this may be attributable to want of consideration alone. I found many instances of this want existing along with an evident delight in cleanliness in domestic arrangements. At a wine farm, in regard to which my opinion was asked as the produce had a peculiar *gout*, I found the building in which the wine was manufactured, and kept, situated on a slope, and immediately behind and above it, the wall of the building forming part of the enclosure, was the *sheep kraal*. And oftener than once have I seen sheep-skins and mouldy harness huddled into the wine-store when not in use for the vintage. In keeping with all this is the picture given of the pollution of the Liesbeek River.

A few months later there was sent to me another number of the *Argus*, with the following letter from another correspondent:—"As there is now again a general 'cry for water,' not only to supply private dwellings, but also to flush the public drains, and as it is suggested to build additional reservoirs, I think it my duty to bring to the notice of those who have taken upon themselves to provide for the wants of the town in this respect that a large supply of this valuable article is to be obtained in the vicinity of Capetown, without buying off the mill-streams, which, under servitude on the properties, formerly had to flow down to the canals that prievously existed in Long and New Streets, the Heeren, Kaizergrachten, and Strand Streets, which were provided with sluices, that were let down in summer and opened every Saturday, so that the collected waters carried down all impurities into the sea.

"About twenty years ago, a private company was formed to purchase the Newlands Spring, and carry the water by pipes over certain properties, above the Rondebosch road, into Capetown, giving the proprietors certain rights to a supply of water from those pipes. Through the selfishness and shortsightedness of one of the owners, whose property would have been *most benefitted*, this project failed.

"Numerous dams and reservoirs could be constructed in the several ravines surrounding Capetown, from which, during the rainy seasons, immense quantities of this, to us, precious element pour down into Table Valley, and find their way into the sea, without any attempt being made to collect them, either by private individuals in constructing tanks on their premises, as at Port Elizabeth, or by the Town Council in retaining some of it for flushing purposes, even by common dams.

"As the subject of *additional reservoirs* is again mooted by the City Engineer, the building of which would, of necessity, entail a heavy expenditure, and perhaps during dry seasons be useless, if the water to fill them cannot be found, as has been the case lately with those now existing, I wish to bring the following facts to his notice:

"In the course of my wanderings I found myself one day on the top of the *Kasteel's Berg*, at the back of Table Mountain, having ascended by the ravine above Camp's Bay. To those who have not been there it will perhaps be interesting to know that on the Kasteel's Berg, about one and a half miles from the poort, rises a strong *rivulet*, which, when I saw it during the driest season, yielded at least six times the quantity of water supplied by the spring on Orangezigt. As far as I could trace, the spring is the property of the public, or rather of the town, and there exists no claim thereto on the part of anyone at *Hout Bay*, to which the water now finds its way, forming a large marsh in the valley, which, when drained, would be the most valuable garden ground.

"From the source of the spring to the Kasteel's Poort the ground is almost level, here and there showing a natural pavement of stone, across which a watercourse could, with little expense, be built, as far as the head of the poort, whence, along the sides of the mountain, iron or earthen pipes could be laid at any elevation, to the battery on the neck of the Kloof Road. When this is reached, the most elevated parts of the town can be supplied with water, and the pressure, which the height from which it comes ensures, would make the flushing of drains or gutters at any time an easy matter.

"I am not an engineer sufficient to give even an approximate idea of the expense the works I propose would entail, but I suggest inspection and inquiry by members of the Town Council, to whom the existence of this abundant supply of pure, fresh water, within—according to my idea—easy reach, may be unknown; difficulties equally great having been surmounted by owners of farms when they were desirous to secure rising water at their homesteads. I may instance the works of the late Mr W. Herman on his farm 'Thornlands,' near Swellendam. I shall be happy to conduct visitors to the spot, or to recommend a competent guide."

The subject has received the attention which it commanded, both in the Municipal Council and elsewhere. In commenting on the discussions of the subject by that body, the editor of the journal named wrote:—"Some discussion yesterday arose in the Town Council on the water supply for Capetown. Some of the councillors remarked that the reservoir question was not forgotten, it was only shelved on account of the press of sanitary work. Mr Louw considered that if the Platteklip stream was purchased by the Town Council there would be an abundant supply of water for the whole of Capetown. But the question arises, what is an abundant supply? People have different ideas on that question. Some think we ought to have enough to flush the streets and alleys, and others that at least every house should have a good supply of water either laid on direct, or at least *within easy reach.* Now this is notoriously not the case at present, and we should recommend the Town Council to make diligent inquiry, by means of a committee or otherwise, as to the water privileges now enjoyed by the people. We have facts to show, not only that the poorer classes lack water for bodily ablutions and house cleaning, but that a good deal of sickness has arisen from the impurity of the wells to which many are compelled to resort. A city like Capetown ought not to be dependent on the flow of a spring for its summer supply of water; it should have an abundant store to fall back upon. We quite agree with Dr Abercrombie that the present reservoirs are ridiculously small for the due supply of a large town like this. If we don't go in for a gigantic reservoir in the Platteklip ravine, large enough to admit of the streets being flushed, two or three smaller ones might be made at the extremities of the town, which would be a great help. The Town Council have done well to purchase the Platteklip stream; but they would do well also to inquire thoroughly into the question of the water supply, with a view more particularly to ascertain whether the poorer classes have *good water, accessible in reasonable quantities.* We are quite aware that a crystal spring in every house would not insure the cleanliness of the inhabitants, and we are well assured that a great deal of the filth and sickness is referable, not to the want of water, but to the dirty habits of the people. Dr Abercrombie was not far out when he said that most of the people washed their faces about once a week, and the rest of their person never. For all that, however, we ought to see that the inhabitants of the town are plentifully supplied with good water. Then they will be responsible for their dirty condition."

In the consideration of the subject thus far, attention has been directed only to the supply from springs and streams; but in a more comprehensive study of the subject, which in process of time will probably be demanded, the basis selected will be the rainfall, and not the fraction of this which is supplied to our hand unsought by these streams and springs. And to this I must refer before taking leave of the Cape district.

In regard to the quantity of the rainfall, information has been given in a previous chapter.

In attempting to determine what proportion or amount of this may be available for storage or irrigation, it will be necessary to attend to the distribution of the rainfall over the days of the year, and even over the hours of the days.

From observations made at the Observatory, it appears that the mean monthly fall, from January 1, 1842, to December 31, 1855, was as follows:—January, 0·880; February, 0·653; March, 0·864; April, 3·576; May, 4·311; June, 2·921; July, 2·951; August, 3·323; September, 2·332; October, 1·014; November, 1·090; and December, 0·576;—showing considerable inequality in its distribution, which inequality would have been still more apparent had the daily observations by which these means were determined been given. And with this inequality of distribution in view it must be borne in mind that, though when the rain falls gently it may be all absorbed by the soil, there are limits to the rapidity with which this can be done; and if the rainfall exceeds these limits, which vary in different circumstances, the excess must stagnate on the surface or run off, and may be lost if it be not collected and retained. This alone is what is available for storage; but this quantity may be affected in various ways, and in every case beneficially for man, by judicially conducted planting of trees and herbage and bush. And whatever might be effected by such means, and by the storage of all waters now escaping by streamlets to the sea, the Cape district, which includes much besides Capetown and Table Mountain, and other interests besides those which centre there, might benefit by the adoption of similar measures elsewhere.

In some districts of the Colony the effects of the existing neglect of such measures is seen in rivers coming down like a wall of water, stretching across the river-bed from bank to bank; and it rarely happens even here that a year passes without a very large rainfall occurring within a brief time. Such a rainfall occasioned, in 1862, a great destruction of bridges and other structures in different parts of the district. In that year there fell at Wynberg, in the month of June, 22 inches of rain—equal to 484,000 gallons of water on every acre. In February, 1866, the last I spent in the Colony, there fell in one day, or twenty-four hours, at Wynberg, 2·9; at Green Point, 3·4; and at Protea, or Bishop's Court, much more. On one occasion there fell at Wynberg, in the course of forty-eight hours, 8 inches of rain, which was equivalent to 176,000 gallons of water per acre.

A very large proportion of such rains might be made available for storage for subsequent use; and this might contribute not a little to the development of agriculture and of horticulture in the district.

But—some one will then say—the expense! Yes; the expense will be considerable, if any measures which are likely to prove efficient be undertaken and accomplished; nor do I know of anything great and effectual which can be done without expense. To provide for the expense of executing such measures as I may see would be effectual does not come within the scope of my present purpose, nor does it do so to indicate how the means of meeting the expenditure might be found without wrong or injustice to any one; but I may direct attention to the losses which accompany the neglect of such measures as are referred to. Within a few months after I wrote the statement in the last paragraph, another deluge of rain carried on the work of destruction, of which the *Cape Argus* reported, in its issue of November 26, 1867:—" Our anticipations of a fine harvest have been somewhat dashed by the up-country news. The south-easter, which blew in Capetown with a violence rarely experienced even in this stormy peninsula, extended up-country, and did great damage to roads, bridges, and, worst of all, to the crops. In Capetown and the neighbourhood the south-easter is a dry wind, but up-country it is usually accompanied with heavy rains. Last week torrents of rain appear to have fallen, and heavy floods have made roads and rivers well nigh impassable. It is many years since the frontier post has been delayed so long. But the delay of the mail, or indeed the destruction of roads and bridges, is of little consequence compared with the loss to the wine and corn farmers. The Government will not probably be hard upon the contractors, who are so uniformly punctual. It does not do to be careless about fines or to admit the validity of every excuse which a contractor makes; but in cases where the whole country is devastated with floods, even a contractor may reasonably plead for indulgence. He must make provision for crossing fords, but he is scarcely expected to find a substitute for a broken bridge in every instance in which one may have been swept away, or to remake a highway that has shared the same fate. But the losses of contractors, even if they be heavy, will not affect the community like those of the farmers. From all sides we hear of vineyards being torn up, and corn crops emptied of their grain. One farmer reported on Saturday that he could scarcely find a grain of wheat in a standing crop covering some sixty acres of ground! Doubtless there are many farms which, from their sheltered position, have not suffered, and we cannot but hope that the grain harvest will not be seriously affected. No reports have yet come to hand of losses of stock, but we shall probably hear of them when the post arrives.

"Perhaps the loss of roads and bridges may prove a blessing in disguise. The necessary repairs and reconstructions will, at least, provide labour for the starving poor. When the fever is over in Capetown we shall have to clear out a number of the population, and send them where they can earn their daily bread. They can be usefully employed for a time in mending roads and rebuilding bridges. The wind, bad as it is to the farmers, may blow some good to the working classes. True, it will be an expense to the Government, and consequently very disheartening to the Retrenchment Committee; but the expenses incurred by Government never afflict the people, unless, indeed, they involve fresh taxation. We believe that there never was a larger number of applications for Government expenditure from the various districts, but apparently never less willingness, at least if Parliament represents the country, to pay a fair share of it. Whatever may be thought of railways for the Cape, or the undertaking of other public works, it is quite certain that roads and bridges must be kept intact. We may do possibly without gas in the towns or railways in the country, but we cannot do without good highways as the arteries of trade and commerce."

And now, while some of the preceding sheets were passing through the press, there has been sent to me a report of a meeting held in Capetown to authorise the Town Council to borrow £50,000 for the construction of a new reservoir, which was agreed to by a large majority! "Where there is a will, there is a way."

It appeared, from the statement submitted to the meeting by the Town Council, that the daily consumption of water at Capetown, as registered at the Town-house, was 482,000 gallons. But the supply from the springs in summer falls as low as 280,000 gallons, leaving a deficiency of over 200,000 gallons a day.

At this meeting Mr Arderne, an intelligent and experienced resident of long standing, said he admitted that the town did require more water for domestic purposes; but for flushing the drains and so on, nature had already given a supply which would not be exhausted in a hurry. There was scarcely any part of Capetown, except in the more elevated portions, where water could not be got at the depth of ten feet or so. He had had stables himself in Barrack Street for the last forty years, in which were from ten to twenty horses, and yet he never used municipal water there. He had a well in Church Square, where there was water all the year round at a depth of ten feet from the surface. There was an abundance of water towards St. George's Church. When they were building the Mutual Insurance Company's offices they had to drain the water continually. In the Botanic Gardens there was plenty of water at a very short distance from the surface.

Of all of this water it may be said that it indicates a subterranean flow to the sea, which, when occasion demands, may be arrested and utilised. And there are yet greater undertakings looming in the distance. Adjacent to the town are extensive flats of sand, stretching from Table Bay to Simon's Bay, and from Table Mountain range to the Hottentot Holland range of mountains. It is many long years —sixty and more—since it it was suggested that this stretch of sand might be utilised. And it seems that some means of utilising them, by conveying to and over them water now lost in the sea, had been previously under discussion; and this was adverted to at the meeting, but only to give occasion for the statement that this was not the scheme which was advanced by the Town Council.

The *Cape Argus* of the following day, October 31, 1876, said:—"The Hamiltonian scheme was nowhere, and we almost regret it, for a little jaunty talk about £400,000 in these hard times would have done us all good. We owe the promoters a grudge for the disappointment, though we commend their judgement in not embarrassing the meeting. By-and-bye, when the population is doubled, and the working classes are elbowed out to Salt River, we will go in for manuring the Cape flats and increasing our cabbage gardens. And the Hamiltonians shall have their scheme, and they shall crow, as they will, over the old foggies of the present day, who opposed their advanced views. Meanwhile we have got another reservoir, and let us be thankful for it. Let us be permitted to say we have got it—for, in our joy at the prospect, we cannot dwell on the interval that separates us from the realisation. We may look forward now to a better era for Capetown, when the streets will no longer reek with filth, nor gutters exhale pestilence at every corner. And possibly in a few years, when the new work is finished, we shall no longer blush for the fact that in a city of 30,000 inhabitants there is not spirit enough to start a public bath or wash-house. In a well-ordered community there ought to be such a place in every poor quarter, for there are hosts of poor households, where the arrangements for washing are so contracted or impracticable as to make washing an act of heroism. Yet the poor may be coaxed into washing, and learn in time to appreciate the comfort of a clean skin. The success of these establishments in the great cities of Europe is remarkable, and the good done in promoting cleanly habits cannot be estimated. No one will pretend that in a semi-tropical climate like this, where for four months of the year the atmosphere is rolled in dust, and human bodies are fried as in a stew-pan, the need for such places is less than in Europe. Plunge-baths are tolerably paying investments in London, where for half the year they cannot be resorted to. In this more favoured climate they could be used throughout the year, and would probably realise a handsome profit. They are a necessity of city life; and, once instituted, soon make their value and importance felt.

"It is difficult for any one not residing in the city itself to understand the suffering to which a large number of the inhabitants are put during the summer months in consequence of the insufficient supply of water. . . . .

"Now that a reforming spirit has asserted itself we shall not let it die out very rapidly. There is much that *must* be done to satify the first conditions of decency and order; there is much that *may* be done in the interest of beauty and taste."

In a preceding chapter treating of meteorological observations there have been given abstracts of observations made at the Royal Observatory near Capetown. The following are abstracts of daily observations made there in 1868; of rainfall at Wynberg Hill during that year; of the same made at Simonstown during seven months of that year; and an abstract of observations at Simonstown, made in the years 1862–65.

Abstract of daily observations made in 1868 at the Royal Observatory, Cape of Good Hope.

From Daily Observations.	Jan.	Feb.	Mar.	April.	May.	June.	July.	Aug.	Sept.	Oct.	Nov.	Dec.	For the year.
	ins.	ins.	ins.	ins.	ins.	ins.	ins.	ins.	ins.	ins.	ins.	ins.	ins.
Mean height of the barometer at the temperature 32 deg. Fahr. for each month, and for the year, from observations at 5 A.M., 9 A.M., 1 P.M., 5 P.M., and 9 P.M., ...	29·906	29·932	30·001	30·055	30·037	30·183	30·157	30·120	30·075	30·055	30·023	29·985	30·044
Maximum observed, ...	30·188	30·298	30·235	30·456	30·340	30·534	30·458	30·368	30·357	30·384	30·349	30·229	30·534
Minimum observed, ...	29·684	29·691	29·671	29·781	29·802	29·865	29·702	29·866	29·889	29·746	29·780	29·642	29·671
	deg.	deg.	deg.	deg.	deg.	deg.	deg.	deg.	deg.	deg.	deg.	deg	deg.
Mean temperature of the air for each month, and for the year (Fahr.), from obervations at 5 A.M., 9 A.M., 1 P.M., 5 P.M., and 9 P.M., ...	70·07	68·18	65·38	61·01	57·93	55·19	55·05	57·72	57·81	60·73	63·93	67·60	61·72
	deg.	deg.	deg.	deg.	deg.	deg.	deg.	deg.	deg.	deg.	deg.	deg.	deg.
Mean temperature of evaporation for each month, and for the year (Fahr.) from observations at 5 A.M., 9 A.M., 1 P.M., 5 P.M., and 9 P.M., ...	62·35	61·35	58·61	55·52	54·49	52·14	50·72	53·23	53·99	54·64	57·92	60·60	56·30
Mean humidity of the air for each month, and for the year (the integer number 100 implies saturation), derived from observations at 5 A.M., 9 A.M., 1 P.M., 5 P.M., and 9 P.M.	64·24	66·91	66·24	71·47	80·31	82·01	75·30	74·89	77·87	68·25	69·09	66·34	71·91
Maximum humidity, ...	96·0	98·0	97·0	97·0	98·0	99·0	94·0	97·0	99·0	98·0	97·0	95·0	99·0
Minimum humidity, ...	25·0	37·0	30·0	26·0	42·0	37·0	35·0	34·0	51·0	32·0	40·0	28·0	25·0
From the self-registering thermometers (Fahr.).	deg.	deg.	deg.	deg.	deg.	deg.	deg.	deg.	deg.	deg.	deg.	deg.	deg.
Mean maximum temperature of the air, ...	79·13	75·91	73·24	69·25	64·94	61·47	61·84	63·61	63·60	67·55	70·71	75·63	68·91
Mean minimum temperature of the air, ...	61·64	60·70	58·48	53·66	51·54	48·71	47·57	52·07	51·78	54·78	56·61	59·11	54·73
Approximate mean temperature of the air, ...	70·39	68·31	65·86	61·46	58·24	55·09	54·71	57·84	57·69	61·17	63·66	67·37	61·82
Mean range of temperature of the air, ...	17·48	15·21	14·76	15·58	13·40	12·75	14·27	11·55	11·81	13·36	14·11	16·52	14·23
Highest temperature, ...	93·5	86·0	95·4	83·0	79·8	70·5	76·7	77·8	69·5	82·1	80·8	90·0	95·4
Lowest temperature, ...	55·0	53·0	53·8	48·0	43·4	40·6	40·3	43·0	44·5	49·0	48·0	49·0	40·3
Extreme range of temperature on any one day, ...	30·5	27·9	34·3	32·7	27·8	23·0	26·6	25·0	18·2	25·6	23·8	30·2	34·3
Least range of temperature on any one day, ...	8·5	6·8	9·0	6·6	5·5	3·0	3·4	4·4	1·5	6·2	4·7	9·8	1·5
Quantity of rain in inches, ...	0·710	1·037	0·467	2·165	1·870	3·359	2·653	0·692	0·904	2·715	2·382	0·999	19·953
Greatest amount of rain on any one day in each month, ..	0·350	0·305	0·230	0·950	0·690	1·204	0·657	0·250	0·185	0·698	1·612	0·552	1·612
Number of days on which rain fell, ...	4	6	4	5	11	9	11	8	10	9	6	8	91

Abstract of daily observations, made in 1868, of rainfall at Wynberg Hill.

From Daily Observations.	Jan.	Feb.	Mar.	April.	May.	June.	July.	Aug.	Sept.	Oct.	Nov.	Dec.	For the year.
	ins.	ins.	ins.	ins.	ins.	ins.	ins.	ins.	ins.	ins.	ins.	ins.	ins.
Quantity of rain in inches, ...	0·927	1·659	0·782	4·556	2·759	4·777	9·032	1·959	2·150	4·626	3·256	0·764	37·247
Greatest amount of rain on any one day in each month, ...	0·712	0·506	0·612	2·487	1·262	1·835	1·6[illegible]4	1·524	0·962	1·780	1·319	0·722	2·487
Number of days on which rain fell, ...	3	7	3	4	5	6	8	4	6	4	5	2	57

Abstract of daily observations made at Simonstown during seven months of 1868.

From Daily Observations.	June.	July.	Aug.	Sept.	Oct.	Nov.	Dec.	For Seven months.
Mean height of the barometer at the temperature 32 deg. Fahr. (not corrected for altitude above the mean level of the sea), for each month, and for the year, from observations at 9 A.M., 1 P.M., and 5 P.M., ...	ins. 30·106	ins. 29·658	ins. 30·063	ins. 29·682	ins. 29·988	ins. 29·974	ins. 29·679	ins. 29·878
Maximum observed, ...	30·490	30·387	30·255	30·279	30·258	30·287	30·668	30·668
Minimum observed, ...	29·874	29·665	29·846	29·817	29·717	29·722	29·720	29·665
Mean temperature of the air for each month, and for the year (Fahr.), from observations at 9 A.M., 1 P.M., and 5 P.M., ...	deg. 60·4	deg. 60·9	deg. 61·3	deg. 60·2	deg. 64·7	deg. 66·8	deg. 69·9	deg. 63·5
Mean temperature of evaporation for each month, and for the year (Fahr.), from observations at 9 A.M., 1 P.M., and 5 P.M., ...	deg. 55·7	deg. 54.1	deg. 55·8	deg. 56·1	deg. 57·6	deg. 60·7	deg. 62·2	deg. 57·5
Mean humidity of the air for each month, and for the year (the integer number 100 implies saturation), derived from observations at 9 A.M., 1 P.M., and 5 P.M., ...	74·0	68·0	71·0	73·0	66·0	69·0	63·0	69·0
Maximum humidity, ...	100·0	100·0	93·0	100·0	100·0	94·0	94·0	100·0
Minimum humidity, ...	34·0	32·0	37·0	37·0	47·0	35·0	32·0	32·0
From the self-registering thermometers (Fahr)	deg.	deg.	deg.	deg.	deg.	deg.	deg.	deg.
Mean maximum temperature of the air, ...	..	67·7	69·0	63·7	68·9	70·8	74·7	69·1
Mean minmum temperature of the air, ...	52·0	52·9	55·0	54·4	56·0	58·9	63·2	56·1
Approximate mean temperature of the air, ...	..	60·3	62·0	59·0	62·4	64·8	68·9	62·9
Mean range of the temperature of the air, ...	..	14·8	14·0	9·3	12·9	11·9	11·5	12·2
Highest temperature, ...	..	87·0	83·0	79·0	84·0	82·0	87·0	87·0
Lowest temperature, ...	48·0	49.0	50·0	49·0	51·0	51·0	51·0	48·0
Extreme range of temperature on any one day, ...	..	32·0	30·0	14·0	24·0	20·0	22·0	32·0
Least range of temperature on any one day, ...	..	8·0	5·0	5·0	5·0	6·0	7·0	8·0
Quantity of rain in inches, ...	0·340	0·299	0·086	0·305	0·320	1·194	0·534	3·076
Greatest amount of rain on any one day in each month, ...	0·164	0·149	0·015	0·116	0·127	1·133	0·460	1·133
Number of days on which rain fell, ...	10	11	5	15	8	6	7	62

Abstract derived from observation at Simon's Town in the years 1862, 1863, 1864, and 1865. Approximate height above sea level, 50 feet.

Months.	Mean height of Barometer at Temperature 32°.	Temperature of the Air. Mean Temperature.	Mean daily range.	Mean of greatest range any one day of each month	Mean of least Range any one day of each month.	Mean of greatest Range throughout each month	Mean Temperature of Evaporation at 9 A.M. 1 P.M. 5 P.M.	Mean Humidity for each month	Rain.	Clouded sky.	Lightning. Recorded No. of Days
	inches.	degrees.	degrees.	degrees.	degrees.	degrees.	degrees.	per cent.	inches.	per cent. of visible hemisphere	
January	29·926	73·99	17·53	24·2	10·1	31·8	68·67	72·3	0·698	36	...
February	29·890	72·99	15·25	22·8	8·4	30·6	68·21	70·7	0·470	37	...
March	29·937	69 98	14·35	23·2	6·2	31·6	66·36	76·4	1·403	42	2
April	29·979	65·34	14·24	22·7	6·4	30·5	62·57	79·2	1·958	56	...
May	30·045	62·18	13·48	21·7	6·3	29·2	59·97	81·4	3·395	55	4
June	30 066	59·17	13·60	22·2	6·1	27·4	57·62	85·2	5·103	60	1
July	30·095	57·46	12·43	19·2	4·2	24·4	56·48	86·5	5·283	57	1
August	30·097	59 02	14·43	27·7	5·8	33·4	57·45	85·0	2·830	57	3
September	30·077	60·77	16·36	27·6	5·9	32·8	58·84	82·6	2·753	55	3
October	29·990	62·72	15·45	26·3	5·3	34·2	60·73	82·4	4·003	66	1
November	30·002	65·75	15·66	24·1	5·7	31·9	62·43	75·7	1·353	56	...
December	29·921	72·34	17·19	26·1	7·6	33·1	67·95	72·6	0·240	44	...
Mean Annual Values.	30 002	65·14	15·00	23.98	6 50	30·91	62·27	79·17	29·489	52	4
	*+·021	...	...	...	...	...	...	...	...	...	...
	30·023	...	...	Greatest 37·9 on Aug 19, 1875.	Least 1 0 on April 18, 1862.	Greatest 46·6 in Aug. 1864.	...	...	...	...	...

* ·021 inch is the reduction to the Observatory Standard Barometer.

SECTION II.—*Divisions of Stellenbosch and the Paarl, and places adjacent.*

The Division of Stellenbosch lies to the east of the Cape Division, between the sea at Hottentot's Holland and the Paarl. It covers about 400 square miles. It lies along the base of the Drakenstein Mountains, and has much beautiful scenery. The Division of the Paarl lies to the south of Stellenbosch. The name is given from the paarl stone, one of three immense granite boulders which are on the top of what is called from it the Paarl Mountain—it has an area of about 600 miles. The soil and climate of the division are similar to those of Stellenbosch; the former is clay, alluvial mould, iron-stone, etc. It is a rich wine district. It is watered by the Great and the Little Berg Rivers, and numerous streams rising in the adjacent mountains.

Zwartland is an adjacent region belonging to the division of Malmesbury, fertile in wheat, to which I find it convenient to advert in connection with Stellenbosch and the Paarl.

Away from the immediate vicinity of Table Mountain the supply of water is somewhat less; but in some quarters it is more so in appearance than in reality.

From the reply of the Van Regneveld, Civil Commissioner of Stellenbosch, to the circular on the subject of irrigation, issued from the Colonial Office in 1862, it appears that the inhabitants of that district residing at Eerste River, especially those near its mouth, suffer greatly at times from the want of water, the supply from the river sometimes almost entirely failing. And from a statement submitted by Mr T. B. Bailie to the Committee of the Legislative Council on Irrigation it would appear that owners of gardens and vineyards at the head of the stream used frequently in former times to draw off the water there for their own use, regardless of the wants of those residing at a lower level, who were thus provoked frequently to take the law into their own hands—provoked thereto by the inconvenience to which they were thus subjected, together with the limited amount of the local supply.

But there are times during which a large supply of water is brought into the district by that river; and this not being there used at the time, or stored up for use in a time of need, passes through the district and is lost in the sea.

I am led, moreover, to conclude that the rainfall is not much less in that district than in Capetown. The only means I have of determining this are records of observations made at Somerset West in 1862 and 1863. From these it appears that the rainfall there was, in these years respectively, 28·707 and 23·630 inches—average, 26·168 inches; while, during the same years, the rainfall observed at the Town-house, Capetown, was 28·750 and 26·900—mean or average, 27·825; and in the following year, 1864, it was then only 17·720. At the Royal Observatory it was, in 1862 and 1863 respectively, 32·008 and 25·603—mean, 28·850; and in 1864 it was only 18·915 inches. It further appears that the average of the rainfall during the six summer months was, at Somerset, 6·8700; and during the six winter months, 19·2984—the latter quantity is equivalent to about 425,000 gallons per acre, which is much more than would be soon evaporated or would be absorbed by the soil.

From Stellenbosch, it was reported by the Civil Commission in 1869:—"The state of the division has been satisfactory, the crops abundant, and of good quality, especially wheat; in fact, owing to the continued depressed state of the wine trade, the farmers have directed their attention to the cultivation of cereals to a much greater extent than hitherto. The *oïdium* made its appearance, but in a much milder form; its progress was, as usual, arrested by the application of sulphur, and the vintage was considered a fair average one."

In 1874 it was reported:—"The state of the division is not so satisfactory, on the whole; and, although rains commenced in due season, and in succession, the farmers expected more abundant crops in grain and wine; but, owing to the strong south-easterly winds, they have sustained a good deal of damage in consequence.

"The oïdium made its appearance again, but by the application of sulphur in time, its progress was arrested, and I am glad to say there was an improvement in the price of wine and other produce.

"Ostrich farming in this division has been quite the rage for the last two years, and has been carried on to great extent and success. Mr Jan Beyers, of Nooitgedagt, near Stellenbosch, especially, has been most successful; the birds seem to thrive exceedingly well on his farm.

"Farmers continue to introduce thrashing machines, which seem to answer their object, and these supply the deficiency of manual labour, which, through emigration to the diamond-fields and railway works, is most severly felt.

"Crime during last year was on the increase."

In this district I have seen indications of water having stagnated so long as to be injurious to the vines growing there, retained probably by an underlying stratum of clay. And there are not wanting in the district facilities for the construction of dams similar to what has been constructed by the Hon. Mr Hare, of Groene-fontein.

In regard to that dam Mr Hare gave detailed information in giving evidence before a Select Committee of the Legislative Council, appointed on 1st July, 1862, "to enquire by what means the Government and the Legislature can promote a more extensive system of irrigation in the Colony." To these details I shall afterwards have occasion to refer. Here let it suffice to say that his dam is calculated to hold 16,000,000 gallons of water; that it has been filled oftener than once; that the district is described by him as "a wine district, a pastoral district, and an agricultural district—it is a little of all;" that he said the tank full of water "would irrigate something like six hundred acres;" and from a calculation made by one of the railway engineers and himself he was led to believe "that every inch of the dam would irrigate 10,000 vines," but that he did not believe irrigation "would be of much use in increasing the quantity of grain produced in the district." He was asked:—"Q. 391. From your knowledge of the Colony, are you not aware that your observation in this respect does not apply to other districts?—Most decidedly not, because I know that most other districts grow their crops with water, and that they would not be able to produce as much without it; but I do not think we would grow a muid more with irrigation. We are peculiarly situated. We grow our stuff mostly on the side of rising ground where you could not make a dam. In a flat country, of course, irrigation could be extensively and beneficially practised.

"392. Is your part of the country also liable to drought?—We always depend upon our rains more or less. There is a difference in the fall now and then; but taking it as a whole, for the last seventeen or eighteen years, I think we have always depended upon our ploughing time and harvest. There have been years when it has been very dry, but we have been always, I think, well satisfied.

"393. *Mr Wood.*] Is the sugar cane at all cultivated in your part of the country?—Not to any extent. I have seen it in gardens and answering remarkably well." The testimony of Mr Hare may be considered indicative of what may be done in the district of the Paarl.

Of the Paarl it was reported by the Civil Commission in 1874:—"The district appears to be in a flourishing state, and the prices of landed property still continue high.

"The vintage does not promise to yield quite so well as was generally anticipated, owing to the exceptionally dry weather, no rain having fallen since November. There is, consequently, a great scarcity of water in the upper parts of the town, the reservoir being quite empty.

"Bush-fires are continually occurring in Du Toit's Kloof, and other parts of the mountain, causing great damage to the bush and underwood. Stringent measures are being taken to put a stop to it.

"Crime does not appear to be on the increase."

The reports in regard to crime I cite because it is alleged that in some parts of the Colony crime increases with drought and diminishes with satisfactory supplies of rain.

Drackenstein and French Hoek, in the division of the Paarl, I visited shortly after a fall of copious rains, when there was not only no lack of moisture, but the moisture was in some places in excess of what was desirable for agricultural purposes; and in both, but especially the latter, the character of the uncultivated vegetation gave indication of the retention of a considerable quantity of water in the soil throughout a great part, if not the whole, of the year.

A fine waterfall in Drackenstein arrests the eye of every one passing that way. This I visited, and though no particular plan suggested itself to me as the best means of utilising this by the employment of the water for irrigation or as a motive power—for either or for both; I saw enough to satisfy me that it might be utilised to a great extent in various ways.

The same might probably be said of a lesser waterfall in French Hoek; but of this I cannot speak from personal observation.

I had heard of irrigation works in Koeberg; and arrangements were proposed for my visiting that district; but, in accordance with the suggestion of friends resident there, my visit was deferred for a few months, and I had to leave the Colony without accomplishing my purpose. I am thus unable to report largely, as probably I might otherwise have been able to do, in regard to this district.

One of my correspondents, a civil engineer, who has given much attention to the subject of irrigation, and the practicability of its application to the agriculture of this country, sent to me a copy of a letter which he had addressed to the editor of the *Cape Monitor*, and which was published in that journal in the beginning of 1861. From this I make the following statement. Speaking of Zwartland

and Koeberg, he says:—"This tract of country—naturally—is perhaps as dry during the summer months as any in the Colony. 'From the Berg River to Capetown millions of pounds of meat and fat are destroyed yearly, for want of water at the resting places.' Fountains are few, and wide apart, and summer streams are entirely absent. The most of the farms are supplied with excavated dams, the waters of which become in summer unwholesome in the extreme, both to man and beast. Now, if this is, so far, a correct description of Koeberg and Zwartland, their proximity to Capetown, the density of their population—annually increasing, and the enterprise and industry of the farmers, points them out as the very tracts of country to show what can be done by artificial irrigation. And it is to be hoped, that when the subject is well investigated—not by 'dealers in impracticabilities,' but by clear headed, earnest men—no obstacle will be in the way of trying the experiment. Two plans are applicable to the irrigation of a large portion of Zwartland and Koeberg—first, raising water by wells; and, secondly, by husbanding the winter streams in artificial lakes. We shall pass over the former, as being too primitive for the enterprise of the Zwartland boer, and draw attention to the latter, as being most worthy of consideration. The districts in question are traversed by periodical rivers, having a large body of water during the winter. Some of them appear to have but little fall, and consequently are well suited for embanking. The aid of the rain-guage is not necessary in this instance, as the years (if any) that the periodicals failed to produce a stream of water can be ascertained within the memory of man. Take, for example, the Malmesbury periodical. It requires no great strain on the imagination to conceive the spirited inhabitants asking the aid of Government in the matter of a few thousand pounds and professional advice, in order to form an artificial lake, near the village, two or three miles in length; and to conceive this water irrigating hundreds of acres, covered with valuable plantations, giving nourishment to hundreds of trees of the finest timber, and fertilising and creating new fountains miles to the rear. A successful experiment of this description and irrigation will gradually spread to the remotest districts of the colony, districts where it can be carried on to a greater extent than ever it can be at Zwartland. 'Wheatlands,' (says Mr Rubidge in his paper on overpasture). 'late the property of Mr Thornton, now Mr Parke's, differs from the millions of square miles around only in having a spring from the Milk River in a position to be led over the fields.' But every one will admit that the ground suitable for irrigation in this colony is unlimited; and water we all know is the only element required to make it highly productive. But what the real value of water is, in connection with farming operations in our different districts, say per thousand cubic feet, is not yet ascertained. To the like of Mr Gird or Mr Hare in the west, and Mr Reitz in the east or midland—with Government aid—we look for this information. Once the value of water is fixed upon by experiments, the construction of artificial lakes, etc., need not be doubtful undertakings, or 'a waste of public money.' *Zot wederziens.*"

Mr Gird, to whom my correspondent referred, in a letter which appeared in the same journal, pointed out different localities in that district where dams for the supply of water at outspan places might be constructed with advantage. After referring to some spots which had been the subject of angry discussion, he proceeds:—"Then there is another outspan place on the same New Kloof route, behind Paardenberg, at the crossing of roads from different directions, and a boundary line between the same two divisions, which is required for a drinking dam, a natural hollow basin, which only requires excavation and channels leading to it for collecting the rainfall, to make it of great service to the slaughter cattle and sheep, as well as to animals of draught of all kinds used in transport on that line: the traffic being very great, more particularly by slaughter stock, nine tenths of the supply for Capetown and the neighbouring district coming down by this route; and this would complete the chain of water supply from the Berg River to the outspan place at Lichtenburg.

"If that dirty, stinking, muddy pool at D'Urban, were properly cleaned out, deepened, and the bottom paved with large stones, as well as the approaches to it, and a stone wall built, to separate the road and road dirt from the water, it would then be not only an ornament and a credit to the village and the divisional council, but be the means of supporting, with the other water stations from the Berg River to this place, the slaughter stock in good condition until it reached Haarde Kraal Station or Outspan Place, near the twelfth mile stone, and facilitate the transport of produce to an extent beyond all belief of the Capetown people, and reduce the price of it in proportion.

"At this Haarde Kraal, Outspan Place, where nature points with unerring certainty to the spot of all others suitable for a large drinking dam, there is only a very limited supply some distance from it, and which at times is barely enough for the tamest horses and mules, but from which the slaughter stock derive no benefit, nor the thousands of oxen used in draught for transport, the poor famished creatures preferring rather to drink from a ditch low down of brackish and deleterious water, from which they purge to death, and scatter the line of road with their bones, mementos of disgrace to the cruel and improvident who thus neglect one of their primary duties to God for the protection of the useful creatures entrusted to their care.

"I trust I have now laid before the public enough to convince them that there is sufficient to be done on this side of the Berg River to furnish slaughter stock with the means of preserving their condition in the frightful heat they are subjected to on the road to market; and that even this instalment of justice for the cattle paths will be the saving of millions of pounds of flesh and fat to the public annually."

The following is an abstract of meteorological observations at Somerset West in the years 1862 and 1863:—

Abstract derived from meteorological observations at Somerset West in the years 1862 and 1863.

Months.	Mean height of Barometer at Temperature 32°.	Temperature of the Air. Mean Temperature.	Mean daily range.	Mean of greatest range any one day of each month	Mean of least Range any one day of each month.	Mean of greatest Range throughout each month	Mean Temperature of Evaporation at 9 A.M. 1 P.M. 5 P.M.	Mean Humidity for each month	Rain.	Clouded sky.	Lightning. Recorded No. of Days
	inches.	degrees.	degrees.	degrees.	degrees.	degrees.	degrees.	per cent.	inches.	per cent. of visible hemisphere	
January	29·817	71·04	24 90	36·9	12·7	43·0	66·97	55·81	0·356	16	1
February	29·801	70·64	22·69	35·2	10·6	45·7	66·34	57·48	0·743	23	...
March	29·840	66 74	22·59	37·9	8·3	42·2	63·72	63 37	1·331	27	...
April	29·862	63·07	22·38	37·3	7·6	48·8	61·17	67·86	1·448	35	1
May	29·971	58·84	20·59	34·0	4·9	43·2	57·63	73·99	2·475	34	1
June	29 945	55·41	17·00	34·0	7·1	41·0	53·86	80·00	5·727	44	1
July	30·007	54·04	17·21	29·9	5·5	39·1	53·03	79·72	4·299	41	1
August	30·021	53·99	16·96	35·7	4·2	41·9	53·55	77·62	3·395	40	1
September	29·981	56 07	19 27	35·1	6·4	39·2	55·56	72·10	2·456	31	...
October	29·897	59·06	18 90	31·6	8·9	35·8	58·63	75 60	2·999	48	...
November	29·888	63·10	22·85	37·3	9·9	45·5	61·61	65·16	0·963	32	...
December	29·811	70·67	25·26	36·1	11·3	49·0	67·11	57·20	0·479	20	...
Mean Annual Values.	29·903	61·89	20·88	35·08	8·12	42·87	59·93	68·83	26·671	33	3
	*+·028	...	...	...	...	...	...	...	...	...	...
	29·931	...	...	39·4 on Aug. 15, 1863.	3·3 on Aug. 8, 1862.	52·6 in Apr. 1862.	...	...	...	...	...

* ·028 inch is the reduction to the Observatory Standard Barometer.

SECTION III.—*Divisions of Malmesbury, Piketberg, Clanwilliam, and Namaqualand and regions beyond.*

Malmesbury is a division lying along the coast to the north of the Cape division. It has an area of 2040 square miles. The soil is fertile, producing both corn and wine; but it is chiefly noted for the production of grain. Its main river is the Berg River, with its affluents the Twenty-four Rivers and the Zout River.

In the town of Malmesbury is a warm spring, to which medicinal virtues are ascribed.

On my arrival at the Colony, in 1863, I heard much of the sufferings of the division from drought. In one year during my residence there I was told that several vleies and dams which had never before failed were dry; amongst others, one which was described as a beautiful lake at Drooge Valley, the property of Messrs John and William Eaton.

In 1858 petitions were presented to Parliament to take into consideration the expediency of constructing a wall across the Berg River, to which proposal I shall afterwards have occasion to advert.

In 1862 queries similar to those addressed to the Civil Commissioners in other parts of the Colony in regard to irrigation were addressed to the Civil Commission at Malmesbury.

The queries were these—(1) Have any, and what, plans for irrigation works within your division been proposed? (2) What irrigation works—such as, by the damming up of rivers or other running streams, by diverting or raising water from such sources, by the construction of dams in favourable localities, by the collection of periodical supplies of water, or by the sinking of wells—appear to you to be practicable in your division? (3) To what extent would the landed proprietors in your division be likely to avail themselves of a law empowering the Government to make them advances of money upon reasonable terms, with convenient periods for repayment, for the purpose of executing works for improving the supply of water on their properties?

Mr Rainier, the Civil Commissioner of the district, wrote in reply to these queries in 1863:—" 1st. I am not aware of any plans for irrigation works having been proposed within my district, with the exception of a plan formed by some of the inhabitants of the Berg, to dam up that river near the mouth, to prevent the salt water from entering; which plan, if carried out, would, in my opinion, effectually destroy the navigation of that river.

"2nd. The Berg River is the only river in this district which could be dammed up; which might be done at great expense, and, I think with advantage, at a distance of forty or fifty miles from the mouth, without injuring the navigation of the river.

"The water of the said river where the tide flows in, on the lower part, might also be made available for irrigation purposes if the inhabitants residing along its banks made large dams with flood gates, which could be closed when the river was salt.

"The greater part of the district is sadly in want of large dams, for the construction of which great facilities are afforded if the farmers would build a strong wall across the narrow parts of valleys.

"The sinking of wells might also be carried on to a great extent, as I feel certain that water will in all places be obtained without going to any great depth.

"During the past dry season many farmers have had to drive their cattle distances of two and three hours to drink water on farms where great facilities offer for making dams, but the owners, from want of funds, have been unable to do so, and consequently the loss of cattle during the past year has been very great. I would also mention that, if the district were well supplied with water, most of the farmers could do without half the quantity of land they now require.

"I believe the farmers of this district would be glad to avail themselves largely of a law empowering the Government to make them advances of money upon reasonable terms, with convenient periods of repayment.

"Any plan for the construction of dams or other irrigation works should be submitted first to some person or board competent of judging of the feasibility of the plan proposed. And the money applied for should be advanced by instalments as the work progressed.

"I may also mention that I know one farmer who has made several dams and is now about to sell a large portion of his farm, finding that, in consequence of the increased supply of water, he can now farm on as large a scale on the remainder of his farm as he could before on the whole farm."

In the report of the Civil Commission for 1865 it was stated:—
"The past year has been a very dry one; and some parts of the

division, particularly that called 'Red Zwartland,' have suffered greatly in their crops from drought, and from want of water for their cattle. The rest of the division has suffered greatly from the rust in the wheat crops, by which it is calculated at least one third of the crop has been lost. Some of the poor farmers are without bread or seed wheat. Had it not been for the rust, the crop would have been nearly an average one.

"As it is, many of the famers have had nothing to sell, in consequence of the failure of their crops; and have neither money to pay nor food to give their labourers. It is, therefore, anticipated that still greater distress will prevail amongst the labouring classes during the year 1866.

"The pasturage has generally been better than it was in 1864; and consequently the stock were in fair condition.

"The vineyards in many places have suffered from the dry weather, so that a short crop is expected.

"There still continues to be a good deal of distress amongst the labouring classes; and pauperism is greatly on the increase."

In 1866 the report was more satisfactory. In this it was stated:—

"In the early part of the year copious rains fell, producing abundance of grass; consequently, the cattle required for ploughing were in excellent condition. The winter set in with beautiful showers, and the farmers generally sowed much more grain than usual; but, owing to the failure of late rains, the wheat crops suffered considerably; notwithstanding which, the yield of wheat is good, while that of rye, oats, and barley is much more abundant than has been known for several years past.

"There have been no improvements in agriculture; the scythe is much used, and is acknowledged by all who use it to be far superior in every respect to the sickle.

"Cattle have been in excellent condition, and large numbers have been purchased by butchers from Capetown.

"The quantity of wool produced in this division is increasing rapidly.

"There does not appear to have been so much distress amongst the labouring classes as was anticipated in last report."

In 1869 it was reported:—"The general state of this division is wholesome, and in every respect highly satisfactory; the grain crops have been unusually good, and free from rust or any other disease. There has been no disease amongst the cattle or horses, and cattle are in excellent condition. Rains fell in abundance; the springs are strong, and the dams better supplied with water than they have been for the last fifteen years.

"There have been no improvements in agriculture. The several fisheries on the coast have been, as usual, successful. The roads in the division suffered much damage from the heavy rains during the winter, but are now in good repair.

"Trade appears to be flourishing, and the general aspect of things is decidedly encouraging."

The following was the quantity of rainfall in inches observed at Groote Post, Malmesbury, in 1867 :—Jany., 0·290; Feby., 0·510; Mar., 1·110; April, 1·525; May, 1·055; June, 2·053; July, 3·623; Aug., 1·350; Sept., 1·000; Oct., 2·990; Nov. 0·330; Dec. 0·175—for the year, 16·011. And in 1868 :—From 1st January till 30th June, 7·107; July, 3 520; Aug., 1·095; Sept., 0·980; Oct., 2·755; Nov., 1·550; Dec., 0·910—for the year, 14·957.

To the north of the division of Malmesbury is the division of Piketberg, which is noted for the large quantity of wheat which is there raised, but wine and tobacco are also grown. It covers an area of 1770 square miles, and includes an isolated range of mountains standing out as it were as a sentinel from the Olifant's River range of mountains, whence its name. The Berg River forms the boundary between the divisions of Malmesbury and Piketberg. From what was stated, in evidence collected by Parliament, in regard to the expediency of constructing a wall across the Berg River, it appears that this district, like that of Malmesbury, might be greatly benefitted by any satisfactory arrangement for excluding the salt-water from the Berg River, or from dams supplied from it near its mouth.

And Mr Schonnberg, the Civil Commissioner, reported in 1862 that extensive irrigation works, such as damming up the rivers and water-courses, constructing dams, and the sinking of wells, are practicable throughout the division, especially in the flats of Voor- and Achter-Piketberg; and that such works would tend greatly to the prosperity of the division.

In the Report of the Civil Commission for 1865 it was reported :—
"The past year was one of which better hopes were entertained than of the preceding one; but unfortunately those hopes have been disappointed.

"Owing to the want of sufficient rain at the proper season, the farmers were not able to sow so much grain as they wished; and the excessive drought, and the attacks of the fly and rust, in wheat, rye, and oats, have destroyed the harvest to such an extent that farmers, who in ordinary years sold from 200 to 500 muids of wheat, are obliged to purchase their bread for this year. The consequence is, that for want of money and food, they have been obliged to discharge nearly all their servants, who are thus thrown on the world without means of support, and with destitution staring them in the face; and unless Parliament or Government adopt some means to meet the general distress among the poorer classes of both colours, it is impossible not to foresee a great increase of crime.

"The only part of this division that produced cereals sufficient to maintain itself, is a part of the Twenty-four Rivers and Great Berg-River.

"The vintage is very fair; and the vine disease showed itself in a very mild form, which the timely application of sulphur removed entirely.

"Stock of all kinds have suffered less from disease, and have maintained their condition better than was expected; but the drought will no doubt affect them during the ensuing three or four months."

In 1866 it was reported:—Piketberg has not suffered so much from drought during the past year as it did during the year 1865; the consequence has been that an increase in the produce of the division, as regards cereals of all kinds, has taken place; but, notwithstanding this, the people have not been able to overcome sufficiently the reverses they suffered in the preceding years to enable them to carry out any improvements likely to benefit themselves or the division. The yield in grain may be estimated at 75 per cent. increase on the census returns of March, 1865. The vintage will not be equal to the last year's crop, owing to the excessive dryness of the latter portion of the year. The water on many of the best wine farms is failing. The vine desease has appeared in many places, but not to any great extent.

"All kinds of stock have done very well; no diminution from disease has taken place. As regards trade, the same may be said of the past year as of the year preceding it; except that in consequence of the increased harvest an increase in export has taken place. It is to be regretted that traders and others at the different ports are so indisposed to afford information as to the quantity of produce shipped out of the division, by which a fair idea of the producing capacity of the division might be obtained.

"In the early part of the year the labouring classes suffered severly from want of employment; but crime has, notwithstanding, not been so prevalent as was expected. It is to be hoped that the ensuing year will be more favourable for this class of the population."

In 1869 it was reported :—"The year 1869 has been a very favourable year for those engaged in agriculture. There has been an abundance of rain throughout the year, and even the present summer is very mild. The crops were good and the yield abundant; the pasturage throughout good. Stock of all description are thriving well. Vineyards have not been subject to oïdium during the past year, and the vines promise a fair yield.

"Fruit of all sorts is in abundance. Tobacco has been much cultivated, and the crops promise well."

In 1874 it was reported :—"If the former year (1873) was one of abundance and prosperity, this one was one of unexampled drought and scarcity. The rainfall was far below the average, the consequence of which was a most severe drought, which lasted through all the winter, and resulting in the almost total failure of the grain crops. Farmers who in former years sold upwards of one thousand muids of grain, are this year obliged to buy for their own consumption. Provisions of all kinds have consequently increased in prices almost beyond precedent. Crime appears to be on the decrease."

And again, in 1875 :—"The year 1875, like its predecessor, has been one of great drought. In some parts of this division more than the half of the cattle and sheep have died from sheer want, and a great many permanent springs have either altogether failed, or greatly diminished in their supply.

"The harvest, although below the average, has, on the whole, been better than that of the former year, on account of the rains having been more seasonable; but it is doubtful whether the rainfall for 1875 was greater than that of 1874, the observations of the same only extending as far back as 1st September, 1874. The rainfall in 1875 amounted to 15·265 inches, but it is impossible to say how far this fall is below the average; there is, however, an excess of 1·085 inches rain during the last four months of 1875, compared with that of a similar period of 1874, the amounts being respectively 4·845 and 3·760 inches.

"We shall continue to suffer from these droughts until men have come to understand that setting fire to the mountains, and thus destroying the trees and vegetation, diminishes the rainfall and water supply; that only a certain and known quantity of water falls upon the earth, and *that* at certain periods; and further, that it is necessary to store the surplus waters to supply the deficiency of the dry seasons."

Reference is made, both in the reports from Malmesbury and from Piketberg, to measures which might be taken to render the Berg River much more productive of good to the district than it at present is. It would require a practical hydraulic engineer to report satisfactorily in regard to the practicability and expediency, or impracticability and inexpediency of carrying out measures such as have been proposed.

The Division of Clanwilliam covers an area of 6000 square miles. It includes the fertile valley of the Olifant's River West, and a vast extent of mountain and karoo on either side of it. The climate is very hot in summer, although cold in winter; and the situation is rather unfavourable for communication with other parts of the Colony, owing to its deep sandy roads and rugged surface.

In regard to the division of Clanwilliam, the Civil Commission reported in 1865:—"Clanwilliam, like most other divisions of the Colony, has suffered severely from the prevailing drought. The grain crops have been completely destroyed, and the inhabitants will be compelled to procure their supplies of wheat from the more southern divisions. It may, however, be remarked that the general failure of the crops is mainly attributable to the great apathy and want of enterprise exhibited amongst the agricultural community; for most parts of the division are well supplied with water, available, with but little trouble or expense, for irrigation purposes; and it is to be hoped that the present lamentable state of things will tend to open the eyes of many of those who are now suffering from want, when they might, with but little trouble, be in the enjoyment of abundance for themselves, and sufficient to supply the wants of their neighbours. At present the division is reduced to a deporable state of poverty: there is no money in circulation; cattle are in very low condition; and perfect stagnation prevails in trade; add to this the appearance of lung-sickness amongst the cattle at the farms *Kookfontein*, near Lambert's Bay, and *Vredensdal*, and the probability that the contagion will spread through the whole division."

In accordance with this report is the following sketch of the position, prospects, and capabilities of the division of Clanwilliam, which appeared at the time in more than one of the newspapers of the Colony:—"It really is most disheartening to see so many thousands of acres of the most fertile soil in this district, in the beautiful valley of the Olifant's River, lying waste and comparatively unproductive, owing to the apathetic and indolent dispositions of the majority of the farmers. It has been calculated from reliable sources that not more than one twenty-fifth of the whole extent of the farms are under cultivation, and in some parts of the division not even that. The usual brawback, namely, drought, does not, or ought not to exist here, as the Olifant's River supplies abundant water at all seasons of the year; and, were there the slightest effort on the part of the landowners, tributaries of the river might be made available for irrigation purposes to almost any extent. As a proof of what energy and enterprise will do, it is only necessary to say that two farmers, some time ago, sold their farm in Upper Oliphant's River for the sum of £250, *because they could not make a living on it* (it was by no means the best in the locality); but it now supports upwards of 60 souls, while the present proprietor is deriving a comfortable independence from his exertions. Some of the finest grain in the Colony is grown in the field-cornetcies of Lange Vlei (Zandveld) and the mouth of Olifant's River; but the only market the farmers have for the sale of their produce is at Lambert's Bay, where the trade is entirely in the hands of certain mercantile establishments. This seems a great pity, as little or no competition is the consequence, and the farmers get but low prices for their grain. Why could not some of the leading men among them take it upon themselves to make shipments direct to Capetown, and thus raise competition. The division generally does not exhibit many signs of progress. The education of the farmers' children, beyond what is required by the Dutch Reformed Church in their confirmation rite, is almost totally neglected. Little or no improvement has taken place in agriculture; and thus the district, with all its natural capabilities, remains almost the same as it was twenty years ago. On the whole, however, the prospects are much more cheering than they were last year. Cattle now are in excellent condition, and the oïdium in the vineyards, owing to the use of sulphur, has almost entirely disappeared. The agricultural prospects are certainly good; rain having fallen in sufficient quantity, and at proper intervals, the farmers have been able to get through their ploughing and sowing in a most satisfactory manner, and it is generally expected that the returns for this year will far exceed the average of the last six or seven years. The village itself is certainly not one of the best in the Colony, and, what is more, does not appear to be improving. It is a fact worthy of remark, that during the fifty-eight years of its establishment, since 1808, the population seems rather to have diminished than increased. The present number of white inhabitants may be taken at about eighty-one. Still the general appearance of the village has been very much improved within the last three years, by the building of two churches : one for the Dutch Reformed community, which was commenced in 1863, and opened the following year, is a handsome and tasteful building, and will cost, when completed £3200; the other, for the English Episcopal community, is built entirely of stone from the neighbourhood, at a cost of £1700. The whole of this sum, with the exception of £400 given by the bishop, has been subscribed for by the congregation—Mr R. Grisold, of Lambert's Bay, having given £400, and the rest by the inhabitants of the village. The building is a very neat one, and the workmanship, both inside and out, of the very best description. . . . The public buildings in the village are in a most wretched state, and have, on the whole, about the most dilapidated appearance it is possible to conceive. During a heavy rain last week the water ran in under the doors, and the clerk's room was a mass of mud up to the ankles. Everything had to be moved out, and another room fitted up for temporary occupation, until the place could be cleared and cleaned. It is high time Government took some steps for the improvement of these buildings, especially in a place where the Circuit Court has to be held regularly."

While such representations are given of the state of prostration in which the population is, the country presents facilities for the application to agriculture of water flowing through the division to be lost in the sea, and patches of considerable extent of fertile soil, which could be cultivated with satisfaction if means of transport, at a moderate expense, could be provided.

I visited Clanwilliam, and formed a much higher opinion, both of the residents in the village and of the farmers in the division, and of the coloured as well as the white population, than the pictures given above could have led me to expect.

Having heard much of the sand which abounds there, I was surprised when I saw the rich soil and productive farms lying immediately beyond the pass by which I entered, travelling from Wellington; and, in travelling to the village, I found the country presenting the appearance of a land well watered everywhere.

Amongst the products were cotton, indigo, and rice; while the returns of wheat seemed almost incredible.

On one farm I was told of 105 muids having been reaped as the produce of one! And on another, of an average of 33 muids having been annually reaped for eleven years from the same ground, without artificial application of manure or other dressing! And this I find borne out by the following evidence, given by the Hon. Mr Hare before the Select Committee of the Legislative Council on Irrigation:—"Q. 396. What is the largest quantity you know of, not only in your district but elsewhere, of grain reaped from one muid—the largest percentage of return? The largest I know of my own was one hundred and one to a bushel, and got from barley. That I am certain of, for it was measured twice under the impression that there was some mistake. I have heard, from men that I could depend upon, that in the Olifant's River and about there they merely turned out the grain in places where the sheep had been standing, and reaped, I believe, upwards of one hundred muids of wheat from one; but, as I told my informant, they must either have had a great many stalks or very long ears to do that.

"397. And have you heard what quantity of barley is returned in some places on the Olifant's River?—A most remarkable one, upwards of two hundred fold they say. But then I may state that where we put three muids they only put one; so that if we sow the same, and they reap two muids to our one, the question is—Is it because they sow thinner, or because the land is better? The people down there say it is both combined; but if we do the same as they do we cannot get that return.

"398. On the Olifant's River the land is irrigated by the rise of the river is it not?—Yes, by the overflowing deposit, exactly the same as with the Nile."

On learning of my intention to visit the district, Mr T. L. Bain wrote me a letter on the subject, which I did not receive until after my return. In this he says,—"I have no doubt you will, in the Clanwilliam district, make some discoveries which may be valuable in an agricultural point of view, though the district is generally so despised. I lived there once, and found cotton growing at many farms—rice, indigo at old Prib Smit's, turmeric, sugar cane, and arrow root, and even dates bear there and get ripe. I distributed among the farmers there, when I was leaving, lots of coffee seeds which I got from Natal; but I never heard the result. The soil near the Olifant's River is particularly adapted for many tropical plants.

The indigo, especially, grows without the slightest trouble, and is very profitable, even in the rude way old Mrs Smit used to manufacture it.

In reply to the queries issued from the Colonial Office, which were as follows:—"(1) Have any, and what, plans for irrigation works within your division been proposed?—(2) What irrigation works, such as by the damming up of rivers or other running waters, by diverting or raising water from such sources, by the construction of dams in favourable localities, by the collection of periodical supplies of water, or by the sinking of wells, appear to you to be practicable in your division?—(3) To what extent would the landed proprietors in your division be likely to avail themselves of a law empowering the Government to make them advances of money upon reasonable terms, with convenient periods for repayment, for the purpose of executing works for improving the supply of water on their properties?"—Mr Ford, then the Civil Commissioner of the district, wrote as follows:—"I have the honour to acknowledge the receipt of your circular No. 16, of the 7th instant, requesting me to report on certain points regarding irrigation works in this division.

"(1) As the two first queries do not appear to require more information than my personal knowledge of the division can afford, I consider it unnecessary to delay my reply to them, and shall therefore proceed to report on the three at once. But before entering upon the subject of each query separately, I may premise by stating that the only portion of this division where the construction of irrigation works to any extent, requiring the investment of more capital than the means of the farmers can afford without borrowing, would be confined to the Lower Olifant's River field-cornetcy. The remaining portion of the division is well, and some parts superabundantly, supplied with water, available for all practicable purposes of irrigation, besides which, the rains during the winter months are generally sufficiently plentiful to rear the crops; an occasional drought will of course occur, such as last year, the effects of which in some parts of the country cannot be averted, irrigation being totally impracticable from the nature of the country and soil. The very fertile valley of the Upper Olifant's River is so abundantly supplied with water, not only from the Olifant's River, but from numerous never-failing streams, issuing from the mountains on the east and west sides, that irrigation to any extent can, if necessary, be resorted to with comparatively little expense.

"(2) I shall now proceed to the first question, and, in reply thereto, beg to state that some years ago an effort was made at the missionary institution Ebenezer, on the Lower Olifant's River, to raise water from that river by means of a pump, but this plan proved a perfect failure. At the suggestion of Mr P. Fletcher, C.E., the system of raising water, adopted in many parts of Central India, by means of a leather bag or bucket worked with oxen, is about to be introduced at that institution; should this succeed, I have no doubt that within a very short time every farmer in that locality will avail himself of similar means for irrigation.

"(3) With respect to the second question, I have already stated that the only portion of this division where irrigation is necessary is in the Lower Olifant's River ward, and as regards this portion, after the detailed and most able report of Mr P. Fletcher, C.E., on the practicability and probable expense of constructing works for irrigating the extensive and fertile valleys along the north and south banks of the Upper Olifant's River, from the confluence of the Great Doorn River to the missionary institution Ebenezer, and other parts of the Lower Olifant's River field-cornetcy, printed and submitted to Parliament at its session in 1860, I should consider it an act of presumption on my part to offer an opinion on the subject. A reference to that report (see papers printed by order of Government, G. 29, page 35, No. 45 of annexures to proceedings of House of Assembly) will afford every particular of information required by this query."

Mr Fletcher's report on the points mentioned was as follows:—

"The river is divided into two portions, the Upper and Lower Olifant's River. The total length of both is about one hundred and fifty miles, and having throughout a general breadth which may be stated at four hundred and fifty feet. The summer stream of the Upper Olifant's is mostly in the form of streamlets rippling over a shingly bed, which is studded with small islets covered with bushes. The river bed, thus, is seldom more than ten feet below the banks in its course through the open portion of the country.

"The bed of the Lower Olifant's River is very different. From about Olifant's Drift, below Kokanaap to Klavervley, a distance of forty-five miles, it varies from fourteen to twenty-two feet below the banks or river deposits, and is sandy, but blends into mud at about Kokanaap, and continues so to the sea. The effect of the tide is felt thirty-five miles from the mouth during summer, but in a straight line this is only about eighteen miles. At Ebenezer, a distance of twenty-two miles by the river course, the tide was observed to rise on one occasion three feet, while at the mouth the average rise indicated by two months' observations was only 2·9 feet.

"The fall of the river for twenty-eight miles above Ebenezer is at an average rate of one foot two inches per mile. At the end of the dry season last year, the nearest point on the river to the sea, where water was fresh during ebb and flood tide, was thirteen miles. In summer there is a channel of six and a half feet at high water for twenty-two miles from the mouth.

"The visible volume of water in the river on the 17th May, at the farm Olifant's River, was four cubic feet per second. On the following day it increased to about twenty-five thousand cubic feet per second, which caused the submersion of nearly half the valley, being the first overflow since 1851. This is about the maximum and minimum of the quantity of water in the river during 1859, at a given point, forty-one miles from the sea.

"The permanent tributaries to the Olifant's River all enter it on the Upper Olifant's River; the principal are the Jan Dissel's River, Hex River, and Roode Hoogte River. Some of those apparently produce (superficially) in the summer, as great a volume of water as the Lower Olifant's River itself. The principal 'periodicals' again enter in the Lower Olifant's River; they are the Doorn, Troetroe, and Holle Rivers. The first of these conveys perhaps larger floods in the winter than the Upper Olifant's River. It issues at a point about three miles from the Olifant's River, from a deep narrow glen, and continues for about nine hours on foot. The water course is a bed of large and polished stones from one ton downwards, showing the nature of the river to be a perfect mountain torrent on a large scale. The surface drainage of the Doorn River is very extensive; it branches off to the northward through Calvinia and to the southward to the Cold Bokkeveld, carrying along with it a large quantity of soil. The other two, Troetroe and Holle Rivers, have floods of a milder character—they both convey large deposits.

"The latter is very much influenced by thunder rains, owing to the wide spread of its branches, which extend as far as Losper's Place in Boschmansland in one case, and in the other circumscribes Calvinia, and consequently may give more or less a discharge of water at any time of the year. I shall have occasion to refer to those rivers again. When all these periodicals and the Upper Olifant's River get heavy rain on their respective surface drainage at certain periods of time, so that the waters will enter the valley of the Lower Olifant's about the same instant, then it is sure to overflow its banks; but this of course seldom happens, which may be given as the principal reason for the uncertainty of the natural irrigation, or overflow of the Lower Olifant's River.

"The river ground commences from the Doorn River, and extends to the sea in alternate patches on each side, a distance of about sixty miles. For fourteen miles, however, from the sea, the ground is too saline for cereals; although, I believe, a considerable portion of it might grow other valuable kinds of produce.*

"As will be seen by the plan, there are several extensive fields, some of two and three miles long and half-a-mile broad. The river-course shows a vertical section of the soil to be from fourteen to twenty-two feet thick, in a series of layers. The origin of each can be recognised at sight, by having a knowledge of the peculiarity of the 'slick' or deposit of the several rivers. Some spots occasionally occur, over which the river flows more or less every year; those spots serve as gardens to the inhabitants. But as such are very limited (if I except the delta of the Troetroe River at Vredensdal), it is not always that they have sufficient to grow their pumpkins, which appear to be the almost only and favourite vegetable in this fertile valley.

"As to the fertility of the soil, the various accounts of the great fertility of the soil of the Lower Olifant's River have not been exaggerated. Under ordinarily favourable circumstances, it yields, at an average, one hundred fold; such is also the established character of the deposit of a few of its periodical tributaries, as well as of many of the Hantam and Bokkeveld Rivers, etc. The soil, however, requires, before it produces to this extent, soaking for about twenty-four hours; consequently, no rain has any productive effect upon it, except when of sufficient amount to produce floods having a sluggish pace, or after the evaporation of standing pools. The 'stooling' of cereals varies from twenty to as many as one hundred and sixty stalks from one grain. I went through the stubble of a patch of wheat at Klavervley, on which a fair average of the stools appeared to be thirty-seven stalks to each grain. Some of the farmers sow as little as one muid to fifteen acres. Sowing so thin as this, however, I believe to be no advantage. I have measured one spot over which

* "I have been informed, by a party who used to grow rice in Clanwilliam, that it grows well in brack or saline soil; cabbage, beet, etc., grow well on the same description of soil."

the flood comes every year, and whose proprietor gets a return of one hundred muids to three schepels, about three bushels or three-fourths of a muid. The size of the ground is 4·8 acres, or nearly five acres. When the river overflows late, 'rust' is expected; in which case from forty to sixty-fold is considered a good return. It is unnecessary to remark any further on the fertility of the soil of the Upper Olifant's River when watered, or of karroo-land in general, as it is already too well known to require any additional comments. The quantity of land more or less of the above nature represented on the accompanying plans amounts to eight thousand seven hundred acres. Between the starting point of the survey at Klavervley and the Doorn River there are an additional one thousand four hundred or one thousand five hundred acres; but being generally detached and in narrow strips, I did not deem it necessary to measure these grounds.

"I shall now briefly state a few conclusions I have come to as to the possibility and expediency of irrigating this invaluable soil, which I hope will suffice, in the meantime, to lead to some practical means of gaining the desired object. Although a section was carried on with the view of leading out the water high up the river, still this mode of irrigation soon appeared to be altogether unsuitable, as the following objections will show:—

"1st. During about three months of the year there is a deficiency of water when most required.

"2nd. The deposit to be irrigated is by the river divided into two portions, so that a distributing canal would be necessary on each side of the river.

"3rd. These canals in many cases would require to be carried through Kranzes; this, together with the construction of a dam, would involve an expenditure to which the result attained would bear a very unfavourable comparison.

"There are, however, other two methods eminently adapted for the case of Olifant's River. The one is raising water by machinery, and the other submerging the valley by means of one or two dams.

"The first of these methods I beg to recommend to your favourable consideration, more particularly owing to the trifling amount required for its introduction. It will act as a stepping-stone to a more extensive system, by infusing a spirit of industry and giving a taste for artificial irrigation to the inhabitants. Refined machinery in the shape of pumps and mills has been found unsuitable for the wants of the rural inhabitants of India, but the Government there gave encouragement to what was most suited to their understanding. The same will hold good to a great extent in this Colony. I have heard myself one of our most enterprising and intelligent farmers condemning in eloquent English the most improved force-pump as perfectly useless; and, after an expenditure of some £70 upon one he had procured for himself, he was obliged to substitute the 'blanket pump' (a slight modification of Verra's machinery).

"A reference to Smith and Cotton's works on India will show to what extent pumps or machines are used in that country for irrigation. I cannot see why they should not be comparatively as numerous in some of our districts at the Cape. One of those simple machines placed on the banks of the Olifant's River, and made to play on a few acres of the deposit, would yield from sixty to a hundred fold. Once introduced, it would be within the means of the native as well as the farmer to make or procure one, and in many localities would ensure bread for families and water for their cattle.

"The second method is that of submerging the valley by means of one or two dams. This appears to be the most effective system of irrigation for the river in question.

"The only objection as likely to be of some disadvantage is that a winter crop might be occasionally drowned, perhaps once in six years. But this, in a measure, could be guarded against by planting the lower portion of the deposit with crops which will be least affected by water, such as several kinds of fruit trees, rice, vine on trellises, etc. The advantange of this system of irrigation is that the soil will always retain its productive quality, as a deposit of any desired depth can be allowed to take place. The whole of the ground within the influence of the dam would be irrigated in the same manner as the present periodical overflows irrigate them—but with this difference, that the artificial process would be more gentle and better regulated.

"One submersion of twenty-four hours would yield a crop; consequently, two or three floodings in the year would be all that will be necessary. The principal considerations connected with this system are: firstly, the nature and expense of the dam, and secondly, the extent of land which it will cause to be flooded. The fact of the valley only requiring submersion two or three times in the year may suggest that the upper portion of the dam should be wood work, which need only be applied when an overflow is wanted. The advantages of this arrangement in connection with high floods is obvious.

"The lower portion would, of course, require to be of *the most substantial character.* Any estimate of the probable expense of such a dam can at best be but an approximation. The more so that I was not able to get such complete sections of the two most eligible sites as desirable; for it was not till I witnessed the comparatively tranquil lake formed by the overflow that this system recommended itself. The best spot in point of cheapness of construction is Blaauwkrantz; although the most extensive fields of the deposits will be embraced by constructing one lower down at Huilkrantz.

"The following estimate will give an idea of the respective merits of the two:—

"Blaauwkrantz—probable cost of dam, £12,000; extent of land irrigated, 2,400 acres; yield, say in corn, 45,000 muids, and one or two other crops.

"Huilkrantz—probable cost of dam, £17,000; extent of land irrigated, 4,500 acres; yield, say in corn, 75,000 muids, and one or two other crops.

"The above outlay for either of the two dams, compared with the extent irrigated, may at first appear unfavourable, but if the extraordinary fertility of the soil is taken into consideration, it will be seen that a very remunerative result may be reasonably expected. Where in the one case forty-five thousand muids and in the other seventy-five thousand would be easily produced, at present (except that the river overflows) not more than twenty muids are grown."

As if in anticipation that a survey would produce some such report, the editor of the *South African Commercial Advertiser* wrote, in the beginning of 1858:—"The year 1858 may be distinguished by a survey of two navigable rivers which traverse two of the most fertile tracts of country in South Africa, being the corn producing portions of Malmesbury and Clanwilliam. The first is the Berg River, the upper course of which, through the Paarl Valley, has long been famous for wines and fruit, cultivated by an industrious and skilful population, and connected at present with Capetown and its port by an excellent road, and about to be brought closer still by a railway. But the lower part, which runs through the division of Malmesbury to the sea in St. Helena Bay, should be the highway for the heavy produce of the land on both sides for many miles inland, and would be so to a vast amount if certain obstacles were removed from the mouth and bed of the river. The second is the Olifant's River, which is at present navigable by boats for upwards of thirty miles, at which distance it is affected by the tide, and if certain obstacles at its mouth were removed it could be entered and navigated by coasting vessels of sufficient burden to establish a traffic between it and the principal markets of the Colony. All who have visited these two tracts of country represent them as unequalled, certainly not surpassed, by districts of equal extent in any part of the world, in their power to produce the most valuable crops with the least labour or expense. From periodically overflowing its banks, and depositing alluvial karroo-mud, the best of manures, this river has been called the Nile of South Africa, and we question whether anywhere in Egypt the husbandman reaps more than a hundredfold from his seed, which is here common, and frequently exceeded. It is, however, cut off from the great markets by a difficult country, or rather the great markets are deprived of the wealth and riches for which nature has provided the materials with so lavish a hand. If these two rivers were made easy for vessels of from 50 to 100 or 150 tons, creating a steady demand, we have been assured that the Cape would immediately become an exporting instead of an importing country for grain and flour—both of which it is well known are of the finest quality, and fetch the highest prices in foreign markets. In the course of the year just ended we have imported flour and grain to the value of £120,000. But for the cost of transport by land the whole of this and many times as much more could have been supplied by Malmesbury and Clanwilliam."

In reference to this, a partriot, whose opinion I have already had occasion to cite, wrote:—"With reference to your late remarks regarding the country bordering on the Olifant's River, I may state that, after careful inspection, I have seen no place within the limits of the Cape Colony which presents such a profitable field for the industrious emigrant as the banks of that river. It is really and truly the Nile of South Africa. At certain seasons the whole country is flooded, and hundreds, nay, thousands of acres of land of the richest description are made fit to receive seed and capable of producing the richest crops, without the least exertion on the part of the agriculturist. Nothing is necessary after the water has receded but to throw wheat or other grain upon the surface and drag a few mimosa trees over it. If the mouth of the river could be opened the produce would find an easy market. At Koknap, some 25 miles from the mouth, there is a rise and fall of the tide equal to three or four feet. The farmers in that neighbourhood appear to be a most apathetic race. On one of my journeys I stopped at the house of a boer to rest and feed my jaded horses. While this was being done I strolled along the banks of the river, and was much struck with the extremely fertile appearance of the soil, and the very little which had been done for turning it to account. It seemed as if the Creator had done everything for the country, and man nothing. Scarcely any rain had fallen for some time past, and the river had not overflowed its banks for more than a year. The stocks of grain and vegetables were getting very low. The *Baas* was complaining much about the long-protracted drought, and when he had finished I took the liberty of pointing out how he could, by leading out the stream for the purposes of irrigation, or by fixing a pump to be propelled by wind on the river's bank, secure an abundant supply independent of the weather. He seemed to listen with some interest to the development of my plans, and I began to hope that he had decided upon doing something to relieve himself of the difficulty; but eventually, after turning round and scrutinising the whole horizon in the direction of the river's source, as if in search of some favourable symptom, he yawned heavily, and merely observed:—'*Ach! wat, dat zal een dag regen.*'"

The publication of this was accompanied by the following editorial remarks:—"We insert with much pleasure the letter signed 'T. W.,' on 'Boring for Water,' and some other methods for converting that precious element into gold. He speaks from personal observation of the Olifant's River; and the fact that at *twenty-five* miles from its mouth the tide rose several feet, and also that at certain seasons the fresh water of this river overflows a large extent of country on both sides, indicate its immense value as the means of both production and transit. From these facts we learn that a dam or weir carried across the stream, near its mouth, would establish a navigable canal twenty-five miles in length; and at the same time would be formed a *lake of fresh water* of equal length, and as broad as the river channel, available at all seasons for cattle and for artificial irrigation; while the natural irrigation in winter prepares the soil for seed, without manure, and almost without the labour of the plough or the spade. As the waters abate you scatter the seed on the rich and moist surface, and in a few months reap from a hundred and fifty to three hundred fold. Nature designed this to be one of the most productive districts on the face of the globe. Up to the present hour it is one of the poorest.

"This description is applicable in every respect to the Berg River. In the channel of this river also a fresh water lake of twenty-five to thirty miles in length could be formed by one or two dams or weirs for the same purposes of irrigation and transit. All that is wanted is water, which now runs to waste. The soil and the sun will do all the rest.

"And we may here remark that when such canals, lakes, or tanks are formed in a level country, the benefit is not confined to irrigation on the surface. The water finds its way underground, or percolates the neighbouring country to an incredible distance, *irrigating from below*, as we soon discover by the growth of plants and trees, and filling the wells many miles off. There is established a constant current, through innumerable pores from these head waters, to supply the consumption by vegetation and the action of the air."

The failure of the inhabitants of the district to secure the full benefit of a position so advantageous as is thus represented may seem to indicate the existence of some counteracting influence, which it may be feared is connected with the physical condition of the country; and the consideration of this seems to me to render it necessary that I should, though I do it with reluctance, introduce into a treatise relating only to the water supply of South Africa the information laid before Parliament, by command of the Governor, in reply to the third of the queries issued from the Colonial Office.

In reply to this, Mr Ford writes :—"This question can only apply to the proprietors of farms in the Under Olifant's River ward, and it is my firm opinion that none of them would feel inclined to avail themselves of any law empowering the Government to make advances for the execution of irrigation works, however favourable the terms of repayment might be made; and I do not express this belief without grounds, but from the fact of my having taken advantage of every opportunity of endeavouring to impress upon the minds of the farmers the necessity of exerting themselves for the improvement of their condition, by incurring some expense to increase the productive capabilities of their farms, where the means are practicable, with a trifling outlay. I regret, however, to say that with few exceptions their apathetic dispositions are so inherent that no argument is sufficient to produce any favourable result; their ideas do not appear to extend beyond the present means of obtaining a subsistence for themselves and families, without a thought for the future prosperity of their children."

And similar to this is the testimony of Mr Fletcher, who wrote in his report :—"The lands of which I have been speaking and the river deposit throughout, however, are all private property in the hands of the farmers or the Ebenezer missionary institution. On these proprietors, accordingly, it must to a great extent depend whether these irrigation improvements can be effected or not. It seems necessary to give you some general idea of the character of the people. Their virtues are many and well known; but, from the nature of the case, I shall be compelled here to refer rather to their faults than their excellencies, that the Government may perfectly understand all the difficulties they may encounter in accomplishing the changes now proposed.

"The farming population along the irrigable portion of the river I estimate at about one hundred and twenty souls. With one or two exceptions, their means are very limited indeed. The total income of the whole valley could not have exceeded fourteen hundred pounds per annum during the last nine years, which will give about eleven pounds thirteen shillings for each man, women, and child. This is I think, if anything overrated. Except when the river overflows, they scarcely grow sufficient bread for their own consumption, and until they got into the way of sowing in the outer or 'buitenlands,' which is not many years ago, they were often obliged to buy bread or otherwise want it. Even last year I know that some of them stored their wheat, owing to the uncertainty of the seasons succeeding.

"For the last eight winters, they have watched with nervous excitement each river flood, and found their watchings ended in disappointment. On the 19th of May last, I was fortunate enough to witness a successful overflow of the river and the excitement it produced among the farmers and natives. All, from the youngest child and upwards, appeared to participate in it. Night and day they watched its progress when it once passed the usual limits. They were ill prepared, however, for taking advantage of the fertile fields ready for the plough, as they had just finished ploughing their 'buitenland,' and their oxen were mere skeletons. They, however, put new life in them as they best could; while members and relatives of the resident families collected from the distant parts of the district with fresh oxen and ploughs.

"For ten or fourteen days the 'apathetic inhabitants of the Under Olifant's River' did not eat the bread of idleness; day and night they worked incessantly; the sun and the moon alike rose and went down upon them at their labours. But to talk to them of artificial irrigation by dams on the river, the idea is ridiculed; and pumps they can satisfactorily demonstrate would prove a dead failure. Those of them who have a vague notion that something might be done in this direction, fear the introduction of taxation; but perhaps the idea most dreaded of all is, that the land will some how or other get into the hands of strangers *(vreemde menschen)*, or that such will be encouraged to settle among them. This is a fact which indicates more deep-seated ignorance than any other; it is one which extends throughout the length and breadth of Clanwilliam. As is the case in the other wards of the district, education is at a very low ebb. There is only one family that has a teacher, to whom several of the neighbours send their children. An inspection of this school shows at once the industry of the teacher and the perseverance of the scholars; there are ten of the latter, the most of them grown-up and some of them even married; they are taught to read their catechism, but the greater portion and efforts of their time is devoted to writing a copy of a manuscript letter placed before them. By the end of six weeks (the time generally supposed requisite to complete their education) they can produce a tolerable copy of the original epistle. It is really astonishing to see the progress they make even under this system. Arithmetic is not taught, and I believe there is not one of the farming population I allude to who can cipher a sum in the simplest rules of arithmetic.

"All the farms along the irrigable portion of the river are partner farms, or what they call *maatschappy plaatsen*; some of them have to the number of seven proprietors, generally brothers or brothers-in-law.

"Those imaginary shares are again subdivided among their families. One of themselves pointed out a farm where seventeen individuals could claim shares. Nothing is more ruinous to the moral as well as material progress of those farmers than the principles of *maatschappy* farms. Quarrels *(roezies)* are created between brothers about cutting a bush, alloting flooded river ground, or keeping a few extra bucks, which extend over years and in some instances through life. In talking to some of the most intelligent of the people, they appeared very desirous of having a portion of land that each shall call his own; but doubted whether they could demand a division. I believe they can. One thing, however, is certain, that until these *maatschappys* are done away with, there will be no harmony or prosperity on the Lower Olifant's River.

"Should His Excellency or the Parliament desire to have direct information from the most extensive and enterprising landholders on the Lower Olifant's River, on the subject of introducing irrigation, I beg to refer them to the following parties :—Messrs Gert du Toit, Tilman Neuwoudt, and Wolfaard, Bakkely Plaats; and Albert van Zyl, or Pieter, his brother, of Langevlei.

"It seems desirable that I should next refer to the present condition of the Ebenezer Institution, for two reasons. In the first place, because it possesses, without occupying or in any respect adequately cultivating, a very large extent of irrigable land granted by the Government of former years. In the second place, because, under the systems which have prevailed there hitherto, it has failed to render its inhabitants, an industrious class of people, available to supply the labour-market for the farmers in the surrounding country."

A description of the state of the people is given, and then Mr Fletcher goes on to say :—" Such is a brief and true description of Ebenezer in 1859. Far be it from me to detract from the good that the worthy missionaries of this benevolent society have done. I need only refer to their labours and successes at Wupperthal and Tulbagh Institutions to confirm this. But at Ebenezer they certainly have failed; not, however, without struggle and hard exertion on the part of the various worthy missionaries who laboured there, or expense to the society—as the existing building, the ruins of a couple of windmills, and a prostrated pump, will testify.

"While the doctrines of the church were promulgated, and education was partially diffused, industry—the lever of christianity—was, and is still, wanting in the native character. When the grant was made to the society, irrigation was evidently contemplated, judging from a clause in the title-deeds. A suction pump of the most improved kind was got, capable of irrigating many acres. I believe it was selected by Mr Smyth, now Astronomer-Royal at Edinburgh, and propelled by wind-power, after a manner of his own invention. It was fixed and set in motion by the hands of the worthy missionary himself, but after giving a few strokes a flange broke, and with this break ended the necessary accompaniment of christianity. To this day the power of raising water for irrigation, by machinery, is not credited by the inhabitants of the Lower Olifant's River.

"Thus, for the want of a knowledge of one of Euclid's simplest propositions, incalculable evil has been done. A great drawback to the progress of the institution is, that the people have no written regulations, and in this respect they are precisely similar to the boers on the *Maatschappy Plaatsen.* This certainly must have been a most serious drawback to the institution, considering that there have been no fewer than eight missionaries stationed there at different periods, all of them good, and some of them even talented, men, but each having doubtless his own peculiar ideas of native character and domestic arrangement. Unless I had seen those natives under different circumstances, I should have concluded that there was no hope for the improvement of the real Ebenezerites; but, as with the farmers so with them, the overflow of the river showed that even they could appreciate 'the sweets of profitable toil.' Even calves were inspanned and the most primitive ploughs were brought into play; in one field I counted fourteen of these implements. Those of the people who had not ploughed for years before, went heart and soul into it, knowing that the returns would be ample and sure; but a better proof still of the prospects of their becoming useful to the Colony at large is, that in one day I engaged sixty able-bodied men at the rate of one shilling per diem and rations; this was a written agreement (*vaste accoord*) that, if required for works of irrigation, they should work at this rate for six months. At a native werf on the Holle River twenty-eight more offered their services upon the same conditions. I shall now show what the society had in their power to do, even with a portion of the means at their command, by simply adopting the rude appliances which have proved so abundantly successful for raising water for irrigation in India. Two men and eight oxen can raise water from a well sufficient to irrigate eight acres of ground; thirty men and fifteen such spans will consequently irrigate one hundred and twenty acres, which would, on the Olifant's River, give at least a return of two thousand muids of corn; this would give more than five muids to every man, woman, and child on the institution, or nearly three pounds per diem to each, for three hundred and sixty-five days, besides one or two other crops for feeding cattle, etc. There would still remain river land to the extent of one thousand seven hundred and eighty acres, skirted for nine miles by a dam of water four hundred feet wide and an average depth of about ten feet, banked at the lower end by the sea water and at the other fed by four or five cubic feet of water per second—even at the end of the dry season. From the above facts it appears that Ebenezer alone could support comfortably a population of at least fifty thousand. To make this statement more palatable, halve it—and say twenty-five thousand; or if you doubt the assertion still, refer to standard works on irrigation—such as Smith's, Cotton's, etc.

"The above remarks are also applicable to the other farms along the Lower Olifant's River."

Mr Fletcher, having given copious abstracts from the report of Colonel Smith, in regard to the physical geography, etc., of Mairwara in Central India, and details given by him of the different kinds of tanks constructed there, and the results which have followed the formation of these, goes on to say:—" Now, were I able to prove to your satisfaction that a large portion of our inland districts comes up to the definition given of Mairwara that their 'rains are extremely precarious, and bad seasons in this respect are the rule, good the exception, that the whole amount of rainfall in favourable seasons rarely exceeds twenty-two inches, and that it often ranges from eight to twelve," then I believe I could satisfactorily show that a large portion of the country I have inspected is even more favourable for cultivation by means of the above artificial irrigation (at least in so far as I am able to judge after a careful perusal of Captain Smith's description) than the district of Central India to which the above extracts refer. But unfortunately, under existing circumstances, I believe no man can say with certainty whether eight or twenty inches is the average rainfall in our inland districts. As an opinion, however, I may state that the Hardeveldt, Kamiesberg, Bushmanland, Hantam, Calvinia, etc., are, with respect to water, *more favourable* than Mairwara seems to have been prior to the introduction of irrigation. Assuming in the mean time this to be the case, our Karroo soil, when it receives moisture, will bear a comparison with any in the world for productiveness; the climate is as good; and the scattered population certainly as prolific.* Those are facts too well known to require proof. It therefore only remains to show that the configuration of large portions of our district is remarkably adapted for the introduction of the variety of dams referred to. As it is inconvenient here to point out the many localities I have observed, I shall only bring a few instances to your notice.

"The Holle or Zout River is a periodical tributary of the Olifant's River, and contributes the most and the best 'Slick' or deposit to that river during the rainy season. By its arteries its brings together the rich Karroo soil of the Hantam and Hardeveldt, and the rich sandy soil of Bushmanland. It rains more or less annually, sometimes once and sometimes as often as ten times in the year. The upper

* "The most of the Clanwilliam farmers have large families; they are notorious for intermarriages; and thus, families of ten bear almost every possible relationship to one another. Medical men know best the physiological results of this practice on a population. I have had a conversation with the late Dr Soek, of Clanwilliam, upon the subject. This talented gentleman had many years' practice among the boers, and he attributes the most of their diseases to intermarriages. This subject, unfortunately, being one of those which a false delicacy forbids being very openly discussed, I shall adduce no farther proof as to the correctness of the assertion on which I have already ventured."

portion of the river, or about fifty miles from its junction with the Olifant's River, overflows more or less every year. I have, however, been informed by one old man, that, as a remarkable and unusual occurrence, he knew two years in which it did not run in his neighbourhood at all. Over the lower portion of the river it now seldom overflows, owing to the increased height of its deposit, which gradually diminishes in depth up the river, and widens in successive seasons to the breadth in some instances of nearly two miles, while the minimum breadth is one or two hundred yards. The river can have but little fall, as I have seen the flood passing over a ploughed patch of ground without moving the soil. On the 18th July, last year, I waded Zout River, the breadth under water was about half a mile, and did not exceed knee deep, except in the water-course, which was crossed by means of a branch of a tree. The depth of the course was about 6 feet and breadth 12 feet. The valley or river deposit of the Zout River is as extensive as that of the Lower Olifant's River, and a considerable portion of it as fertile. The best crop of oats I have seen in Africa was in the deposit of this 'periodical.' Other portions are of a very saline character. At a rough guess, I believe that in many spots a dam might be constructed three or four feet high, and a couple hundred feet long, which would flood several hundred acres, thereby rendering them richly arable. I have measured some of last year's 'slick,' two feet deep; this, of course, was under the most favourable circumstances, but by the use of dams, the deposit might be regulated, the fresh slick might be allowed to deposit to its full extent, so that in a few years the lands would be out of the reach of ordinary floods—if desirable that they should be so. By this system of irrigation, even the most saline basin would become available to agriculture,† and about nine or ten thousand acres on the banks of this one periodical river might be brought under cultivation, which would even excel the richest soil in the 'Baland.' This is at least my confirmed opinion.

"Several of the tributaries to the Zout River have extensive Karroo deposit. Some of their basins reaching to nearly one mile in breadth, and their fall so little, that, standing in their delta, a person cannot sometimes judge with the eye which direction water would flow. Their water-course, which winds through the middle of the deposit, is always well-defined and shews a longitudinal section of the plain.

† "The whole of the deposit of the Olifant's River rests upon a saline stratum. A crust of salt may be observed in the summer, along the water's edge. The water, in wells sunk in the deposit, is perfectly salt, although the surface soil is fresh.

Except in ordinary heavy rains, those channels carry off all the water without overflowing, while a few pounds would leave them in a condition to produce fifty, eighty, or even one hundred-fold. Such is the nature of the Farsh River, Geelbek River, portions of the Troetroe River, Oorlogs Kloof, the Zak River at Amandelboom, and other tributaries of the Hartebeeste River. I have not seen the latter, but have been more than once informed that it has, in some places, a deposit of an hour on horseback in breadth (about five miles), and that when it does overflow there is abundance of grass for all the cattle that visit that quarter. If this description of the Hartebeeste River be correct (and I have no doubt it is—see Mr Moffat's report on a portion of the Orange River, dated 1st February, 1857), the products it may be able to yield either in the form of grain or pasture for cattle, would appear to most people fabulous.

"We have here, and not here only, but an extensive portion of the whole colony, the richest soil in the world, lying at present for two-thirds of the year utterly unoccupied, waste and worthless; and if I have failed to impress on you the abundant and beneficial results which may reasonably be expected from the introduction of irrigation, it must certainly be attributed to my own inability to do sufficient justice to the argument. I trust, however, that I have at least said enough to direct earnest attention to the subject, and to induce others to follow up the same inquiry to a practical and prosperous issue."

It need excite no surprise if I state that such a report did not give universal satisfaction in the district; but I find it stated in the journals of the day,—"The majority of the landed proprietors altogether coincide with said report, and consider it a most able, correct, and *truthful* document. Moreover, they feel sincerely thankful to Mr Fletcher for bringing to the notice of Government one of the most important and really necessary works required in this long-neglected district." But nothing was done!

In 1866 the Civil Commissioner reported:—"This division is still suffering from the effects of protracted drought. The state of the crops is, nevertheless, more encouraging than at a corresponding period of last year, owing to a fall of rain during the early part of the winter. A road from Clanwilliam to Rondegate is in course of construction. . . . It is in contemplation to make a causeway through the morass at Rondegat, which has been successfully drained for that purpose."

Of Clanwilliam, in 1874, it was reported:—"The past year has been a trying one for this district. The drought has been unprecedented; no farmer now residing in the district can remember so great a drought. The wheat and rye crops have in very many instances entirely failed, numerous farmers not having reaped a grain, and others not sufficient for their own consumption. Cattle and other stock are in very low condition, and great losses of stock have been sustained in consequence. . . .

"Ostrich farming is on the increase, and birds, old and young, are in demand at high rates.

"A water-mill is in course of construction in the village by an enterprising farmer and capitalist, who, it is said, intends erecting several dwelling-houses in the village—a very great want, there not being a single vacant dwelling in the place.

"Crime is, unfortunately, on the increase, the result, doubtless, of the very severe drought."

And in 1875:—"The drought for nearly two-thirds of the year was unprecedented. The usual rains for the ploughing season, in April, May, and June, did not fall in those months, consequently the usual extent of the land was not cultivated,—very many farmers, especially in the ward under Olifant's River, having no cattle to plough with.

"The loss of stock, in horses, cattle, sheep, and goats, since the census was taken in March last, has been enormous, fully one-third, if not more.

"Crime, especially thefts of goats, has been on the increase, owing, no doubt, to the high price of provisions."

Namaqualand is a division which derives its importance from its copper mines, which are estimated to be of great extent and value. It has an area of 21,000 square miles; but of some two and a half millions of acres there are only about 11,000 under cultivation. With the exception of a tract of country in the valley of the Olifant's River, it is very barren and devoid of moisture. It is described as "a vast expanse of barren and rugged country, perfectly waterless, with an almost rainless climate, and chiefly valuable for its immense mineral wealth, as yet but partially developed. Along the coast the country is covered with deep sand, making transport difficult; but a railway across the desert conveys the copper ore of the mountains to the port of shipment."

It comprises the country lying between the division of Clanwilliam and the Orange River. It is thus described by Mr A. G. Bain in a narrative of a journey made by him, in 1854, from Clanwilliam to the Copper Mines, submitted by him in accordance with a desire expressed, by the Lieutenant-Governor at the time, to be furnished with a written account of the geology of that district, and presented to Parliament by order of His Honour :—

"Leaving the Village of Clanwilliam, I proceeded *via* Zeekoe Valley and Heerelogement, to Bakkeley Plaats, on the lower Olifant's River, the road being, all the way, exceedingly heavy and sandy. At the last-mentioned place the sandstone rocks, such as those that cap the Table Mountain, Stellenbosch Mountains, Bain's Pass, etc., entirely disappear, and a mica schist of a very bright silvery appearance, with large milk-white quartz veins, take their place. After crossing the river here, some hills, nearly entirely composed of pure white quartz and decomposed clayslate, attract the attention of the geologist, and I have been told that, in this neighbourhood, some excellent specimens of graphite or plumbago, and other valuable minerals, have been discovered, but I did not see any of them myself, nor can I vouch for the correctness of the report.

"The whole of the distance from this place to Kokenaap (16 miles), is a flat country covered with blown sand, under which is a substratum of coarse blown calcareous tuffa resting on mica slate.

"From Kokenaap we enter what is called the 'Hardeveldt,' which forms a marked contrast to the sandy country, on the Southern side of the Olifant's River, for here the roads are good and hard, and hence the name of 'Hardeveldt.'

"An hour's ride beyond Kokenaap, the first granite hill begins to show its head above the extensive Karroo Flats, which here bound the horizon on all sides; and as we advance to the northward these hills increase in number and altitude, all of them rich in veins of white quartz. Water begins to get scarce and brackish, though there are many situations on private property, well adapted for the formation of dams, if taken advantage of, that would contain water the whole year through.

"At Zout River, I found some fine specimens of porphyritic granite, and at Louis Fontein a beautiful red-spotted rock, resembling in appearance the Egyptian porphyry of Zuurberg, in the Uitenhage division, but much softer.

"Our route still continues through granite hills with broad white veins of quartz (but no appearance of gold), till we reach the foot of the Kamies Bergen, when they suddenly become a confused mass of rugged and lofty mountains,—

'Hills over Hills and Alps on Alps arise.'

"We are now in Klein Namaqualand and fast approaching the copper regions. The roads become steep and undulating, though of such a nature as easily to be converted into good ones. Leaving the Kamies Bergen behind, we descend by a very tortuous route to the missionary station of Kamaggas, a wretched-looking place, as far as buildings go, there being only the missionary's house and one or two other small walled houses, the rest being merely a Namaqua kraal, consisting of a few mat huts. There is, however, a beautiful little stream of fresh water, the first I had yet seen in Namaqualand, which will doubtless fix this locality as the future capital of the mining regions.

"Twenty miles' ride over a good level road, in a North-easterly direction from Kamaggas, brings us to Messrs Phillips & King's celebrated mine 'Spectakel,' situated near the confluence of the Kousie and Schaap Rivers. Here the surface indications are very strong, the green carbonate of copper boiling out, as it were, through the gneiss rocks in all directions. . . .

"It has generally been reported that no other rock but granite was to be found in Namaqualand, but from this mountain I could distinctly see three different geological formations, viz: granite, gneiss, and mica schist, in regular sequence, the latter being totally different from any of its contemporaries in other parts of the colony, as containing massive beds of cellular sandstone, filled with small quartz pebbles, and very distinctly stratified.

"In speaking of the different formations to be found in Namaqualand, I may here observe that all the copper mines I have met with are invariably situated about the junction of the granite and gneiss rocks, and never, so far as I am aware, has any indication of copper being found there in the mica schist. Unfortunately, the lots of many persons whose leases of 'Mineral Centres' from Government were chosen because they were near to those known to be rich, are in the mica schist, and consequently good for nothing.

"Springbok Fontein is situated about four hours' ride on horseback beyond 'Spectakel,' the road being merely a bridle path, very steep and stony, and the view not always pleasing to the eye, from the great quantity of bare amorphous granite mountains, which everywhere rear their naked heads above their rugged bases of gneiss, the former without a fissure or crack, and perfectly destitute of vegetation, if we except some patches of red and brown lichens, which only add to their dreary appearance.

"The mines are situated in rather a pleasant valley (an oasis in the desert); and here we already behold the nucleus of a rapidly rising village.

"It has been said that Namaqualand is perfectly destitute of water, but by sinking here to the depth of a few feet the best of water has been procured; and from the geological construction of the country, I feel certain that, by digging to a very slight depth, good water will be sure to be found in every valley. . . .

"From 'Springbok' I took a trip to the borders of the Bushman-land, where several copper mines have also lately been discovered.

"The mountains here begin to diminish in size, until the whole country stretches out into one interminable bare granite plain, upwards of 3,000 feet above the sea level. To this country the Namaqualand farmers invariably *trek* with their flocks and herds after the first rains of February, March, or April, where they find rich pasturage for their cattle, which fatten in an incredibly short time. They return again in May or June to plough and winter at their homes. . . .

"I regret that circumstances would not permit me to visit Steinkop and the Orange River, though so near to both, as many of the richest mines are, I am told, about those localities, some of which at the latter place are being worked to great advantage.

"Namaqualand has hitherto been considered by the colonists as an arid, worthless waste; but fortunately for us, Sir Harry Smith, a few years ago, extended the boundary from the Kousie to the Orange River, and thus was this contemned country included within the Colony.

"In an agricultural point of view, no part of the Colony surpasses Namaqualand, for in favourable seasons sixty and seventy-fold are common returns of its cereal produce, whereas in Zwartland (called the granary of the Colony) twelve to fifteen-fold are reckoned good returns.

"For several months in the year good grazing is also to be had for cattle and sheep, except in very dry seasons. Water, it is true, is scarce, and generally bad, but, as above stated, it is almost everywhere to be obtained, of good quality, by digging to a small depth. Many favourable sites are also to be had for dams or tanks, which will doubtless be appropriated for that purpose when the country becomes sufficiently populated.

"When a boer has from 30 to 40,000 acres of land, as is now commonly the case, little improvement can be expected. Let the farms be cultivated and the mines opened, and it will be found that Namaqualand contains more real wealth than the whole Colony put together; for I have little doubt that all the country from the Olifant's River to the Orange River abounds in mineral wealth, and only wants to be thoroughly explored to develop its riches.

"One of the greatest drawbacks to Namaqualand is the intense heat of summer, with the present scarcity and badness of the water. Some parts of the country, such as the Bushman Flats, for instance, are entirely abandoned in the summer for want of water, till the rains fall, when the boers (as above mentioned) immediately *trek* thither.

"The general physical appearance of the country is, on the whole, not unlike that of the neighbourhood of Capetown, only the mountains of the former are granite and gneiss, while the latter are clay-slate and sandstone: still we have the same ranges of mountains running parallel to the coast, with the same sandy plains intervening."

Valuable details are given in regard to the mines of copper ore, in regard to the richness of them, and in regard to working of them; but these I find it necessary to omit.

Somewhat similar, but somewhat more discouraging, in regard to any immediate prospect of agricultural operations being carried on with a probability of remuneration and profit as the result, is a report of the Surveyor-General, C. D. Bell, Esq., made after an official visit to the district, in accordance with instructions from the Colonial Office, towards the end of 1854. Writing of the physical character and productions of the district, he says,—"In Namaqualand, as in many parts of South Africa, the rivers are so useless for transport, and their beds supply water of drinkable quality so scantily, that, as influencing the distribution of population, their course is of little importance in comparison with the run of the great mountain barriers. I will therefore confine myself principally to the latter, there being no stream in the whole country excepting the Orange River. To make a brief general description intelligible, I propose, as the leading idea, a rugged back-bone of mountains, about 120 miles long, and from two to four or five thousand feet high, stretching with a slightly eastern curve from T'Konolosip, near Missionary Drift, on the north, to Spectaakle, on the south,—these two points being best known to those interested in the country. This back-bone is merely part of a longer range in the same direction. It is dislocated at a point a little the southward of half-way, the vertebræ having there an abrupt twist of 12 or 13 miles, due eastwards, towards Kookfontein, and back again to the general line, which is nearly coincident with the present magnetic meridian, namely, 28° to 30° westwards of true north. I wish, now, to be understood as speaking of topographical, not geological, features; I have only had a very distant view of that part of the country where the valley or flat, left by this twist, occurs. It opens one pass, viz., from Steinkopf to the low lands, the only others being that near Byzondermeid to the Spectaakle and Komaggas field, and that through where I have marked Bellkloof, across the Fannin's mine plateau, from the T'Kodas country to Nabas.

"On the west the sea coast runs nearly parallel to, and about 40 miles distant from, the back-bone mountains, from which, on this side, ribs of irregular magnitude and length project generally more or less in a true southerly direction, and with, perhaps, an inclination to a final bend south-westwards at the points, until they are lost in the Dunes. This is merely a general view of what seemed the prevailing characteristic. There is a considerable irregularity, and on the south all is in confusion. This country, from the mountains to the sea, I propose to call the low lands.

"On the eastern side, beginning at the north, as before, the great central range breaks down into a belt of rugged and mountainous country, fifteen or twenty miles broad, which I shall call the high lands. For about one-third of the distance southward from T'Konolosip the high lands are bounded by the Orange River, which sweeps in from the eastern interior and flows smoothly, in long tortuous bends, channelled deeply through fearful-looking masses of rock, in a north-westerly direction, being also parallel in the whole to the great back-bone, until it rounds it at T'Konolosip; further south, the TkNabies and TkGams mountains may be taken as the eastern limit. The only distinct cross ranges or spurs I observed were the Old Tataps and Seven Sisters range, on the north, raising their heads in line to the true eastwards above the lesser peaks, and bounding on each side the elevated hilly region that slopes down to Nabas. On the south of the Seven Sisters, passing the valley of the TcOuleroop River (figuratively, but impracticable in reality), there is an elevated, mountainous country, little known, up to Kookfontein. Southwards of Kookfontein there is a comparative flat, and another to the south of Concordia.

"Bushmanland stretches away to the east of the high lands. It may be described as a vast sandy flat, highly elevated, on which isolated hills and ranges of hillocks, hills, and mountains are scattered. It slopes gently north-eastwards, so far as I saw it, from the level of the Kookfontein and Concordia plateau, down to the Orange River. I now remark in my sketches here, in Bushmanland, a tendency, previously unobserved, to true south-easting and north-westing in these lines of hills rising through the sands, too remarkable to be left unnoticed, seeing that light may be thrown on the original structure of the country through a correct knowledge of such facts. As on the western side, I find that these ribs, if they are ribs, are lost in the confused labyrinth of hills, into which the back-bone itself spreads out at the southern end of the part now under description.

"The vegetable productions of the whole country are curious in a scientific point of view, and most interesting to a botanist. The resintjeboom and one or two others are worthy of the attention of horticulturists; but I must pass mere curiosities of science, and only notice those plants and trees which, in their natural state, may be found useful either for export or the sustenance of any population requiring better food than do Bushmen and Hottentots. It only occurs to me to mention the Zandveld-grass of the low lands, rich in a South African point of view, and the nutritious pasture afforded by the Tcwaa-grass of the hill valleys and the TcGaa and other grasses of the Bushmanland. I have seen tracts of country in Bushmanland, far as the eye could reach, covered with waving seed-grass, like corn-fields, with the stalks growing in separate sheaves,—and if on trial, that grass, cut at the proper season, be found to retain its substance as hay, tens of thousands of tons may be annually reaped for the cattle at the mines, at a very trifling cost. The black-ebony tree, the mimosa, and the willows of the Orange River, are worthless, beyond a very limited extent; and I know of no great promise of useful gum in very considerable quantity.

"I would here refer, on these and other points, to the graphic journal of Mr Charles Pillans, published in the local newspapers in October and November last. It will supply my deficiencies in description of the northern part of the country. I saw Namaqualand at the best time of the year, and its aspect, even then, was far from inviting. The sandy flats and rugged mountains on the south present a general character of wild sterility, only relieved here and there by scanty corn patches in the nooks of the hills; but it is in the northern tracts, bordering on the Orange River, where its desert assumes its most appalling appearance. Twenty-four years of occasional travel has led me over most parts of Southern Africa; I have been in the Kalihari desert, and across a considerable portion of the

Island of Ascension, and therefore ought to know what a wilderness is; but until I saw the country near the banks of the Orange River, in Namaqualand, I had no adequate idea of *utter desolation*."

The description given by Mr Bell embraces in part the districts of Great Namaqualand to the north and of Bushmanland to the east; but it is the supply of water in Little Namaqualand with which we have at present to do.

Here the first requirement is water to enable the miners to subsist while prosecuting their work, and to sustain drivers, oxen, and mules when engaged in the transport of ore to the coast, and to support those who are there engaged in the shipping of that ore for exportation.

At present water is sent thither by ship from Capetown, and clothes are brought to Capetown to be washed—a distance by the coast-line of some 300 miles!

In reply to inquiries which I addressed to a gentleman previously resident there, and personally engaged in the transport of ore, I was informed that he knew of nine springs in the traversed district, which he described as follows:—

On the direct road from Hondelip to Springbok there was the first, at De Reit, a spring which he believed to be permanent, but brackish; and next at Riet house, a spring of which he gave a similar character. At Ovesi was a spring of fresh water, alleged to be permanent, but of this he was doubtful. At Kikvias was one, the water of which was quite fresh. At Koorn Huis was another of fresh water. At Springbok Fontein was a spring of fresh water from which the place takes its name. The flow of water from this was permanent, but not strong; and about a hundred yards from this, at a lower level, was a dam. These seemed to be the only sources of supply in the country traversed in a journey of many miles. At Concordia, fifteen miles further on, was a strong and permanent spring of fresh water, which was raised by a pump. And at De Reit, on the border of Bushmanland, on the road to the Orange River, was a weak spring of water, which was slightly brackish, issuing from blue rock.

Imagination may picture what must be the aridity of a district in which sources of such supply are noted and are valued, and determine the route of a traffic amounting to thousands of tons per annum.

It appears, from an incidental allusion in the reply of the Civil Commissioners to the queries issued from the Colonial Office, that Messrs King and Co. had sunk wells at most of their mining stations, and had obtained water, which is only in accordance with what might have been expected. All these springs are outlets of subterranean rills, and the rainfall though small must bring to the division a supply which should not be despised, but carefully husbanded.

To this supply, Mr Fletcher, whose report in regard to Clanwilliam I have quoted at such length, thus alludes in a letter to me from Keerom, in Namaqualand, under date of 8th September, 1865 :—"I fully expected to have been able to send to you a collection of Namaqualand plants this season, but hardly any of the vegetation in the district has come to flower, owing to the drought. You can judge of this better from the rainfall at Keerom—a height of say 2,000 feet above the level of the sea—which has been as follows :—

1865.—11th Feb.,	... ...	1·00, thunder shower.
23rd Mar.,	... ...	0·50
10th and 11th Apr.,	...	0·20
11th May,	... ...	0·68
16th „	... ...	0·58
17th and 18th May,...	...	0·32
25th and 26th „	... ...	0·37
20th June,	... ...	0·05
1st Aug.,	... ...	0·14
29th „	... ...	0·13
Total rain up to 8th September,		3·97 inches.

The oldest inhabitant in the district does not remember a year so void of rains."

I shrink from penning a word which might be construed by any ingenuity into a reproach of those who are subjected to such drought; and it is only as a consideration calculated to show the importance of anything possible being done in more favourable years to husband the supply, and at all times to secure the treasure hid in the sand, that I remark, that in accordance with what has already been stated this rainfall was equivalent to nearly 100,000 gallons of water to every acre upon which it fell. Of this much would be absorbed by the soil, and much would be quickly evaporated, in both cases perhaps becoming subservient to the maintenance of vegetation in the district, but a portion would also percolate the soil to some little depth, where it would accumulate, and whence it might possibly, by appropriate appliances, be drawn for the supply of the wants of man and of beast.

In a memorandum as to mineral lands, the property of the Crown within the Colony of the Cape of Good Hope, prepared by Mr Bell, and dated 23rd February, 1854, looking at it chiefly in regard to its adaptation to the support of a comparatively large mining operation, says:—"The climate of Little Namaqualand is not unhealthy; but the heat in summer is very distressing to the European constitution; and in winter the cold is very severe, especially in the mountains. The rainy season is nearly the same as that of the Cape division. It may be described as continuing from the latter end of March to the middle of July—the prevalent rainy wind being from the north-west; at other times there are occasional thunder-showers; but the district is subject to long-continued drought, so that frequently the rivers do not run for years. I am informed that even the Buffels River has not continued to run a fortnight at a time for the last three years; but a tremendous torrent sweeps down its bed when much rain has fallen on the hills.

"In the brief notice of the physical aspect of Little Namaqualand, I may divide it into three sections: the first being a tract of about 40 miles in breadth, along the coast, which probably consists of a base of clay-slate, on which may occur, in a few places, strata of much more recent formation, the granite projecting here and there. It is a tolerably level, although slightly undulating, plain, mostly covered with a deep layer of loose heavy sand, on which there is a scanty vegetation. It is quite destitute of fresh water throughout its whole extent; but dams might be found in various places sufficient to mitigate this evil, except in cases of very long protracted drought. The second section may be taken to be the belt of mountains, about 40 miles in width, adjacent to this tract, along the coast. These mountains probably consist of granite; very old sandstone, and perhaps trap gneiss, is said to occur frequently. The present supply of water, and that which may be obtained by digging, is, comparatively speaking, much better there, both in quantity and quality, although still scanty and nauseous to those unaccustomed to it. And lastly, that section behind the mountains occupied by the people of Steinkopf and Pella, also undulating and hilly, where there are a number of small springs, and where, in many situations, water may be found a few feet below the surface. Khamghaap's country, though properly Bushmanland, may be described as a part of Little Namaqualand, and it probably partakes of the characteristics of the two last-mentioned sections. I believe that throughout the whole of Namaqualand there is not one running stream, the Orange River excepted; and after some months of drought, the pools, or water holes (or graaf-water), in the beds of the torrents, become so very brack and bitter, as to produce most disagreeable and debilitating effects on all those who are not thoroughly inured to their use. The immense boulder-like, and of course I need not say unstratified, masses of granite are not favourable to the retention of water near the surface, and its delivery in springs; and it has been observed that when ever a fountain rises from clay-slate it is invariable brack.

"The capabilities of the country for the maintenance of people and working cattle, employed on the mines, must depend principally on water. Other absolute necessaries can be brought from districts better adapted to their production, that is, slaughter cattle and sheep may be driven without insuperable difficulty, and grain to a limited extent conveyed by sea and by return waggons over the heavy sands from Hondeklip Bay, where, as may be mentioned, in illustration of the state of the country, the supply of water is imported from Capetown. Many parts of the Kamiesberg and the Steinkopf and Pella country are, however, very fertile, but uncertain, for in dry years the growth of cereals is stinted, and the crops often burnt up; on the other hand, if they have enough of rain, the heat and moisture burst the eggs of the locust, and immense swarms or flights, many miles in length and breadth, and hundreds of yards in height, devour everything green within their reach. Altogether, I should say, there will be much difficulty in maintaining at present a sufficiency of men and working cattle at all the mines; but this difficulty may be entirely overcome, if those concerned in the enterprises work together, and time their efforts, instead of plunging too far a-head at the commencement.

"In conclusion, I may state that I have received intelligence from a surveyor that water has been discovered at Hondeklip Bay, and although its quality is not described, it appears that it is such as men and cattle can drink. A notice as to discovery of water on a farm about eight miles from that place also appears in a newspaper this day published."

In regard to the Kamiesberg, the Rev. Mr Haddy, long resident in that region, wrote in a letter to Mr J. Fox Wilson, under date of 21st December, 1862, that that designation is a mere imitation by the Dutch of the name given to it by the Hottentots. It was translated by Sir James Alexander as meaning Lion Mountains, but this Mr Haddy alleges to have been an error. "The lion," he says "does not inhabit these high and often snow-covered elevations." The district he describes as less a ridge of hills than a mountainous region, covering little less, perhaps, than 4,000 square miles of territory, capped here and there with conical hills of different altitudes; and the name given by the Hottentots is *Camies di Qumghu—Multitudenans,—Clustered*, or, as he renders it, *Associated Mountains*. He says,—"The very best water is found here in considerable quantities; and there is also much good arable land."

A similar rendering of the name is given by the Rev. Mr Tindall, some ten years or more of whose life, including boyhood and youth, was spent among the people. The name, he alleges, should be written *Chamies*, and not *Kamies bergen*—the meaning of which word he gives as Multitude; and from the summit of these the eye, he says, surveys a sea of mountains—the scene he describes as grand and imposing.

The division of Namaqualand, so sterile, so arid, so scorched, so uninviting as most people would consider it, is not without its attractions—and it is not without its promise of reward to those who will make it what it might become. It is not the benighted Namaqua, who had no opportunity of bettering his position, nor the miner, who for the sake of present hope of future comfort, inspired by what he reckoned tempting wages, has made it his temporary home, who alone have taken up their residence there, but there are numerous bodies of farmers and others in the division; and in many places there are indications that a larger supply of water than is actually used might be found.

But it is with the supply of water available for pastoral or agricultural operations that we have at present more particularly to do. And the statement of Mr Bell, as well as the narrative of Mr Bain, together with the statements of Messrs Hardy and Tindall, show that the land, though arid everywhere, is not, throughout its extent, incapable of culture; and, from the same and other sources of information, it appears that not a little might be made of this land.

Mr Bell, in an interview with Paul Lynx (a Hottentot chieftain, resident in the district) and his councillors, which he had while there, said to them, he reports,—"I have seen that in this land there is grass where there is no water, and water where there is no grass. I have seen that there are numberless places where water could be found in the ground, or where dams could be made to preserve the rain."

In the same report he says,—"Water has been incidentally mentioned in § B 48. I believe that much of the existing supply is intentionally concealed by the natives. I have seen numberless places where water could be found, and I am under the impression that a proper set of regulations for the preservation of dams and wells, and security of return for outlay theron, will lead to such an improvement in this respect, that ere long no man need ride two hours in Namaqualand without letting his horse drink. The quality may not always be first rate, but still drinkable for horses and men accustomed to its use."

The incidental reference is to the following :—" Right to the use of fountains by natives presents a most difficult question, which can only be settled in detail by the local authorities. In all cases where better water can be found by labour, the copper miners are of course not content to drink out of the small filthy puddles left open to the sheep and cattle by the Namaquas. They will blast the rocks, they will dig, they will make dams at much cost; but this gives them no legal right to preserve their work, and they are liable to the mortification of seeing it destroyed by the wanton carelessness of others who, if they can satisfy a present want, have no interest in the permanent supply.

" This question might be settled on the understanding that, if a supply equal to that formerly afforded by the unopened and unpreserved fountain be run off for the public, no one shall be entitled to the residue who is not willing to share or repay a fair portion of the original outlay on improvements, besides somewhat for original risk, especially if the consent of the commissioner to experimental dams or wells has been obtained. I can see no other general principle which can be equitably put in practice. Of course, it will follow that if the former Namaqua occupant obtains a supply equal to what he always enjoyed, he has no right to claim compensation for the overplus."

Mr Bell evidently saw that water was to be had in many localities, and that there were others in which it might be retained; and he says of the Buffels River, "a tremendous torrent sweeps down its bed when much rain has fallen on the hills." Mr Bain, also in a communication to the Honourable the Colonial Secretary in the same year, also bears testimony to the fact that water is to be had. In this communication he says,—" It has been said that Namaqualand is perfectly destitute of water, but by sinking here [Springbok] to the depth of a few feet the best of water has been procured; and from the geological construction of the country, I feel certain that, by digging to a very slight depth, good water will be seen to be found in every valley." And again,—"Many favourable sites are also to be had for dams or tanks."

In replying to the queries issued from the Colonial Office, Mr Ford, Civil Commissioner of the district, writes, under date 30th May, 1862:—"As far as I have been able to learn, there have been as yet no plans proposed for irrigation works within this division.

"It is the opinion of practical men, to whom I have spoken on this subject, that much may be done for the good of this district by the sinking of wells and by the damming up of our periodical rivers. Up to the present time few attempts have been made by the farmers of this district to improve their places by sinking wells, although it is evident from the experience of Messrs King & Co., who have sunk wells at most of their mining stations, that water may generally be obtained in this way. Nor has any advantage, except in a solitary instance here and there, been taken of the many favourable localities that exist for collecting supplies of water by damming up our periodical rivers."

In a letter which I received from Mr Fletcher, and from which I have already quoted his statement of the rainfall at Keeron, during the first nine months of 1865, he says,—"Namaqualand will have no harvest this year worth speaking of. Bushmanland may possibly save our cattle, but I fear the natives will be very badly provided for." And it proved to be the case.

The Rev. Mr Hardey, the superintendent of the Wesleyan Missions, had to visit the stations of the body with which they are connected; and on his return attention was called by one of the leading journals to his reports:—"The Rev. Mr Hardey, the esteemed superintendent of the Wesleyan body, has just returned from an extended tour in Namaqualand. If any one doubts the existence of dire distress in the country, five minutes' talk with Mr Hardey will dissipate all his scepticism. We hasten to place before our readers the result of his observations, for it is not a matter in which time can be lost. He describes Little Namaqualand as a dried-up, parched land, literally without crops or water supply. On the Kamiesberg mountains, where some two thousand persons are wont to live on the grain which they sow from year to year, there are literally not fifty muids of wheat in the whole district. The cattle, too, are all dying for want of water, and are being driven northwards, into Bushmanland, as a last remaining hope. Even there, however, no water can be found within reach of what pasturage is available. So dry and desolate is the whole country, that even the goats refuse to give their milk, and, as a consequence, the last refuge for the children and the poor is taken away. The wells and water-courses are all as dry as if water had not touched them for years. From Piketberg to Olifant's River, Mr Hardey and his party found not a drop of water for man or beast. The endurance and unselfish kindness of the people, as described by Mr Hardey, is like a green oasis in the midst of this gloomy wilderness of sorrow. They go about from house to house sharing with each other whatever they may have to keep life within them. If a kid or a goat be killed, no one dreams of keeping the feast to himself, but it is carried to those who are known to be in want. The last resources in the shape of provisions are being fast used up, and, after the end of the year, Mr Hardey thinks there will be literally nothing on which the people can lay their hands to keep them from starvation. Already many hundreds are trekking to Bushmanland with what little stock remains to them, but there seems little chance of their finding a livelihood there. Though the pasturage is better, water is not to be found."

The Rev. Mr Godman, the Wesleyan Missionary at Lilyfontein, in Namaqualand, addressed to the editor, on the distress in that country, the following:—"Sir,—Permit me to thank you most sincerely for your very seasonable appeal to the public on behalf of our suffering people, which has just come to my notice in the *Mercantile Advertiser*. To the general accuracy of the statements I can bear testimony, confirmed, as they daily are, by saddening observation. I would, however, beg to be allowed to make a few remarks, not to weaken the statement, but by way of explanation. Of the 2,000 persons you mention as residing on the Kamiesbergen, 1,283 are living on the Wesleyan Institution at Lilyfontein. Among these, instead of upwards of 2,000 muids of grain being gathered in, as in ordinary years, there will not be more than twenty muids this year. Through the drought, the oxen of the people are so weak that they have not been able to ride copper to half their usual extent, and the few who have ridden have lost more by the death of their oxen than they have gained by their work. In some instances, they have lost more than double the amount received. The almost entire failure of the crops among the farmers deprives the people of another kind of employment; for by harvesting on the surrounding farms, those who had not been able to sow for themselves have usually earned a little corn;

but this year most of the farmers have none for themselves. Then, so little has been done at copper riding this year, by reason of the drought, and there is so much ore lying at the mines, that the company are not anxious to raise much at present, and are, therefore, employing less men than usual; so that every kind of employment fails. And, as you justly say, the goats and cows have to a great degree ceased to give milk, on which, in good seasons, adults as well as children subsist in great measure, so that the people are indeed in much distress; and failing assistance from without, the worst results may be anticipated. I can bear testimony to the endurance of the people, and that whatever they may have is fairly divided among all who come to their houses; consequently, as soon as it becomes known that any one has slaughtered, or has obtained a little food, he soon has plenty of visitors, until all again are reduced to the same level; and hence, when they obtain a little food, they are often obliged to eat it under cover of darkness to prevent others from depriving them of a part of what, perhaps, is scarcely enough for a meal for themselves and families. It is true many of our people have gone to Bushmanland to try to live, though we hardly know how they are to subsist on account of the drought; and it is equally true that all the poorest and most helpless are left here, with no one to assist them, excepting Mr Morris and the minister. We are doing what we can, but the work is overwhelming, and many, I fear, must be lost unless assistance come from some other source. We are not unmindful of the many demands made on the liberality of those living in Capetown; but our necessities compel us to state our case, and we trust we shall not be left without aid. Should means be placed at our disposal, it is not our intention to support those in idleness who are able to work. The old and infirm we must assist; but we purpose rather employing the able-bodied men in such works as improving the roads on the institution grounds, which will benefit the public as well as ourselves, for they are constantly used by others, although they are not recognized as district roads; attending to the dams so as to secure a better supply of water in future, if possible; and other such works of general utility. While they are so employed, we intend to give them their food, and, by way of payment, supplying their families with, say, half a bucket of corn per week, that they too may live. The carrying out of these intentions must, of course, depend on the assistance we receive; but we purpose acting upon the apostolic principle—'That if any would not work, they should not eat.'"

Of Namaqualand the Civil Commissioner reported in 1865:— "Namaqualand has, during the past year, in common with other parts of the Colony, suffered seriously from the drought. The harvest has been a failure. In most instances, the yield has been less than the quantity of seed sown.

"Sheep and cattle have pined away in large numbers. Amongst the latter, lung-sickness has also been very destructive.

"The carriage of copper ore has not been so profitable as usual to the riders. Their losses in cattle on the roads must have been nearly, if not quite, equal to their receipts. This is attributable partly to the badness of the roads, and partly to the drought."

In 1869 it was reported:—"The year opened very favourably; fine rains fell, which were hailed with great joy by the farmers, the division having suffered from a continuous drought of three years. Grain was sown to a great extent, as far as means would admit of. Rains did not continue to fall around Springbok-fontein during the season, and consequently the harvest did not reach the anticipations of the people; in some instances it was a perfect failure. However, the Kamiesberg and Hardeveld were more favoured with rain, and there the crops have turned out pretty well. The farmers on the borders of Bushmanland and around Springbok-fontein are very poor, —in some instances almost reduced to beggary, owing to their loss of stock by the continued drought they had to contend with, and the daily depredations of marauding Bushmen, who infest Bushmanland. Several murders have been committed, and thefts of stock by wholesale (as many as forty head of cattle at one time), by these bands of marauders, who are led by one Roman Brandrug, the ring-leader of the murderers of the constables on their way to Clanwilliam.

"There is no municipality here, and no market, and all articles of consumption (including vegetables), with the exception of beef, mutton, and goat's flesh, are imported from Capetown; and those who indulge in luxuries have to pay dearly for them.

"Crime is greatly on the decrease. In the early part of the year the gaol was filled with prisoners,—in fact, the gaol could not hold them all."

In 1874 it was reported:—"1874 has been a very bad year for Namaqualand; the entire division has suffered from a very severe drought. There has, so to say, been no corn in the land; the only part of the district where a little was gained was on the Kamiesberg, and I believe one person there, after thrashing out his corn, obtained three (3) pannicans. The poorer farmers and natives will severely feel the drought, as not only will they be famished by having no crops, but the loss in cattle and sheep will also be felt, and the more so in 1875, as they will, in consequence of having no bread-stuffs, be compelled to maintain themselves by slaughtering more freely their live-stock. The more well-to-do farmers have ordered their grain from Capetown. Many farmers have, up to the present, saved their stock by sending them to the Orange River and Bushmanland for water and pasture, and, should timely thunder-showers fall in Bushmanland, these will save their stock; but it is feared that the loss will be great to those who did not do so. Living is very expensive, and were it not for the tram-line we would have been on the verge of starvation, as there were no cattle in the country in a fit state to ride transport from the coast. The old adage, 'troubles never come singly,' has been verified with us, as the mortality amongst the kids and lambs, 'by crimp sick,' has been most severe; but we live in hopes of better times coming.

"There were ninety civil cases, and two hundred and thirty-five criminal, tried during this year.

"The O'kiep mine is steadily progressing in size and importance. The yield of ore during the year has been ten thousand tons. There have been many improvements in the way of machinery and buildings. The new pumping engine, a thirty cylinder steam jacketed condensing beam pumping engine, on the Cornish principle, costing, including two boilers, engine, and boiler-houses, £3,800, started to work on the 5th of March; it pumps water from a depth of four hundred and eighty feet. The yield from the Spectacle mine has been on the decrease. The village of Springbok is going down-hill, and it is hoped that the company will again ere long renew their trials for ore here."

And in 1875:—"The same drought of the previous years did not break up until about the middle of 1875, and the poorer farmers and natives, who were much reduced thereby, had no grain for sowing; the former were, however, assisted by the Cape Copper Mining Company with grain, and the latter, in the Kamiesberg (where need was most required), by the Government. The rains came rather late in the year (August), and the yield in corn was very small, as, owing to the latter rains not falling, water has been scarce; the fact is, the country has not recovered from the severe drought, nor will it until we have good soaking winter rains. It is almost impossible to describe the poverty and miserable way in which the poorer classes exist in this country in a severe drought; in good seasons they live principally on corn, milk, and roots. That part of the population in the neighbourhood of the mine, will, as long as there is copper in the country, be well cared for in drought, as the tram-line between the Port and the Mine has been opened, and they can always be provided with the necessaries of life from Capetown. Typhoid fever has been very prevalent at the mine, and has carried off a great many, both whites and blacks, and this appears to be the case every summer. The O'kiep mine has been steadily progressing, its yield being now 1,000 tons per month; the trial mines, I believe, have not paid expenses. The Concordia mine is still being worked by the Concordia Copper Company, but the ore at present raised is of low percentage."

The following is a tabulated statement of meteorological observations made at Concordia, in Namaqualand.

The rainfall, trifling as it is, is not to be despised. Taking each inch as equivalent to, say, 25,000 gallons on every acre, the multiplication of this number by the number of acres in Namaqualand and Bushmanland would give the measurement of a quantity of water almost incredible. The evaporation in so dry a climate would be very great, but even this would not be without its advantage; and that not only in creating a moister atmosphere, favourable to vegetation, but also in moderating the temperature of the district, which, from one of the reports quoted, appears to be characterised, like that of some other districts of the Colony, by extremes of heat and of cold; it would at once moderate the heat of summer and the cold of winter, the heat of the day and the cold of the night, to the benefit of agriculture.

Tabulated statement of meteorological observations made, 1867 and 1868, at Concordia, in Namaqualand.

From Daily Observations.	Jan.	Feb.	March.	April.	May.	June.	July.	Aug.	Sept.	Oct.	Nov.	Dec.	For 9 months.
1867.	degrees	degrees.	degrees.	degrees.	degrees	degrees.	degrees.	degrees.	degrees.	degrees.	degrees.	degrees.	degrees.
Mean maximum temperature of the air,	93·5	86·6	79·97	74·97	72·07	62·83	59·13	59·32	71·43	...	...	...	73·31
Mean minimum temperature of the air,	69·4	65·8	62·90	57·17	54·68	48·90	47·40	45·52	52·50	...	...	...	56·03
Approximate mean temperature of the air,.......	81·4	76·2	71·44	66·07	63·37	55·87	53·27	52·42	62·97	...	...	...	64·67
Mean range of temperature of the air,	24·1	20·8	17·07	17·80	17·99	13·93	11·73	13·80	18·93	...	...	...	17·28
Highest temperature,......................	95·0	96·0	92·0	86·0	86·0	74·0	74·0	76·0	87·0	...	...	...	96·0
Lowest temperature,......................	54·0	55·0	50·0	45·0	42·0	41·0	36·0	36·0	39·0	...	...	...	36·0
Extreme range of temperature on any one day,................................	32·0	32·0	26·0	28·0	30·0	20·0	21·0	25·0	27·0	...	...	...	32·0
Lowest range of temperature on any one day, ...	13·0	4·0	4·0	6·0	6·0	6·0	2·0	4·0	4·0	...	...	...	2·0
Quantity of rain in inches,..............	0·02	1·78	1·26	0·705	0·00	0·45	1·11	0·78	0·17	...	...	...	6·275
Greatest amount of rain on any one day in each month,.....................	0·02	1·21	0·75	0·28	0·00	0·27	0·78	0·52	0·17	...	...	...	1·21
Number of days on which rain fell,....	1	2	4	5	0	2	3	4	1	...	...	...	22
1868.													For 12 months.
Quantity of rain in inches,..............	0·00	0·00	0·760	0 63	0·00	0·78	1·485	0·75	0·00	0·955	0·155	0·97	6·485
Greatest amount of rain on any one day in each month,.....................	0·00	0·00	0·350	0·22	0·00	0·78	0·985	0·225	0·00	0·440	0·155	0·07	0·985
Number of days on which rain fell,....	0	0	3	4	0	1	4	5	0	5	1	1	24

Mr P. Fletcher, C.E., in a report of surveys presented to both Houses of Parliament, by command of His Excellency the Lieutenant-Governor, in May, 1860, referring to Captain Smith's valuable work on irrigation, printed by order of the Court of Directors of the East Indian Company, which had been placed at his disposal by the Hon. Sir Rawson W. Rawson, then Colonial Secretary, cites his description of Mairwara, in Rajahpootan, describes the form of dam there used, and shows, in a passage I have quoted (ante p. 267), the applicability of the system also to the division of Clanwilliam. But the district described by Mr Fletcher embraces a vast extent of country, and his statements indicate his conviction to have been that the whole of the district from the latitude of Clanwilliam and Fraserburg, about 32° north, to the Orange River, say 28° 30′, arid as it is, presents greater facilities for irrigation than did then Mairwara, and that the district of Namaqualand on the west of this is apparently not more hopeless than was that district in India prior to the introduction of irrigation.

On the practicability of irrigating of this arid region I have dwelt at greater length, influenced in some measure by the consideration that if it can be shown that so hopeless a district can be made productive, it will probably be admitted that in regard to none other need we foreclose the discussion in despair.

Reference having been made in the report by Mr Fletcher relative to the character of the native population in the vicinity of Ebenezer, in the division of Clanwilliam, and as this may be considered an important element to be taken into account in considering the expediency of commencing the construction of dams in the district, I consider it proper to introduce the following statements in regard to the changes produced upon the native population of Namaqualand by the introduction of mining operations into that district. They were made in a lecture delivered in the Mechanics' Institute, Capetown, in 1856, by the Rev. H. Tindall, who, as has been intimated, had spent some ten years, if not more, in the adjacent country to the north of the Orange River.

Speaking of that time, and what was true then is true still, he said :—" As might be expected, the events of the last five or six years have produced great changes in Little Namaqualand. Before that period it was a very quiet pastoral district. A few mission stations formed centres of attraction, and places of refuge, for a people who had once roamed at large in undisputed possession of all they surveyed, but who were now glad to accept any security by which their claim for protection and property might be recognised. The farmers were living in easy and listless enjoyment of their extensive grounds, rich corn fields, and numerous flocks; there was nothing to attract the notice of the world, or to excite the spirit of enterprise; trade was advancing with a sluggish pace; and there was no prospect of any other change in the circumstances of the people, than such as would ultimately result from their being gradually brought into closer connection with their more civilized fellow-colonists. The first intimation of the approaching excitement was the purchase of a farm by some English gentlemen—it was rumoured that they were going to dig for copper. After a while, travellers began to pass through the country in rather unusual numbers, ostensibly to learn the opinion of the inhabitants about a certain 'Constitution,' which was, as well to farmers as to natives, an undefinable, unknown something;—these were soon multiplied into a host, who scaled every mountain, crowned every height, peered into every gully, and ranged every plain in search of hidden treasure. The people looked on with wonder; the natives, fearing that they would lose their last foot of ground, and their last spring of water; and the boers alternately blaming and praising as they vacillated between their love of seclusion and their thirst for gain. Wherever a speck of green was discovered on the rocks, beacons were erected which marked the spots where wealth was to be sunk instead of raised, where hopes were to be buried and disappointment exhumed. Vessel after vessel arrived upon the coast and discharged its cargo; men, tents, tools, and provisions were transported to different parts; stations were formed, and the sound of the pickaxe, the hammer, and the blast began to echo among the hills. New roads were opened and old ones worn out. Provisions rose to an enormous price; every waggon and draught animal in the country was in demand for transport service; and, by the introduction of intoxicating drink, many parts of the country were thrown into a state of lawless excitement and confusion. The reaction came, the country awoke as from a vision; the curtain rose and displayed another scene in the drama; the results are,—a magistracy; three or four good mines; numerous unsightly excavations; the ruins of forsaken establishments; and another page in the volume of experience, upon which lessons of wisdom stand inscribed. . . . .

"These events have exercised various influences upon the inhabitants of the Namaqualand district. Several of the Dutch farmers, unaccustomed to the turmoil of the times, and fearing increased difficulties in obtaining labourers, have disposed of their farms and removed to the Sovereignty. Others have suffered severely by the facilities with which intoxicating liquors have been obtained, not being able to resist the temptations which they presented; whilst many have caught the spirit of the age, and risen a step higher in the scale of humanity. The natives have been both benefitted and injured. The high prices which they have obtained for their labour and produce, the employment they have had for their waggons and draught cattle, and their intercourse with Europeans, which has 'brought civilized habits under their immediate observation,' are all circumstances which have had a tendency to elevate them and to promote their advancement. Many of them have also profited by their visits to the mines to improve their knowledge of various trades, and have become very efficient and respectable workmen.

"On the other hand, numbers have suffered irreparable injury; their waggons have been worn out, their oxen and horses killed by excessive work, and their flocks diminished for the sake of a few of the luxuries of life. More desolating still have been the effects of strong drink. It has not only impoverished, but also degraded. It has produced habits of irregularity and vice, and, in some parts, almost entirely destroyed the good which had previously been effected. To this we must add the effect produced on the minds of the natives by the vices of some civilized men, which have so seriously counteracted the influence which others' virtues have exerted. The rule of the native chiefs or corporals was even formerly feeble; it was almost totally destroyed by the influx of those over whom they had no control—their own subjects learnt to throw off restraints which they saw others despise. It will be with Little Namaqualand, as it has been with all other heathen countries over which the civilized rule of a free nation has extended; those of its aboriginal inhabitants who acquire habits of temperance and industry will gradually rise in intelligence and station, and will ultimately become amalgamated with the people by whom they are governed; whilst those who cling to their native habits of indolence, and add to them the extravagant vices of more refined wickedness, will either emigrate to lands where they may enjoy their barbarous freedom and indulge their evil propensities unmolested, or will soon be engulphed in a vortex which will consign their very name to oblivion. . . .

"The present population of the destrict is composed of farmers, natives connected with the mission stations, and a few wandering half-castes and bushmen. The natives at a rough computation may be estimated at three thousand (3,000) souls. This number would be distributed as follows:—Xkanghap, 50; Paul Lynx's tribe, 450; Steinkoff, Kookfontein, and Komaggas, 1000—stations of the Rhenish Mission. Lily Fountain, 900; NChorap, 250; stations of the Wesleyan Mission. Pella, 200. And parties having no fixed location, including a few wandering border boers, 150.

"The natives of the mission stations have made considerable progress during the last few years. Many of them stand in high repute for their comparative intelligence and respectability. This is not only true of the half-castes, who are further removed from the Namaqua than from the European in birth and appearance, but also of those who are purely Namaqua. The mission station of Lily Fountain may be regarded as affording an encouraging example. Only a few years ago it was occupied by a poor and ignorant people, but now they are both comparatively enlightened and enriched. They possess large numbers of waggons, horses, and cattle; they have built and purchased several cottages; they cultivate almost all the corn land which their grounds contain, or which can be spared from the pasturage; they have built a handsome and spacious church, at a cost of nearly one thousand pounds; they have laid out good roads over their mountain ranges and passes, in the face of many difficulties, and at considerable expense. Several of them have acquired a practical knowledge of useful trades; they are respectably clad in European clothing; the Dutch language has entirely superseded their mother-tongue, and that, in its turn, is being gradually supplanted by the English. Their services are very valuable to the adjacent farmers, many of whom greatly envy their prosperity. The internal affairs of the institution are well regulated by the corporals, who are elected annually. The causes which operated much to the disadvantage of other stations, being more remote from this, did not produce similar effects on the social condition of its inhabitants, so that good order and subordination to local regulation have been uniformly maintained. Collectively, the conduct of the people of this institution—and for anything I know to the contrary, that of the other mission stations—is irreprehensible; though it is not attempted to be denied that there are frequent individual cases of irregularity and moral delinquency, when the persons are absent from the station, and especially when exposed to the debasing influence of ardent spirits. Such cases, however, are more to be pitied than condemned. The same things excite little surprise when they occur in civilized communities.

"I have no doubt that Little Namaqualand will now advance both in internal prosperity, and in importance to the rest of the Colony. Its formation into a separate district and magistracy will provide for its efficient government, and will afford important facilities for the development of its various resources. One of the best ends to which the energies of its Government can be directed, and one that calls for immediate attention, is the construction of roads and dams. It has justly been observed that 'the force of Government—general and local—should be mainly directed to the works that tend directly to encourage rural production, and to diminish the expense of transport. Open the country by roads and bridges, facilitate export by harbour improvements, and the instincts of human industry will do the rest. The importance,' and, we may add, the internal prosperity, 'of isolated districts is determined much more by their relative position and means of intercourse with the rest of the country than by their special character. The naturally poorest district, having easy access to markets and seaports, may greatly exceed in value the richest in natural advantages of soil when cut off from communication with the world by mountain barriers or bridge less rivers.'"

Of the country and of the climate Mr Tindall writes:—"Much of the country along the sea coast and the Orange River, as well as the Bushmanland, is of no value for agricultural purposes. The winter rains fall very lightly, and it is only in those years when heavy thunder rains fall in the summer, that the plains are covered with the beautiful long waving Towa and Qga grasses. The southern part, on the contrary, is an excellent grain country, and is capable of yielding a much larger quantity than has ever yet been produced. One of the principal disadvantages is the scarcity of water during the summer season; but this difficulty, at least as far as the southern part is concerned, might in a great measure be overcome. In such a country numerous tanks and reservoirs might be constructed, and probably many new fountains might be opened. It is not like other parts where the rains are so uncertain and irregular that there would be little encouragement or remuneration for such projections. . .

"The climate of Little Namaqualand is in general salubrious, but in the summer, except on the high mountain ranges, the heat is excessive. It would differ little in temperature from the country north of the Orange River if it were not for the cool and refreshing breeze which comes up from the sea in the evening."

The Orange River, passing towards its mouth, flows through the territory formerly peopled by the Namaquas, a tribe of Hottentots, dividing it into two very unequal portions—the lesser only of which has been taken into the Colony, and the other, still inhabited by Namaquas, is called, in contradistinction, Great Namaqualand. The southern boundary of this is the Gariep, Qgariep or Orange River; the western boundary is the Atlantic.

It is a land of drought—of drought as severe as that of Little Namaqualand, within the division of the Colony last spoken of; and the region of drought extends far beyond. It is worth while just to look at it, were it only to lead us to appreciate, as this ought to be appreciated, the water supply and the facilities for the storage of water at command within the bounds of the Colony.

Great Namaqualand extends along the coast from the Orange River to Walwich Bay, lying between the sea and the Kalahari desert. I have already reported that, from a missionary of the Wesleyan body, who had been stationed in Great Namaqualand, I learned that while he was there it was only by a six weeks' journey that he could procure flour for his family, and *that* a journey made with difficulty. The statement with which this was followed is now recalled, well nigh thirty years after it was made, and there may be slight inaccuracies in my report of it, though I do not see this to be possible. It was substantially this—that they could not grow wheat and had to send into the Colony for flour, and in doing so their waggons had to be got across the Orange River, where there was neither ford nor ferry. They sent on the waggon a boat; the waggon was taken to pieces on the bank of the river; the boat was launched; the waggon was carried across piece by piece, and when reconstructed on the colonial side of the river, the journey to the nearest village was resumed. Supplies having been obtained, in returning the same operations had to be gone through in re-crossing the Orange River, and the supplies carried across in a similar way; six weeks in all being in general consumed in the journey to and from the Colony.

Dr Moffat writes :—" The entire country, extending in some places hundreds of miles on each side of the Orange River, and from where it empties itself into the Atlantic to beyond the 24th degree east longtitude, appears to have the curse of Gilboa resting on it. It is rare that rains to any extent or quantity fall in those regions. Extreme droughts continue for years together. The fountains are exceedingly few, precarious, and latterly many of these have been dried up altogether. . . .

"One morning, after travelling several days in those Karroo plains, Mr Campbell stood still, and remarked with great emphasis to Mrs Moffat and myself,—'Sirs, it would require a good pair of spectacles to see a blade of grass in this world.' . . .

"As an inhabited country, it is scarcely possible to conceive of one more destitute and miserable; and it is impossible to traverse its extensive plains, its rugged, undulating surface, and to descend to the waterless beds of its rivers, without viewing it as emphatically 'a land of droughts,' bearing the heavy curse of

> 'Man's first disobedience, and the fruit;
> Of that forbidden tree, whose mortal taste
> Brought death into the world, and all our woe.'

"Meeting with an individual on my journey thither who had spent years in that country, I asked what was its character and appearance. 'Sir,' he replied, 'you will find plenty of sand and stones, a thinly scattered population, always suffering from want of water, on plains and hills roasted like a burnt loaf under the scorching rays of a cloudless sun.' Of the truth of this description I soon had ample demonstration. It is intersected by the Fish and 'Oup Rivers, with their numberless tributary streams, if such their dry and often glowing beds may be termed. Sometimes for years together they are not known to run; then, after the stagnant pools are dried up, the natives congregate to their beds, and dig holes, or wells, in some instances to the depth of twenty feet, from which they draw water, generally of an inferior quality. They place branches of trees in the excavation, and, with great labour, under a hot sun, hand up the water in a wooden vessel, and pour it into an artificial trough; to which the panting, lowing herds approach, partially to satiate their thirst. Thunder-storms are eagerly anticipated, for by these only rain falls; and frequently these storms will pass over with tremendous violence, striking the inhabitants with awe, while not a single drop of rain descends to cool and fructify the parched waste.

"When the heavens do let down their watery treasures, it is generally on a partial strip of country, which the electric cloud has traversed; so that the traveller will frequently pass, almost instantaneously, from ground on which there is not a blade of grass, into tracts of luxuriant green, sprung up after a passing storm. Fountains are indeed few and far between, the best very inconsiderable, frequently very salt, and some of them hot springs; while the soil contiguous is generally so impregnated with saltpetre as to crackle under the feet like hoar frost, and it is with great difficulty that any kind of vegetable can be made to grow. Much of the country is hard and stony, interspersed with plains of deep sand. There is much granite; and quartz is so abundantly scattered, reflecting such a glare of light from the rays of the sun, that the traveller, if exposed at noon-day, can scarcely allow his eyelids to be sufficiently open to enable him to keep the course he wishes to pursue."

Describing his journey from Bysondermeid, he says,—"My oxen being somewhat rested, I bade farewell to my companions in travel, and proceeded with a guide through a comparatively trackless desert. Having travelled nearly the whole night through deep sand, the oxen began to lie down in the yoke from fatigue, obliging us to halt before reaching water. The next day we pursued our course, and on arriving at the place where we had hoped to find water, we were disappointed. As it appeared evident that if we continued the same route we must perish from thirst, at the suggestion of my guide we turned northward, over a dreary, trackless, sandy waste, without one green blade of grass, and scarcely a bush on which the wearied eye could rest. Becoming dark, the oxen unable to proceed, ourselves exhausted with dreadful thirst and fatigue, we stretched our wearied limbs on sand still warm, from the noon-tide heat, being the hot season of the year. Thirst aroused us at an early hour; and finding the oxen incapable of moving the waggon one inch, we took a spade, and, with the oxen, proceeded to a hollow in a neighbouring mountain. Here we laboured for a long time, digging an immense hole in the sand, whence we obtained a scanty supply, exactly resembling the old bilge-water of a ship, but which was drunk with an avidity which no pen can describe. Hours were occupied in incessant labour to obtain a sufficiency for the oxen, which, by the time all had partaken, were ready for a second draught; while some, from the depth of the hole and the loose sand, got scarcely any. We filled the small vessels which we had brought, and returned to the waggon over a plain glowing with a meridian sun; the sand being so hot, it was distressingly painful to walk. The oxen ran frantic, till they came to a place indurated, with little sand. Here they stood together, to cool their burning hoofs in the shade of their own bodies; those on the outside always trying to get into the centre. In the evening, when about to yoke them in order to proceed on our journey, we found that most of the oxen had run off towards Bysondermeid, An attendant, who was despatched in search of them, returned at midnight with the sad tidings that he was compelled by thirst, and terror of meeting with lions, to abandon his pursuit.

"No time was to be lost, and I instantly sent off the remaining oxen with two men, to take them to the next fountain, and then proceed to solicit assistance from Mr Bartlett, at Pella. Three days I remained with my waggon-driver on this burning plain, with scarcely a breath of wind; and what there was felt as if coming from the mouth of an oven. We had only tufts of dry grass to make a small fire, or rather flame; and little was needful, for we had scarcely any food to prepare. We saw no human being, although we had an extensive prospect; not a single antelope or beast of prey made its appearance; but in the dead of the night we sometimes heard the distant roar of the lion on the mountain, where we had to go twice a day for our nauseous but grateful beverage. At last, when we were beginning to fear that the men had either perished or wandered, Mr Bartlett arrived on horseback, with two men having a quantity of mutton tied to their saddles. I cannot conceive of an epicure gazing on a table groaning under the weight of viands, with half the delight that I did on the mutton, which, though killed only the preceding evening, required no keeping to make it tender. Oxen had been sent for, which were to arrive in two days. This time was spent in mutually refreshing intercourse; but Mr B., although inured to Namaqua heat, remarked, that what we experienced was enough to set the grass on fire."*

The ludicrous was not awanting. He tells,—"We had troubles of another kind, and such as we did not expect in so dry and thirsty a land. Rain had fallen some time previous in the neighbourhood of Kamiesberg; the loose soil, abounding in limy particles, had become so saturated, that frequently, as the oxen and waggon went along the road, they would suddenly sink into a mire, from which they were extricated with difficulty, being obliged to unload the waggons and drag them out backwards. One river was so swollen and rapid, that

* Inferior as the tribes inhabiting such a region may be supposed to be to nations more favourable conditioned for intellectual development, the men are not all devoid of intellect or capability of reasoning. The Rev. John Campbell, when he penetrated the regions lying beyond the Colony as the pioneer of missionaries, on one occasion drawing a bottle of ginger beer had the mortification to see it squirt into in the air a continuous curve till the bottle was emptied. But in his disappointment he saw a Namaqua looking on in amazement, and enjoying this. Mr Campbell desired the interpreter to ask him if he was surprised to see so much come out of a little bottle. "No," was the reply; which led to the question, what was it then that had astonished him? The reply was to the effect that he was surprised to find that white men could *put so much* into a little bottle; and he could not conceive how it could be done! No; get it in, and no wonder it comes out! And a savage who could perceive this was not a man whose intellect was to be despised.—J. C. B.

Mrs K. preferred being carried over to going in the waggon. Being rather more robust than Mr K., this duty devolved on me, and it was not an easy one, as the stones in the river were as slippery as butter, and the whole party standing on the bank, all in a titter, expecting every moment that we should both have a plunge, which, though not unattended with danger, excited the risible faculties in no ordinary degree."

Again,—"At last Magerman consented to take me to the other side of the river; and the good man, hoping to gain his point, conducted me to a ford, opposite which a village of his people lay, who he expected would take me by force. The waggon and its contents were swam over piecemeal, on a fragile raft of dry willow logs, about six feet long, and from four to six inches in diameter, fastened together with the inner bark of the mimosas, which stud the banks of the river, which is at this place 500 yards wide, rocky, with a rapid current. The rafts are carried a great distance down by the stream, taken to pieces every time of crossing, each man swimming back with a log.

"When, after some days' labour, all was conveyed to the opposite shore, the last raft was prepared for me, on which I was requested to place myself, and hold fast. I confess, though a swimmer, I did not like the voyage, independently of not wishing to give them the trouble of another laborious crossing. I withdrew along the woody bank, and plunged into the river, leaving my clothes to be conveyed over. As soon as they saw me approaching the middle of the current, terrified lest evil should befall me, some of the most expert swimmers plunged in, and laboured hard to overtake me, but in vain; and when I reached the northern bank, an individual came up to me, almost out of breath, and asked, 'Were you born in the great sea water?'"

Mr Tindall, writing of the country of Great Namaqualand, proceeds to say:—"The various countries of the world are marked by great diversities. Some are provided with all that could constitute them charming and happy abodes for man; all that could make life sweet, and earth attractive; and all that could favour the onward march of civilisation and enlightenment: whilst others are so sterile, so dreary, so uninteresting, and, by their natural incapabilities, so adverse to industry and advancement, as to inspire a sigh that ever human foot should have trod their soil. Great Namaqualand more properly belongs to the latter than to the former class. There are but few who would envy its nomadic tribes, their rugged mountains and boundless plains. Though it is clad at some seasons with beauteous verdure, it presents but little of the lovely or the sublime. There are times when the eye of a quadruped might glisten at the sight of its rich pasture; but the eye of man soon tires of a scene so lonesome and monotonous. . . .

"The line of coast is the most sterile that can be imagined. There are but few spots where water can be found, and wherever it is procured is of the most inferior quality. Angra Pequina, Sandwich Harbour, and Walwich Bay, are the principal havens. The last named is the most capacious, and affords a safe and excellent anchorage. It was formerly much frequented by whaling vessels. From the bay nothing is to be seen but a succession of low sand hills, the very pictures of desolation. The valleys are covered with the dabee tree, and the vnaras bush, which yields a delicious fruit, about the size of an orange. The bush and the fruit are both covered with long sharp thorns. The skin is soft, and encloses a yellow pulp which is of a rich flavour, and contains many nuts, larger than the seed of a water-melon, and similar in taste to an almond. This fruit, and such fish as they can succeed in spearing among the rocks of Pelican Point, form the staple food of the natives living near the bay. The sand hills occupy the bed of the Qkhuisip River, which is here three or four miles wide. They are covered with a dense fog early in the morning, which seems to favour the growth of the dabee and vnaras, but is unfriendly to other vegetation. Beyond the river stretches an immense plain, on which rain seldom, if ever, falls. When we have travelled about eighty or a hundred miles, we enter a mountainous region, which is tolerably fertile, though water is uniformly scarce. The northern boundary of Great Namaqualand is undefined. Perhaps the Qkhuisip River may be considered a just line of demarcation between the territory of the Namaquas and that of the Damaras. The former have, however, considerably crossed their bounds, and invaded their neighbours' rights. Most of the large rivers take their rise in the mountains sorrounding Jonker Africaner's place. The Zwaghop and Qkhuisip run for about 200 or 250 miles due west, and fall into Walwich Bay. The Fish River runs in a southerly direction, intersects Namaqualand, and empties itself into the Orange River, a little below Vnabas. It is the finest of all the periodical rivers of Namaqualand. Its bed is rocky, hence it contains in many places, large pools of water, which, in a land of droughts and deserts, is the most valuable commodity to the traveller. Its banks are covered with large mimosa and camel thorn trees. Wild fowl are abundant, and in many places fish is plentiful. A few years ago vast herds of game resorted to its sparkling pools to quench their thirst, but now they have fled from the face of man, and sought other abodes where they roam undisturbed by the sharp crack of the murderous rifle, and the fleet pursuit of the hunter's steed. The country between the Fish River and the sand plains of the coast is characterised by high and rugged mountains. Eastward are undulating plains which lead to the trackless wastes of the desert. The Qnosop and its principal tributaries take their rise in Damaraland, and form the eastern boundary of Namaqualand. After winding through the sandy desert, and uniting with the Kuruman and another river, it falls into the Orange a little below the Waterfalls. Its bed is sandy, so that water can seldom be procured from it.

"The principal mountains between the Fish River and the Qnosop are the Xgharas range, situated about 150 north of the Qgariep. From these rise the Lion River, which runs in a westerly direction, and joins the Fish River, the Xhum, Xham, and Kei Qaap; which run south and fall into the Qgariep, and the Bak River, which takes an easterly course, and loses itself in the sand hills.

"All the rivers of Namaqualand, except the great stream of Southern Africa, are merely periodical water-courses, their beds are generally dry and covered with glowing sand, and, except the Fish River, even in the best seasons, seldom continue running for more than eight or ten days successively. Yet they are the great reservoirs of the country. In many parts they contain large pools in which water may be obtained for many months after rain; occasionally a spring oozes out from their burning sands, and wherever a ledge of rocks crosses their course one may dig in hope. They are also valuable to the natives on account of the trees which in most parts cover their banks. The quantity and the quality of the timber, however, only serves their scanty wants; it would not be worth exporting even if greater facilities of transport existed. Throughout the vast region which we have traversed but few fountains are to be found, and of those few some are so situated, and so weak, as to be useless for all the purposes of irrigation.

"Some tribes do not possess, in all their territory, a single fountain which will admit of irrigation. The principal springs are at Nisbet Bath, the Hoole's Fountain, Bethany, Beerseba, Rehoboth, Elephant Fountain, Rhinoceros Vlei, and Wind Hoek; a short list to enumerate for so many thousand square miles as Great Namaqualand contains. Some of these are warm and hot or thermal springs. The most remarkable are at Windhock, Jonker Africaner's place. They are situated in two localities. Cold, warm, and hot springs issue from the ground within a few feet of each other; they are happily on elevated ground, and are capable of watering a large tract of garden ground. There are two baths, which are celebrated for their sanative qualities, to which many of the natives resort. One is situated at the junction of the Xhams with the Orange River, and the other in the Fish River, some distance above its union with the Orange. In some parts large '*vleis*,' or natural ponds, are found on the plains, which, after heavy rain, contain water for two or three months; to these the natives invariably resort with their herds, so as to save the pasturage near the fountains and rivers for scarcer times. The general appearance of the country and its geological features have been described by Moffat; what he says of the part he traversed will hold good with reference to the whole. In describing a tour which embraced the greater part of south-eastern Namaqualand, and gave him an opportunity of forming a correct idea of its general features, he says :—'The country over which we passed was sterile in the extreme, sandy from the abundance of granite. Iron stone was also to be found, and occasionally indications of copper. Slaty formations were also to be met with, and much quartz, filling up large fissures occasioned by former convulsions, and the hills in some places presenting a mass of confusion, the strata bending and dipping from the perpendicular to the horizontal, and in others extending in a straight line from one hill to another. Native iron, in a very pure state, is procured in these regions, and from the account given by the natives, I should suppose some of it is meteoric. The plains are invariably sandy, and there are even hills of pure sand.'

"If I were to say anything of the mineralogy of the country, it would be merely on speculation. I feel justified, however, in venturing an opinion that whatever mineral wealth there may be, it is principally confined to the neighbourhood of the Orange River, to the mountainous tract between the sea cost and the Orange River, and to the mountains of Southern Damaraland. It would be folly in me to say more, when practical men require such a searching investigation before they can hazard an opinion. The nature of the country, its scarcity of water, its mountain barriers, and its barren coasts, would always be serious difficulties in the path of enterprise, however rich its bowels might be.

"The climate of Great Namaqualand is extremely hot in summer and cold in winter. It is, however, very salubrious, except at the northern extremity, where its contiguity to moister regions, occasions fatal intermittent fevers. The rainy season commences in November and ends in May. The rains are generally late and more uncertain, the nearer we approach the south. The country about the Orange River, the vicinity of Nisbet Bath, the sea coast, and some parts of the Fish River region, are especially subject to continued and oft-recurring droughts. The only rain is from thunder clouds; they rise from the N.E., and are always hailed with anxious delight. . . There is something sublime in a Namaqualand thunder storm. The air is sultry and oppressive, not a breath of wind is felt, a faint rumbling is heard, coming from a dense black cloud in the distance, which momentarly becomes louder, the threatening mass gradually rises, lighted up with the quick flashes of forked lightning. At length a cloud of dust approaches, a storm of wind rushes over the plain, overturning trees, uprooting bushes, and sweeping everything before it in its tumultuous course. A few large spattering drops are heard, and then with the almost simultaneous lightning's glare and thunder's crash, torrents of mingled hail and rain descend. In a few minutes the country is flooded, rivulets flow where one would think water had never run before, and the ear is charmed with the sweet strains of a long silent music. Perhaps in less than an hour the cloud has passed over, and may be seen speeding onwards to pour out its treasures over many a sunburnt plain and parched mountain. The sun shines brilliantly forth, nature seems to rejoice in a resurrection, and everything that has life appears to accord new praise to a bounteous God.

"Great Namaqualand is only sterile from the scarcity of water, and the frequent recurrence of drought. If the rains were more certain, and fell at the proper season, it would amply repay the agriculturist for all his toil, but as it is, nothing can be done. In a good season the plains are covered with long waving grass, like corn-fields, inviting the reaper's sickle. This grass will stand for nearly two years without perishing. Along the Fish River, and the mountainous regions of North-Western Namaqualand it is more tender, and is sooner scorched by the burning heat, so that those parts suffer most when visited by a drought. There are but few native fruits. The principal are the '*vnaras*' of Walwich Bay, and the '*vaûs*.' The latter is a berry about the size of a large pea; it grows on a large bush and is very sweet. The natives distil very strong brandy from this fruit, the process of which will be afterwards described. Various kinds of roots and a few bulbs are also articles of consumption, but they would only suit a palate which stern hunger had deprived of all fastidious taste. In some parts wild melons grow, but nowhere so abundantly as in the eastern desert. For cattle and game, and natives who are accustomed to them, they form a sufficient substitute for water for a time. If it were not for the grateful relief they afford, many an adventurous hunter would have perished of thirst, and many a swift courser of the plain would have failed in the chase.

"There are not many varieties of trees in the country under consideration. The mimosa, camel thorn, wild ebony, a yellow wood tree, and the koker tree, are the principal. Timber, for building or other purposes, has often to be brought from a great distance. There are no forests, unless the lines of trees which ornament the banks of the periodical rivers deserve the name.

"In a country which is so noted for barreness we cannot expect to meet with much richness or variety of flora. There are few bulbs and plants which would interest the botanist. Even in the more fertile regions of the north a flower is almost a miracle. The golden blossoms of the mimosa give a little charm to the scenery. In the south the hills and valleys are covered after rain by the common yellow flowers of the 'opslag.' . . .

"If Namaqualand can boast at all of its feathery tribes, it is of the richness of their plume, and not of the sweetness of their song. Ostriches may be seen on the plains in considerable numbers. Vultures and crows abound, and are the bushman's guide to the spot where the victim of his venomed shaft has fallen. There are several beautiful varieties of the hawk tribe, and a few handsome cranes may occasionally be seen. In Northern Namaqualand immense flocks of the locust bird follow the flight of Afric's scourge. Their evolutions in the air present a beautiful spectacle, they travel onward in a revolving circle, and when they alight on some open plain they have the appearance of a vast army drawn up in battle array, though their raven attire aud snow-white breasts (if you will pardon the sacrilegious allusion) resemble more an assembly of grave divines. There are handsome varieties of paroquets and woodpeckers; the hoopoo may also be found in some parts. Guinea fowl, Namaqua partridges and pheasants, wood pigeons, turtle doves, bustards, with wild geese, ducks, and other waterfowl, are most plentiful along the Fish River and the borders of Damaraland."

To the east of the lands inhabited by the Namaquas and the Damaras is the Kalahari Desert, where there is not only a want of water, but there has been a great diminution of the supply within the lifetime and memory of its present inhabitants.

Mr J. Fox Wilson, in a paper on this subject, wrote :—" That great expanse of wilderness called the Kalahari, remarkable for few inhabitants, little water, and considerable vegetation, seems to be gaining in extent, gradually swallowing up large portions of the habitable country on its confines, and slowly, but surely, assimilating their fertile character to its own sterile one. It has become matter of notoriety that springs, which a few years ago supplied a sufficient quantity of fluid to irrigate considerable breadths of garden and fields, have diminished in their flow and dwindled away, causing the migration of the inhabitants to a more favourable dwelling-place; while desert sucking-places and well-filled pools, such as that at Serotli, described by Livingstone, are at present either completely dry, or afford only a small quantity of liquid after much digging, where formerly existed a large piece of water.

" At Lopepe and other places on the road to Lake 'Ngami, this is the case, as well as at Tunobis, in Damaraland, and elsewhere, but it is most conspicuous in the territory of the Bakwain tribes, in which, as one of the many evidences of the growing desiccation of the country, streams—*e.g.,* the Mahalapis River, that at Lopélole and at Porapora Pass—are pointed out where thousands and thousands of cattle formerly drank, but in which water never now flows, and where a single herd could not find fluid for its support.—(Livingstone, pp. 14, 150).

" When Mr Moffat first attempted a settlement at the Kuruman, forty years ago, he made a dam six or seven miles below the present one, and led out the stream for irrigation, where not a drop of the fountain-water ever now flows; and other parts, fourteen miles below the Kuruman gardens, are pointed out as having contained within the memory of people now living hippopotami and pools sufficient to drown both men and cattle.—(Idem. p. 110).

" The fountain at Griquatown, which a few years ago yielded a sufficiency of water to irrigate four square miles of corn and garden-ground, has of late years, and in the most marked manner, diminished its supplies, almost ceasing to flow, and occasioning the emigration of many of the Dutch-speaking inhabitants to other and more fertile localities, not subject to the absence of water.

" As this diminution of water has been coincident with the failure of fountains over a wide extent of territory in Bechuanaland, it is evident that from some cause, more or less obscure, *a great change in the external physical characteristics of the entire region between the Orange and the Ngami Lake has taken place since the country was first explored by Europeans.*

"This great change has not, however, been confined to the comparatively short space of time during which missionaries have been in the country. On the contrary, the traditions of the natives point to more remote periods, when the country was far more fertile and much better ordered than at present, when the Kuruman and other rivers, with their impassable torrents, were something to boast of. Moffat says, that accounts of floods of ancient times, of incessant showers which clothed the very rocks with verdure, and of the existence of giant trees and forests which once covered the brows of the Hamhana Hills, are wont to be related by garrulous elders, to the utter astonishment of their younger listeners. In those ancient days, the lowing herds walked to their necks in grass, and filling their owners' milk-sacks with milk, made every heart to sing for joy. —(Moffat's South Africa, p. 86.)

"But, independent of this oral and traditional testimony, travellers have before their eyes, in the immense numbers of stumps and roots of enormous trunks of the acacia giraffe, where now scarcely a single living specimen is to be seen raising its stately head above the shrubs, and the ancient beds of the dried up rivers—Matlaurin, Mashaua, Molapo, and others—positive demonstration of the departed former fertility of the lands of the Bechuana nation. In fact, the whole country north of the Orange River, and lying east of the Kalahari Desert, presents to the eye of a European, to use the words of the missionary just quoted, 'something like an old neglected garden or field.'"

This desiccation is attributed by him to the destruction of forests, and herbage, and bush; and many facts corroborative of his views have come under my consideration. By Dr Livingstone and others the desiccation of these regions has been attributed to an upheaval of the land; and a study of the hydrology of South Africa will show that no small influence has been thus excited in producing the aridity of the continent, and of the southern extremity of it not less than of any other portions of it whatever. This, as I have shown in another volume, I consider to have been the primary, and the destruction of forests and herbage the secondary, cause of the changes indicated by the geological formations and physical geography of the Colony and of extra-colonial districts, and noticed by the observers on whose statements Mr Fox Wilson bases his views.

If we have correctly diagnosed the evil and its origin, the ascertained causes of it are suggestive of the appropriate remedies to be employed; but here we take into consideration only the facilities presented for the adoption of one of these—the utilisation of the water supply, such as it is.

What has been reported does not promise much for the region described; but I do not consider that there exists reasons why even the Kalahari should be considered a land doomed to perpetual sterility and barrenness.

Mr Fox Wilson's hope of any amelioration of the condition of the country seems to have rested primarily on the expectation that an extra supply of water might be obtained by Artesian Wells. The hope is not unreasonable; but, whatever may come of it, I do not know of any reasons why the measures deemed by Mr Fletcher applicable to Namaqualand might not prove applicable also to these regions beyond the river.

Mr Wilson writes:—"According to the latest geological observations on the neighbourhood of the Kuruman, there is no perennial fountain in that part which does not come from beneath the quartzose trap that constitutes the filling up of the ancient valley, the igneous rock resting on Silurian schists, from the surface of which the water appears to rise. We may remark, by the way, that this association of rocks is favourable, if we may rely on the dicta of experienced observers (*teste* Sir Roderick J. Murchison), to the probability of the existence of gold. But confining ourselves strictly to our subject, let us notice a few facts in connection with the most accurately described fountain in South Africa, that called Gasigonyane, which gives rise to the River Kuruman.

"This remarkable spring issues from caverns in a little hill which is composed of blue and gray limestone mixed with considerable quantities of flint, but not in nodules as found in beds of chalk. From the appearance of the caves and the irregularity of the strata, one might be led to suppose they have been the results of internal convulsions. The water, which is pure and wholesome, is rather calcareous. It is evident that its source must be at a very great distance, as all the rains which fall on the hills and plains for forty miles round in one year could not possibly supply such a stream for one month. Although there are no sandstone formations nearer than thirty miles, great quantities of exceedingly fine sand come from it, and silicious particles appear to boil up out of the smaller springs in front of the larger, and are found in deposit in the bed of the river for miles distant, a proof this of the subterraneous passage of the water from the sides of the elevated basin to the centre, and a strong argument in favour of the probability of the success of Artesian Wells, if attempted. Large, however, as is the body of water rising from the Kuruman fountain at its origin, like so many others in South Africa, it is largest at its source, and after running a course of ten miles, becomes evaporated in its bed to the north-west of the settlement.

"From the thick deposits of tufa visible on the stems of plants and elsewhere near fountains, the supply of water must in former times have been much greater than at present; and next to the occurrence of ancient river beds and water courses in the desert, perhaps the most conclusive proof of the former well watered condition of Bechuanaland is to be found in the numerous eyes of ancient fountains partially filled with sand, rushes, and tufa, which are to be met with in the neighbourhood of the Kuruman River and elsewhere. Many of these old fountains, by means of long and deep canals made by the Dutch boers from lower levels up to spots that indicate the former presence of water, have been resuscitated, and now form permanent streams of great value for irrigation purposes. From the occurrence also of voluminous fountains like the Kuruman at an immense distance from the sources of water supply, as well as from the fact of the rising of water from the surface of the Silurian schists which constitutes the bottom rocks of the South African central valley, it is pretty certain that the supplies of fluid under consideration flow beneath the plains from the well watered table-headed mountains of the eastern ranges, in the same way that the waters of the Algerian Sahara flow beneath the surface of the soil from the mountains of the Atlas Chain and re-appear in the noted Bahr-taht-el-erd, or underground sea of the plain of Tuggurt. There are no fountains in that part of Algeria, but the inhabitants dig through several layers of sand and gravel till they reach a flaky stone-like slate known always to be above the Bahr. This layer is easily broken through, and the water thereupon rushes up so quickly that the man who digs through it is sometimes drowned. (*Morell's Algeria*, pp. 229 and 248).

"Made aware of these facts, the French Government, some years ago, conceived the idea of forming a chain of stations across the

Sahara, or a portion of it, from the Tell to Senegambia, each station to be provided with an Artesian Well. (*Revue de Paris*, April 24, 1845; *Chambers' Journal*, Vol. 4, p. 92; *Davies' Algiers in 1857*; *Annee Scientifique*, 1863; *Almanack des bons conseils*, 1864). In 1856, French engineers dug a well of this character at Zamerna in the Oued-Ris. After forty days labour, at the depth of sixty *metres*, a supply of water yielding 4,500 *litres* per minute was struck, which so surprised the Mogrebbin Arabs that they organised a grand festival in honour of the fountain. In the oäsis of Sidi Rached, the formerly extensive verdure was perishing, and with it the supply from the spring, when the French sappers constructed a deep well yielding 4,300 *litres* per minute. At the sight of this abundance of water the tribe was beside itself for joy, and the old sheik, falling on his knees, is said to have thanked God for sending the French to be the means of restoring the oäsis to the possession of the tribe. During the last five years fifty wells have been pierced in the Algerian Sahara, yielding 36,766 *litres* of the fertilising element per minute,—an amount of water equal to the flow of several rivers. Notwithstanding this pleasing success, which has been followed by the settlement in villages of several hitherto nomadic tribes, and the planting of palm groves, it is matter of grave doubt whether the whole chain can be successfully formed to the Senegal. Sublime visionaries have not hesitated to assure us of the possibility of reclaiming even the Sahara, while more sober intellects have coolly argued the soundness and practicability of their bold theories; but if the digging of deep wells in the great desert should be followed by the growth of date and Doum palms, the French have already taken a first and most energetic step towards the accomplishment of its reclamation. And, reasoning from analogy, there can be but little doubt that if Artesian Wells have been found successful in the Algerian Sahara, they may also be practicable in the Lesuto or the Bechuana country. The time may be nigher than we anticipate when the advance of railroads from the ports towards the interior, the increased civilisation of the aboriginal tribes, and the gradual development of commerce, may occasion the experiment to be made as to whether deep wells may or may not afford perennial supplies of fluid from subterraneous reservoirs of water beneath the central plains of South Africa, those supplies being derived from the gravitation of the rains from the mountains towards the centre."

Mr Wilson having no object in view but the eliciting of the truth, that it may be known and made known what can and what cannot be done, generously placed in my hands, on my return to Europe, the whole of his correspondence on this subject. Amongst other documents thus supplied to me I find a letter from Mr Francis Galton, who had been one of the first travellers who had thoroughly explored those regions, to Sir Roderick Murchison, who had forwarded to him a letter of inquiry which he had received from Mr Wilson.

In this letter Mr Galton writes, under date of 20th Nov., 1862 :— "Often, when in Africa, I looked for conditions that might hold out hopes for Artesian Wells, but saw none; and from the similar character of parts I know of by report, but have not seen myself, I doubt whether they exist anywhere in Namaqualand or the Kalahari.

"In the first place, there is *no* high well-watered range whence underground supplies might regularly percolate below the plains; on the contrary, the hill tops are a very abomination of desolation at all times, except during the rainy seasons, when they are no better off than the plains.

"Secondly, there is a remarkable absence of clay or other impervious strata (other than solid rock). The natives are just able to find a very limited supply of argillaceous earth (of the whereabouts of which they make great mystery) sufficient to make a rotten kind of pottery. A pot is a highly prized article with them, and a very scarce one, owing to the absence of raw material. Consequently, the rains sink directly through the porous sandy earth, till they reach a solid bottom, and then drain rapidly into the usual arid river bed. In a day's time, a turbid current, half-a-mile broad, will roar through a ravine that has not shown a drop of water for years previously. There are, however, thin strata of mud here and there, which retain, with more or less completeness, the wet above them. It is on these, and these almost alone, that the habitability of Damaraland and Namaqualand depends. Either it makes possible the formation of a pool or lagoon which receives the drainage of a large area, and, sheltered by the rushes and trees about it, occasionally holds out until the next rains; or it forms the basis of a layer of wet sand. In one of the reaches of a sand river bed, the sand is dry on the surface, but a bushman's eye discovers indications of subterranean moisture; he digs and finds wet sand. Then, and *this is a bushman's law throughout South Africa*, let him abstain from digging any further than is necessary for his wants, or he may tap the basis of clayey soil on which the water reposes, and the contents of his well will run through like that of a bucket with a hole in the bottom.

"I believe the land to have very little more water in it during the dry season than the natives know of. Often the rocky beds of the rivers are exposed, and no streamlet ever courses along them. In fact all the water lies in stagnant reservoirs either as shallow lagoons above ground or as circumscribed wet patches of sand with a precarious bottom below ground.

"The only indications of anything else lies in less than six small hot water-springs, in places where, I believe, limestone rocks crop out. These springs are fairly constant, but the largest of them, that at Barmen is, as I know well, far from being wholly so.

"Lastly, I should have said, as a further proof of the circumscribed nature of the underground reservoirs, that a limited number of cattle will soon exhaust any one of them.

"The Australian practices are, I believe, those that would improve the lands we are speaking of—viz., damming the smaller river courses, so as to make artificial reservoirs, then encouraging a judicious vegetation to check evaporation, and finally trusting to the stamping of the feet of sheep to harden by 'puddling' the surface of the land in the neighbourhood of the newly-formed lagoon."

This reply was considered by Sir Roderick Murchison to shut out all hope of Artesian Wells being dug with success. But he considered the suggestions of Mr Galton, in regard to other measures for gathering and maintaining supplies of water, sensible and good. With all deference to such authority, I must add I fail to see that the corectness of the views of Mr Galton are incompatible with the views of Mr Fox Wilson being also correct.

Mr W. C. Baldwin, to whom also application was made for information, gave it as his most decided opinion that throughout the greater portion of the interior over which he had travelled the successful sinking of Artesian Wells was impossible; assigning, as reasons for his opinion, the great distance of any mountain range whence water could flow, and the rare and limited and uncertain rainfall. Of Damaraland and Namaqualand he said he knew nothing personally. In the Kalahari he saw no chance of such wells supplying water.

But in mountainous districts he considered that tanks and deeper wells, for surface and superficial draining, if means for their formation were found, would be a priceless boon.

The acting secretary *pro tempore* of the South African Irrigation and Investment Company, with whom Mr Wilson had also entered into correspondence on the subject, wrote to him also somewhat discouragingly.

In his letter he said,—"The geological formation predominating in the Colony is unfavourable, as I understand, to the sinking of Artesian Wells. I have only heard of one attempt being made to sink such a well, and that in the neighbourhood of Capetown, where the rock arrived at, after passing through a comparatively thin bed of clay, proved extremely hard and incapable of being worked in a country where skilled labour is excessively dear. The instruments employed, I am informed, were immediately broken.

"I can give you no information under this head as to the extra-colonial or desert tracts you refer to."

Queries were addressed to Rev. Mr Haddy, who travelled extensively through the district in the years 1843 and 1844, with a view to establishing a mission station on the conterminous borders of Great Namaqualand and the Damara country, and who resided seven years at the northern limits of the country, at about the longitude of the Cape, in the course of which time he made frequent journeys both through the districts adjoining his station and districts far remote. In answer to these Mr Wilson received the following replies, with an intimation, however, that they must be understood as relating chiefly, and almost exclusively, to Great Namaqualand:—

"1. Rains fall annually, with more or less regularity; they are, however, akin to tropical thunder-showers, instead of seasonable periodical rains. Yet considerable tracts of country are left without rain enough to revive vegetation (small and tender plants) for a series of years. Such parts must be deserted by man and beast during such intervals.

"2. Clay may be procured, probably, but at distant points; but would be difficult of carriage, involving an enormous outlay.

"3. The formation of pools or reservoirs is no doubt practicable, but it would be of difficult execution and very expensive.

"4. Reservoirs of water, if covered, would probably soon become centres of animal and vegetable putrifactions, utterly unfit for the use of man. If uncovered and well supplied with water they would be of little use to a pastoral people, unless the rains were sufficient to supply the amount of vegetation required, in which case they would not be needed.

"5. Willows and all such trees as line the banks of the Gariep.

"6. The 'Nqara of Walwisch Bay' is not required to bind the sands beyond the limits of the belt of marine sands that lines and borders the western coast to a considerable depth inwards from the sea. There is no necessity for cultivating any plant for binding the soil of the interior, which is, for the most part, covered with vegetation.

"7. It is quite likely that the palm might grow and would flourish near the Orange River.

"8. Water from the springs is stored by some of the nations, as far as it suits their purpose, and large dams might be constructed; but, as before observed, water without pasture is of no use.

"9. Probably in some parts of Great Namaqualand and Damara country Artesian Wells might be bored, as the abundant pasturage, numerous trees (chiefly varieties of thorns), and plentiful supplies of game, indicate the presence of moisture not far from the surface."

In accordance with this is a letter from the Rev. H. Tindall, to whom Mr Wilson had subsequently written on the subject, at the suggestion of Mr Haddy, as a man than whom, perhaps, there was not another better able to describe the country. In this letter, Mr Tindall wrote, under date of 15th July, 1863:—"I think I can do no more than endorse Mr Galton's views. They apply to all Namaqualand and Damaraland that I have seen. 1st. There very little rain falls. Two or three thunder showers lasting an hour or so in a year is about the average quantity. The rain comes down in torrents and finds its way in a very short time to the beds of the periodical rivers. 2nd. There are no elevated regions in which moisture might be collected in order to furnish a supply to lower tracts of country. 3rd. Sand and stones prevail to the exclusion of clay. 4th. Nearly all parts of the country slope down to periodical rivers, which drain into the Atlantic or the Qgariep. I should suppose that these conditions are generally unfavourable for Artesian Wells. I have, however, occasionally seen tracts of land which suggested the idea of Artesian Wells to my mind, though not sufficiently acquainted with practical geology to form a correct opinion.

"The Namaquas are not in the habit of making wells, but among the Damaras it was a general practice. Most of those I saw when travelling in Damaraland yielded water at a depth of six or eight feet, though some were thirty feet in depth.

"The warm springs at Nisbet Bath are constant. I have never known them in the least affected by long continued drought. I was under the impression that those at Barmen, Concordia Ville, and elsewhere in Damaraland, were equally so, though Mr Galton says they are not. With regard to the planting of trees, etc., to protect water, the difficulty in those arid countries is to get the water. More might be done to collect it in good seasons, but the incorrigible sloth of the natives presents a great barrier to such enterprises, and the fall of rain is generally so heavy when it does come that very strong works would be required to stem the impetuous torrents that come rolling down."

And again, under date of 10th December, 1863 :—"With regard to Namaqualand, the water supply very much depends on rain. Many of the fountains are situated in the beds of periodical rivers, and are strong or weak as the rivers run frequently or seldom. Other fountains, such as Nisbet Bath, Hooles Fountain, Concordia Ville, etc., which are few and far between, and which are mostly warm, keep their strength, and, during the 25 years that I have known them, have suffered no sensible diminution.

"The natives in those regions very seldom burn the field, the sun seems to perform that office for them. A summer without rain effectually consumes a vast proportion of young shrubs and trees that sprang up in a wet season. So that by ever recurring droughts the vegetation is kept down, and thus the effect of drought becomes in its turn the cause of drought. I have seldom if ever known the Namaquas set any tract of country on fire, and fires rarely occur by accident. The exception is when they are in pursuit of young locusts, when the bushes in which the insects rest at night in dense masses are fired, and in this way millions are caught for food; but then the conflagration is very partial. . . .

"Vast tracks of the country are not swept by any current of air from the sea coast, consequently they have no rain in winter. Their rains fall in summer from thunder-clouds. Perhaps they have three tremendous showers in the course of three months; and these, lasting on an average two or three hours each, suffice to cover them with the richest vegetation. But these are the regions which, cut off from the sea and its prevailing winds, suffer most from drought.

"As far as I have been the thunder-showers are brought up by a wind from the N.E. A steady breeze from this direction is sure to bring rain, but no other will. In this part of the Colony the rain winds are in the winter from the N.W., and in the summer from the N.E.; but all thunder-rain nearly comes from the N.E. point. Very occasionally it seems to rise from the S.W.

"I think a very close study and accurate observation for some years of atmospheric currents will be required before the frequent recurrence of drought will be satisfactorily accounted for,

"Some years the heavens seems so hermetically closed that one would think it never could rain in certain regions; and then again, in other years the rains are so generous and frequent, and seem to be brought up with so little difficulty, that one would suppose a dry season could not possibly occur. . . .

"As for the planting of trees, I have never known natives think of such a thing. . . . The trees in Namaqualand and Damaraland are principally acacias; very few attempts have been made to raise others there." Such is the country bordering on the Kalahari.

The Kalahari itself is described as "a vast extent of dry and sterile country, extending north from the Orange River, in lat. 29° S., to near Lake Nyassa, in lat. 20°, with an average width of 6° of longitude, or an area of nearly 200,000 square miles. It is a region of sandstone resting on a bed of tuffaceous lime, or perhaps Dunn's glacial conglomerate, and covered with a dense low bush. It has no running streams, although the few bushmen who inhabit it discover water here and there in the dry beds of streams, forming what are known as 'sucking-places.' Owing to this want of water it is seldom traversed even by the natives."

Baines and Andersson, and Moffat and Livingstone, have all told of its sterility and its drought.

Moffat, skirting that country, tells:—"A party of Bechuanas, who had accompanied us thus far, now proceeded north-west, towards the Kalagare, and we journeyed towards the east.

"The country became rather more interesting, being interspersed by hills of lime-stone, covered with trees and shrubs, with an abundance of game, some of which must travel at least twenty miles to obtain water. I found, however, that many of the antelope kind could remain two days without water, while rhinoceroses and quaggas were in the habit of frequenting it daily. Some of the company killed two elephants during the day, and we were compelled to bivouac in a plain without water, as it was dangerous to proceed, owing to the number of lions, whose roar we heard in the distance. Next day we proceeded in a more northerly direction over an undulating country, covered with a considerable quantity of timber, but of the poorest quality. We halted at two natural wells of rather an extraordinary description, an iron schist formation, about 100 yards from each other. One is about sixteen feet deep, with four feet of water—they are both nearly perpendicular, and about two feet and a half in diameter. The hill in which they are is composed of a conglomerate mass of iron schist, and near the mouth, as well as in the sides of these holes, are appearances as if the whole had once been in a state of fusion, and that these were the apertures of some internal fires, but nothing like lava appears in the neighbourhood. From the older natives who have resided near these wells all their lives I learned that they were once much deeper. The water was excellent, and to obtain sufficient for ourselves and horses, we fastened a vessel to the end of a rope; the oxen we sent to a water at a distance, called Khuari."

Of the wretched condition of the Balala or Bechuana bushmen inhabiting this desert, he says, "I had frequent opportunities of convincing myself during the present journey. They are generally spoken of in the same manner as pack-oxen or beasts of burden, being employed for that purpose. While we were here a Mochuana met some of these people carrying meat which they had procured at a great distance, and were taking to their families, when he ordered them to take every ounce of it to his own abode. If the wounded game happen to fall at a place remote from water, these people are collected, especially the females, and compelled to carry the meat perhaps a distance of thirty miles; and, to prevent their elopement when their services are required the following day, they are sometimes hedged into a fold made of hook-thorn bushes, precisely like so many sheep, and there they must pass the night. Many of the poor women came to the water, particularly when they found there was a stranger there who took their part. The Bechuanas, who were travelling with us to the Barolongs, did not object to my interference on their behalf, and only laughed at my foolishness in making such 'lincha,' dogs, the objects of my sympathy. They, like the natives in general, live at a distance from water, which they visit at most once a day. As they never wash themselves, little of that precious beverage serves; their vessels consist of sacks made from skins, also the entrails and paunches of animals. They use also ostrich eggs for the same purpose, corked with grass, and of which a woman can carry thirty."

And again,—"The landscape was somewhat pleasing to the eye; many clumps of trees were scattered around, and on the plain to the north, between us and the Molapo river, appeared a forest, but the timber, chiefly acacia, was of small dimensions. On the distant horizon, hills in the Bauangketsi country were seen, apparently covered with timber, indicating a more fertile region. Lions abounded in this neighbourhood, but they did not disturb us, except by an occasional roar. Some of the horsemen having visited the Molapo, and found the bed of it dry, it was necessary to alter our course.

One evening we came to a pool of rain water, which was surrounded by fires, in order to prevent the game from approaching. This was to us a most providential supply, as there was no water for two days after leaving this. The few natives who visited us, finding that we were very friendly, brought the whole village to our encampment; and as we had plenty of meat, they were, to their no small surprise, liberally supplied and rewarded for allowing our cattle to drink at their guarded pool. The soup in which our meat was cooked, and which contained an ample share of mud, was swallowed with avidity: a dozen would surround a pot, and having no spoons, and not allowing time for the soup to cool, they used the right hand to take out a little, threw it quickly into the hollow of the left, thence into the mouth, and afterwards licked both that nothing might be lost. The following day we travelled over a dry and sandy plain, and halted, without water. Early next day we resumed our journey, and it was distressing to see the suffering of the poor cattle from thirst, running into the shadow of a tree or bush, from which it was difficult to remove them. We at last descended into the bed of the Molapo, but it was as dry as the neighbouring plains. We proceeded eastward along the bed of the river, but could not meet with an individual to give us information as to where we might find water. The valley becoming rocky, we were compelled to lead out our waggons to the open country. We had scarcely done this, when two lions passed along the spot we hed left, roaring furiously. After some miles' jogging over a rough bushy country, we descended again into the river bed, where it was discovered the reeds were on fire. Nearly the whole party ran, expecting water, but found none. Men and cattle being worn out, we halted for the night, every one feeling as if this night was to be his last. Two very hot days' travelling over a dusty plain, with a dry and parching wind, had reduced mind and body to a state of great exhaustion. A camp of eleven waggons, upwards of one hundred and fifty oxen, and nearly a hundred human beings, generally, make a terrible uproar, especially when there is plenty of meat; ours was silent as the desert around, interrupted only by an occasional groan from the wearied, worn-out cattle.

"Thirst aroused us at an early hour, and examining the footmarks, we found that the horsemen who had left us on the previous day in search of water had passed eastward. Before we had proceeded far, a buffalo was discovered in a thicket of reeds. The men, seizing their guns, fired upon him, but as he concealed himself in the middle of the reeds, it was difficult to reach him. I entreated the men to desist, as from the character of the buffalo when wounded an accident appeared to be inevitable; however, they persisted, saying, 'If we cannot get water, we must have raw flesh.' In order to dislodge the animal, they set fire to the reeds, when the enraged buffalo rushed out through the fire and smoke; and though his gait seemed as awkward and heavy as that of a great pig, he instantly overtook one of the men, who escaped with merely being thrown down, slightly wounded, and having his jacket torn open. Had not the dogs at the same moment seized the animal from behind, the man would have been killed on the spot. The buffalo returned to the flaming reeds, from which he would not move, but was shot after his skin was literally roasted in the fire. About noon we came unexpectedly to the stream, into which men, oxen, horses, and sheep rushed promiscuously, presenting a scene of the most ludicrous description. One man is pushed down by an ox, pleased with the refreshing coolness of the water; another, in his haste, tumbles head-foremost over the bank, followed by a sheep or a goat. One crawls between the legs of oxen; another tries to force himself in between their bodies. One shouts that a horse is trampling upon him, and another that he is fast in the mud. But while all this was going on, there was no disposition for merriment, till every one was satisfied and withdrew from the water; when wet, muddy-looking spectacles presented themselves, which would have caused even gravity itself to laugh. While the meat was preparing over the fire, a quaff of the tobacco pipe unloosed every tongue, and made all eloquent on the hardships of the past. Correctly to conceive of such a scene, it is necessary to have witnessed it."

Of the wretched inhabitants of the Kalahari Livingstone states that he considers the Bushmen as probably the aborigines of the southern portion of the Continent and the Bakalahari—or people of the Kalahari—the remnants of the first emigration of the Bechuanas. The Bushmen, says he, live in the desert from choice, the Bakalahari from compulsion, and both possess an intense love of liberty. Of the Bushmen he says:—"Many are of low stature, though not dwarfish; the specimens brought to Europe have been selected, like costermongers' dogs, on account of their extreme ugliness; consequently English ideas of the whole tribe are formed in the same way as if the ugliest specimens of the English were exhibited in Africa as characteristic of the entire British nation. That they are like baboons is in some degree true, just as these and other simiæ are in some points frightfully human."

Of the Bakalahari he says:—"The dread of visits from Bechuanas of strange tribes causes the Bakalahari to choose their residences far from water; and they not unfrequently hide their supplies by filling the pits with sand and making a fire over the spot. When they wish to draw water for use, the women come with twenty or thirty of their water-vessels in a bag or net on their backs. These water-vessels consist of ostrich egg-shells, with a hole in the end of each, such as would admit one's finger. The women tie a bunch of grass to one end of a reed about two feet long, insert it in a hole dug as deep as the arm will reach, and then ram down the wet sand firmly round it. Applying the mouth to the free end of the reed, they form a vacuum in the grass beneath, in which the water collects, and in a short time rises into the mouth. An egg-shell is placed on the ground alongside the reed, some inches below the mouth of the sucker. A straw guides the water into the hole of the vessel, as she draws mouthful after mouthful from below. The water is made to pass along the outside, not through the straw. If any one will attempt to squirt water into a bottle placed some distance below his mouth he will soon perceive the wisdom of the Bushwoman's contrivance for giving the stream direction by means of a straw. The whole stock of water is thus passed through the woman's mouth as a pump, and when taken home is carefully buried. I have come into villages where, had we acted a domineering part, and rummaged every hut, we should have found nothing; but by sitting down quietly and waiting with patience until the villagers were led to form a favourable opinion of us, a woman would bring out a shellful of the precious fluid from I know not where.

"The so-called Desert, it may be observed, is by no means a useless tract of country. Besides supporting multitudes of both small and large animals, it sends something to the market of the world, and has proved a refuge to many a fugitive tribe—to the Bakalahari first, and to the other Bechuanas in turn—as their lands were overrun by the tribe of true Caffres, called Matebele. The Bakwains, the Bangwaketze, and the Bamangwato all fled thither; and the Matebele marauders, who came from the well-watered east, perished by hundreds in their attempts to follow them. One of the Bangwaketze chiefs, more wily than the rest, sent false guides to lead them on a track where, for hundreds of miles, not a drop of water could be found, and they perished in consequence. Many Bakwains perished too. Their old men who could have told us ancient stories perished in these flights. An intelligent Mokwain related to me how the Bushmen effectually baulked a party of his tribe which lighted on their village in a state of burning thirst. Believing, as he said, that nothing human could subsist without water, they demanded some, but were coolly told by these Bushmen that they had none, and never drank any. Expecting to find them out, they resolved to watch them night and day. They persevered for some days, thinking that at last the water must come forth; but, notwithstanding their watchfulness, kept alive by most tormenting thirst, the Bakwains were compelled to exclaim, 'Yak! yak! these are not men; let us go.' Probably the Bushmen had been subsisting on a store hidden under ground, which had eluded the vigilance of their visitors."

By Hall it is mentioned that in Great and Little Namaqualand and the Kalahari Desert, whole years may elapse without the phenomena of a running stream. But in Ovampoland, lying between the Kalahari Desert and the sea, and to the north of Damaraland, which again is to the north of Walwisch Bay, which marks the northern boundary of Great Namaqualand, it is otherwise. And Andersson writes of a second journey through it in the beginning of 1867:—"The like magnificent trees, both forest and fruit, were observable everywhere; and the landscape, as heretofore, was dotted in all directions with patriarchal-looking hamlets, and further enlivened by groups of men, women, and children, occupied in their several vocations. The cheerful twitterings and warblings of numerous birds, moreover, added an additional charm to the scene. I have seen many lands and places—some, perhaps, strickly speaking, more beautiful; but amongst the extensive savanahs, or the interminable forests of South Africa, which so long has been my home, there was but one Ondonga! It is a beautiful, open, and, for Africa, highly cultivated country, environed on all sides with interminable and dense forests; in short, a perfect oasis in the desert. In the fall of the year, indeed, the landscape presents a boundless expanse of yellow, waving corn, studded with primitive-looking homesteads, whilst here and there, whether alone or in clusters, arise gigantic, widespreading, and dark foliaged trees. The population of Ondonga is considerable. By counting the houses in a given extent of country, and taking the average number of individuals to each, we came to the conclusion that there were about 100 people to the square mile. There is but little pauperism; the crippled and aged, moreover, seem to be carefully tended and nursed. What a contrast to their near neighbours, the Damaras!"

CHAPTER III.

Supply of Water and Facilities for the Storage of it in Inland Western Divisions of the Colony of the Cape of Good Hope.

Section I.—*Divisions of Calvinia, Tulbagh, and Worcester.*

Falling back from the territories of the Namaquas and Damaras and the Kalahari Desert, and re-crossing the Orange River, we re-enter the Colony by Bushmanland—a very thinly peopled district, though visited at times by numbers of farmers with their flocks and herds, trekking from Little Namaqualand, when this has been watered by thunder-showers, while they at their homes suffered from drought. It has already been described, in connection with Namaqualand, as stretching away to the east of the high-lands of that district—a highly elevated sandy flat slipping down gently to the Orange River, with isolated hills, and with ranges of hillocks, hills, and mountains scattered over its surface.

It is some 3,800 feet above the level of the sea—about the level of the summit of Table Mountain, of which some geologists consider it to have been a continuation before violent denudating currents swept away the intervening material. It extends southwards from the Orange River to Calvinia and Fraserburg; and "is a barren and waterless region, except after heavy rains, when a large crop of tall grass, said to be very nutritious, covers it. Its only inhabitants are a few families of migrating boers, and a few wandering Corannas, Namaquas, and Bushmen. Near to the banks of the Orange River the plateau gets broken up into fearful precipices; to the south it slopes away into the elevated region of the Roggeveld and Hantam district."

Of the whole of the region situated on the southern slope of the Nieuweveld and Roggeveld mountains, as of Great and Little Namaqualand and the Kalahari Desert, Captain Hall states that whole years may elapse without the phenomenon of a running stream, and yet, says he, "the magnitude of the dry water-courses of the Buffels,

Hartebeest, and Oup, or Borradail River, all tributaries of the Orange River, show how immense must be the torrents which sometimes sweep along them."

Of the luxuriance of the Towa and the Qga grasses growing here and in Namaqualand, I have received the most glowing accounts. By one of my correspondents I was informed :—"The *Zuid-Afrikaan* has now in its office for inspection a bundle of grass from Namaqualand, from three to nearly four feet long, the growth of a few weeks. 'The country, which, before the late rains, was a barren waste, is described as being covered over with such grass—giving it the appearance of one unbroken waving corn-field. If a country so prolific were blessed with irrigation works, what could it not produce? We only hope that the inhabitants of Namaqualand will have made a good use of the scythe and sickle to lay in a large stock of this grass as fodder for their cattle during the ensuing winter. To allow it to rot on the field would be the basest ingratitude towards a bountiful God.'"

And this is in keeping with statements made to me in regard to the growth of these grasses in Bushmanland.

To the south of Bushmanland, and bordering on the division of Clanwilliam, is the division of Calvinia, which includes a vast extent of the barren region known as Bushmanland.

It has an area of about 2500 square miles. "It is a vast grazing country, liable to severe drought, and devoid of any constantly running streams. That portion of it near the edge of the Roggeveld, called the Hantam, is better watered, and considered healthy for horses and cattle. From thence the country slopes down along the dry plateau of Bushmanland to the Orange River, which forms the northern boundary of the division. The large water-courses marked on the map of South Africa in this neighbourhood are generally dry channels, seldom with any current, unless after heavy thunder-storms. The Roggeveld spoken of is a rocky and barren table-land on the top of a long escarped mountain range, about 5,000 feet high, which extends northwest and southwest from the Komsberg (5,300) nearly to the village of Calvinia. From a distance it looks like an immense wall, and in an extent of nearly 200 miles there are only one or two available passes."

Mr Truter, Civil Commissioner, wrote, in 1862, that the damming up of some of the rivers in the districts which were not subject to excessive or violent overflowing from the effect of thunderstorms, and the construction of dams in favourable localities, were quite practicable, and in some cases would be inexpensive. Such he intimated would likely be the case with one if formed in the vicinity of, and immediately above, the village of Calvinia, which then was, and I presume still is, supplied from wells; but he states, "that no plans for irrigation works in this division had been proposed."

In 1865 the Civil Commissioners reported:—"There has been an entire absence of prosperity during this year. In the early part, the horse sickness prevailed; and made great ravages in Achter Hantam, and the Achterveld generally; while the other field-cornetcies suffered little from this scourge. The early winter promised favourably; but, since the 18th May, 1865, no rains have fallen. Although they were heavy on that day, the rivers were not sufficiently flooded for irrigation. From the continued drought, the harvest has been, so to say, an entire failure. Breadstuffs, sufficient exclusively for home consumption, have been raised only on some four or five farms. The oldest inhabitant does not remember such a drought. The prospects are dreary, and starvation imminent. The cattle are impoverished, and dying by hundreds. Most of those that survive the drought will be carried off by the first cold rains, if no relief come in the meanwhile. The lung-sickness and *bloed ziekte* have also appeared among the rough cattle. Fruits and vegetables are meagre and scarce. The cold of winter was moderate, but the summer heat intense and insupportable."

One of my correspondents has furnished me with a list of the fountains or watering places known to him between Calvinia and the Karroopoort on the route to Capetown. He stated that there was one in the bed of the Doorn River, but the precise spot was unknown to him. At about three or four hours' ride from Doorn River, on the direct road to Fraserburg, by way of Vloksberg, are two springs within a short distance of each other, one of which is on a rising ground, and the water is warm to the feeling, attributable to its coming from a great depth, while the mount may have been formed by a deposit of matter from its waters.

At the foot of Yoke Hooghte, on the further side, is Great Fontein, the water of which is also warm, and tastes of sulphur; the flow appears to be considerable.

On the top of Tuis Hooghte is a permanent spring; and about half an hour's ride from this, at the foot of Tuis Hooghte, is a fountain of cold, pure, fresh water, issuing from blue rock.

At Tuin place—a preaching station of the Dutch Reformed Church, connected with the parish of Ceres—which lies a little out of the direct road, is a very fine spring in the bed of the Tanqua River, which is used to some extent for irrigation. And at Zie Place, which is about six or seven miles further up on the course of that river, and so out of the direct line of the road, is another spring of the same character.

About two hours' ride from the Tanqua River is a spring which appears to be permanent. And at the foot of the Roggeveld Range, at a spot which my informant could not exactly describe, but near where the route is crossed by a road known as the Vlok-Berg, is a good spring of fresh water, which my correspondent believed to be permanent.

From this it appears that fountains exist upon an average throughout the route, at about three hours' ride, or eighteen miles apart.

I have in some other districts traced, or thought I could trace, something like a linear direction perceptible in the relative position of many springs. It may be admitted by me frankly that routes through arid regions may have been, and probably were, determined by fountains existing there; but I advance a step further, and suggest that when the route thus formed follows something like a definite line, there may possibly be found upon examination some dyke or fault, or other geological peculiarity occasioning this disposition of the springs. And should some such fault or dyke be found existing along this route valuable indications of where subterranean water is likely to be found may thence be obtained.

To the south of Calvinia is the division of Tulbagh, lying in the mountains forming the buttresses to the plateaux of the cold and warm Bokkeveldt—the former a region about 4,000 feet above the sea level, bounded on the west by the Olifant's River Mountains, and on the east, like the Roggeveld Mountains, by the Great Karroo,—the latter a fertile region lying behind the second range of mountains at the head of Mitchell's Pass; in the former English fruits ripen to perfection, in the latter the grape and the cereals are produced in abundance, whence has been given the name Ceres to the village in the district. The division has an area of 6,090 square miles. It is well watered, and contains a large proportion of fertile soil.

Many who have visited this district have been struck with the facilities which it offers for remunerative irrigation. Mr Munnik, the Civil Commissioner, reported :—" Irrigation works, in damming up rivers and running waters, and the construction of dams in favourable localities for the collection of periodical supplies of water, could be effected with great advantage to this division;" but "no plans for irrigation works have been hitherto (1862) prepared in my division."

Before I left the Colony, however, works on a large scale were begun by Mr Bennet on the farm of Noodgedacht, in Tulbagh Kloof. It was before they were undertaken that I visited the district. It was the first rural district of the Colony which I visited on my return to it in 1863, and the impression produced on my mind was that it was a district which could scarcely be surpassed in facilities for irrigation. A waterfall not far from the Kloof supplies water entering the valley at such an elevation as might admit of its being led on to any part of the valley, and the more tortuous the course by which this might be done the more advantageous would it be for the interests of the community, as the water, percolating and evaporating more extensively, would thus more extensively promote the vegetation of the district.

Of the dam constructed on the farm of Noodgedacht I received, while the work was in progress, the following account:—"The dam is constructed across a shallow kloof at the base of the mountain range, in which there is a permanent spring. Proceedings were commenced by excavating, along the site of the embankment, a trench three feet wide, and deep enough to reach the underlying pot-clay. This trench was then filled in with well worked puddle, and a puddle dyke was carried up through the heart of the embankment two feet in thickness.

"The material for the embankment was obtained by excavating the soil, chiefly shale, from the site of the reservoir. The dimensions of the embankment, when finished, will be—Length, 700 feet; height in centre, 40 feet; width on top, 9 feet; slope on lower side, 2½ to 1, and on the upper side, 1½ to 1, the latter being pitched with stone. The embankment wants about 6 feet in height to complete it.

"There are three discharge pipes laid in puddle through the embankment, each 4 inch in diameter, and provided with proper sluice valves. When full, the reservoir is calculated to contain 25 million gallons of water.

"The work has been severely tested during the preceding winter, when the water rose and over-topped the unfinished part of the bank, which sustained no injury beyond the removal of a little of the top soil.

"There were, on an average, eighty men employed for four months in the construction of the embankment, under the direction of Mr J. H. Corbett, C.E."

A little later, a Tulbagh correspondent of the *Zuid-Afrikaan*, giving an account of this dam, reported :—" We have seen the extensive waterwork which Mr Bennett is constructing under the superintendence of Mr H. Perrott, on the farm ' Nooitgedacht. It is an unusually large dam, is to collect all the water on the place, has been scientifically begun and carried on, and will be completed within a couple of months, should the fall of rain not prove too great and consequently obstruct its progress. At this moment there are upwards of 40 men employed on the work; some weeks ago there were upwards of seventy. There is no doubt that such undertakings give bread and meat to the poor, and, if reproductive, tend to benefit the community in general, for the wages paid there may be termed fair, considering that there is a universal cry of want of employment. The men are paid from 2s to 2s 6d per diem. The dam which is to contain the water, and runs from south to north, has a length of 640 feet, and from east to west a width of 300 feet, forming therefore an oblong square of 192,000 square feet, and the depth in diameter being 20 feet, gives a body of 3,840,000 cubic feet, or about 10,520 cubic feet per diem throughout the year. Now, admitting that the dam can be filled once a month from a strong spring on the place and by others, we get 12 × 10,520—126,240 cubic feet of water per diem—a mass with which a good deal can be done. The dam which keeps the water within bounds has a bottom width of 123 feet, and will consequently have a surface width of at least 60 feet; which dam or dyke will, in our opinion, judging from the manner in which it is being constructed, be able to resist the billows of the roaring ocean, and therefore remain extant for ages."

And subsequently a correspondent of the *Eastern Province Herald*, himself from the Eastern Province, reporting a visit paid by him to the Breakwater, the Tulbagh Kloof Railway, and this dam, reported of it :—" On my return towards the Kloof, I was induced to visit what will be, when completed, one of the largest dams in the Colony, now in course of formation on a farm lately purchased by Mr Bennett, the contractor for the Kloof works. The farm slopes gradually from the range of hills at its back down into the Breede River valley, and the position selected for the dam is at the bottom of a water-bearing kloof, which, supplied by a spring in the mountain, ensures a continual flow. But were this not the case, its position and vast area would always guarantee its full complement of water of fourteen million gallons from the rainfall in the winter; for nothing less than this quantity would satisfy the energetic projector. The total length of the embankment is 640 feet in a straight line; its section is like a railway embankment, 9 feet wide at its top and 114 feet at its base; the inner slope, or that which is next the water, being somewhat steeper than the outer one, for certain reasons of construction. The width of the dam at the water level is 250 feet, and its depth in the centre is 35 feet. In its construction, not an ounce of material has or will be carted or conveyed, the embankment being formed solely of earth or clay puddle, both of which are dug out of the bottom of the dam, and thrown up to form the embankment; by this means the bottom is levelled, and its contents increased. The water cannot percolate through, because Mr Bennett has constructed a backbone, so to speak, of well-puddled clay, about four feet thick, running lengthwise through the centre of the embankment, like a wall, dividing it into two equal portions. This is far more effectual than lining the dam with clay; because, in the latter case, when the water in the dam is lowered, the sun at once cracks the exposed clay, and the water must of necessity escape; in the former, the clay, being buried in the centre of the embankment, must always remain moist, and therefore quite impervious to water. Overflow dykes are provided, at either end of the embankment, to prevent the possibility of a heavy rainfall overfilling and bursting the dam. Three cast-iron pipes, each four inches in diameter, pass through the embankment for distributing the water over the farm, with valves to regulate the supply. These are laid in at the bottom of the dam, so as to draw off from time to time all silt or sediment, which would otherwise collect. Many other ingenious methods were pointed out to me for preventing the pipes ever getting choked with bush or other matter, and much other useful detail which your readers will scarcely follow. I have endeavoured to describe clearly and faithfully a work which I think is calculated to be of immense advantage to its originator, and I hope, not without its useful lesson to the neighbouring farmers, and perhaps the Colony at large. But the great feature of the work, in my estimation, is its wondrous cheapness, the total cost being very considerably under one thousand pounds—fourteen million gallons of water for £1,000! Oh! inhabitants of Graaff-Reinet; and ye, dwellers in Somerset, report says that ye are entertaining praiseworthy schemes for preserving your water in dams and reservoirs—take my advice, and find out something of Mr Bennett's works in this way, and also of Robt. Hare's property at Klapmuts, which has been containing 20 million gallons for the last three years, and has enabled its worthy proprietor to supply Capetown with strawberries and butter, and indulge himself occasionally with a cruise in his yacht! You may think that I am romancing a little. Pay him a visit, and judge for yourself."

In 1869 it was reported in regard to Tulbagh :—" The abundant rains in the early part of the year promised a plentiful harvest. Although this expectation has not been fully realised, the crops have been better than for many years past. In some parts of the division the 'stink-vliegen' did considerable damage to the oats and wheat, and the caterpillars in others destroyed the young mealies and beans. The fruit suffered considerably, too, from the very strong northerly gale in December; and in the basin of Tulbagh disease has shown itself in the orange and peach trees; but in spite of these drawbacks the yield has been satisfactory, and a few such years would go far to raise the depressed state of the agriculturist.

"At one time horse-breeding was carried on to some extent in the Bokkeveld. The removal of the Indian Remount Agency, the introduction of the rail, and the substitution of the mule, a cheaper and hardier draft animal, have had the effect of reducing the demand for horses; and the losses resulting from periodical attacks of horse-sickness have led in a great measure to the discontinuance of horse-breeding. Lately, an experiment has again been made of shipping horses to India, many of which were procured in this district. Should this attempt to re-open the trade prove successful, there is no doubt that horse-breeding will again be engaged in.

"Linseed has this year been grown successfully at Witzenberg, and there is no reason why it should not be cultivated to a large extent in parts proved to be so favourable to its growth. The white mulberry plantation at the mission station Steinthal now numbers 800 trees, all thriving well. During the year a few castor oil seeds obtained from Mr Charles Barry were distributed in the ward of Tulbagh; the plants from them are growing very luxuriantly, and as the soil and climate appear suitable, it is proposed to procure larger quantities of the seed, and to have the tame plant substituted for the wild, which is found in many parts of the valley. In earlier times, the oil was pressed from the berries of the wild tree, and many of the inhabitants remember that sufficient was made for family consumption. The growth of linseed, the white mulberry, and castor oil may be carried on to some extent on very many of the farms without interfering with the more important agricultural operations, and, while supplementing the farmer's income, be made the means of affording employment to women and children, and work for the male population to fill up the intervals between the busier seasons on a farm.

"Improved agricultural implements are being gradually introduced. The mowing-machine and the scythe are doing a great deal of what was done exclusively by the sickle. Several properly constructed dams are to be found in the district, and the sufferings from drought have made many a one resort to dam-building and other works for irrigation. Owing to the number of farms lying along the Kleinberg River, flowing from the Winterhoek mountains, the supply of water coming to those lying on its banks in the ward Twenty-four Rivers, at so long a distance from the source of the river, became very scant last summer, and was useless for irrigating purposes, exposing these farms to considerable loss, if not absolute ruin. Just then large quantities of water were led out of the Breede River at Mitchell's Pass, across the Tulbagh Basin into the Kleinberg River, restoring to the residents on its lower banks what they had been deprived of by the farmers living higher up. The timely supply of water thus procured has prevented much distress and averted a good deal of bitterness of feeling which was beginning to exhibit itself. This change has been brought about by the energy of one man; and is a pleasing instance of what may be effected in a quiet way by individual effort, while grander plans are being matured or discussed, or rather talked of, as has too long been the case in the matter of the Colonial irrigation works.

"The prospects of successful farming are still not over-bright. Several farmers from other districts have purchased farms here, the original owners of which had become bankrupt, and the new proprietors have been enabled, even in the bad seasons, not only to hold their ground, but to make marked improvements on the farms. On the other hand there are instances where no effort is left untried; but, either from the lowness of the price of produce, the want of facilities of reaching a market and turning everything produced to account, or from some other unexplained cause, success is not attained.

"The state of the roads is good, the expenditure upon them for the year being £1,150, and the divisional council is now in a fair way of getting out of the difficulties in which it was involved, and responding more satisfactorily to its proper functions.

"An aqueduct, erected by the municipality of Ceres at a cost of

£300, was blown down by the gales last winter. This has been replaced by a more substantial one, and in connection with it a water-mill has been erected, which will enable the inhabitants to have their corn ground on the spot, instead of sending it a distance of seven miles to the mill at Prince Alfred's Hamlet.

"The chief crime of the district is sheep-stealing, and is still carried on to a large extent, although it may be said to be somewhat on the decrease."

Of Tulbagh, in 1875, it was reported:—"The year 1875 has been characterised by an event of great importance to this division, the effects of which have shown themselves in increased trade, increased traffic, and the prosperity incident on all such undertakings.

"Allusion is made to the opening of the Worcester Extension Railway Line first to Tulbagh Kloof and then to Ceres Road Station, reducing the ordinary journey between Ceres and the Metropolis from twelve to eight hours, and as a particular benefit, rendering unnecessary to heavy traffic the wearisome, and at times dangerous, passage of Bains' Kloof. As one result of this extension three new villages have sprung up along the line—viz., Gouda, Berghville, and Wolsely, and although their sites have not been most happily selected, a few years may see them entering into serious competition with the older townships.

"A census of the Colony has been taken during the past year. As far as this Division is concerned its results must be accepted with much reserve. It was a time of great depression, after a short wool clip, great losses of stock, drought, etc., similar to the previous decennial period of 1865, and as in this Colony seasons move in almost regular cycles, it is not impossible that the next one may prove alike misleading.

"The district has been singularly free from crime, notwithstanding the temporary presence in it of from 700 to 1,000 Railway employés of different colours, nationalities, and creeds. . . .

"Considering the untoward circumstancs of bad crop in 1874, the drought in the first half of 1875, the past year has, owing to counteracting reasons, not been on the whole a bad one for agriculturists, wine growers, and stock farmers. High rates have compensated for small returns. The late rains of last year have given a crop much above the average; the summer rains from December up to this date (Feb. 12, 1876), augur well for the prospects of the present year."

Thus far I have followed the order in which the replies received from the several civil commissioners to the queries issued from the Colonial Office were presented to Parliament; and finding it convenient to do so I shall continue this course.

Next in this order is the district of Worcester.

The division of Worcester is separated from the division of Tulbagh by the Dwyka River; it is situated behind the first or coast line of mountains through which flows the Brede River in the upper part of its course. It has an area of 5,420 square miles, and is traversed by several ranges of lofty mountains. The climate is agreeable; but it is subject to heavy thunderstorms.

The reply of Mr Le Sour Civil Commissioner of the district to the queries issued from the Colonial Office in 1862, was as follows:—

"In compliance with the request contained in circular No. 16, 1862, dated the 7th instant, I have the honour, with reference to the several queries therein submitted, for the information of his Excellency the Governor to state:—

"1. The only plan for irrigation works now proposed in this division consists in constructing or forming dams in rivers at such places where it may be practicable.

"There are five or six rivers in this division which are led out at convenient places by the proprietors of the farms contiguous to such rivers, for irrigating their farms, and to which the proprietors of the same claim or have a certain right by virtue of their quit-rent grant.

"The rivers above alluded to are:

"The Breede River, which, having its source in the Warm Bokkeveldt, in the division of Tulbagh, traverses the divisions of Worcester, Robertson, and Swellendam.

"The Smalblader, or Du Toit's Kloof River, and Louwshoek River, in the Goudinie.

"The Hex River, the Wilge River, Norma, Doorn River, Wagebooms River, Jan du Toit's River, Hartebeeste River, etc., streams of minor importance, are all dammed up at convenient places by the proprietors of contiguous farms.

"The plan of irrigation may be much extended or improved in cases where the proprietors are desirous to sub-divide their properties.

"2. At the present time there are, to my knowledge, only three dams in this division, in which the water from springs originating in the immediate vicinity are collected, and afterwards used for irrigating certain extents of land. There are, however, a great many places, particularly at the bases of the several mountain chains intersecting this division, where dams may be constructed with some success, for collecting periodical supplies of water, and from which portions of land may be irrigated.

"Sinking of wells appears to me also practicable in many parts of this division, but, I consider, not to such an extent that the water obtained by these means will be sufficient for irrigating purposes."

In 1865 I had an opportunity of visiting the district. On the day following that of my arrival I was taken up the Hex River Kloof. There I found water led out upon different farms at which we rested, but not in such a way as to secure the full benefit which might have been derived from the water which was thus emoloyed in irrigation. None of the water so led out was lost to the community, but much of it was lost to the farmer, and had it been more economically applied the results would probably have been more remunerative of the expenditure of money or of labour which the work had entailed.

An approximation to the Marcine irrigation of Italy had been introduced by Mr Heatly npon his farm, which was in some measure adapted for that form of irrigation. By laying the ground under water he could effect without difficulty the destruction of bushes; by subsequent appropriate treatment, produce rich meadow grass; and by enclosing a large plot of ground with wire fencing, he kept his cattle from the cultivated and the irrigated ground. Though only comparable to a rude and rough imitation of the Italian irrigation referred to, it was in advance of anything of the kind which I had seen in the Colony; and I found he was enabled by it to engage largely, and with satisfactory results, in the production of butter, for the Capetown market, some 80 or 90 miles distant.

At a lower level a water-leading had been constructed by Mr P. J. de Wet. The cost of this was about £600. On asking him if he had found the outlay remunerative, appealing playfully to the friends with me, he said—"All know that if I get 600 muids of corn from my irrigated field I will be repaid my outlay in a single year!" He was engaged at the time in executing similar works near Robertson, and at my desire he promised me more explicit information in regard to the expense, and the returns received. At the close of the year I received from him the following statement:—

"According to my promise, I can now tell you of the outlay and the pecuniary return connected with the water-course on my Hex River farm. I had to lay out £585. I have worked it four years, by which I have cleared £760; and I have enclosed some 28 morgen for vineyard and garden, which will also clear me £100 at least per annum.

"On the other farm, that in the Robertson district, I only finished the water-course last September. There I have not sown largely as it was too late, the season was past; but I hope to do so next year. Thirty morgen of splendid land, however, is enclosed and planted with maize, beans, pumpkins, potatoes, lentils, and all sorts of vegeables; every bit of it looks uncommonly well and promises a good return, which will certainly be more than the interest of the capital laid out.

"A great many more such improvements might be made in the district, but very few of the farmers have the means. I think the Government, for its own sake, should come forward in aid of such undertakings,"

In accordance with this statement is the following, in regard to another enterprising farmer in the district, which appeared in the *Worcester Courant*:—"Those who have had anything like a harvest are well off this year, on account of the high prices. We have heard that the income of Mr H. A. du Toit, Over Hex River, from his farm this year will be about £2,000. This is partly owing to a dam or dams—on a small scale, we believe—which he lately erected himself; and partly, no doubt, to obtaining his fair share of the Hex River water; but principally, to *his own industry!* This amount includes crops of all kinds—cereals, vintage, and, we believe, increase of stock. But it is a proof, if any were wanting, that money is to be made by farming in this Colony, if pursued with intelligence and perseverance."

I found that a large stream of water was led off from Hex River for the supply of the town of Worcester, with its cereal crops and its gardens and vineyards. The supply was at that time abundant; and much was allowed to go to waste. In the orchards, I found that, contrary to the practice very general in the Colony of leading the water by the base of the trunks of the trees, it was led midway between rows of these, by which arrangement it probably effected more good with less evil; but still, diseases of fruit trees told of stagnation and of waste of the valuable element. In the spacious green in the centre of the town were plants growing which were indicative of water having also stagnated there; and the protracted exposure to the sun's rays to which the water in the aqueduct was subjected must have occasioned the loss of much by evaporation, which might have been prevented by the planting of trees along the banks of the aqueducts and water leadings.

On farms at a still lower level, intersected by the main road to Robertson, I found that irrigation had been carried out upon a much more extensive scale, and apparently with great advantage, but also with great waste, both of water and of land. Valuable portions of the latter were lying under water, and in several places the high-road was impassable to passengers on foot, without great inconvenience, from the depth of the water retained in valleys which it crossed. In one of these the water came to near the axle of the cart in which I travelled, and the impassable swamp on both sides seemed to extend so far as could be seen from the cart. Men and women alike, if unprovided with a conveyance, must partially undress to prosecute their journeys on what is, I presume, the Queen's highway; and this apparently was occasioned by the system of irrigation adopted.

A day was spent by me in a visit to Goudinie, a fertile tract of land lying along the Breede River. There I found the soil rich and productive, and the place well adapted for the culture of the grape for the preparation of raisins, which is carried on extensively by the inhabitants of the district, and for the fine quality of which the place is celebrated. Water is not deficient; but, in addition to the other, supplies, which are suitable to the products now raised, there is a warm spring, which, after supplying baths, as at present, might possibly be utilised as a means of supplying the bottom heat necessary for the production of tropical fruits.

Another day was spent in visiting the Brand Vley, where a large body of water, of the temperature of 150°, rising, fills a spacious basin, and runs off by several large artificial channels, to be cooled by exposure to the air before being led on to irrigate the fields lying below. This might apparently be employed to an indefinite extent in the culture of tropical fruits; and, for anything I saw leading to a contrary conclusion, it might also be utilised as a means of producing green crops at all seasons of the year; but my confidence in suggesting this is kept in check by the reply given by the Honourable Mr Hare to queries put to him while giving evidence before a Select Committee of the Legislative Council. He was asked :—

"394. *Chairman.*] Do you not think that in those parts of the Colony where you can irrigate with advantage there is another advantage in being able to do so above what you have here—namely, that in very wet seasons they may sow later, whereas in very wet seasons you perhaps have a bad crop on that account?—From what cause it arises I do not know, but in our part of the country we have never any certainty, even if we had all the water in the land, of raising a crop sown after the 15th of August. For instance, I know that in two successive years I have tried to reap a piece of green stuff in February, for a trial of reaping machines, at our great shows, but I have never yet been successful.

"395. Why?—They say it is out of season and will not grow. Now at Brand Vlei there is just the sort of soil for it, so that if it would grow nowhere else, it would grow late there. It is a light loam with plenty of water. An experiment was made, and I took a little trouble to inquire from Mr de Wet why it was it would not even grow there; but the only answer I got was: '*Ik weet nie, maar dit wil nie groei nie,*'—I don't know, but it wont grow. We have never been able to succeed with late crops. They come up and seem to die off."

Apart from such sources of supply as the warm springs at Goudinie, and the warm springs at Brand Vley, it appears that the Breede River brings into the district, carries through the district, and drains off from the district of Worcester a great quantity of water which might be stored up for the promotion of agriculture; but that comparatively little is done to improve the advantages which are thus afforded.

Mr J. H. Munnik gave the following evidence, bearing upon this subject, before the Committee of the Legislative Council on Irrigation, in 1862:—

"308. I am well acquainted with the district of Worcester.

"309. I know that in Worcester the greatest proportion of the crops is raised entirely by irrigation.

"310. Have you any idea of the difference of the quantity reaped on lands not irrigated, and on lands that are irrigated?—The lands not irrigated are chiefly in the division of Tulbagh, in and about the village of Tulbagh; there the return would be from ten to twelve fold in wheat. I am not certain as to the light crops, but the yield in oats, I think, would be about twenty-five fold. In the Cold Bokkeveldt, which is a very high country indeed, perhaps 1,500 or 1,600 feet higher than the parts I have just alluded to, the crops are raised without irrigation, and the yield is very good. The soil is rich, and they are enabled to sow wheat known as Baartkoorn, a very delicate wheat, which is subject to rust, and therefore not sown about Tulbagh; and the yield there, I should say, is from eighteen to twenty-four fold. Round about Worcester and in the town of Worcester, where the crops are raised entirely by irrigation, and also by the Hex River, the yield is the most prolific of any part of the colony that I am acquainted with; I have heard of one hundred and twenty fold having been reaped in barley. In wheat, the yield is from twenty to thirty fold, and it is but an average crop when the yield is twenty-four or twenty-five. There are two Bokkeveldts, the Warm and the Cold, and I may mention that the yield is not so prolific in the Warm as in the Cold.

"311. Do you ascribe that to the difference in climate?—To the natural moisture of the soil; which moisture arises from its elevation, and its being more abundantly watered.

"312. Do you mean by springs or by rain!—I should think from both, being situated at such an elevation and between mountains, the clouds are attracted, and the rains fall more extensively than on the lower ground.

"313. *Mr Wood.*] Are you acquainted with much of the Colony? —I am acquainted with a good deal, but not with the Eastern Province.

"314. Do you know much of the Western Province?—I am acquainted with a portion of Clanwilliam, Worcester, Tulbagh, Caledon, Malmesbury, Piketberg, and Saldanha Bay.

"315. Then, from your extensive knowledge of that country, would you have the kindness to state to the committee what your opinion is of the whole of that as a grain-producing country, without irrigation?—In every part of it where the crops are raised without irrigation, the yield is not nearly so good as in those parts where irrigation can be carried out.

"316. Is it as certain?—No; where they are dependent on the rains it is naturally more uncertain.

"317. I mean with their annual crops?—Yes.

"318. And do you know whether there are facilities in that country for husbanding the water?—If you will allow me to confine myself to Worcester, which will perhaps give us a more definite idea, I may state that in certain portions that I have mentioned there are facilities for husbanding the water and making it available with much greater advantage to the country than at present, because we are all aware that everything is now almost left in its natural state. There is, in fact, very little done at all in the way of the adoption of artificial measures for such useful purposes as irrigation; in fact, I hardly know of anything except the water-courses that were taken out, perhaps to some distance, many years ago, during the slave time. Since then, I believe, but little has been done.

"319. Can you give the committee an idea why so little improvement in this way has been accomplished in the district you are speaking of—in other words, is it a wealthy or a needy district?—Taking it in general, I should say that the people are well to do, but agriculture has been accompanied with such difficulties, in consequence of changes that have taken place, that parties are almost disposed to limit their operations to a great extent.

"320. Will you explain what the difficulties are?—They consist chiefly in labour.

"321. The want of labour?—I would not say so much the want of labour as a change of circumstances; an increased rate of wages in comparison with remunerating prices of articles of production.

"322. Do you think there is any possibility of stimulating these people to take such steps as would increase the amount of cultivation? —I am disposed to think that at present changes are beginning to force themselves upon the people, and that during the last two years there has been an increase of cultivation, which is now progressing, and, with such a stimulus as you have alluded to, would progress more extensively.

"323. *Chairman.*] Do you think if the railway were carried on to Worcester that would act as a stimulus?—I should think it would be the greatest stimulus that could be, for a great obstacle in the way of improvement is the difficulty of transport. I may state to the committee that wheat can but seldom be conveyed to the Cape-town market, from that long distance, across bad roads, to remunerate the producer; and with light crops, such as barley and oats, it is quite out of the question.

"324. What produce is mostly raised in your district,—grain?—In Tulbagh and towards the Winterhoek the land produces grain of every description, and might produce wine in great abundance, vegetables also, and in fact any agricultural produce. Worcester and the Goudinie, and the surrounding parts, are well known to be a very rich and fertile country, capable of producing anything. The Warm and Cold Bokkevelds chiefly produce grain; and during the last six years their attention has been turned to wool, the produce of which has increased to a marvellous extent in that district, it being a very excellent grazing country.

"325. *Mr Wood.*] You stated that you think a railway extension would act as a great stimulant to the district: do you believe that with such extension the farmers would be induced to apply artificial means for irrigation purposes?—I think if they could be placed in a position to see their way clearly, as to how it may be attained, they would gladly embrace any opportunity of extending irrigation.

"326. Has any system, do you know, been adopted for artificially feeding stock in that district?—Not to any extent worth mentioning.

"327. Have you any practical knowledge of farming?—I have some knowledge.

"328. What is your opinion as to the artificial feeding of stock?—I have not the slightest doubt that if once the attention of the farmers was properly directed to the subject, it might be carried on with the most beneficial result to themselves and the colony at large.

"329. *Chairman.*] Do you know anything about farms on the Hex River, where they irrigate pasture land; and could you give the committee any account of how it is done?—At the mouth of the Hex River Kloof, and before it enters the Breede River, the Hex River is diverted, and four or five farms are brought under irrigation by it, and the limited extent that is so brought under irrigation is what these farms chiefly depend on. These are the farms which we hear of in the division of Worcester and Hex River that supply a number of slaughter oxen annually to the butcher. Upon one of these farms a dairy was established by an English gentleman, Mr Hack, who afterwards went to Australia, when Mr Andries du Toit took up the dairy; and he himself has told me that that dairy was one of his most profitable occupations, and that the income from his dairy was more than from any other of his farm pursuits.

"330. By going higher up the Hex River, would it not be possible to lead out the water at a considerable distance above where it is at present led out?—The water of the Hex River, to my knowledge, belongs properly speaking to the parties to whom I have just alluded, and to the village of Worcester. Just at the mouth of the kloof of the Hex River there is a division made of the water into two portions, one of which flows to the farms already mentioned, and the other to the town of Worcester.

"331. *Mr Wood.*] You state that you know of four farms in the locality of Hex River which have the advantage of irrigation: are there any farms adjacent?—There are other farms adjacent, but they obtain the water from another river, called the Noillege Rivier, running parallel with the Hex River, and coming from behind the mountains.

"332. Do they irrigate also?—Yes, and are equally productive, also providing slaughter oxen for the butcher.

"333. But are there farms in the immediate vicinity that you know of which have not the facility of irrigation?—Yes, beyond that; from there on to Robertson the farms have not that facility.

"334. And do the farmers reap as large returns?—No, the farms are not considered of equal value; and to give you an idea of the difference, I may mention with regard to the farm of Charl Plessis, which is in extent only 1,300 morgen of meadow land, that the owner would, I believe, have no difficulty in getting £10,000 for it, whereas an adjoining farm, Kloppersbosch, or Fink Rivier, of four times the extent, would not be of one-third the value, certainly not one-half.

"335. And is that about an average of the difference between the value of a dry and irrigable farm; or could you give the committee an idea of the difference?—The lowest proportion would always be as three to one.

"336. *Chairman.*] Can you name any large stream which might in your district be led out or dammed up so as to bring under water a larger tract of country than is now done?—There are several streams; one coming from Ceres which is the origin of the Breede River, and is a very formidable stream at the mouth of Mitchell's Pass, where it might be turned westward to come through New Kloof, and so to irrigate on this side of the New Kloof a large tract of land; or it may be taken out higher up, to run parallel with the range of mountains which extend to the back of Worcester, and so also bring an immense tract of country under irrigation in the valley between Mitchell's Pass and Worcester, a distance of twenty-six miles. The river, the committee must remember, runs parallel with the mountain east and west, and after it is past the town of Worcester it may be taken out and turned off south, over a rich fertile part of the country, namely, Moortkuil, Scherpeheuvel, and other parts which could be brought under irrigation by it. I must now turn to a different part of the country, and take the well-watered region of the Cold Bokkeveld. The farms along there are all situated on the spur of the mountains, and the river running along it, in a northern direction, receives all the waste water from the different farms, and passes through a range of hills called the Zwarte Ruggens, where it falls into the Doorn River. But before it reaches that river, there is a vast open flat of great extent, and if the water were brought over that flat it would be impossible to overrate the value of the produce that could be obtained from it. Higher up, towards Karroo Poort, there are three streams, but they are entirely dependent on the fall of winter rains; they do not run all the year round, but are dry from about December or January. While they continue to run, however, they are very formidable, and large quantities of water that flow along these mountains collect together at about Baviaan's Kloof, and enter the Karroo on an open plain of ever so many square miles. If that water were collected at proper seasons of the year, the quantity which might so be gathered would be quite sufficient to supply the farms from one year's end to another.

"337. You said that the Breede River might be led out from the southern bank, why not from the northern; that is to say, why not on both sides?—The range of mountains which extends from Worcester to the back of Robertson and Swellendam causes a natural declivity towards the river, which is in the lower part.

"338. Might not a canal be made round the spurs of these mountains?—They are intersected by so many kloofs and other rivers; for instance, those I have alluded to, the Hex River, etc.

"339. But you would not consider one river crossing another an impediment, but rather an increase of the supply of water?—Yes, an addition of water; but what I principally allude to is the intersection of the kloofs.

"340. You have given us an account of the district generally, and as to the quantity of water it possesses at present. Now, if the water were husbanded and led out artificially in the way you have mentioned, could you tell us, at a rough guess, what quantity of land could be put under tillage, or used for pastoral purposes, more than can be done at present?—I think that would entirely depend upon the extent of the work and the quantity of water that could be so husbanded. The proportion of ground under cultivation at this moment is very trifling in comparison with what could be brought under cultivation by irrigation. I have no hesitation in stating that there is not at this moment one-tenth part cultivated of what could be with a sufficient quantity of water.

"341. It has been said here that one great impediment in the way of adopting any system of irrigation is, that our rivers are not permanent; now even in the rivers which are not permanent, but in which there is a superabundance in some seasons of the year, do you not think that the pasture land could be improved in the summer season by storing water which would otherwise run into the sea?—Very materially so; because in many parts it may be seen that where the water collects on low grounds during the winter, that ground in summer produces luxuriant grass, and is, perhaps, the spot upon which the farm is mainly dependent. . . .

"349. *Mr de Wet.*] In the event that water running in the rivers were led off, would it not interfere with the rights of the proprietors of the land now lying alongside the river and entitled to a share of of the water?—I must state, as far as concerns the Breede River, to which I have alluded, that in no part is it turned out or made available as yet for irrigation purposes, and that no farms are at present dependent upon it for cultivation. As regards the other river towards the Karroo that I have also alluded to, there is a farm which is dependent upon it, but to what extent it is impossible to say, inasmuch as it is situated about twenty miles away from where it enters the Karroo, and where I mentioned it could be made available. The smaller rivers belong to the farms that have been established upon them to a certain extent. I have already alluded to the Hex River and how that water is disposed of. I happen to know that Noillege River, the next river to that, is exactly the same; but when you go back to the Goudinie, the rivers running from the Du Toit's Kloof, the Bastiaan's Kloof, etc., all these are large and very strong streams,—in fact the strongest in the whole district,—are not made use of by any farmers at present.

"350. As an instance, I believe the town of Worcester is supplied by water led out of the Hex River. Taking, then, the case of the Hex River, in the event extensive irrigation were to be carried out, and the water were led out of the river for that purpose, would it not interfere with the supply to the town of Worcester?—I have already stated that at the mouth of the Hex River Kloof the water is divided into two parts; that one-half is taken to the town of Worcester, which was originally established on two farms that had a better right to it, and that the other half goes to supply the four farms I have spoken of. I have been resident at Worcester for nearly ten years, and I may state that during that period I have hardly known at any time that the whole of the water was required by those at present entitled to it, the stream being more abundant than was necessary to supply the wants of all entitled to it.

"351. That is in the present state of the river; but suppose irrigation were to be carried on to a large extent, will that not altogether alter circumstances?—I think that further irrigation could not be properly applied to that stream, to which I consider that the town of Worcester and the other farms mentioned by me have a claim."

The Honourable Mr Hare, in giving evidence before the Committee of Legislative Council, gave it as his opinion that in the district of Worcester there were facilities for irrigation greater, perhaps, than are to be found any where else in the Colony—"because it is such a level country." He goes on to say:—"It is a common saying among the farmers about the Breede River that you cannot take the water out of it, which clearly proves that there is very little fall, and that thus there would be very little difficulty or danger in making a dam there." This is a measure he considered could be adopted on comparatively few of the other rivers of the Colony with safety. And in accordance with this is the report of Mr le Seur, in 1862, quoted above. He also gave the following evidence:—

"406. Are you acquainted with the farms on the Hex River, where they irrigate pasture lands?—Yes, I know two or three such farms.

"407. Do you know du Toit's?—Quite well.

"408. Could you give us some account, as nearly as you can, of the manner in which they proceed there in regard to the irrigation of this pasture land?—Their proceedings are conducted in a very extraordinary way. Wherever they can get the water they lead it out, but they have no system of irrigation; none whatever. They could get twice or three times as much done if they went to work in a proper sort of way. They merely turn it out, and it runs over. It runs over a sort of ground neither a valley nor a veld, and, after it has gone through that land, you see a beautiful rich crop of clover.

"409. They do not attempt artificial irrigation?—Not a bit. I made the remark when I was at the farm of Mr Plessis, that if he had English clover with rye grass mixed he would have seen his cattle not dying, but in a different condition."

In 1865 it was reported by the Civil Commissioner:—"This division is suffering terribly from the drought. Springs and dams are reported to have failed and dried up in every direction, in consequence of which hundreds of sheep have died in the ward of Klein Roggeveld and in the Great Karroo, as well as horses and cattle, for want of pasturage and water. Many farmers were therefore compelled to abandon their homesteads, and go to other parts, so as to keep their flocks alive.

"It is surmised that, should this drought continue much longer, it will be extremely difficult for traders and others to proceed through the Karroo with cattle and sheep, as most of the springs near to or in the proximity of the trekpaths and main roads have failed, or are drying up. The accounts received from the interior and distant parts of this division, regarding the state of the country and the effects of the drought, are distressing in the extreme. It is stated that cattle, sheep, and horses are in such a state of starvation that, should it commence raining heavily at once, large numbers will perish."

In 1866 it was reported:—"The prospects of the farmers have considerably improved during the past year. The crops have been generally very fair, although the wheat and oats have in some places suffered from rust. The barley crops are, however, very fine. The vintage was also a very fair one, and a large quantity of brandy and raisins were obtained.

"The cattle, which, at the commencement of the year, were in very low condition, have greatly improved; but, in consequence of the low condition of the cows in the beginning of the year, the calving season has been a bad one; the lambing season has, on the contrary, been very prosperous.

"No material improvement has taken place in agriculture, but there is a very wide field for it; very few thrashing machines are used, and the grain is still generally cut down with the sickle. The farmers here have not yet found out the immense saving to be made by the use of the scythe. There is at present no agricultural society in this division; but there is every prospect of one being formed in a very short time, which it is to be hoped will be the means of inaugurating many improvements.

"Several of the farmers have begun to see the importance of making dams to husband their surplus water, and a few extensive ones have been formed during the past year. Those individuals who have had the enterprise to make dams, have already reaped great benefits from them, in having been able to cultivate much more ground. It is to be hoped that their good example will be speedily followed, which will effectually put a stop to the continual litigation that prevails about water. The municipality have it in contemplation to erect a large dam above the town, which will, it is to be hoped, be a model for the rest of the division."

In 1868 it was reported:—"A large dam is now being constructed for the municipality of Worcester, by convict labour, which, when finished, will prove a great benefit to the lower part of the town, and the locations especially; for, although Worcester has a large supply of water, more could be used with great advantage. This dam will not be dependent on rains, as it can be filled in winter from the same source that supplies the town."

From Worcester it was reported in 1869 :—" The year 1869 has been marked by very seasonable and refreshing rains, far above the average quantity, and naturally followed by abundant crops, which, I am happy to be able to add, have been almost entirely free from rust. The vintage has been a very good one; but brandy and raisins, owing to there being no demand, were sold for hardly remunerative prices. The general health of the town and district has been on the whole most satisfactory. Crime has not increased."

In 1874 it was reported :—" Having had no opportunity of visiting any considerable portion of this district during the past year, I can only write from hear-say of its condition. I believe, however, that facts will bear me out when I report that, although the whole of the district suffered more or less from drought, the crops were much better than in the previous year. The vineyards, too, are unusually healthy, and the yield promises to be greatly above an average."

The following is an abstract derived from meteorological observations at Worcester :—

Abstract derived from meteorological observations at Worcester Station in the years 1862, 1863, 1864, 1865, 1866, 1867, and 1868.

Months.	Mean height of Barometer at Temperature 32°.	Temperature of the Air. Mean Temperature.	Mean daily range.	Mean of greatest range any one day of each month	Mean of least Range any one day of each month.	Mean of greatest Range throughout each month	Mean Temperature of Evaporation at 9 A.M. 1 P.M. 5 P.M.	Mean Humidity for each month	Rain.	Clouded sky.	Lightning. Recorded No. of Days
	inches.	degrees.	degrees.	degrees.	degrees.	degrees.	degrees.	per cent.	inches.	per cent. of visible hemisphere	
January	29·146	73·51	29·14	42·5	12·8	52·5	65·96	43·48	0·095.	18	15
February	29·138	72·53	26·55	38·8	11·3	47·4	65·39	50·15	1·189	31	13
March	29·199	68·71	26·37	42·8	11·9	52·1	63·21	50·10	0·398	26	9
April	29·222	62·74	26·51	43·7	10·2	51·2	58·50	54·33	0·622	29	13
May	29·379	58·40	25·00	41·5	8·4	49·9	55·34	60·45	1·149	34	5
June	29·306	54·22	22·61	40·8	8·1	42·2	51·66	65·53	1·861	37	9
July	29·343	53·11	21·46	38·3	7·6	45·6	50·77	66·44	1·806	33	9
August	29 341	54·93	21·62	39·7	7·0	46·9	52·09	63·96	1·431	34	4
September	29·307	58·25	22·26	40·1	8·9	47·9	54·05	56·37	0·981	34	6
October	29·227	61·44	23·41	42·0	9·4	49·8	56·95	54·96	1·320	42	5
November	29·216	65·35	26·13	40·5	11·0	50·2	59·82	49·59	0·661	31	9
December	29·152	71·39	27·89	42·5	13·2	51·9	64·26	44·48	0·106	21	3
Mean Annual Values.	29·248	62·88	24·91	41·10	9·98	48·97	58·17	54·99	11·619	31	14
	*+·017	...	...	...	...	...	...	...	...	...	...
	29·265	...	...	...	...	...	...	...	...	...	...

* ·017 inch is the reduction to the Observatory Standard Barometer.

Abstract of daily observations made in 1867 at the Worcester Station. Approximate height above the sea level, 776 feet.

From Daily Observations.	Jan.	Feb.	Mar.	April.	May.	June.	July.	Aug.	Sept.	Oct.	Nov.	Dec.	For the year
	ins.	ins.	ins.		ins.	ins.	ins.	ins.	ins.	ins.	ins.	ins.	ins.
Mean height of the barometer at the temperature 32 deg. Fahr. (not corrected for altitude above the mean level of the sea) for each month, and for the year, from observations at 9 A.M., 1 P.M., and 5 P.M., ...	29·123	29·155	29·346	Incomplete.	28·977	29·143	20·290	29·349	29·358	29·267	29·256	29·221	29·310
Maximum observed, ...	29·328	29·430	29·301		29·267	29·386	29·571	29·676	29·845	29·561	29·594	29·453	29·845
Minimum observed, ...	28·927	28·896	28·737		28·782	28·847	28·909	28·902	29·047	29·027	28·949	28·912	28·737
	deg.	deg.	deg.	deg.	deg.	deg.	deg.	deg.	deg.	deg.	deg.	deg.	deg.
Mean temperature of the air for each month, and for the year (Fahr.), from observations at 9 A.M., 1 P.M., and 5 P.M., ...	83·2	78·8	74·5	68·7	62·7	58·6	59·1	60·2	64·2	66·7	73·0	81·6	69·3
	deg.	deg.	deg.	deg.	deg.	deg.	deg.	deg.	deg.	deg.	deg.	deg.	deg.
Mean temperature of evaporation for each month, and for the year (Fahr.) from observations at 9 A.M., 1 P.M., and 5 P.M., ...	67·5	65·1	64·3	60·5	56·5	52·1	52·2	53·4	54·2	58·0	61·3	65·0	59·2
Mean humidity of the air for each month, and for the year (the integer number 100 implies saturation), derived from observations at 9 A.M., 1 P.M., and 5 P.M., ...	42·8	46·9	55·8	58·8	68·4	65·6	66·3	66·3	53·5	56·1	50·8	42·6	56·2
Maximum humidity, ...	70·0	83·0	88·0	93·0	100·0	100·0	100·0	100·0	93·0	94·0	94·0	82·0	100·0
Minimum humidity, ...	14·0	21·0	35·0	28·0	30·0	30·0	25·0	30·0	30·0	30·0	23·0	23·0	14·0
From the self-registering thermometers (Fahr.).	deg.	deg.	deg.	deg.	deg.	deg.	deg.	deg.	deg.	deg.	deg.	deg.	deg.
Mean maximum temperature of the air, ...	91·10	92·9	74·1	77·2	70·4	65·2	65·9	66·9	70·7	73·3	80·2	85·8	76·1
Mean minimum temperature of the air, ...	59·17	59·3	56·4	52·7	47·2	44·6	42·7	43·6	54·2	48·1	50·6	57·9	51·4
Approximate mean temperature of the air, ...	75·13	76·1	65·25	64·9	58·8	54·9	54·3	55·2	62·4	60·7	65·4	71·9	63·7
Mean range of temperature of the air, ...	31·93	33·6	17·7	24·5	23·2	20·6	23·2	23·3	16·5	25·2	29·6	27·9	24·7
Highest temperature, ...	105·0	103·25	95·5	91·0	84·0	78·5	81·0	85·0	94·7	90·0	97·0	99·0	105·0
Lowest temperature, ...	52·0	51·0	48·0	40·5	34·0	33·5	34·0	36·0	36·0	37·0	44·5	49·0	33·5
Extreme range of temperature on any one day, ...	46·0	45·25	42·5	40·0	37·5	39·0	40·0	47·0	44·7	42·5	40·0	41·0	47·0
Least range of temperature on any one day, ...	13·5	10·5	5·5	10·0	9·0	7·0	8·5	8·0	10·0	6·7	7·2	12·0	5·5
Quantity of rain in inches, ...	0·165	1·600	0·895	1·275	1·520	1·100	1·880	1·290	0·130	0·900	0·445	0·100	11·300
Greatest amount of rain on any one day in each month, ..	0·145	1·140	0·430	0·945	0·615	0·790	1·630	1·150	0·105	0·320	0·250	0·065	1·150
Number of days on which rain fell, ...	2	3	4	2	8	6	5	5	3	7	6	2	55

Abstract of daily observations made in 1868 at the Worcester Station.

From Daily Observations.	Jan.	Feb.	Mar.	April.	May.	June.	July.	Aug.	Sept.	Oct.	Nov.	Dec.	For 11 months.
	ins.	ins.	ins.	ins.	May return is incomplete.	ins.	ins.	ins.	ins.	ins.	ins.	ins.	ins.
Mean height of the barometer at the temperature 32 deg. Fahr. (not corrected for altitude above the mean level of the sea), for each month, and for the year, from observations at 9 A.M., 1 P.M., and 5 P.M.,	29·139	29·167	29·207	29·238		29·382	29·314	29·343	29·266	29·268	29·208	29·169	29·245
Maximum observed, ...	29·350	29·534	29·500	29·500		29·770	29·743	29·616	29·544	29·656	29·548	29·440	29·770
Minimum observed, ...	28·950	28·919	28·880	28·939		29·111	28·961	28·997	29·041	28·960	29·011	28·928	28·880
	deg.	deg.	deg.	deg.		deg.	deg.	deg.	deg.	deg.	deg.	deg.	deg.
Mean temperature of the air for each month, and for the year (Fahr.), from observations at 9 A.M., 1 P.M., and 5 P.M., ...	81·2	76·3	73·4	69·3		59·7	57·4	61·9	64·6	66·5	72·8	76·6	69·0
	deg.	deg.	deg.	deg.		deg.	deg.	deg.	deg.	deg.	deg.	deg.	deg.
Mean temperature of evaporation for each month, and for the year (Fahr.), from observations at 9 A.M., 1 P.M., and 5 P.M., ...	69·5	64·6	63·2	59·8		52·9	50·9	54·3	57·4	57·4	61·0	63·6	59·5
Mean humidity of the air for each month, and for the year (the integer number 100 implies saturation), derived from observations at 9 A.M., 1 P.M., and 5 P.M.,	44·0	56·0	54·0	56·0		66·0	65·0	63·0	59·0	57·0	50·0	45·0	56 0
Maximum humidity, ...	94·0	83·0	93·0	87·0		100·0	93·0	100·0	87·0	93·0	81·0	86·0	100 0
Minimum humidity, ...	21·0	28·0	22·0	21·0		30 0	22·0	22·0	27·0	27·0	27·0	22·0	21·0
From the self-registering thermometers (Fahr.).	deg.	deg.	deg.	deg.		deg.	deg.	deg.	deg.	deg.	deg.	deg.	deg.
Mean maximum temperature of the air, ...	89·4	83·5	80·3	74·9		67.4	62·0	71·4	70·4	72·7	79·6	83·8	75.9
Mean minimum temperature of the air, ...	58·9	57·3	55·3	48·0		43·3	42·6	46·7	46·7	48·0	54·6	56·0	50·7
Approximate mean temperature of the air, ...	74·2	70·4	67·8	61·5		55·4	52·3	59·0	58·5	60·4	67·1	69·9	63·3
Mean range of the temperature of the air, ...	30·5	26·2	25·0	26·9		24·1	19·4	24·7	23·7	24·7	25·0	27·3	25·2
Highest temperature, ...	106·5	97·3	102·5	89·3		79·8	86·5	84·5	87·0	86·0	96·0	97·0	106·5
Lowest temperature, ...	50·0	48·0	42·5	40·5		33·0	33·5	36·5	42·0	38·5	37·5	44·5	33·0
Extreme range of temperature on any one day,...	43·5	38·8	42·5	47·7		88·5	41·5	40·5	35·7	41·0	38·0	41·8	47·7
Least range of temperature on any one day, ...	14·5	10·8	13·0	13·7		7·5	6·5	5·2	8·3	9·8	11·2	12·0	8·3
Quantity of rain in inches, ...	0·165	1·055	0·775	0·865		0·840	2·500	0·910	0·375	2·020	1·385	0·420	11·310
Greatest amount of rain on any day in each month	0·165	0·800	0·435	0·665		0·475	0·850	0·645	0·330	1·100	1·070	0·420	1·100
Number of days on which rain fell, ...	1	4	4	3		2	7	4	2	6	4	1	38

Number of Observations of the Wind at the several Points of the Compass during each Month of the Years 1867 and 1868, at Worcester.

Points.	Jan.	Feb.	Mar.	Apr.	May.	June.	July.	Aug.	Sept.	Oct.	Nov.	Dec.	Total	Jan.	Feb.	Mar.	Apr.	May.	June.	July.	Aug.	Sept.	Oct.	Nov.	Dec.	Total
N.	-	-	-	-	-	-	1	-	-	-	-	-	1	-	-	-	-	-	1	-	-	-	-	-	-	1
N.byE.	-	-	-	-	-	-	-	-	-	-	-	-	-	-	-	-	-	-	-	-	-	-	-	-	-	-
N.N.E.	-	-	-	-	-	-	-	-	-	-	-	-	-	-	-	-	-	-	-	-	-	-	-	-	-	-
N.E.byN.	-	-	-	-	-	-	-	-	-	-	-	-	-	-	-	-	-	-	-	-	-	-	-	-	-	-
N.E.	-	-	-	-	-	-	4	1	-	-	-	-	5	-	-	-	-	-	-	1	1	-	2	-	-	4
N.E.byE.	-	-	-	-	-	-	-	-	-	-	-	-	-	-	-	-	-	-	-	-	-	-	-	-	-	-
E.N.E.	3	-	-	-	-	-	-	-	-	-	-	-	3	-	-	-	-	-	-	-	-	-	-	-	-	-
E.byN.	-	-	-	-	-	-	-	-	-	-	-	-	-	-	-	-	-	-	-	-	-	-	-	-	-	-
E.	-	-	-	1	-	-	3	-	-	-	-	-	4	-	-	-	-	-	-	4	3	-	-	-	-	7
E.byS.	-	-	-	-	-	-	-	-	-	-	4	-	4	-	2	2	-	-	-	-	1	-	-	-	-	5
E.S.E.	3	3	3	2	-	3	1	2	-	4	3	3	27	1	-	-	-	-	-	-	-	-	-	-	-	1
S.E.by E.	-	-	-	-	-	-	4	1	1	-	-	-	6	-	-	-	-	-	-	-	-	-	-	-	-	-
S.E.	33	41	51	34	25	30	18	34	29	27	42	25	389	29	33	38	24	6	22	32	29	25	46	28	21	333
S.E.by S.	2	1	-	2	-	-	-	1	-	-	1	-	7	-	1	-	-	-	-	-	3	-	-	-	-	4
S.S.E.	-	1	-	-	-	3	-	3	6	7	2	2	24	12	10	16	4	-	-	-	3	4	6	9	8	72
S.by E.	-	-	-	-	-	-	-	2	2	-	-	-	4	6	3	-	-	-	-	-	-	-	2	3	-	14
S.	3	-	-	-	-	-	-	-	-	1	1	-	5	2	-	2	-	-	-	-	-	-	-	-	-	4
S.byW.	-	-	-	-	-	-	-	-	-	2	-	4	6	-	-	-	-	-	-	-	-	-	1	-	-	1
S.S.W.	-	1	-	-	-	-	-	-	-	1	-	-	2	-	-	-	-	-	-	-	-	1	-	-	-	1
S.W.by S.	1	-	-	-	-	-	-	-	-	-	-	-	1	-	1	-	-	-	-	-	-	-	-	-	-	1
S.W.	13	3	1	-	1	1	-	5	-	6	1	4	35	9	-	1	5	2	1	-	3	-	3	-	1	25
S.W.byW.	-	-	-	-	-	-	2	-	-	-	-	-	2	-	-	-	-	-	-	-	-	-	-	-	-	-
W.S.W.	3	-	-	-	-	-	3	5	-	-	1	-	12	2	-	-	1	-	-	-	-	-	-	-	4	7
W.by S.	-	-	-	-	-	-	-	-	-	-	-	-	-	-	-	2	-	-	-	-	-	-	-	-	-	2
W.	-	-	1	-	2	1	-	2	-	-	-	1	7	1	-	-	-	-	2	-	-	-	-	-	1	4
W.by N.	-	-	-	-	-	-	-	-	8	4	-	4	16	-	2	6	-	-	2	-	6	8	4	-	3	31
W.N.W.	7	17	7	-	21	25	4	20	15	7	11	23	157	7	7	5	5	3	15	15	14	25	9	8	23	136
N.W.byW.	-	-	-	2	-	-	5	-	-	-	-	-	7	-	-	-	-	-	-	-	2	-	-	-	-	2
N.W.	21	17	19	45	36	21	41	12	25	32	24	22	315	24	18	20	29	30	30	31	15	19	19	27	25	287
N.W.by N.	-	-	-	-	-	-	-	-	-	-	-	-	-	-	-	-	-	-	-	-	-	-	-	-	-	-
N.N.W.	2	-	-	-	-	-	-	-	1	-	-	-	3	-	-	-	-	-	-	-	-	-	-	5	-	5
N.byW.	-	-	-	-	-	-	-	-	-	-	-	-	-	-	-	-	-	-	-	-	-	-	-	-	-	-

N.B.—No observations in 1868, from May 13 to 28 (inclusive).

SECTION II.—*Divisions of Beaufort, Fraserburg, and Victoria West.*

Beaufort is described as one of the finest districts of the Western Province, "especially noted for its extensive sheep walks, upon which are raised a greater number of woolled sheep than in any other of the western districts; and, consequently, wool is the great staple of the district." "It has an area of 9,600 square miles. It includes a vast expanse of mountain and karroo country, but is an excellent sheep farming region, and by the formation of dams the scarcity of water is fast becoming remedied. It is watered, or rather drained of its waters, by the channels of the Dwyka, Gamka, Blood, and other water courses, by which millions of tons of much required water now run to waste, which, however, present in many places fit sites for dams, or rather artificial lakes. All water-courses north of the Nieuwveld and Roggeveld mountains flow into the Orange River; those south of the mountains, into the Indian Ocean;—the watershed line of this part of the continent running over their summits."

In the Introduction to "Hydrology of South Africa," it is stated, that, in making the tour of the Colony, in 1847, "while passing through the Karroo, I witnessed the privations to which the inhabitants were subjected through the aridity of the climate.

"My recollections of the journey call up vividly even now oft-recurring visions of bones of oxen at varying distances along the road—the bones of oxen which had succumbed by the way travelling in a land where no water is. And they call up incidents more definitely declarative of the characteristic of the land, amongst which are the following:—

"At one place at which we arrived on a Saturday we learned that beyond that place there was no water to be obtained within a distance of 84 miles on the road to Beaufort—whither we were bound—and we found this to be the case. Resting the horses on Sabbath, when we resumed our journey we started before bay-break, and managed by night-fall to reach the fountain, but water the horses touched not by the way.

"The day following we had at mid-day to send our horses six miles off the road to slake their thirst while we rested, letting them brouse by the way in going and coming, the *achter reiters* driving them slowly, very slowly along, that they might not be unfitted for resuming the journey on their return.

"At a farm house, at which towards evening we were, in accordance with colonial hospitality, welcomed and served with tea, I, inconsiderately perhaps, but stay-at-home travellers will say very naturally, said I would be obliged if they would give me also a little bread. 'Bread,' said the farmer, 'we have not seen bread for nearly three years.' 'Why, how is that?' said I. 'Because of the drought,' was the reply, 'we cannot raise corn' (the name given in the Colony to wheat). 'Then what do you raise?' I asked. 'Nothing,' said the farmer, 'we have occasionally had showers, and after these we have sown beans and they grew; but scarcely were they above the ground when they died away.' Then what do you eat?' 'Mutton.' 'But what do you eat with the mutton.' 'Mutton.' 'What do you mean?' 'I mean what I say, we eat the fat with the lean, and the lean with the fat, and so do the best we can.'"

"Karroo is an old Hottentot name signifying dry or barren, and has been from the earliest times applied to that arid and barren plateau in the Western Province lying between the Roggeveld and Nieuwveld mountains on the north and the Lower and Great Zwartberg on the south, drained by the dry channel courses of the Dwyka and Gamka Rivers in the centre, the Buffels River to the east, and the Doorn and Tanqua Rivers to the west. . . . As a generic name, Karrooveld is applied to all barren lands of the same nature as the Great Karroo in every part of the Colony, composed of hard-baked red clay, lying in a bed of blue schistose slate."

"The deserts of South Africa," says the writer quoted, an esteemed collaborateur, in Messrs Silver's Handbook for South Africa, in writing of these, which he describes as physical features which give a distinguishing stamp to this part of the earth's surface, "if we may call them such, are not, like the Saharas of the northern portion of the continent, sandy wastes shifting their surface under the breath of every gust of wind, or cold inhospitable plains like the Gobi of Central Asia; on the contrary, they are generally composed of shallow beds of the richest soil, which only want the fertilising power of water to render them, not only as rich but much richer than any other parts of the surface. Hence the importance of adopting a systematic mode of irrigation for these vast plains, which comprise nearly two-thirds of the surface of the Cape Colony, by storing the water up in the kloofs of the mountain ranges which bound them, not merely with extemporised dams, but in reservoirs with scientifically constructed retaining walls, capable of resisting the pressure of the enormous floods of water that pour down when the thunder-storms of the interior burst over them—floods that in the pre-historic era, when our Karroos were no doubt vast lakes, were sufficient to tear passages through the weak points of the coast range of mountains, and still form the passages of rivers such as the Gamka, the Groote or Gamtoos River, the Great Fish, etc., etc.

"Lichtenstein and Pringle describe in eloquent terms the beauty of the Great Karroo after the vivifying effect of a few thunder-storms. In like manner Bushmanland after the periodical rains presents the appearance of a vast field of grass; but owing to the formation of the surface the rains that fall soon run off, and the summer sun speedily reduces it to its previous burnt and barren appearance. Dr Livingstone's theory, which is adopted by all our best authorities, is that our Karroos and other large plains once formed the beds of immense lakes (an opinion strongly corroborated by the fossil remains found there), and of which the fast diminishing Lake Ngami now presents the only remains. We may also note that on the immense though gradual slope from the Nieuwveld Mountains to the Orange River, when the thunder-storms are more violent than usual, hundreds of square miles of the country along the line of the dry watercourses or natural lines of drainage are converted into what are locally called Vleis, or large shallow lakes. The whole of these waters it is supposed were let off by fissures or cracks in the subtending slopes rent by the upheaval of the country. The fissures thus made at the Victoria Falls, discovered by Livingstone, let out the waters of the great Zambesi Lake. The fissure through which the Orange River pours itself at the Falls of Aukrabees probably drained off the waters of the desert of Kalahari and the table-lands of Bushmanland. The Warm Bokkeveld valley and Kannaland, as well as the Great Karroo itself, were evidently lakes at one period, their waters escaping by the fissures of Mitchell's Pass, the Gauritz, and the Hex River valley; and indeed the rugged and fearful kloofs through which their surface waters still escape show the evident traces of some violent convulsions of nature. The basins of Cradock and Queenstown, evidently old lake-beds, are now drained by the watercourses of the Great Fish and Kei Rivers. A proof of the elevation of the land at a comparatively recent geological period is seen in the beds of shells along the coast, varying in thickness from 50 to 500 feet, from which is derived the greater part of the lime for industrial purposes. Indeed many appearances point out that the mountain ranges nearest the coast formerly formed the coast line of South Africa.

"The soil of the Kalahari desert is generally a light-coloured, soft, and pure sand, resting on a bed of tufaceous conglomerate, and probably on a substratum of granite or gneiss. Although destitute of running water, and so called a desert, it is covered with a dense vegetation of bush, and water is generally found by digging a few feet down into the native marked watercourses. Towards the north, indeed, the whole region seems to be covered with a hard sunbaked alluvial deposit, and to have been the bed of an ancient lake. Towards the east the sandstone formations, capped with trap-rock, appear.

"The substratum of the Great Karroo is a blue shaly rock, comprising part of Mr Bain's dicynodon strata, covered with thin beds of red argillaceous soil; both rock and clay appear to contain much soluble saline matter, which causes the general brackish nature both of the springs and rivers found in the Karroo. The term Karroo is a geographical term peculiar to South Africa. Barren, as these Karroo plains may appear to the eyes of a stranger, they form the most valuable sheep farms and pasture lands of the Cape Colony, and under the influence of a skilfully-arranged system of irrigation will be found capable of producing both wine and cereal crops to any amount. The market value of Karroo land has actually quintupled within a period of a very few years."

Mr Kinnear, Civil Commissioner for the division of Beaufort, wrote in 1862, in reply to the queries issued from the Colonial Office :—

"1. No plans for irrigation works have been proposed within this division. It was generally deemed useless to do so, as there was no prospect of the necessary funds being forthcoming.

"2. It is seldom that any river in this division runs longer than a few days at a time; no well that has yet been sunk yields a sufficient supply of water for irrigation. But there are many places along the Salt River, Zak River, Gamka, Karriega, and other rivers, where irrigation works on a large scale, by the collection of periodical supplies of water, are practicable. I may mention, among others, Nelspoort and Saltriverspoort on the Salt River, Botpoort on the Zak River, and Beaufort on the Gamka. For such works on a small scale there are innumerable favourable situations on the banks of rivers, and throughout the country generally."

Since that time, however, if not before, not a little has been undertaken and accomplished by public spirit and private enterprise.

The present Premier, the Honourable J. C. Molteno, then representative of the district in the Legislative Assembly, gave in the course of the same year the following evidence before the Select Committee of the Legislative Council on Irrigation :—

"1. *Chairman.*] You have had a good deal of experience in farming, particularly in the Beaufort district, I believe. You have also erected dams on a large scale, and will perhaps favour the committee, by giving us some idea of the increased value of Karroo and other fruitful soil when it can be irrigated ?—Yes, I have had such experience, and have erected some dams. The increase of value you may say is from very little indeed; for a piece of land which you would scarcely value at all without water may be made by the formation of dams useful, in the first instance, for grazing purposes; and where the works are sufficiently extensive and circumstances favourable, may become quite as good for garden and sowing land as any other portion of the district with a natural supply of water. I have myself erected several dams on portions of my own property, which was formerly only of use for a month or six weeks in the year, after rains fell, by which means I have been able to use these lands as separate farms, without the stock depending on any other water than that secured by the dam; and have likewise been able to spare sufficient water for the purpose of irrigating garden and sowing ground to a limited extent. . . .

"5. Are there any, or many, so-called dry rivers in Beaufort which could be partially dammed up, either at their heads, or at their feeders, so as to prevent the water which fills them in rainy seasons from escaping to the sea ?—Very many. And not only rivers but *leegtes*, which are filled with water after rains, and where, by damming up a small *poort*, or kloof, in a suitable situation, you might retain water to a great extent, and thus render a large quantity of land which is now of very little value for want of water, or which has only pools of water now and then for a short time, as good as any farm where there is permanent water. In many cases you could dam up to such an extent that irrigation could be carried out and large tracts of land brought under cultivation.

"6. These *leegtes*, are they not the feeders of the rivers, by which their water, if not stopped, runs off to the sea ?—Yes, I think in many instances damming up these *leegtes* would be preferable to damming up the rivers; except in some places where you might lead the water into a *vley*, or other suitable place for making a dam. These things have been done to a considerable extent by private individuals; but

I think the time has come to go beyond anything that is within the means of private individuals, and also for rendering valuable large tracts of land still in the possession of Goverment, which are at present scarcely used at all. . . .

"23. Can you speak as to the practicability of reservoirs being made which can be filled during the rainy season?—I think there are many places where you can collect the surface water and store it, which now runs away, without going to the expense of artesian wells. In Karroo the soil is very fertile; all you want is water.

"24. Of what capacity are those dams which you say have been already erected by private individuals?—Some of them are so extensive that, as I said before, a piece of land which before was scarcely worth anything has now become a farm, and people are living there with cattle and stock—to all intents and purposes as good a farm as one with a permanent fountain of water. The dam has, therefore, created a value in the land which it had not before.

"25. When you told the committee that parts of the country could be used for grazing ground if supplied with water, I think you merely suggested that they should be supplied with water to give drink to the cattle, not to fertilise it?—I think it would be better to consider them as grazing farms at present, as they are too far from a port to raise grain for export.

"26. But they would serve that purpose also?—Yes; but where you construct works to lead out water for irrigation, they require to be on a much more extended scale than for watering cattle or sheep.

"27. You have not seen the mode of irrigating in Italy?—No.

"28. In what way would you collect water for irrigation?—By damming up, in most instances, in suitable situations, and perhaps excavating to a certain extent, as well in others.

"29. But would that not be to the detriment of parties lying lower down the river?—Not in such cases, because this is water that would otherwise never be made use of; it is only obtained after the rains, and if not stopped would immediately flow off.

"30. Then probably there are many places where parties could raise their own fruits and vegetables, and what they want for their own household supply, which they are now prevented from doing by want of water?—Quite so. There are also many valuable tracts of Goverment ground where there is excellent pasturage, but useless through being destitute of water, which land would fetch a very high price if water could be obtained, not merely for grazing purposes, but for growing vegetables, fruit, forage, etc., to such an extent as would suffice for the consumption of the farm residents.

"31. In this way you could fertilise the whole Karroo, for the Karroo ground is considered very fertile?—Yes; such works of course would be highly reproductive.

"32. What is the rainy season in the Karroo?—Generally speaking, rains fall in the spring and summer months.

"33. It does not coincide with our winter months?—No. The rains come on sometimes unexpectedly, and change the whole appearance of the country in a few days. You might pass through and find it as dry as possible, and returning eight days afterwards through the same country your horses would be able to get a good feed of grass.

"34. Mr *von Maltitz.*] Do you think the value of Goverment ground would be increased to any great extent by the construction of such works as you speak of for irrigation?—No doubt it would. A piece of ground with a dam erected upon it would sell for much more than it otherwise would, and for very much more than the cost of erecting the dam.

"35. You said that you would prefer to dam up the *leegtes* rather than rivers: what was your reason?—I spoke chiefly of the Beaufort district, with which I am most acquainted.

"36. You are probably of opinion that if you dâmmed up the rivers they might become blocked up with sand, which would not be the case with *leegtes?*—It might be so; but in leading out rivers you have to lead them through the banks which are often high above the surface; it would depend upon the locality; in some cases it would be better to dam up a *leegte* or a *poort*, and in others to take the lead out of the river. Both plans might be adopted, according to the nature of the ground and other circumstances."

In the report of the Civil Commissioner for 1865 it is stated that in consequence of the continued drought very little had been done in the formation of dams, but that the municipality of Beaufort had "commenced the formation of a dam on a very large scale, the length of the embankments to be 1,500 feet; breadth, 100 feet; and height at the highest part, 30 feet. It is being formed at the northern extremity, above one of the fountains which supplies the town. When full it will contain a very large body of water, not only sufficient to irrigate a great extent of ground but greatly to strengthen the spring."

From time to time intelligence reached me in regard to the progress of the work. All was encouraging, and at length it was brought successfully to completion. The first report received spoke of the "successful progress made by the municipality in the construction of a splendid dam for the water supply of the town and neighbourhood. By forming an embankment of 800 yards in length, they are damming up the valley of the Kuil's River for a distance of three or four miles, and will thus produce what may really be called a lake sufficient not merely to supply the town but to irrigate all the garden lands about it during the driest months and the hottest seasons. And yet the estimated expense of the undertaking is exceedingly reasonable. The whole of the funds available amount to £1,500, which has been raised on ample security from the Savings' Bank. The works are superintended and managed by the municipal commissioners themselves, each member taking his fair share of the task, and thus the expenditure is reduced to the lowest and most economical scale possible."

In regard to the work at Beaufort I find it stated in one of the journals of the day, published while the work was in progress:—"At Beaufort the municipality are busy erecting a splendid reservoir above the town, by closing the run of the Kuils and Springfontein Rivers, which is about ninety feet in breadth. It is to abut on the stony ridge in the river, and then run westward about five or six hundred yards. It is to be entirely an earthwork, and will be, at its greatest height, about twenty-six feet, with a base in some parts of one hundred feet in width. The original estimated cost was not to exceed £1,500, and this amount was raised by a loan from one of the institutions in Capetown. About half the work is reported to be completed at a cost of £800, and it is now evident that to complete it in the substantial and safe manner originally contemplated, a further loan of £500 will be required. The water will be distributed by pipes running through the wall of the reservoir; and the original plan was to include a small water mill, and there is a splendid fall in the bed of the river suitable for the purpose. In addition to giving an increased supply of water for irrigation purposes to any of the inhabitants willing to pay for it, it is proposed to clear a very large tract of rich arable land at the lower end of the village, which is to be subdivided into water erven, and let, to enable the commissioners, without any additional taxation, to pay the interest annually upon the loan raised, and with every hope that the town revenue will also be largely increased. It is being superintended and carried out by Mr D. G. de Villiers, one of our most enterprising, substantial, and successful farmers, and who deserves every praise for the sacrifice of his time and ability for the good of his fellow citizens. Other members of the reservoir committee, who are not taking such an active part, deserve much praise. It is a noble work, and we all hope it may be a success, as it will enable the inhabitants to grow their own bread, for which they have always been dependent upon other parts of the country to supply, and often at very high rates."

The following graphic narrative tells of the severe test to which it was subjected on its completion. It occurs in a letter from Mr Molteno to Mr Porter, then Attorney-General in the Colony, dated 11th November, 1867:—"It affords me the greatest pleasure to be able to inform you that the Beaufort reservoir has received an immense supply of water, far beyond what, judging by the experience of years, people were led to suppose at all probable in so short a space of time. As you are aware, they have been busy for some months or more closing up the bed of the Springfontein River, which is at the end of the embankment, by carrying up with cart and oxen an immense mound of earth and stone, about 150 feet broad at the base to 15 at the top. To expedite this part of work the clay had been loosened beforehand, as it was necessary to complete it with all despatch, as a run of the river before it was finished might have done serious damage. The weather had been threatening for rain some time, when on the 7th instant, the work having progressed so far as be out of danger, a heavy thunderstorm passed over the flat country a good many miles above the reservoir; but as it was so distant, few thought that any strong run of the river would take place. However, on the weather clearing up, it could be seen that a great extent of country was flooded, and that the river would soon be down. From the circumstance that no rain had fallen near the town, and all the convicts and workpeople having left, it being about sunset, but few were on the embankment. It, however, so happened that I was there, and had an opportunity, which I would not have missed for any consideration, of witnessing the angry water of the river reaching the embankment, running along its old bed, as near as it could find it (for the banks had been all cut down for some distance, and the clay used to form the embankment), until checked and thrown back by the great mound which lay in its path. For one, like myself, who had so often stood by and seen this river in full flood, roaring and tearing down its rocky bed at this point, to see it checked and thrown back, and then quietly fill the dry bottom of what is now a good-sized lake, it was a most interesting sight, and I consider myself most fortunate in being on the spot just at the right time, for I had frequently said that if it were possible I would willingly make a special trip to see the reservoir fill. The news that the river was down soon spread, and the townspeople, men, women, and children, quickly mustered on the long embankment to see the water gradually rise. This it did for some time rapidly, until the water, by degrees, getting a larger space to flow over, the rise was less perceptible. The excitement, as you may suppose, was now very considerable; some champagne was brought up on to the embankment, and 'Success to the undertaking, and may it never break!' was drunk. It was a fine moonlight night, and many people remained enjoying the sight until a late hour. A watch was set to keep a constant look-out for any slight flaw which in such a mass of new work might be expected. The river continued running strongly all night, and the sight next morning was a magnificent one; but still the quantity of water seemed but little compared to the capacity of the reservoir. However, it was soon to receive a much larger supply, for the weather again became threatening, and in the afternoon a tremendous fall of rain commenced, the reservoir steadily filling, and extending far and wide, so that the river bed for a couple of miles up could no longer be traced. Good-sized mimosa trees were submerged, and the whole look of things changed. People began to grow timid, and to say there was enough, at any rate for a time, until the bank could get soaked and more or less consolidated. But it was not to be so. The weather was more or less rainy all Friday, and in the evening set in heavier, with thunder and lightning, flooding the town from the water which fell in and round about it. Night set in, the reservoir gradually rising. Considerable alarm prevailed. Such a fall of rain in such a space of time was most unusual, and had scarcely been calculated upon. The post-driver from Victoria reported having had to swim his horses though an otherwise insignificant tributary to the Springfontein River. All contributed to make it a night of much anxiety, for a sixty hours' rise of the river, and all the surrounding country flooded, so that water was running in from every direction, told every one that an enormous body of water would be arrested. Not but that the reservoir, or lake, as it might more properly be called, would contain it, but so large a supply all at once, with a new untried work, was alarming. However, on Saturday morning all was right, and the weather inclined to clear up. The water had again risen about five feet, and as it rose gradually extended over a much larger area, so that now it covers a very large extent of ground. Some leakages have appeared here and there in the upper and less dangerous parts of the embankment, but a constant watch and force have been at hand night and day, and the greatest care and attention paid to anything of this kind, which is always to be expected more or less in new works of the sort. I look upon the work as sound and secure. Still, if possible, it would be desirable not to retain more water *for the present*; and as this view is generally entertained, the convicts, free labourers, and volunteers, to a good number, perhaps 100 men altogether, are vigorously at work blasting the rock and clearing away stone and ground in the *uitloop*, in the rand, or stony hill, which forms the one side of the reservoir; so that if another rise of the river takes place, the water may be let off, at any rate for the present, until the work has become more consolidated, when a much larger body of water can be stored; although, in my opinion, what there now is is almost inexhaustible for the use of the town as it now is, or until more ground is laid out and prepared for cultivation. I should think the water at the top, where it becomes more shallow, cannot be less than three miles round. Boats, both rowing and sailing, are now the whole talk, and immediate efforts will be made to get some up. Bathing and swimming are being indulged in by all classes to their hearts' content. The embankment is a favourite place of promenade; and altogether it is clear that this work, if no mishap takes place, will entirely alter the character and position of things. . . .

"I am glad to think that a district in which I am so much interested, and with which I have so long been connected, should be the first to complete a work of so much importance, and can only hope that it may be the forerunner of many others; for the more I see of, and think about, this Colony, the more I am persuaded that the storing up upon a large scale of the waters which now run to waste is the one thing to which we must look for progress."

The reservoir was completed and filled with water in November, 1867. A correspondent of the compilers of the Cape of Good Hope Directory for 1868 wrote:—"The large Beaufort dam, or reservoir, is completed, and the waters now collected give it the appearance of a small inland lake, with two of its sides belted with mimosa trees of luxuriant growth. So pleasant is the vicinity that the reservoir has become the favourite resort of the villagers. . . . The climate of this neighbourhood has undergone a great change for the better. In lieu of the dry, hot summer air, there is now a coolness which is pleasant to all; and the Kuil's River, lower down than the dam, where formerly there was nothing but marshy puddles, the evaporation from which frequently produced cases of fever, now contains always fresh water." The successful carrying out of this work was spoken of as of inestimable value to the Colony, "as a pioneer work to show what can be done by enterprise." And it was such.

But while the embankment stood the test to which it was thus subjected immediately after its completion, within two years, a deluge of rain washed away a large portion of it. And in the following year (1870) the town suffered from a deluge of rain, from which the river Gamka, on the one side, overflowed its banks, carrying away parapet walls constructed to prevent it from inundating the town; and at the same time the Springfontein was full, and not having vent sufficient through the gap in the embankment made by the breakage which had occurred in the preceding year, both rivers flowed uncontrolled into the town, flooding many houses, blocking up the water furrows, ploughing up the street, and causing great loss to the inhabitants.

Still the dam had given demonstration of the quantities of water which might be stored up and utilised; and what was done revealed with this what were defects against which it was necessary to take precautions.

In 1869 it was reported by the Civil Commission:—"The bursting of the Beaufort reservoir occurred on the 23rd of October last, to the great loss of this town. A large extent of land had been brought into cultivation, owing to the command of water supplied by the dam, and was capable of much extension, giving employment to a number of people, and adding very materially to the good supply of the town. I had on many previous occasions expressed my utter want of confidence in its stability. It is most essential that the dam should be reconstructed under the plan and proper supervision of a qualified engineer.

"Another site has been suggested by myself as being safer, giving deeper water, and able to be constructed for a sum not exceeding £700, while to re-make the dam at Kuils River, the present dilapidated reservoir, would entail an expenditure of at least £3,000, besides the labour of one hundred convicts to carry out the new plan submitted by Mr Brand, the town engineer. It will be a question, therefore, for the consideration of the public, whether they are prepared to approach Parliament for another Loan Bill, and thus saddle the town with a debt of £7,000, or sanction the making of a new dam on the *site* proposed by the civil commissioner, which has the great advantages of small cost in construction, deeper water, and freedom from danger to the town, besides enabling the municipal commissioners to lease, with the sanction of His Excellency, a larger extent of land, and securing at the same time ample water-leading to the present lands. . . .

"Crime is much the same as last reported, and is not caused in most cases by want of food,—that is, if the natives are willing to work. This they shun, and would rather live in idleness, trusting to chance for their food; and hence the sheep-stealing.

"In agricultural and farming pursuits generally the past year has been a severe trial to many, the drought having been very severe; many of the farmers being obliged to leave their homesteads to seek pasture and water elsewhere. That there is much improvidence and want of foresight among these people is patent to any observer. They have few dams for storing water, and again, although the general experience is such as to warn them of a coming drought, they still hang on with large stocks, instead of selling their overstocked numbers of sheep at market value, and thus reducing their risk when the calamity of drought comes. Thousands of sheep are lost in this way, whereas, if sold at any market and any price, whatever they realised would at all events, not as now, be a total loss."

In 1874 it was reported:—"The drought mentioned in the last report lasted until November of this, when the district was favoured with beautiful rains; they came, however, too late for the sheep-farmer, and the lambing season has been, therefore, a very poor one,—worse even than that of the former year.

"The harvest, on account of the long and continuous drought, has not turned out well. Since the fall of the recent rains, however, the prospects of the division are bright for the coming year.

"Mr Devenish, of Salt River, has been successfully using a reaping and thrashing machine. This, I am informed, answers admirably, and has effected a great saving in labour to the owner during the past season.

"A wool-washing establishment, on a small scale, has been erected near the town by Mr Maddison. This improvement is of so recent a date that it can hardly be said to have had a fair trial.

"The re-construction of the Beaufort reservoir is in progress. About one hundred men (convicts) are engaged on the work, under the supervision of Mr Brand, C.E. The strengthening of the embankment and outlet ought to ensure the success of this important and much needed undertaking. The estimated cost is £5,500. It is encouraging to note that many of the farmers are becoming alive to the necessity of dam-making; no fewer than ten new dams have been finished during the year, while many are in course of construction.

"The 'Xanthium Spinosum,' owing to the prompt measures taken by the Divisional Council, is disappearing. . . .

"Crimes of a serious nature have been of less frequent occurrence than during any previous year."

In 1875 it was reported:—"The western portion of this division has, during the past year, suffered from severe drought. The veld became so parched up, and water so scarce, that several flock-masters had to leave their farms, for the neighbouring districts, in search of pasture and water for their flocks. The Nieuwveld farmers, however, were more favoured. Rain fell early in the year, the veld became most luxuriant, and sheep and cattle have not been known to be in such excellent condition for many years. In spite of these favourable circumstances, however, the lambing season was but a poor one, and this was doubtless owing to the severe drought of the preceding year.

"This district being essentially a pastural one very little attention is given to agricultural pursuits;—there are a few farms in the Nieuwveld wards, upon which crops are raised, but the quantity produced is barley sufficient for the requirements of the producers themselves. The heavy rains which fell in November last produced a destructive flood in Salt River valley. The water has not been known to be so high for upwards of 25 years. A portion of Mr Charles Devenish's large dam, built of solid masonry, in the bed of the river, and which was all but completed, was swept away; and I have since ascertained that no fewer than eight other dams, within the same drainage area, burst simultaneously."

The following is an abstract of observations made at Camfers Kraal, near Beaufort:—

Tabulated statement of meteorological observations made at Camfer's Kraal, near Beaufort—1868.

FROM DAILY OBSERVATIONS.	Jan.	Feb.	March.	April.	May.	June.	July.	Aug.	Sept.	Oct.	Nov.	Dec.	For 4 months.
From the self-registering thermometers (Fahr.).									degrees.	degrees.	degrees.	degrees.	degrees.
Mean maximum temperature of the air,....................	...	...	...	...	...	...	...	...	81·30	82·01	89·90	94·32	86·88
Mean minimum temperature of the air,........	...	...	...	...	...	...	...	...	48·09	43.71	50·03	53·87	48·93
Approximate mean temperature of the air,........	...	...	...	...	...	...	...	...	64·70	62·86	69.97	74·10	67·91
Mean range of temperature of the air,............	...	...	...	...	...	...	...	...	33·22	38·29	39·87	40·45	37·96
Highest temperature,............	...	...	...	...	...	...	...	...	95·0	94·0	104·0	111·0	111·0
Lowest temperature,..............	...	...	...	...	...	...	...	...	35·0	29·0	35·0	43·0	29·0
Extreme range of temperature on any one day,......	...	...	...	...	...	...	...	...	46·0	57·0	54·0	50·0	57·0
Lowest range of temperature on any one day,...........	...	...	...	...	...	...	...	...	13·5	24·0	23·0	25·0	13·5
Quantity of rain in inches,......	0·00	1·69	0·30	0·94	0·35	0·00	0·86	0·18	00·2	0·23	0·95	0·61	6·130
Greatest amount of rain on any one day in each month,......................	0·00	1·16	0·20	0·94	0·18	0·00	0·38	0·12	00·2	0·11	0·50	0·45	1·160
Number of days on which rain fell,.......................	0	4	2	1	4	0	4	2	1	4	2	3	27

When the report by Mr Kinnear, as Civil Commissioner of Beaufort, was made in 1862, and the evidence of Mr Molteno was given before a Committee of the Legislative Council on irrigation, the division of Beaufort extended to the Orange River, including then what now constitutes the division of Fraserburg—a great extent of country north of the Roggeveld and Nieuwveld mountains, with an area of 29,600 square miles, a vast inclined plain sloping from the mountain ranges down to the Orange River, intersected by the dry channels of the Hartebeest, Zak, and other periodical streams. It borders on Bushmanland, and includes a region sometimes called the Trekveld, inhabited by migrating boers, who roam with their flocks up and down seeking fresh pastures. It is alleged that the formation of dams would wonderfully increase the productiveness of the rather barren uninviting region.

The Hydraulic Engineer reported to the Commission of Crown Lands and Public Works, under date of 27th May, 1876, in regard to Bushmanland after passing through it on his return from Klaver Vley by way of Calvinia and Karroo Poort to Capetown, after a tour of observations embracing the course of the Orange River:—"The most important thing to be done with regard to Bushmanland is to push on with the surveys, so that longer leases may be granted. Farmers would then sink wells and make reservoirs. They will also have an interest in protecting trees and herbage; and might be induced to plant trees in the kloofs so as to preserve the springs.

"It seems very likely that if the Bushman-grass or other herbage is too closely eaten off before its seed time, sufficient provision may not be made for reproduction, and the less useful plants which are not eaten and so allowed to remain till seed time, will increase largely.

"It is a common belief among the Boers that the country has got drier of late years. This may possibly be due to some secular, or perhaps periodic, change of climate, but there are causes at work that may somewhat account for it. One of these is the multiplication of little paths by the sheep and goats. Along these the water can more easily find its way off the land than it did formerly. Another cause is that the reeds and ranker herbage of the valleys is trodden down or eaten up by stock, and a quicker outlet thus given to the storm-water. Places where in times past hunters found a *rietvley*, or a thick and almost impassable jungle, are now for the greater part of the year *vloors* of cracked mud with a scanty growth of small bushes.

"Till the surveys are finished, little improvement will be made, unless Government could see its way to granting some compensation to the yearly tenant for wells and reservoirs. I understand that the giving of compensation for houses and other improvements was tried once, but did not succeed. I am however inclined to think that compensation might be given for such important and obvious improvements as wells and dams, but refused for cottages or kraals.

"The suggestion has often been made, and is, I think, reasonable, that some assistance should be given, by way of loan, to farmers who propose to make extensive reservoirs, or otherwise improve their water supply. Such a loan should not be made unless the opinion of some engineer or practical man is taken previously as to the capability of the site, and the strength of the design. The loan should, I think, not exceed half the value of the work, and the interest on such loan should be a first charge on the land benefited.

"The chief faults of the existing dams or reservoirs are:—

"1. The overflows are generally too small.

"2. The position of the supply pipe or culvert is often bad. At one dam I visited the irrigation water was actually supplied through a leak in the bank. I cannot too strongly recommend the use of siphons in small reservoirs so as to avoid having pipes through the bank.

"3. The slopes are generally too steep.

"4. Sufficient provision is not made for puddling the bank; or at any rate for trampling in the layers of material as they are put on.

"5. The original ground is not properly broken up or otherwise prepared before beginning the embankment."

In regard to Fraserburg, the Civil Commissioner reported in 1865:—"This division has shared with others in the effects of the great drought so prevalent over the whole Colony, and has perhaps suffered more than some others on account of the open and arid nature of the soil. The Roggeveld, so well known as the agricultural part of the division, from its high position and from receiving, as it has always hitherto done, regular periodical rains, has produced literally nothing this season. All the crops have failed, and the herbage is entirely parched up. This, together with the partial failure of the springs, renders it no surprising thing to hear of the numbers of sheep that have died and that are still dying. It is asserted by the oldest and most respectable inhabitants that a drought so severe has not been known since 1824.

"In consequence of this calamity, affecting, as it does, all classes of the community, transport has been obtained with much difficulty, and that only in a few instances. Thus every common necessary of life has become extremely scarce and dear, to the village people especially, and it is the general expressed opinion that the sufferers will not be able to regain what they have lost for the next two or three years to come, even should the seasons prove favourable.

"Owing to these drawbacks, and to the low price offered for wool, that article has neither increased in quality nor quantity.

"The village, nevertheless, shows a marked improvement, in a neat and well-built water furrow, along which trees have been planted, which are thriving very fast."

And in 1866 it was reported:—"Since the severe drought of last year this division has been progressing favourably, both in agricultural and pastoral farming. Much grain has been sown, and many merino sheep introduced; and it is the general expressed opinion that when the Crown lands are disposed of, under the Act No. 19 of 1864, they will be chiefly occupied by men of capital and enterprise, and that the division will in a few years become second to none in the Colony as a sheep-walk, for which it is already so proverbial."

In 1869 it was reported from Fraserburg:—"The prospects of the Division have not improved during the past year. Nearly the whole district having suffered as usual from drought, preceded as it was by so many previous years consecutively, has gone far to retard progress, and indeed brought many to the verge of insolvency. Rains of any consequence have been confined to the northern and eastern portions of the division in the early part of the year, and it was not until June and August, and again in October, when the Beaufort West dam was carried away, that the southern and western extremities were inundated to an extent unknown for many previous years. As an instance of the severity of the drought at the present time in the middle, eastern, and northern portions of the division, the lessee of a Government, and proprietor of four quitrent farms, notwithstanding the extent of country which they embrace, and that generally they are well supplied with water, has been compelled to seek pasturage for his flocks elsewhere. The heavy north-west gales of wind felt in Capetown during December, also blew here with much violence, and were followed by severe frosts and a heavy fall of snow in the Roggeveld, causing severe losses in stock and considerable damage to the late crops and gardens. Countless herds of springbuck have followed in the tracks of the late rains, to the incalculable loss of the farmers; in fact, these animals in this respect are as great a scourge as locusts or drought.

"Several farmers, discouraged by the low price of wool during the last couple of years, have given up breeding merino sheep, and again taken to Cape sheep, which they erroneously expected would yield larger profits; the present low price of slaughter sheep, however, and the late rise in wool have caused them to repent of this step. The breeding of Angora goats seems to be attracting some attention, and ostrich farming has been successfully carried on by a couple of our most enterprising landowners during the last few years. . . .

"As regards crime, there were, during the year 1868, 133 criminal cases, including police cases and cases under the Masters and Servants Act and Pound Ordinance, and during the past year 128. The poverty among the coloured population caused by the drought gave reason to suppose that many more cases of theft of stock would have been brought on. That this did not occur must be accounted for by the fact that the farmers are willing to submit to the small pecuniary loss of a few sheep or the like, rather than incur the expense and great personal inconvenience attendant on a journey to the circuit court held at Graaff-Reinet or other distant towns,—the patriotism and public spirit of the pastoral population generally not being of sufficient warmth to overcome self-interest. . . .

"The Floksberg and Komsberg passes on the Roggeveld range were much damaged during the heavy rains which fell there in June and October."

In 1874 it was reported of Fraserburg:—"In the early portion of the year indications of an approaching drought retarded much the general progress and advancement of this division. Farmers are, however, now becoming fully convinced of the dire necessity for making some provision to meet the exigency of these periodical droughts, and more attention has therefore been bestowed upon the construction of dams and augmentation of the water supply generally. . . . The ostrich appears to be gradually superceding the sheep, and almost as much attention is now devoted to the former as to the latter—a precarious foreboding for the future.

"Owing to the drought during the early part of the season, as also to that malignant infection, 'the scab,' there has been a decrease in wool in this division to the alarming extent of upwards of twenty-five per cent."

In 1875 it was reported:—"There is nothing of much interest to report since 1874. The past year may be regarded as one of prosperity, and advancement. The magnificent rains during the early portion of the year, as also those more seasonable later, have entirely blotted out the ill effects of the droughts of former years.

"The lambing season has been a most prolific and successful one —comparatively few, or no losses on this score have been reported; there has, however, been a decline in the supply of wool. This decrease, it is thought, may be attributed to the large numbers of sheep bought up for the Capetown and surrounding markets. A rise in the price of meat in the local markets may also be ascribed to this.

"Ostrich-farming has made rapid strides, and enclosures, stretching for miles, may now be seen on several farms in this neighbourhood. This enterprise may, however, be said to have reached its climax; the instability, and downward tendency of the feather-market has created some alarm among this class of farmers, and ostriches are now changing hands at wonderfully reduced prices. This change may, however, be regarded rather with satisfaction, as, taking the best view of it, ostrich-farming cannot be of much ulterior benefit to the Colony—indeed, its introduction has impeded the culture of many other and more staple products.

"The storage and impounding of waters is receiving deserved attention. The absence of skilled labour, as also of scientific knowledge in the construction of dams is still a great impediment to their durability. Abundant proof of this was manifested during the heavy rains at the commencement of the year, when several large dams, entailing no small outlay of capital in erection, were completely destroyed.

"With sufficient irrigation, this district could easily produce bread-stuffs enough and more than is required for its own consumption. On the few farms that can command a water supply, the harvest has been all that could be desired."

From my son, who has lately returned from Fraserburg, where he was district surgeon for many years, I have learned that the water supply of the division of Fraserburg seems to be less abundant than it was some years ago. "The fontein, or spring, at the missionary station of Amandelboom, for instance, used formerly to supply water, which ran and watered the gardens of the missionary and his people, while now, to get it to run in the old furrow, the water has to be pumped up, or baled up with buckets, a height of some two or three feet. The supply pumped up is still sufficient for the cultivation of the ground, but the level of the under-ground water has sunk two or three feet.

"Many other instances are to be found of springs, where there was formerly an extent of ground cultivated, now standing too low to let the waters run along the surface.

"The farmers attribute this, among other causes, to the very much greater number of sheep that are now to be found, and which, by their walking, beat the surface of the ground into a harder, less porous mass, from which the rain water runs off, in place of, as formerly, sinking in. This effect on the soil is helped to a large extent by the fact that formerly grass grew much more abundantly in these parts where now the schaap-bosch and other bosches have taken its place. These bosches are sown apparently by the sheep and goats, the seeds passing out with their droppings.

"The water supply at the farms is mainly from permanent springs; these are very often situated on 'aars' (aders), which seem to be geological breaks, extending over miles and miles of under-ground rocks; they are marked by straight lines of more luxuriant vegetation. The one out of which the village spring comes runs for some ninety miles, along which line numerous springs and, of course, farms are to be found. At Fraserburg the water contains a very large quantity of bi-carbonate of lime—all the lime used in the village is obtained from soil round the spring and above the present level of the spring, impregnated with this lime by the evaporation of the water. There may thus possibly be large underground reservoirs in the limestone. The occasional appearance of fish (small but true fish) in the fountain may possibly be connected with these reservoirs; the fish, however, have eyes which fact militates against this.*

* The reference to the existence of eyes in these fishes, militating against their having come from subterranean streams, is founded on the observation that from disuse animal organs dwindle, and in some case disappear; and an interesting animal, inhabiting subterranean streams and reservoirs in Illyria, from which all light is excluded, are found to be destitute of eyes. It is the *Proteus anguinus* of which Sir Humphry Davy, in his "Consolation in Travel, or The Last Days of a Philosopher," in 1828, wrote as "a far greater wonder of nature than any of those which the Baron Valvasa detailed to the Royal Society, a century and a half ago, as belonging to Carniola, with far too romantic an air for a philosopher." And he gives of them the following description:—"At first view, you might suppose this animal to be a lizard, but it has the motions of a fish. Its head, and the lower part of its body and its tail, bear a strong resemblance to those of the eel; but it has no fins; and its curious

"There are one or two fountains or springs which, after a series of dry years, give up altogether; but again, after good rains, yield a fair supply.

"There is a reservoir at the village of Fraserburg. It has but a very small area of collection. It is formed by a wall some fourteen feet high at the highest, and tapering away at the sides; the water will hardly stand nine feet deep at the side of the wall in the middle, and a few feet from the wall it shallows rapidly.

"When filled it will hold water such as cattle can drink for, say eight months in winter, or perhaps five months in summer. It lies above the level of the spring and has a very marked effect

branchial organs are not like the gills of fishes; they form a singular vascular structure, as you see, almost like a crest, round the throat, which may be removed without occasioning the death of the animal, who is likewise furnished with lungs. With this double apparatus for supplying air to the blood, it can live either below or above the surface of the water. Its fore feet resemble hands, but they have only three claws or fingers, and are too feeble to be of use in grasping or supporting the weight of the animal; the hinder feet have only two claws or toes, and in the larger specimens are found so imperfect as to be almost obliterated. It has small points in place of eyes, as if to preserve the analogy of nature. It is of a fleshy whiteness and transparency in its natural state, but when exposed to light, its skin gradually becomes darker, and at last gains an olive tint. Its nasal organs appear large; and it is abundantly furnished with teeth, from which it may be concluded that it is an animal of prey, yet, in its confined state, it has never been known to eat, and it has been kept alive for many years by occasionally changing the water in which it was placed. . . They were first discovered here by the late Baron Zois; but they have since been found, though rarely, at Sittich, about thirty miles distant, thrown up by water from a subterraneous cavity; and I have lately heard it reported that some individuals of the same species have been recognised in the calcareous strata in Sicily. . . . In dry seasons they are seldom found here, but after great rains they are often abundant. I think it cannot be doubted, that their natural residence is in an extensive deep subterranean lake, from which in great floods they sometimes are forced through the crevices of the rocks into this place where they are found; and, it does not appear to me impossible, when the peculiar nature of the country in which we are is considered, that the same great cavity may furnish the individuals which have been found at Adelsburg and at Sittich. . . .

"I cannot suppose that they are larvæ. There is I believe in nature no instance of a transition by this species of metamorphosis, from a more perfect to a less perfect animal. The tadpole has a resemblance to a fish before it becomes a frog; the caterpillar and the maggot gain not only more perfect powers of motion on the earth in their new state, but acquire organs by which they inhabit a new element. This animal, I dare say, is much larger than we now see it, when mature in its native place; but its comparative anatomy is exceedingly hostile to the idea that it is an animal in a state of transition. It has been found of various sizes, from that of the thickness of a quill to that of the thumb, but its form of organs has been always the same. It is surely a perfect animal of a peculiar species. And it adds one instance more to the number already known of the wonderful manner in which life is produced and perpetuated in every part of our globe, even in places which seem the least suited to on the spring, rendering its flow very much greater, even double what it has been in droughts, when there was no dam, or the dam was dry.

"There are two large dams in the district: one at Waterfall, Mr C. F. Marrais'; the other at the Ayofontein, Mr A. S. le Roex's. At both of these places the dams once filled have never been emptied. Water is led out for irrigation purposes to a very large extent, and the cost has doubtless been repaid by the produce thus raised in the course of three or four years at any rate."

My son had a well sunk at Fraserburg. In regard to this he says:—"In sinking my well I was guided by the accounts I saw of Norton's tubular wells, used in the American war and the Abyssinian

organised existences. And the same infinite power and wisdom which has fitted the camel and the ostrich for the deserts of Africa, the swallow that secretes its own nest for the caves of Java, the whale for the Polar seas, and the morse and white bear for the Arctic ice, has given the Proteus to the deep and dark subterraneous lakes of Illyria,—an animal to whom the presence of light is not essential, and who can live indifferently in air and in water, on the surface of the rock, or in the depths of the mud.

"It is now ten years since I first visited this spot. I was exceedingly anxious to see the Proteus, and came here with the guide in the evening of the day I arrived at Adelsberg; but though we examined the bottom of the cave with the greatest care, we could find no specimens. We returned the next morning and were more fortunate, for we discovered five close to the bank on the mud covering the bottom of the lake; the mud was smooth and perfectly undisturbed, and the water quite clear. This fact of their appearance during the night seemed to me so extraordinary, that I could hardly avoid the fancy that they were new creations. I saw no cavities through which they could have entered, and the undisturbed state of the lake seemed to give weight to my notion. My reveries became discursive, I was carried in imagination back to the primitive state of the globe, when the great animals of the sauri kind were created under the pressure of a heavy atmosphere; and my notion on this subject was not destroyed, when I heard from a celebrated anatomist, to whom I sent the specimens I had collected, that the organisation of the spine of the Proteus was analogous to that of one of the sauri, the remains of which are found in the older secondary strata. It was said at this time that no organs of reproduction had been discovered in any of the specimens examined by physiologists, and this lent a weight to my opinion of the possibility of their being actually new creations. . . .

"But on such grounds eels might be considered new creations, for their mature ovaria have not yet been discovered, and they come from the sea into rivers under circumstances when it is difficult to trace their course. The problem of the reproduction of the Proteus, like that of the common eel, is not yet solved; but ovaria have been discovered in animals of both species, and in this instance, as in all others belonging to the existing order of things, Harvey's maxim of '*omne vivum ab ovo*' will apply. . . .

"This inquiry has been pursued by Schreiber and Configliachi; my researches were made upon its respiration and the changes occasioned in water by its branchiæ. . . . They proved to me at least, that not merely the oxygen dissolved in water, but likewise a part of the azote was absorbed in the respiration of this animal."—J. C. B.

campaign, and by the fact that at very many of the farms, more especially where dams were above the fontein, the farmers got a largely increased supply by boring holes about one and a half inch in diameter through the rock in the bed of their springs. From these holes the water would rise to sometimes two inches above the level of the surrounding spring—boiling up, as they say.

"There were three wells within about 900 yards of where I fixed on boring, which was just at my kitchen door. I bored down through 18 feet of sandstone, getting bores welded on as far as necessary, at that depth we came on water which rose four feet in the hole. I went four feet deeper and got a supply of water which never failed me. By vigorous work the hole could be pumped dry in little over five minutes, but two minutes was enough for the water to collect again; and the servants not pumping so quickly a continuous supply was got by them. The boring cost me about £7 10s, but labour was then very expensive indeed. At least half-a-dozen other holes were sunk; four of these were even more successful: they could not be pumped dry; two cases were failures, no water being obtained, though in one case on a hill a depth of thirty feet was reached.

"At two neighbouring farms wells were made, and at Victoria one was made, all three answering thoroughly.

"In the district known as The Flats, extending to the north of Amandelboom there are no springs, or very few. There seems a vast deposit of soil here—not a thin scanty covering, as over the higher part of the district; and by digging pits, as they are called all over this region, water is reached at depths varying from two to twelve feet. In these parts there is less bush and more grass. After good rains the country looks lovely, and the grass is cut by the farmers with the sickle—almost like corn, as they say.

"There seems to be any amount of water to be had, but below the level of the surface, and with no chance of getting it to run on the surface, so as to raise crops. In these regions, the people go for months at times without having any bread to eat."

When the Hydraulic Engineeer was at Beaufort West, in accordance with instructions from the Commissioner of Crown Lands and Public Works, he gave his advice in writing to the engineer in charge of the dam then being made there, and he reported to the Government:—"About two hours ride from Nel's Poort near Beaufort, is a 'poort' called Salt River's Poort, where I think an embankment might be made and a large quantity of water stored. The poort is about

400 yards wide, and the river runs, when it does run, in a channel, the sandy bed of which is thirteen feet below the general surface of the alluvial deposit in the poort. The hills on both sides are very suitable for abutments. It would be necessary to bore or sink shafts to find out at what depth under the alluvium the solid rock can be reached; and a complete survey with levels would have to be taken before any designs or estimates could be made. There is a 'nek' or 'col' not far off which might possibly be used as an overflow channel. The land is, I understand, all in private hands."

Formerly was included in the division of Beaufort also what is now the division of Victoria West. In soil, climate, etc., it resembles Fraserburg. To the north of it is a vast expanse of barren country, thinly inhabited, in which a few Bushmen and Korannas still wander. It has an area of 9,000 square miles.

The Civil Commissioner of this division, in his report for 1865, stated that "a pump had been introduced into the division by Mr Jesse Park, to work by the hand and by horse power. The quantity of water thrown out by it is reckoned at between 9,000 and 10,000 gallons per hour. Mr Park had encountered great difficulties, but his endeavours were at length crowned with success, and visitors who have seen it speak in high terms of its efficiency."

In 1866 it was reported:—"The scarcity of water at many farms is still felt; but the farmers are busy in making dams, and some fountains have been opened. The crops are much better than they have been for several previous years." And again,—"This division may be said to have steadily progressed in many ways during the past year, although there is a very wide field for improvement still open; for most of the farmers have as yet mainly confined themselves to wool growing." In the previous year mention had been made of a dam being constructed by the municipality of the district town. This year it was reported:—"The municipality of Victoria West completed the dam mentioned in last year's report, and this year the water furrow leading from it to the town has been brought to a proper water level, and 250 yards paved with stone. The intention at present is to continue the paving whenever the funds admit of it."

In 1871, the year after the disastrous floods at Beaufort, a similar flood occasioned great loss of life and property in Victoria West. Of this, details have been given in "Hydrology of South Africa," (pp. 241–243); and the same deluge of rain proved similarly destructive to property in the divisions of Oudtshoorn, of which details are also there given (pp. 243–244).

Of Victoria West, in 1874, it was reported:—"The year, 1874, was ushered in by very welcome showers of rain, and a more favourable year than 1873 was then anticipated, but from February, or the beginning of March last, until the first week in November the district has had no rain. Several farmers were in consequence heavy losers in stock, transport from the sea-board almost entirely stopped, and the common necessaries of life rose in proportion. About the commencement of November it rained in gentle showers for five or six days almost continuously, saturating the parched soil; later in the month we had heavier rains. No damage worth mentioning was sustained by the people in the immediate vicinity of this town, but farmers residing to the north and north-east of this division lost considerably, some in live-stock, and the greater number of them by the destruction of dams. It is reported that from Doornberg to the Orange River, a distance of about one hundred miles as the crow flies, some twenty or thirty dams have been carried away, and in that portion known as the 'Vlakte,' not one is left. The ever-recurring destruction of these dams, which appear to be built to serve as sport for the elements, suggests that there must be some defect in their construction, and as the question of husbanding a sufficient supply of water is of vital importance to the pastoral and agricultural population, the subject demands serious attention. It has been proposed that Government should engage the services of qualified engineers to inspect the different dam works in the Midland Province, note the defects in their construction, as well as profit by any useful hints of a local nature they may obtain from farmers experienced in the making of dams, and then proceed to construct model dams on Crown lands, sufficiently strong to withstand the periodical heavy rains, and with due regard to economy and local adaptability. Specifications of such dams should be lodged in the offices of the Civil Commissioners, for the guidance and information of parties desirous of effecting such works on their private properties.

"It would be desirable, also, if a reservatory clause were inserted in future conditions of lease, retaining to the Government the right of at any time constructing a dam on leased Crown land, the lessee to contribute a fair and reasonable sum (in proportion to the advantages likely to be derived by him) towards such work.

"Crime has been on the decrease. No report of stock thefts on the northern border of this division have reached the magistrate during the past year; this is attributed to the fact that during 1874 a detachment of the Northern Border Police has patrolled that part of this division, and the presence of the police at uncertain intervals has not only put a stop to thefts, so prevalent a year ago, but reduced the number of squatters in that neighbourhood.

"Locusts and 'Rispers' have entered an appearance in this district; the 'Voet gangers' particularly are in myriads, covering large tracts of country in dense masses, leaving no vestige of green herbage behind them, and defy every endeavour to stop them in their devastating course. . . .

"Public health, on the whole, has been good. The mortality in the district has been, perhaps, higher than during 1873."

And in 1875:—"The past year has been on the whole a prosperous one for the division. During the first half it was visited by abundant showers of rain. From August to December it remained dry, but during the latter month, gentle, continuous, and saturating rains fell, and the veld just now abounds in pasturage, and stock of all kinds is in good order.

"The extent of land brought under cultivation during the past year has been in excess of previous years, the crops in most instances unusually abundant. . . . The high rate at which every description of grain is being sold here is accounted for by the enormous traffic through Victoria West. . . .

"The census taken during 1875 places this division in a position favourably to be compared with any other pastoral district in the Colony, and shows an enormous increase in the number of woolled sheep, despite the 'Vomeerziekte' which carried off a large number of sheep in the Doornberg and Karrebergen wards, and in the consequent increased supply of wool during the last ten years; the quality of which latter is daily being improved upon by the introduction of the best thoroughbred rams.

"Owing to the inferior quality and scarcity of water, wool is seldom washed in the division, but sent to the seaports in grease; recently wool-washing establishments were started at Zoutpoort and Boschduwar Fontein, and the result of the enterprise has proved most satisfactory, and will probably encourage others who may have the means of so doing, to go and do likewise. It is said that each bale of unwashed wool contains from 60 to 70 per cent. of dirt, on which carriage has to be paid at the rate of from 8s to 15s per 100 lbs.

"The Colonial Hydraulic Engineer has proceeded to Prieska, after inspecting the dams and some waterworks in this neighbourhood, and will on his way examine some of the Crown farms, with a view of noting the facilities there may be of constructing dams or reservoirs, —Rooipoort, in the Doornbergen ward, has been suggested as one such. . . .

"Crime has not decreased as compared with the year 1874."

The Hydraulic Engineer visited, as stated, Victoria West on his tour of observation, and reported to the Government:—"On my way between Victoria West and the Orange River, I visited a poort near the farm of Schilderspan, called Rooi or Roode Poort. It was pointed out to me that an embankment might be made across this poort, and a village laid out below it. The Brak or Ongars River passes through the poort, and, like the streams of that part of the country, runs very rarely, but when it does run it discharges a very large volume of water. The farm Roode Poort was leased in December, 1873, for five years; its area is about 5,000 morgen. The village could be laid out in such a position as to be quite out of danger in case of any accident to the embankment, and at the same time water could be taken in pipes or furrows from the reservoir to the village. There is good commonage on the hills round. The farmers who live near have as a rule wells or weak springs on their farms, but the water available is only just enough for stock. It is said that they would gladly purchase lots in the village, both building lots and arable land. The situation is about 3,500 feet above sea level, and is about 14 hours' ride from Hopetown; 18 hours' ride from Richmond; 15 hours' ride from Victoria West; 16 hours' ride from Carnarvon. Could sufficient water be stored, a wool-washing establishment might be set up, which would be a great benefit to the sheep farmers of the neighbourhood.

"I would venture to suggest that some competent surveyor be instructed to make an accurate survey, to a scale of not more than 400 feet to the inch, of the land above and below the poort, with longitudinal and cross sections. It would then be possible to estimate the amount of water that could be stored, the area of land that could be irrigated, and the cost of the work.

"The surveyor should take levels of any neks or places where an overflow could be made, and examine the subsoil by borings or by shafts. At a depression close to the poort which was suggested to me as likely for an overflow, there would have to be, I fear, some very expensive rock-cutting. If works are carried out, great care would have to be taken to allow a free passage for the river, and to protect the embankment against storm-water while the works are in progress, but I do not anticipate any insuperable difficulty from this cause if the works are entrusted to a competent engineer to carry out.

"A considerable amount of deposit in the reservoir is, I fear, inevitable."

SECTION III.—*Divisions of Prince Albert and Oudtshoorn.*

To the south of Beaufort and the east of the division of Worcester is the division of Prince Albert. It lies along the north face of the Great Zwarteberg range, and extends some distance into the Karroo. The farms at the foot of the mountains are said to be very fertile. It has an area of 3,980 square miles.

In Prince Albert, it is stated, £75 is paid for the right to one hour's run of water supply weekly from the town canal. And while a large piece of fertile ground, having no water right, but with a chance of sometimes getting water when the river is running to such an extent that the water cannot be used in the town canal, may be purchased in that town for £75 or £100; such a piece of ground, with the ordinary supply of water, limited as that is, allotted to it, would probably fetch £1,000.

Mr Molteno, being asked, in reference to the first of these statements, if he knew the quantity of water supplied from the source mentioned, said—"It is a good stream of water; I could not define it exactly; but that is the price frequently paid by one proprietor to another for the right to one hour's run per week." He was asked further:—

"9. Can you give us an idea of the difference of value between dry and water erven?—In the town of Prince Albert you might buy for £75 or £100 a large piece of fertile ground without any water right, but taking its chance of sometimes getting water when there is superabundance, when the river is running to such an extent that the water cannot be used in the town canal. Such a piece of ground with the ordinary supply of water, such as it now is, allotted to it, would probably fetch £1,000.

"10. *Mr de Wet.*] I suppose it depends upon the quantity of water: what is the rate they usually have?—I have mentioned the price paid for one hour per week; this is for one hour's run from the town canal, into which all the water running into the river is turned, except when the river is running stronger than usual.

"11. You cannot say how much water?—Sometimes the stream is stronger and sometimes weaker, according to the season. Each proprietor has a fixed time when he is at liberty to turn the stream on to his property."

From the reply of Mr Borcherds, Civil Commissioner, to the queries issued in 1862, there does not appear to be any very striking natural features of this district suggestive of the adaptation of the country for facilitating irrigation; but from the reply it may be gathered that there are localities in which water might be husbanded, by "damming up rivers or other running waters." And the importance of this being done when practicable receives some illustration from the evidence given by Mr Molteno before the Select Committee of the Legislative Council on irrigation, which I have cited.

Mr Deacon, one of my correspondents, residing in the vicinity of Meiring's Poort, writing to me of measures adopted by his father-in-law at Vondeling, on the Great Zwartbergen—a range of mountains with an average height of 5,000 feet, which is penetrated by the passes of Seven Weeks' Poort and this of Meiring's Poort—said:—"On each side of his dwelling-house is a large dam, the work of his own hands: the one, now about 1,500 yards in circumference and averaging twelve feet deep—the other a little smaller; when full, the former will be about a mile round. He has other large dams on his property; but his time is too much occupied with a large run of business, being a store-keeper, so that his extensive property of 40,000 morgen of most beautiful soil is, except that it is grazed by a flock of sheep, almost neglected. It is true it is being kept in reserve for a numerous rising family; but I do not hesitate to say, that, with the facilities in this part of the country for constructing dams, it might be made, with little outlay, capable of containing and enriching a population of hundreds of farmers."

Of the division of Prince Albert, the Civil Commissioner wrote in 1865:—"Prince Albert in common with a large portion of the Colony, is now passing through a very severe trial from drought.

"No crops have been grown, except in a few localities favoured with moisture, and the result has been considerable scarcity, if not actual distress, among many of the inhabitants of this division.

"Farmers have been reduced to great extremities to procure water for their stock; and the scarcity of that article, and the frequent removal in search of better pasturage, has caused heavy loss both in cattle and sheep, particularly in the latter. In addition to this, the condition of the flocks has been so poor that the clip of wool has been considerably below the usual average.

"Crime has greatly increased, principally in sheep-stealing, another consequence of the great drought."

In 1874 it was reported:—"Prince Albert, which, in common with other portions of the Colony, suffered from prolonged drought in 1865, has during the last year been efficiently relieved. The rains have fallen generally throughout the division, and there is every reason to hope that an abundant harvest will be reaped. The lambing season has been productive, and the increase, should no untoward circumstances arise, will go very far towards making up the loss in stock sustained in the two years' continued drought.

"The soil is in most parts adapted for extensive cultivation; but the want of water, or the deficient economising of water, is a serious drawback. The agriculturists have lately begun to see the utility of making dams; and it would be well if the farmers generally would follow the example of a few individuals who at little expense have made large dams. Energy, combined with no large amount of capital, thus applied, would render the division capable of producing a large amount of grain for export. At present there is much apathy on this subject.

"Fruit this year is abundant; but it is feared that, owing to the oïdium in the vineyards, the crops will be below the average. . . .

"But the favourable condition of things above alluded to has had no corresponding effect upon the amount of crime, for a comparison of the last with the three preceding years shows an increase in the number of offences punished, and especially those of theft of stock. Police cases have diminished, both relatively and absolutely, a circumstance which has been mainly owing to the disposition shown by the Licensing Board to lessen the means for that excessive indulgence in drink to which the coloured races in these parts are so notoriously addicted. The measure adopted for this purpose, and which has been suggested by the paucity of the police and their inability to suppress riotous drunkenness, has been the limiting of the retail of ardent spirits to quantities of not less than a quart bottle, which is not to be drunk on the premises. The following comparative return will show the number of criminal cases, etc., in the specified years:—

	1871.	1872.	1873.	1874.
Criminal cases,.........	117	118	156	150
Police cases,............	34	31	61	32
Thefts of stock,.........	12	21	8	17

And again in 1875:—"This district has been blessed with a good and abundant harvest, high prices have been maintained for all sorts of produce, and it has reaped the full benefit of a main thoroughfare passing through a considerable extent of it, which affords a ready outlet for all its superfluous productions.

"The vintage of the past year, as well as the crops of maize, dried fruit, etc., has been excellent, while, as affecting the future, the vineyards and fruit trees are at present well laden and look promising. Good soaking rains have fallen at intervals, extending nearly over the whole district; the 'veld' has been generally good, and, as a natural consequence, the flocks have been maintained in a fair condition, and the lambing season has been favourable."

In the report made by the Hydraulic Engineer of his tour of observations in 1875, in regard to Prince Albert he reported:—"At Prince Albert there is little or no Crown land except what is high up on the mountains, and is not suitable for irrigation. The inhabitants of the village however have a fair supply of water, and they are anxious to improve it. I understand that the run of the furrow for one hour once a week, which was worth as much as £75 per annum in 1862, is now worth £100 per annum. I do not know how many gallons this represents as I had no time to gauge the furrow, but I understood that such an amount was barely enough for a very small garden.

"And here I venture to note the inaccurate way of estimating water common in this colony. People continually talk about a fountain or a furrow giving as much as a pipe of such and such a diameter would deliver, forgetting that the discharge of pipes depends not only on the cross section but also on the 'fall' or 'head' as the case may be. Thus a pipe of any given dimensions laid on an incline of one foot in the mile will only throw half the quantity that the same pipe laid with a fall of four feet in the mile will throw.

"I visited the kloofs behind the town, and I think there are one or more suitable sites for making storage reservoirs.

"The Civil Commissioner, Mr J. N. P. de Villiers, was of opinion that if the Government would take the initiative and make a large reservoir, it would enlighten the farmers, and bring under cultivation five or six times the land they work now. I am inclined to think that as the whole of the land is in private hands, such a work would be done much more economically by a local company, or at any rate by a board of local proprietors, of course with proper professional advice. I think that Government might advance a sum towards such purpose, such sum to be always less than half the value of the work done; and the interest on such loan to be a first charge on the properties benefited.

"At present I can form no estimate of cost. A careful survey of the valleys above the town with longitudinal and cross sections should first be taken. The best positions for embankments can then be chosen, after which a careful examination of the ground on the lines of the proposed embankments by borings or shafts would have to be made."

Much of the territory in both of the divisions described in this section is included in the Great Karroo; and, as has been intimated, in the Karroo the scarcity of water is great. I am told that along one or more of the principal routes by which it is crossed, tanks for the retention of water have been dug at several outspan places. When I crossed it in 1847 it was otherwise.

In the Karroo, as has been stated, thunder-showers are frequent; and the passage of one of these across the country causes, in a short time, a district to be clothed with verdure and flower, which shortly before was barrenness itself—such as called forth, some fifty years ago, the remark by the Rev. John Campbell, cited elsewhere—"Hech, Sirs! It would take a good pair of spectacles to see a blade of grass in this world. Its as bare as the back of my hand!" Yet that soil apparently will yield, under cultivation, with moisture abundant, crops of fruit and of culinary vegetables of a superior character.

Throughout the districts which have been described much of the water is brackish. On this subject, Captain Hall writes:—"These brackish waters occur in all parts of South Africa, both along the coast and in the interior, and are often a source of bitter disappointment to the thirsty traveller, being very often of a beautifully clear and crystal appearance. Habit, however, reconciles the Cape colonists to these bitter waters, so that often, like the use of salt, really fresh and sweet water appears insipid to a farmer who has been quaffing the saline draughts of some Pekel Fontein or Brakke River. On some soils it is believed brackish water can be usefully employed in irrigation, especially those formed by the decomposition of the old quartzite rocks."

To the south of the division of Prince Albert, lying between that division and the division of George, which extends to the sea, is the division of Oudtshoorn. It "is situated at the west extremity of the valley of Olifant River East; bounded on the north by the Great Zwarteberg range, on the south by the Outeniqua Mountains and Lange Kloof, on the west by the Guaritz River, and on the east by the Antonies Berg. It has an area of 1,980 square miles. It possesses a most fertile soil, producing the best brandy and tobacco in the Colony, and has great facilities for irrigation."

The inhabitants of the division are mostly engaged in agriculture, and wheat and tobacco are the chief staples. By the census of 1865 Oudtshoorn ranked second only to Malmesbury in the quantity (172,085 bushels) of wheat produced; and of 1,560,875 lbs. of tobacco raised in the western division, 966,641 lbs. was returned from Oudtshoorn. A few miles north of the town of Oudtshoorn are the celebrated Cango caves, which have been described in a previous chapter (ante p. 141).

In reply to the queries issued in 1862, Mr Scholtz, then Civil Commissioner, reported that meetings had been held in each field cornetcy of the district, in order to ascertain the sentiments of the inhabitants upon subjects connected with irrigation, and the reports received from the field cornets had been laid before the divisional council; and that the replies given by that body were—that no plans for irrigation works had been proposed in that division, and that the council was unable to state what irrigation works—such as the damming up of rivers or other running water, by diverting or raising water from such sources by the construction of dams in favourable localities, by the collection of periodical supplies of water, or by the sinking of wells—appeared practicable in the division.

But by the Select Committee of the Legislative Council on irrigation much additional information was elicited. By the Honourable Mr Barry, a member of committee, it was stated, in proposing one query, that there are localities on the banks of the Olifant's River in which farms with frontages to the river are valued at £5,000; while at the back of them a large extent of the same kind of land is valued at £1,500. And the following testimony was borne by Mr Walter, M.L.A., resident at Oudtshoorn :—

"115. *Chairman.*] You are a member of the assembly for the district of George, and have long been resident there?—Yes.

"116. Could you give us an account of the district about Oudtshoorn and the Olifant's River?—It is a very productive district. Much produce is raised there, especially wheat and tobacco. The vine is largely cultivated, and a considerable quantity of brandy made. Wine is also made of a very superior quality, but the farmers have not, as a general rule, given attention to wine-making, though I have had opportunities of seeing most delicious wine made there. It is a country wholly supported by irrigation, and the farmers are enabled, when they can obtain water, to get two crops a year. In fact, they all depend upon two crops; but, during the last two or three years, the *na-oost*, as they call it, has been very inferior as regards returns. There are some fine rivers in the district—such as the Olifant's River, the Kamnasie, and the Grobbelaar's River, and irrigation is carried on very extensively on their banks. There is also a very large population in that part of the country.

"117. Do you know the difference in value between land which can be irrigated and other land?—Well, I should say, as a general rule, land that can be irrigated is worth about £50 per morgen. Of course, it varies, in proportion to its quality, and other circumstances. Since I have been down here I have heard of the thirty-second part of a farm being sold for £900. I do not know the exact number of morgen, but I should imagine about two thousand or two thousand five hundred; but, I must observe, it is a divided farm. Ground has risen considerably lately in that part of the country. Erven in Oudtshoorn, which, about ten years ago, could have been purchased for £20 or £25, are now selling at £300. These erven are about two morgen in size. There are other water erven in the town, about the size of one morgen, and these have been readily sold at £200. It is a curious fact that formerly the farm upon which Oudtshoorn is now established was so much in want of water,—as I have been informed, though I do not know it personally,—that the proprietor had occasionally great difficulty in raising his crops; but the abundance of water we now have there is mainly owing to the damming up of the river, which is done to a very considerable extent between Oudtshoorn and the Cango. The whole distance between those places, and, in fact, in the Cango itself, the river is dammed up very considerably. This prevents the rush of water to the sea, and it has been found that the more you dam up a river the more water there will be.

"118. What is the value of the other land not irrigated?—Many circumstances must be taken into consideration; but, perhaps, generally speaking, you may say ten shillings a morgen, and from that to a pound.

"119. I mean land of the same quality?—Yes.

"120. *Mr Wood.*] What would you value this land at, which you speak of as being £50 a morgen, if there were no water accommodation? —I should not consider it of much value, £2 or £3 a morgen. Irrigation might be carried on to a much greater extent in that part of the country, provided parties who have not the advantage of river frontages were allowed to lead water over their neighbour's land. . .

"125. *Chairman.*] Is there much land situated behind the farms having river frontage; and can you form any idea in what proportion to the land at present irrigable?—I should say that one per cent. is irrigated.

"126. *Mr Wood.*] Is the land, not now irrigable, of equal quality with that which is irrigable?—Quite so.

"127. *Chairman.*] Have you any idea of the quantity of grain, tobacco, wine, etc., obtained from irrigable land in proportion to the quantity sown?—That will depend, of course, on the nature of the soil.

"128. Have you any idea what quantity is obtained in the Olifant's River valley from one muid of wheat sown!—Not exactly.

"129. Nor have you any idea of the quantity of wine made from 100 vines!—No.

"130. Could you obtain some information in these respects before the committee finishes its labours?—I will endeavour to do so.

"131. *Mr de Wet.*] Have you an idea, however, that in the event a party went to the expense of having his land irrigated the increased produce would amply repay?—There is no doubt about it; water is the only requisite.

"132. *Chairman.*] What are the sources from which the Olifant's River derives its water?—I have not been as high as the sources of the river; but of course the main feeders are from the Beaufort district. There is the Touver Water and Meiring's Poort River. These are the principal ones, though there are several minor streams.

"133. Are these feeders made use of for irrigation?—I think not, with regard to the Touver Water.

"134. What, in your opinion, impedes further cultivation?—The farms being held in partnership, and the difficulty I have already mentioned in regard to leading out water from the rivers. In my opinion, however, the main cause is the partnership properties. I have known of instances, lately, where properties have been divided and improvements carried out to a very considerable extent; large dams have been constructed and fine buildings erected, which could not have been done without such sub-division.

"135. *Mr Barry.*] But these sub-divisions are increasing daily?—Yes, though not to any great extent.

"136. *Chairman.*] Do you not think that want of capital is an impediment?—In many cases, no doubt.

"137. Do you think engineering works could be constructed so as to increase the supply of water to any great extent?—I do not know of a farm where dams might not be easily made. I know of no large works that could be constructed in any part of the division of George; all would be works of minor importance.

"138. *Mr Barry.*] If dams were erected so as to furnish water on each of these farms on the frontage of the Olifant's River, do you think there is a sufficient supply of water to extend down the whole line of frontage?—At times there would be, no doubt; but in the summer months of course the supply would be materially diminished by leading out the water in the upper part of the river.

"139. But by making much larger embankments on the river than the present insignificant ones, you would increase the water to each farm?—Yes, of course; the larger the embankment the more water is thrown upon the farm; but that would not increase the quantity of water. . . .

"148. Do you happen to know whether the water is harvested to any extent in the district in which you reside?—No, it is not, to any great extent.

"149. Do you think, if it were properly harvested, a much greater supply would be secured?—Undoubtedly; but here again there is a difficulty, owing to partnership property. If a man has a divided share of a farm, of course he can make such improvements as he finds necessary, but in some cases where a man is only entitled to his water once in six, eight, or ten days, as is frequently the case, he is obliged to husband his water for that period.

"150. Is it expensive to keep these dams and water-leadings in order?—No; I do not think there is any great expense. Of course some water-courses are larger than others, and therefore the expense will be greater in cleaning them and keeping them in repair.

"151. Can you give the committee an idea what the annual cost is, taking into consideration the making of the dams and the water-courses?—The expense is not material, I think. Now and then the farmers are employed two or three days with their people in cleaning the water-courses.

"152. From what you have stated as to the manner in which the farms in your district are situated as regards the river, it appears that a large number of farmers are only able to irrigate a small portion, as a rule?—Yes.

"153. Do you think, from your experience, that if these farms could be watered to a much greater extent by means of irrigation, the owners would object to a tax for such an advantage?—I do not think they would. They would be very glad, in my opinion, to pay a reasonable sum for an additional supply.

"154. You have also stated that the district is largely peopled?—Yes.

"155. How do you account for that?—It is mainly owing to the fertility of the soil.

"156. And the products raised?—Yes, the rich returns yielded.

"157. And, as a colonist, is it your opinion that if larger amounts of that country could be put under irrrigation population would increase in the same ratio?—There is no doubt of it. . . .

"169. Would you give us some information with regard to the other parts of your district; the farms situated between the Gamka and the Dwyka?—I am acquainted with that part of the country, and the observations I have already made equally apply to it. With regard to that part of the country south of the Outeniqua Mountains, formerly, you are aware, drainage was necessary in some parts, owing to the great quantity of rain we had then; but of course during the past season that part of the district has also suffered very materially from drought. When I first went into the district, several years ago, sheep farming was not carried on to any great extent in Outeniqua-land, on account of the various sicknesses they were liable to. If a farmer increased his flock in one year, he lost it the next; but now sheep get on better since there has been less rain.

"170. But even where drainage is necessary, in the moist climate of England, for instance, you are aware that irrigation for meadows is practised?—Yes.

"171. *Mr de Wet.*] You are of opinion that dams could be made in the Karroo?—Yes. There are some few farms, perhaps, where you might have difficulty in leading out the water; but there are deep kloofs, where the water might be stored for the supply of cattle. I do not mean to say that on every farm irrigation could be carried on; still there are few farms indeed where it could not, even in the Karroo.

"172. *Mr Barry.*] Is there still a large quantity of Government land about the Olifant's River to be disposed of?—Not much; there is a considerable extent between Georgetown and the Knysna. . . .

"180. *Chairman.*] Could good tobacco in large quantities be produced in the George district?—Large quantities are grown at Olifant's River. Tobacco might also be raised in Outeniqualand, that is George Proper; but Oudtshoorn is the great tobacco-producing country. Some very good specimens were produced at the recent agricultural show at George."

In connection with this testimony, I may again introduce the statement already made to the effect that from notices in journals of the day it appears that Mr W. C. Guest, of Oudtshoorn, has discovered and opened up a stream of water on his farm of Leeuwinblad capable of giving 400,000 gallons of water every twenty-four hours. It appears to be a permanent stream, and has increased the value of his place at least £1,000.

And for further information in regard to what has been done I quote the following letter, which appeared some time since in one of the Capetown journals, illustrative of the extent to which this one district of the Colony might be benefited by a system of irrigation. The writer says:—"The object of the present letter is to direct attention to a whole division, which in its whole length and breadth could be made the finest field for irrigation, and become a real source of prosperity for the Colony at large. The great granary for the surrounding parts of this division is Oudtshoorn, situated between the ranges of the Outeniqua Mountains and the Zwartbergen—adjoining the district of Prince Albert to the north, George and Mossel Bay to the south, Riversdale to the west, and Uitenhage towards the east. Until the making of Montagu Pass and of Meiring's Poort, the access to and outlet for this important and fertile part were much impeded, and the cultivation naturally much more restricted than even at present; yet its great value was known. It is solely an agricultural district; and the whole actual produce is dependent upon and secured by 'irrigation,' conducted in a most primitive state, and impeded and encumbered by many drawbacks of various sorts. The Olifant's River runs from east to west in the middle, through the division, for a length of about 100 miles, and is fed by numerous smaller rivers or streams running into the main channel from both sides of the mountain ranges, thus forming an inviting field for large irrigation purposes. Along those rivers, and in the flats, between the same, a large number of farms have been established; and, for upwards of fifty or sixty years, the slow but gradually increasing cultivation of those farms has been going on, and water has been led out for carrying this out. The lands along those rivers consist for the greater part of a rich clay soil, the result of the overflow of the rivers for hundreds of years; ground where the mimosa luxuriantly grows, and where thousands and thousands of acres are yet found to be uncultivated. To give somewhat an idea of the fertility of those grounds, and of the extent to which cultivation may be brought, if only the great want of preserving water is secured, I submit the following particulars, the result of local experience and information :—

"For about forty or fifty years the district was only occupied by a few farmers, each the proprietor of one or more farms, easily obtained from the Government on the then usual terms, and for many years only cultivated to provide for each farmer's wants. Gradually those farms have become more and more populated; grown-up children have erected homesteads and cultivated lands; and there are now places thus sub-divided and made so valuable as to be appraised for road purposes at from £4,000 to £18,000.

"One farm, well known to the writer of this memorandum, was some twenty-five or thirty years ago the property of a single farmer —since bequeathed to his children, and yielding, for some ten or twelve years, every year 2,000 muids of wheat, besides having large vineyards and gardens.

"Another farm, above alluded to, now valued at about £18,000, was found some years ago to have brought up in one year—in brandy, wine, wheat, tobacco, whipsticks, raisins, and other fruits—not less than £6,500.

"A third farm, abandoned some fourteen or fifteen years since by its owners, who left for the Free State, and at that time uncultivated, was bought by the energetic and active missionary, Mr Anderson, with eighteen of the best men of his congregation, and now provides for perhaps upwards of 300 to 400 inhabitants.

"The village of Oudtshoorn itself, now a large place, fit for cultivation all along the borders of the Grobbelaar's River, a tributary of the Olifant's River, was some twenty or thirty years ago occupied by only one family, and cultivated to a very limited extent.

"These few instances could be multiplied *ad infinitum*; and the history of every farm in the extensive division of Oudtshoorn, now bearing a population of 12,000 inhabitants, would be conclusive evidence of the fertility of the locality, provided water is properly secured. And yet this very same division has been suffering during the last year, perhaps, more than any other, certainly more than ever before; and although some of the more fortunate farms have produced abundantly, others, and some of those above mentioned, have not produced sufficient to provide for their own wants."

In 1865 it was reported in regard to Oudtshoorn :—"The harvest of wheat, barley, and oats in January was exceedingly good and plentiful; and the wheat sold at low prices, from 12s to 10s per muid; but no rain fell from April to July, and the field-cornetcies of Upper Olifant's River, Lower Olifant's River, part of Grobbelaar's River, part of the Cango, and Gamkas Vlakte, were unable to plough. Consequently very little wheat, barley, or oats were sown; and upon some of the farms the crops were lost. The prices have already risen, —for wheat to £1 5s a muid, barley and oats to 15s a muid, and for the 100 bundles of oatsheaves, £1 10s to £2. No mealies, beans, or pease have been sown. No potatoes, sweet potatoes, or tobacco have been planted, and the rivers are actually dried up.

"The farm Armoed and the surrounding farms at Kandelaars River have been without rain and without water for upwards of twelve months. The well-known farms Oude Muragie and Riet Vallei, in the Lower Olifant's River, near the Meiring's Poort, some of those of the Upper Olifant's River, the farm Welbedagt, and all those along the Wynand River, and in the ward Gamkas Vlakte, including the village Calitzdorp, have no water to irrigate their gardens or trees, and the drink water is obtained by digging in the rivers.

"The *oïdium* is not so bad as last year; but the grapes have been destroyed by the heat of the sun, and have suffered from the want of being watered. The fruits on the trees are small, and cannot come to perfection; and thousands of birds, seeking for food, are destroying the half-ripe grapes and other fruits. The orange trees are dying; the vegetables in the gardens of the village have actually been burnt by the heat of the sun, and are destroyed; and if the drought continues much longer there will be very great suffering.

"The Grobbelaar's River, which supplies the village of Oudtshoorn, failed in December; the water-courses are dry, and people have recourse to wells for obtaining water to drink. A number of wells have already been dug by the farmers, for obtaining water for themselves and for the poultry. Some have succeeded in getting good water; and others not even at the depth of thirty-four feet.

"A large number of oxen, horses, goats, and sheep have died for want of pasture, and by drinking bad water in the rivers, which has become salt. A large number of goats have also died from the disease called brandziekte. Those farmers who can afford it have sent their oxen, horses, and mules for pasture to the George and Knysna divisions.

"The cattle are in so poor a condition that the farmers are unable to proceed on a journey with them; and there is no food on the road for them."

In 1866 it was reported:—"The drought in the year 1865 was so severe, that scarcely any wheat, barley, or oats were sown, in consequence of which wheat was sold from £1 to 16s 8d the bushel, barley at 7s 6d the bushel, and oatsheaves from £1 to 16s 8d the hundred pounds.

"The Camnatie and Grobbelaar's Rivers were the only two rivers which had running water in 1865.

"The number of days in which rain fell during the year 1866 were:—In January, 4 days; in February, 2 days; in March, 1 day; in April, 3 days; in May, 2 days; in June, 2 days; in July, 2 days; in August, 1 day; in September, 2 days; in October, 3 days; in November, 3 days; in December, 1 day. Therefore this year a great deal of grain was sown; but, owing to the rust having appeared both in the wheat and oats, and to some of the farmers having already tramped their wheat, it is feared that the latter will rise again to a high price. The oats are more affected, and the sheaves are unfit for forage for horses.

"The *oïdium* has again destroyed the grapes, so that very little brandy and wine were raised.

"Tobacco was in abundance, as also whipsticks. The tobacco fetched good prices, but the whipsticks were cheap.

"The fruit and orange trees suffered much for want of water.

"The cattle were in good condition. The sheep were fat, and selling at the usual prices. A great number of horses were brought in from other divisions, and sold at reasonable prices.

"Landed property has fallen in value, particularly the erven in the village. There is no building going on."

Of Oudtshoorn in 1869 it was reported:—"The rains, which in the commencement of the year were rather scanty, increased as the season advanced, and culminated in the destructive flood of November last, common to the whole Colony, and which in this division resulted in the destruction of the Karroopoort and the Toverwaterpoort roads. The former, constructed by the Divisional Council of Oudtshoorn some eight or nine years ago at an expense of upwards of £2,000, led into the fertile valley of the Cango, and gave access to the celebrated Cango caves. This road has been entirely destroyed by the flood; but has been rendered passable by the exertions of the Cango farmers. The Toverwaterpoort has just been completed at the expense of the Council, and opened up a short means of communication between this division and the northern and eastern divisions of the Colony. It is feared that from the depressed state of the country, and the difficulty experienced by the farmers in paying their road rates, the Council will not be in a position to restore these two very useful roads for some time to come.

"The newly-constructed Robinson's Pass (Ruyterbosch), which connects this division and Mossel Bay, was also partly destroyed by the flood, and the traffic is still interrupted.

"Meiring's Poort, which connects this division with Beaufort, also suffered seriously; but by the exertions of the Divisional Council of Beaufort, was soon rendered available for traffic.

"Serious damage was done to the fruitful lands in the valleys of the Grobbelaar's River (village of Oudtshoorn), Weynand's River and Buffel's River (Calitzdorp), the soil being washed away, and the crops destroyed and carried off. Several lives were also sacrificed to the fury and suddenness of the flood.

"The seasonable rains, however, just before harvesting time proved a great boon to the farmers.

"The yield of wheat, oats, and barley has been large, and of excellent quality. Prices, however, range very low. Rust hardly made its appearance, and that at a late period, doing comparatively little damage. The late crops bid fair to turn out well, though it must be observed that the rains have ceased, and that water for irrigation is becoming scarce.

"The temperature has been, up to the end of the year, unprecedentedly low, such a mild summer not having been experienced for the last ten or twelve years. A light fall of snow took place on the Zwartberg range on the 15th of December, and there was frost in the Cango at the same date.

"*Oïdium* exists generally in the vineyards in the division, but not to a serious extent, and the farmers anticipate an average yield.

"The cattle are at present in better condition than is generally found at this time of the year. A move is being made to breed angora goats; several thoroughbred rams and ewes have been introduced into the district; and the Divisional Council are making strenuous exertions to eradicate the *Xanthium spinosum*, which has unfortunately gained a footing in the district. This division is admirably adapted for the grazing and breeding of goats, and it is hoped that the time is not far distant when Oudtshoorn will add to its already valuable and varied products, mohair.

"Fever has been very prevalent among the coloured people, leaving whole families quite destitute, and the Government has had to support them with pauper rations.

"Crime has decreased to a great extent, owing to the plentifulness and low price of provisions, and in a great measure also to the Government regulations regarding spare diet."

In 1874 it was reported:—The report for the year 1873, in reference to this division, may be applied, with a few alterations, to the condition and circumstances of the past year (1874), which has been more prosperous than its predecessor, inasmuch as there has been an absence of drought, and crops of every kind have been abundant, and produce has maintained high prices.

"Crime has been less, notwithstanding the high price of the necessaries of life. There has been less drunkenness and not a single case of sheep stealing or cattle lifting."

And in 1875:—The year 1875 has, on the whole, been a prosperous one; the grain crops have turned out well, although the wheat has in some parts of the division been slightly affected with rust. The rainfall has been considerably more than during the previous year, and has proved most beneficial to farming operations in general, and stock is in good condition owing to the abundant pasturage.

"In consequence of the heavy rains which fell in August and September the roads, both main and divisional, were very much damaged, and a large expenditure has been incurred to restore them to their previous condition.

"The present rate of assessment on the value of fixed property for road purposes, which for several years past has been levied—viz., one halfpenny in the £, is insufficient to keep these thoroughfares in anything approaching the condition they should be in to meet the requirements of the large and increasing traffic carried on, and little improvement may be expected unless recourse is had to increased taxation, a measure not likely to be received with favour by the community generally, nor with any support in the Divisional Council.

"The works at Caledon's Kloof, which had been temporarily stopped through want of funds, are now being proceeded with under the superintendence of a committee appointed by the Divisional Council of Oudtshoorn, a grant of £200 having been obtained from Government in aid of the cost of construction.

"The opening of this mountain pass for traffic, it is hoped, may prove of great benefit to the community generally, more particularly to the inhabitants of the Ladysmith District, and will also be the most direct road between Oudtshoorn and Capetown, *via* Ladysmith, Robertson, Worcester, etc."

In the report of the Hydraulic Engineer of a tour of observations made by him towards the close of 1875, it is stated:—"At Oudtshoorn and at other places on my route up country, I visited various works of irrigation, and I am of opinion that in most cases more water might be made available, and that considerable improvements might be introduced into the reservoirs, furrows, and other works. It was pointed out to me that the great drawback to making improvements is the want which the lower proprietors experience of some legal means of obtaining a right of water-way through the lands of upper proprietors. The storm-water seemed especially neglected for this reason.

"In my letter to you of December 28th I offered some suggestions on this subject. In your reply of January 12th you were good enough to point out some of the legal difficulties involved. You mentioned that irrigation works on the scale likely to be undertaken in this colony are of a private nature, and consequently have not the same claim to a right of way as railways. This is very true, still, in Italy, Spain, and other countries, there is a power of expropriating lands required for irrigation channels and other works, even in the case of works which, in the first instance, only benefit private individuals. You have pointed out that the Italian *diritto d'acquedotto* is only given to persons who have legally the right to the water.

But there is no difficulty in settling to whom the water belongs, for if no definite lease or grant can be shown, the water-rights belong to the Government.

"Of course, if the upper proprietor can show a legal right to the whole of the water from a stream, then it would not be desirable to take the water from him unless some great public benefit could be gained, say the water-supply of a village already established. In the case of the lower proprietor having the right to the water, I believe you intend to propose that he should also have a right of way through the upper proprietor's land for that water. But I venture to submit that there are probably many cases in which neither the upper nor the lower proprietor has any legal claim to the whole water. I think this a case in which perhaps the Italian practice might be followed, and Government have the right to dispose of the surplus water by lease or otherwise.

"In this matter associations of persons with local knowledge and local interests would be of great service. As you mention, such associations have existed in Northern Italy for a long time, they were also common among the Moors in Spain six hundred years ago. I quite agree with you that such associations, if recognised by the Government, would give a considerable impetus to irrigation and land improvement generally. They would be almost indispensable in the case of works that affected a long line of river or a large area of land.

"The considerable fall of the Grobbelaar's River at Oudtshoorn naturally favours the taking out of the water; and I think sites could be found where reservoirs for the storage of surplus or storm-water could be constructed. There are I believe no Crown lands that could be benefited. I understand that the inhabitants do not ask for pecuniary aid from the Government, and I am told that capital would be found wherewith to carry out considerable improvements if the difficulties as to the right of water-leading were removed.

"One point which I noticed about irrigation at Oudtshoorn and elsewhere, was that the use of inverted siphons for passing water under roadways, such as are very common in North Italy, is apparently unknown here. Instead of which, whenever it is necessary for water to cross a street or road, an open ditch is left, causing, among other inconveniences, an uncomfortable jolt to every carriage that passes. At some towns these nitches are covered over with planks, but a brick or pipe culvert at sufficient distance below the roadway would be more lasting and more convenient."

CHAPTER IV.

Supply of Water and Facilities for the Storage of it in Divisions along the South Coast of the Colony of the Cape of Good Hope.

Section I.—*Divisions of Caledon, Bredasdorp, and Swellendam.*

The divisions of the Colony which have been under consideration may be called the western and adjacent divisions. Leaving these to follow the south coast-line of the Colony, or the great eastern road from Capetown to the Eastern Province and Kaffraria, the first division coming under consideration is Caledon, situated behind the first mountain range, and, with the division of Bredasdorp, comprehends the country between the ridge of the Zonder End mountains on the north and the ocean on the south. The division of Caledon is a fine undulating country, especially adapted for sheep-farming, and timber appears to thrive well. It has an area of 1,700 square miles. In certain parts near the coast are several large vleys, or natural collections of water. The principal river is the Zonder End, flowing east from the coast mountains into the Breede River.

From the reply of Mr Haw to the queries issued from the Colonial Office it appears "that a very considerable portion of this division consists of what is termed *ruggen veldt*, which is very scantily supplied with water." And in a subsequent reply he quotes and answers the queries thus:—"1. Have any, and what, irrigation works within your division been proposed?—Although irrigation works within this division would be of great advantage, I am not aware of any having as yet been proposed as a public measure. Those farmers who own land where water is available generally turn to the best advantage the facilities within their reach, to the extent of the means at their disposal. On one or two farms along the River Zonder End machinery has been made use of to a small extent to raise water, which has proved very successful. I may instance the farm of J. C. Linde, Esq., J.P., where water has been raised to a very considerable height from the bed of the river just named.

"2. What irrigation works—such as by damming up of rivers or other running waters, by diverting or raising water from such sources, by the construction of dams in favourable localities, by the collection of periodical supplies of water, or by the sinking of wells—appear to you to be practicable in your division?—By damming up mountain streams, which do not generally run very deep, and thus diverting their course so as to be made available upon land below, I have no doubt that a very considerable additional extent of land might be brought under cultivation in this division. From the mountain range known as the Zonder End numbers of these streams issue, and in some instances flow to a distance of two or three miles before they become lost in the River Zonder End. These are the streams to which I allude as capable of being turned to great advantage. Most of them are permanent, excepting in droughts, such as we are now suffering from, but which may not occur again for several years, and the expense of diverting their courses would be but trifling.

"Several small rivers in the division might, at certain places, be also dammed up, and the water turned to advantage. I may instance the Bot River, Klein River, and River Zonder End. But as these rivers sometimes become flooded to a large extent, and run with great rapidity and force, great care would be necessary in making selections of spots at which to erect dams. The waters in these rivers are in most places far below the level of the adjacent lands, and, consequently, the formation of suitable dams would become a matter of great expense. I am more disposed to think that machinery for raising water, worked either by wind or horses, would be found more practicable, much less expensive, and less liable to destruction, along the rivers just named, than the damming up of the entire streams.

"By the construction of dams in favourable localities, of which there are numbers in this division, great quantities of water, now allowed to run to waste in rainy weather, would of course be collected, and be turned to good account. I think the want of means among the farmers has been the cause why dams of sufficient extent have not been formed. Upon many farms in what is termed the 'ruggens' in this division, you find small dams or tanks formed by the farmers for their own household use, but not of sufficient extent to contain water enough for half the year. This appears to me clear proof that if their means were larger they would gladly form larger dams. The sinking of wells might also in some localities be found to answer. It is, however, frequently very difficult to hit upon spots where water might be obtained at a reasonable depth from the surface. I am not of opinion that much benefit would be derived from this mode, as most of the water obtained far below the surface in this division is found to be brackish, excepting along the course of fresh-water streams.

"Before any action is taken in this matter, I am of opinion that it would be necessary to obtain the report of an officer of engineering qualifications."

And the Civil Commissioner reported at the close of 1865:—"The severity of the drought, during 1865, has been far beyond anything of the kind experienced in this division for many years. The few dams that have been made have become in some cases quite dry, and in others nearly so. Springs which have never been known to fail have ceased, and others have become so weak as to be of little use for irrigation.

"The want of dams is much felt. Farmers, either from want of funds or inclination, make them so small that they are incapable of holding sufficient water for times of drought. Endeavours have repeatedly been made to impress upon them the necessity of increasing the size of their dams, but with little success.

"No rain of any consequence has fallen for some time past; and the veld is now (8th Jan. 1866) so dry, that in the course of a month or two, should no rain fall, the little grass which now remains will be carried away by the winds. Altogether the prospect is bad for the farmers.

"The crops of 1865 have been far below the average of any recent year. In some instances the farmers have barely recovered the seed they have sown, especially in oats. The barley crop is equally bad.

"The wheat crop is likely to turn out better than those of oats and barley; but this is also very poor, in some cases giving a return of not more than six bushels to one. It is the general opinion that during 1866 the supplies of wheat and other cereals will have to be procured from Capetown."

Of 1866 it was reported:—"During 1866 a greater quantity of rain fell than had fallen for several years previously. The sowing season appeared most propitious, and those farmers who could command seed sowed to a much greater extent than usual. There were many small farmers in the division, who, from poverty, were unable to purchase much seed; these were, however, assisted by others, and thus enabled to plough and sow enough for their own consumption.

"Owing to frequent rains in the early part of the season, the rust has in several parts of the division done a great deal of damage to the wheat crop, and whole fields have been lost thereby; but, all things considered, the farmers have not much reason to complain.

"The rust extended also to the oat crop. It damaged the straw, but the grain itself has not been much injured. The yield of oats has been far beyond the average.

"The barley crops everywhere have been good; the grain has been full and healthy, and the quantity expected is likely to be far beyond the yield of several previous harvests.

"The veld has been good throughout the year; and stock, especially sheep, have remained in good condition. The lambing season has been very prolific; and few deaths have occurred, excepting from ordinary causes.

"It does not appear that there has been any sickness among the horned cattle during the year. The lung-sickness seems to have vanished entirely at the commencement of the year (1866). Horse sickness made its appearance from the east, and carried off a good many horses and mules, especially in the field-cornetcy of Uilenkraal. Instances also occurred here and there in other parts of the division.

"The farmers have had reason to congratulate themselves upon the result of the wool season. The yield has been good, and the prices obtained fair; and many farmers who were drifting into debt have been enabled to recover their former positions.

"The vine disease has again appeared in several places, but not with that virulence which raged some few years ago; and no doubt the timely use of sulphur will, in a great measure, shield the fruit from material damage.

In 1869 it was reported:—"The present harvest, including all kinds of grain, has been most abundant, and the yield in many instances far above average, and all things which depend upon natural causes are prosperous and encouraging. Two or three more seasons like the present will tend much to advance the prosperity and well-being of our farmers.

"The criminal calendar shows a decrease of crime. Owing to the abundant supply of food much temptation to steal has been removed, to which an indigent population is often exposed through want, as the majority of thefts committed in this district, I believe, emanate from this source. Pilfering is a crime inherent in the Hottentot and all the coloured races in this Colony; but crimes of any magnitude are of rare occurrence.

"The general health of the inhabitants has been good, and it is constantly remarked that Caledon is one of the healthiest spots in the Colony, owing, no doubt, to the mildness of the atmosphere and even temperature.

"The Caledon hot baths, noted for the efficacious and beneficial qualities in all cases of cutaneous and rheumatic disorders, as well as complaints of a chronic nature, are kept in excellent order by the proprietor.

In 1874 it was reported :—"The district of Caledon, owing to the fine rains that fell in due season during the past year, yielded a larger amount of grain than for many years previously, and hopes were entertained that food for man and beast would be supplied at considerably cheaper rates than during the preceding year; but, unfortunately, just as the crops had been gathered in, and before the 'stacks' had been completed, a continuous rain set in, lasting for several days, saturating the unfinished hay-stacks, and damaging to a large extent a great quantity of wheat and oatsheaves, so that instead of obtaining oats and barley at six shillings and seven shillings a muid, as was anticipated, double that sum will have to be paid for these articles of consumption, which those keeping horses for hire or pleasure will feel to their cost.

"As this is acknowledged to be one of the best sheep-walks in the Western Province, it is satisfactory to state the season was favourable for stock-breeding, and large additions were made to the flocks, as the ewes were able to rear nearly all the lambs that were dropped.

"The farmers were also well satisfied with the prices obtained for their wool, and the younger members of the family, who two or three years ago could scarcely afford to buy a decent suit of clothes, now turn out in fashionable attire, riding sleek and prancing horses, which they think nothing of asking £40 or £50 for. An animal of the same description could have been purchased five years ago for half those amounts. . . .

"More cases have been disposed of in the magistrate's court during the last year, both civil and criminal, than for some years previously, but mostly trivial offences. No crime of any magnitude was brought to light, and only three cases of sufficient gravity to justify their being sent for trial before the Circuit Court."

In a valley north of the River Zonder End is situated Gnadenthal, a Moravian mission station, the first founded in South Africa, and the missionaries have done much successfully to train the natives to industrial habits, instructing them in various handicrafts. At another of their stations, Hemel-en-Aarde, a station for lepers, so early as 1823, the brethren had commenced the construction of irrigation works. In the report for that year, Brother Leitner writes:—"With the assistance of those patients who are able to work, I have constructed a little aqueduct, about 600 paces in length, by which the water from the River Unrust, proceeding from the Altaquas Kloof, is brought to our place, and made to pass between the houses and garden; there is enough of it to turn a mill." *

I cannot withhold the following evidence that the enterprise at Hemel-en-Aarde was only in keeping with the endeavours made by these brethren to develop self-reliance and self-help in those who were brought under their influence. In a letter from Gnadenthal, under date of 24th October of the same year (1823), Brother Hallbeck writes:—"We have lately had plentiful rains, which is rather unusual at this season, and unless the rust, or mildew, which is, alas, already making its appearance, or some other misfortune should spoil the corn-fields, there is now every reasonable hope of their escaping the hurtful influence of the drought. How should we rejoice if the Hottentots were rewarded for their diligence in agricultural labours. We have this year planted about 600 oaks, and are, as much as possible, enlarging our plantations on the hills, which will be of great advantage to the settlement. The Hottentots are now very busy in their gardens, and we assist the widows and infirm people to sow and plant, that they may have a garden-crop, in case the harvest crop should fail. Many have been obliged to seek work at the farms, and of the adults belonging to us nearly half the number are abroad, so that our church is not so much

* At one mission station, either within or beyond the bounds of the Colony, the missionaries dug a water leading whereby they were enabled to irrigate their gardens and corn-fields at a lower level; and some of the natives seeing the beneficial results which followed set about to do the same for their gardens and corn patches on a hill at a higher level; but when all was finished, to their great disappointment, the water which flowed so copiously to the lands of the Christians would not flow up the hill to theirs, let them do what they would!

I have been amused with the mother wit with which a substitute for a theodolite or a spirit-level is devised, in carrying on such operations, by farmers and others. A hat is filled with water to overflowing, and by looking across the surface of this the level at a distance is determined!

crowded as formerly, but the schools are well attended, and we do all in our power to promote the improvement of the children, and find that our labour is not in vain. We have many clever readers and some who write tolerably well. My wife employs some of the girls in needle-work, which they do very well. Orders are numerous.

"I can easily believe, that some kind, benevolent persons, as you state, wonder that, amidst the severe distress of last year, we caused the poor people to do some public work before they partook of the meal of charity, and I ought perhaps to have explained our plan more fully. I would certainly have done it long ago, if I had ventured to anticipate that general participation in our sufferings which it has pleased God to excite among your countrymen. For the satisfaction of those kind friends I therefore beg leave to observe, that we did not expect the people to fatigue themselves by work, nor were the aged and infirm called upon to take any share in the labour; and it was evident that it was a gratification to the people themselves to be favoured to do something for their own benefit. Our view in adopting the measure was to guard against any possible abuse of the gifts entrusted to us for distribution, and to prevent that degrading spirit of pauperism from creeping in amongst our Hottentots, which is so very pernicious in its consequences. We felt ourselves, as we still do, under great responsibility to those who have stepped forward to our relief; and, with a view to make their donations go as far as possible, we have from the beginning endeavoured to use them in such a manner, that whilst they relieved the needy, they at the same time acted as an incitement to useful exertion; and I have no doubt but this view of the subject, if properly understood, will be approved of by our kind benefactors. Another question, which probably has arisen in the minds of some of our friends, and which I ought to have answered ere now, is this: how does it happen that Gnadenthal, the most flourishing of our settlements has suffered most? The very same circumstance which rendered it more prosperous than the other settlements in good years has plunged it in greater distress at this season, not to take also into consideration the heavier affliction during the storms. Gnadenthal is the only missionary station which depends chiefly upon agriculture for its outward prosperity; and because agriculture in general affords more certain resources than other branches, such as carrying of goods, rearing of cattle, etc. (the resources of Enon, Groenekloof, and other stations), therefore Gnadenthal prospered more than its sisters, as long as agriculture flourished. It was always the favourite spot of the Hottentots, and capable of maintaining more inhabitants than any other station. Hence it happened that this place could not fail to suffer more than the rest, by the successive calamities which befel agriculture; and for this same reason, the farmers of our neighbourhood, whose only resource was agriculture, were so much reduced that Government came forward to distribute alms among the most needy of them, which was done in no other district. One circumstance proves to every reflecting mind the intensity of last year's distress amongst our Hottentots, viz., that within the last twelvemonth only half the number of children were born, as, on the average, in every preceding year. It being now about a year since we were so severely afflicted, I have in these days had much conversation with Hottentots on the subject, with a view to encourage them to gratitude towards our Saviour for His wonderful help. I asked an old industrious man, who has been accustomed to better days, how it was possible that he could maintain life by the little nourishment contained in the leaves of herbs. His answer was: 'I had some old slips of an oxhide, and a couple of worn shoes (made of raw hides). These I first laid on hot ashes, till they were half burnt, then I cut them in pieces and pounded them; after which they were boiled with the grassy substances, and this made them more nourishing.' A noble feature in the conduct of this very man, during the distress, deserves to be noticed. As long as nothing but grass and herbs were to be found in the fields, this man (he is a widower, half blind) and his two little daughters appeared with the rest to partake of the meal prepared for the poor. But, as soon as the bulb season came, he disappeared. Meeting him once, accidentally, in the village, I asked him the reason. He replied: 'Thanks to God; I can now, with the assistance of my girls, find a sufficiency of bulbs in the fields, which are both nourishing and palatable; and therefore I do not think it right for me to be troublesome, and to take away what ought to be given to those only who are not able to provide for themselves. It is true, I have neither bread nor meat; but I will rather forego these luxuries than deprive those who are more infirm than myself.' . . .

"Alas! the mildew has again shown itself in some parts of the Colony, and in our neighbourhood; and I hear that some early-sown fields are quite destroyed by its pernicious influence. We have however of late had very plentiful rains for six or seven successive days; and I therefore still cherish the hope that a general destruction of the harvest may thereby, under God's kind providence, be prevented. We can do nothing but pray; and you will easily believe, that after what we have gone through, we do it from our inmost soul. May God have mercy upon us!"

The adjacent division of Bredasdorp lies to the south of Caledon, extending to the most southern point of South Africa, Cape L'Agulhas, near Struis Bay. With the exception of the range of mountains which, in common with those of Caledon, run parallel to the coast, there is a good deal of flat country in the division of Bredasdorp, but the soil in some parts is very dry and sterile. The inhabitants are principally engaged in sheep farming. This part of the Colony was formerly called the Strandveld. Its eastern boundary is the Breede River. This and the Zonder End are the only rivers of any importance in the division. A few isolated mountains form some remarkable groups, especially that called Babylon's Tower, or the Tower of Babel, the Potteberg, and some others. The area of the division comprises 1,200 square miles.

Mr Hoffmeyer, Civil Commissioner, reports, in reply to the queries issued from the Colonial Office in 1862,—"I am of opinion that the landed proprietors in my division are not likely to avail themselves of a law empowering the Government to make them advances of money for the purpose of executing works for improving the supply of water on their properties." He engaged to bring the subject before the Divisional Council at their first meeting, to obtain their views upon the subject; but I do not know with what results this was done.

From Mr W. H. van Breda, at Zee-Koe-gat, I learned that at Strandveld, near Cape L'Agulhas, in that division, the soil is of clay intermixed with limestone, with plenty of water at a depth of ten feet below the surface, which in some places is fresh, but in others perfectly salt.

In regard to Bredasdorp, the Civil Commissioner reported in 1865:—"Bredasdorp has had to experience much adversity in consequence of the continuing drought during almost the whole year. The harvest of wheat, barley, and oats has been very bad, and the prices for these articles have risen more than 100 per cent. since 1864. With very few exceptions, the inhabitants will, as well for their own consumption as for seed in the next sowing season, have to provide themselves either from Capetown or some other distant part which may have been blessed with a better harvest. There will be some people who from want will not be able to provide themselves with seed and what they require for consumption, and who will be left to the sympathy of their more fortunate neighbours. Although the supply of water in the public spring of Bredasdorp, which supplies the whole town and population, has not diminished, yet nearly all other springs in the division have come to a stand, and some have even dried up. There is scarcely a blade of grass or plant visible in the fields.

"The Kars River, which rises about six hours' distance from the sea, and flows through the whole division, and which in ordinary years runs into the sea near Struis Bay, has not been so far filled that it has flowed at all; so that at present almost all the dams are dried up.

"The oldest inhabitants do not recollect ever to have witnessed such a continuously dry year as the year now past. It is calculated that one fourth of the whole quantity of horses, cattle, and sheep have died of want."

Mention is made in this report of Kars River, which in general flows into the sea near Struis Bay, though that year it had been but a river bed. From a description of the coast between Cape L'Agulhas and the Breede River, given in a report by Mr Daniel John May, Master, R.N., dated 10th July, 1866, and published by the Governor, I cite the following account of Struis Bay and of the country adjoining it:—"Northumberland point, in 34° 49′ south and 20° 4′ east, is about three miles eastward from Cape L'Agulhas; it is a low, scant-covered continuation of the higher land of L'Agulhas, with a shore of shingly beach and rocks, which extend south-eastward for a mile, on which the sea only sometimes breaks; it forms the western limit of Struis Bay. . . .

"Struis Bay is comprised between Northumberland and Struis Points; from the former to the outer visible rock of the latter is, N. 60° E., eleven and three-quarter miles; it is not so deep an indentation as other bays on the south coast, but affords shelter in westerly and north-westerly winds. There are two stone white houses and a flag-staff near one of them, near a landing cove close round Northumberland Point, which serves to distinguish this bay from that eastward of Struis Point, which it is otherwise very much like.

"This landing cove is sandy, but is fast filling up; a wooden jetty, at the end of which a vessel drawing seven feet used to lie, is now high and dry. From opposite these two houses the beach is clean sandy to within two miles of Struis Point, where flat jagged rocks, backed and mixed with sand, commence. Behind the sand beach, throughout the whole bay, is a line of sand hills, the highest 100 feet above L.W.S., some bare, and others more or less covered with bush. Behind these, again, is a green-covered ridge, attaining a height of two hundred feet, which, at an average distance of three miles from the beach, drops suddenly away to a dead level, which extends for six or seven miles to the base of the Driefontein range of hills, which run east and west from westward of Bredasdorp into the sea westwards of Cape Infanta. The white sand is fast encroaching, in consequence, it is affirmed, of the destruction of bush by fishing parties who outspan and locate, which is a legal privilege.

"Fresh water is scarce and brackish; it is procured only (besides rain accumulations) by digging in the sand beach above high-water mark.

"The village of Bredasdorp is sixteen miles, or three hours' journey, distant from the landing place in Struis Bay, where supplies can be obtained, and which is in postal communication with Capetown.

"The Matjies River runs into Struis Bay, about its centre, from between bare sand hills. It is the outlet of many streams flowing from the hills northward and westward for many miles, which form often into lagoons in the flat described above as lying behind this coast, flowing sluggishly, and under various names, with its tributaries, as the 'Honing-nest,' 'Kars,' and 'Nieuwjaar,' until they unite and form this river. Is is unimportant, and usually fordable three-quarters of a mile from the mouth, and often at the mouth, but dangerous.

"Struis Point is formed by a conspicuous mass of bare sand-hills, two hundred feet high, sloping southward for nearly a mile to low-water mark, where it is rocky. Seaward from the point are detached patches of rock in a S. 50° to 60° E. direction, the outermost, visible at low-water, being one mile and three-quarters from low-water mark, or two miles and four tenths from the summit of the sand-hills. This reef is known to extend considerably further, the local fishermen say from three to four miles."

In 1866 it was reported:—"The year 1866 has been a less trying year than the two last; and may even be said, in many respects, to have afforded relief generally.

"The effect of the rain, which had fallen plentifully in the beginning of February, and occasionally thereafter, had soon occasioned a verdure and an abundance of food for sheep and cattle in every part of the division; so that, soon after, the sheep particularly appeared in so splendid a condition as has not been observed for many years past. The lambing was, consequently, during the whole season, very prosperous; and scarcely any complaint was heard of disappointment.

"The production of wool fully answered the expectation, both as regards quantity and quality; and it is believed that the division never before produced a larger quantity, nor of better quality, than in the year just now ended.

"The prices of the wool were also encouraging, and were from 2d. to 3d. per ℔. higher than the year before.

"The grain harvest was also much better.

"It is true that the crops of wheat, sown late in the ploughing season, have on some farms failed in consequence of the rust, which made its appearance in November last, but there are farmers who have gained even more than they require for their own consumption and seed, whilst others who are deficient in wheat have been so fortunate as to reap a good quantity of oats and barley; the latter they will partly apply for mixing with wheat, which produces good bread.

"The supply of water, for drinking and for irrigation, from the public fountains in the town, has been undiminished; and there has been scarcely any complaint of want of water for cattle or sheep during the year.

"Scarcely a report has been heard of cattle or sheep having died last year from want."

Of Bredasdorp in 1869 it was reported:—"The year 1869 has been a year of plenty, compared with the last three previous years. The losses of sheep were greater, and the number of lambs less than in 1868, yet these losses were fully compensated by an abundant harvest of grain, wheat, as well as barley and oats; so that no importation from any other division, nor from abroad, will be required this year. Some farmers will even be enabled, after fully providing this division, to dispose of the surplus to other places.

"The production of wool—the staple produce of this division—has not been so good as in the year 1868, owing to the loss of so many sheep, on account of the drought during four months continuously, viz., from the beginning of December, 1868, to last of March, 1869."

In 1874 the report bore:—"The past year has been a disastrous one for the sheep-farmers. In consequence of the great drought during the greater part of the year, thousands of sheep died of the 'dik kop' and 'brand ziekte' (scab).

"As regards horse-breeding, however, it has been a very favourable one, very high prices having been realised for young horses from thoroughbred stock, the latter having this year been increased by several splendid importations, both at Fairfield and Nachtwacht.

"The wheat crops, which were very promising until the heavy rains which fell in November, commenced to show rust, and several very promising fields were completely blighted by the disease.

"Ostrich farming is on the increase, and prices for both old and young birds are nearly double those of the previous year. . . .

"The criminal records show a decrease in crime,—not a single case of sheep or cattle stealing is on record during the year. The bulk of cases were for drunkenness, and offences committed through its effects."

The following is an abstract of observations on the rainfall at Bredasdorp in the years 1867 and 1868 :—

From Daily Observations.	Jan.	Feb	Mar	Apr	May	Jne.	July	Aug	Sep.	Oct.	Nov	Dec.	Year.
1867.													
Quantity of rain in inches,....	0·47	0·74	3·07	2·14	2·08	1·30	1·66	1·92	0·48	1·27	0·99	0·20	16·32
Greatest amount of rain on any one day in each month,	0·29	0·40	2·43	1·03	1·43	0·25	0·52	1·01	0·30	0·40	0·86	0·20	2.43
Number of days on which rain fell,..................	2	4	4	4	6	8	7	4	3	5	2	1	50
1868.													
Quantity of rain in inches,....	0·42	3·02	0·88	0·25	1·28	0·79	0·88	1·47	0·48	1·24	1·10	0·00	11·81
Greatest amount of rain on any one day in each month,	0·42	1·04	0·80	0·25	1·28	0·31	0·60	0·62	0·23	0·66	0·60	0·00	1·28
Number of days on which rain fell, ...	1	5	2	1	1	4	2	3	3	3	2	0	27

The following tabulated statement shows the directions of the winds observed :—

Number of Observations of Wind at the several Points of the Compass during each Month of the Years 1867 and 1868, at Bredasdorp.

Points.	Jan.	Feb.	Mar.	Apr.	May.	June.	July.	Aug.	Sept.	Oct.	Nov.	Dec.	Total	Jan.	Feb.	Mar.	Apr.	May.	June.	July.	Aug.	Sept.	Oct.	Nov.	Dec.	Total
N.	-	-	-	10	17	12	29	16	8	9	6	2	109	2	2	2	16	2	19	16	21	6	2	4	4	96
N. by E.	-	-	-	-	-	-	-	-	-	-	-	-	-	-	-	-	-	-	-	-	-	-	-	-	-	-
N.N.E.	-	-	-	-	-	-	-	-	-	-	-	-	-	-	-	-	-	-	-	-	-	-	-	-	-	-
N.E. by N.	-	-	-	-	-	-	-	-	-	-	-	-	-	-	-	-	-	-	-	-	-	-	-	-	-	-
N.E.	-	-	-	-	-	3	-	-	2	1	-	-	6	3	-	1	-	-	3	1	-	-	-	-	-	8
N.E. by E.	-	-	-	-	-	-	-	-	-	-	-	-	-	-	-	-	-	-	-	-	-	-	-	-	-	-
E.N.E.	1	-	-	-	-	-	-	-	-	-	-	-	1	-	-	-	-	-	-	-	-	-	-	-	-	-
E. by N.	-	-	-	-	-	-	-	-	-	-	-	-	-	-	-	-	-	-	-	-	-	-	-	-	-	-
E.	-	-	1	1	-	2	-	-	1	-	-	-	5	-	-	-	-	-	1	-	-	-	-	-	-	1
E. by S.	-	-	1	-	-	-	-	-	-	-	-	-	1	-	-	-	-	-	-	-	-	-	-	-	-	-
E.S.E.	1	3	2	1	-	3	-	-	1	-	2	-	13	-	-	-	-	-	-	-	-	1	1	2	-	5
S.E. by E.	-	-	-	-	-	-	-	-	-	-	-	-	-	-	-	-	1	-	-	-	-	-	-	-	-	-
S.E.	49	40	40	23	2	8	8	19	27	29	54	35	334	41	48	49	17	9	4	5	21	29	40	30	29	322
S.E. by S.	-	-	-	-	-	-	-	-	-	-	-	-	-	-	-	-	-	-	-	-	-	-	-	-	-	-
S.S.E.	-	5	7	-	-	-	-	3	-	5	-	2	22	1	-	-	-	-	-	-	-	-	4	-	3	8
S. by E.	-	-	-	-	-	-	-	-	-	-	-	-	-	-	-	-	-	-	-	-	-	-	-	-	-	-
S.	-	-	-	3	1	4	3	10	5	3	-	5	34	-	5	1	4	1	9	7	5	1	2	1	5	41
S. by W.	-	-	-	-	-	-	-	-	-	-	-	-	-	-	-	-	-	-	-	-	-	-	-	-	-	-
S.S.W.	-	2	2	1	1	-	-	-	-	-	-	1	7	3	-	4	-	-	-	-	-	-	1	1	-	9
S.W. by S.	-	-	-	-	-	-	-	-	-	-	-	-	-	-	-	-	-	-	-	-	-	-	-	-	-	-
S.W.	13	8	10	13	10	10	2	9	10	12	2	11	110	12	6	3	18	20	3	5	9	8	9	16	12	121
S.W. by W.	-	-	-	-	-	-	-	-	-	-	-	-	-	-	-	-	-	-	-	-	-	-	-	-	-	-
W.S.W.	7	2	5	3	4	2	2	2	2	7	-	1	37	3	3	7	-	2	7	5	3	11	2	8	-	51
W. by S.	-	-	-	-	-	-	-	-	-	-	-	-	-	-	-	-	-	-	-	-	-	-	-	-	-	-
W.	4	6	-	3	5	1	12	3	4	7	5	15	65	7	5	2	5	38	10	15	13	9	12	6	1	123
W. by N.	-	-	-	-	-	-	-	-	-	-	-	-	-	-	-	-	-	-	-	-	-	-	-	-	-	-
W.N.W.	-	4	3	10	14	7	6	-	4	1	3	5	57	2	4	7	13	-	2	5	-	4	2	4	13	56
N.W. by W.	-	-	-	-	-	-	-	-	-	-	-	-	-	-	-	-	-	-	-	-	-	-	-	-	-	-
N.W.	1	1	10	10	26	24	21	18	13	10	7	7	148	7	6	5	6	-	19	24	10	11	10	7	8	113
N.W. by N.	-	-	-	-	-	-	-	-	-	-	-	-	-	-	-	-	-	-	-	-	-	-	-	-	-	-
N.N.W.	-	-	-	-	-	-	-	-	-	-	-	-	-	-	-	-	-	-	-	-	-	-	-	-	-	-
N. by W.	-	-	-	-	-	-	-	-	-	-	-	-	-	-	-	-	-	-	-	-	-	-	-	-	-	-

N.B.—No observations in 1868, from December 25 to 28 (inclusive).

To the east of Caledon and Bredasdorp is the division of Swellendam, with an area of 1,500 square miles, including the country between the Langeberg Mountains and the sea. It is a productive land, well adapted for either agricultural or stock farming. The climate resembles generally that of the coast region between the sea and the first mountain range. The Langebergen rise in some points to an altitude of 5,000 feet. The kloofs of these mountains are represented, by Latrobe and by Lichtenstein, as in many places well wooded. The deep river-beds crossed by the high-road from Capetown to the frontier occasionally make travelling dangerous.

In 1862, Mr van Oudtshoorn, the Civil Commissioner of the division, reported in reply to the circular issued from the Colonial Office:—"Having gathered all the information in my power, I beg to submit the following points, in answer to the questions contained in your circular of the 7th instant:—

"1. Proposals of various kinds have, from time to time, been made for supplying the villages of Swellendam and Heidelberg with water, by damming up the Cornlands River and the Duivenhoks. This latter river is, in the opinion of some, capable of being led out to Port Beaufort; but Mr Atmore, a professional man, thinks the distance far too great for an open water-course, but practicable with the aid of pipes. In that case, however, the expense would far exceed the estimated cost of carrying water across the Breede River from Potberg, which is calculated at from six to eight thousand pounds. The Buffeljagts River has been partially dammed up, and, as far as the work goes, with complete success; and many dams have been constructed and have answered very well.

"2. On the south of the Langebergen, proceeding from east to west, the following streams seem capable of being diverted or dammed up, with every prospect of success, and at a moderate cost:—

"(1) Duivenhoks River, to water Heidelberg and the plains between that village and the sea.

"(2) Buffeljagts River, to water the valley in which Zuurbraak stands, as well as a plain about twenty square miles in extent.

"(3) Appelbosch River, to water about four square miles.

"(4) Cornlands River, about the same area, including the village of Swellendam.

"(5) Leeuw River, Bruintjes River, and Nooitgedachts River, to water valleys, each of which may be estimated at about five square miles, lying along the southern face of Langebergen.

"(6) River Zonder End, to water a large area, which it would be difficult to estimate without some more reliable data than can be obtained without a proper survey. This stream should, however, be turned off in the Caldon district, probably as far up as Gnadenthal.

"(7) On the northern slope of the Langebergen there are a large number of small streams which may very well be dammed up at a small outlay, as they mostly run in rocky beds and through small kloofs, many of which form a kind of basin, with the materials for closing the mouth, and so converting it into a reservoir, close at hand.

"(8) On the coast, the Potberg stream, which I have noted above, would probably supply a large tract of country on the west of the Breede River, even after diverting enough for Port Beaufort.

"Plans and sections of the beds of the Cornlands and Buffeljagts Rivers have been prepared, and I would suggest that this should be done in all cases before any further action, or rather that a general survey should be made; and in this case, particular regard should be had to the white waters, which are from great depths, and far more permanent than the red (or surface) springs. Before leaving this question I would add that dams, particularly those formed by closing the lower ends of kloofs, are of great benefit, and generally easy of construction; while wells are not advisable, the ground being so thoroughly impregnated with salt that they soon become brack.

"In this report I have not noted the Breede River, because the expense of damming it up would far exceed the means of any private person, and probably even any sum that Government would be willing to allot to any one division, putting out of the question the difficulty of leading out the water, when secured, by side cuttings along banks of rock and sand.

"3. The farmers would, I am led to believe, be glad to avail themselves of a loan for this purpose, provided the terms of repayment were easy, say extending over a period of ten years, and by instalments, and the interest did not exceed six per cent."

In 1866 it was reported:—"To frame a report of the state of a division impoverished by successive years of drought, followed by what at one time threatened to become general insolvency, is no pleasing task; and although there are apparent symptoms, weak and feeble it is true, of a quickened pulsation in the veins of trade, the features characterising the past year's history will mainly exhibit a distress and depression so great that it requires every confidence in the resources of the division to believe it can successfully rally.

"Some of the crops, which came up with astonishing promise suffered so severly from rust as to be scarcely worth harvesting in some instances. The early forage suffered most. The yield of barley was abundant; but the wheat crops were a great disappointment, especially a new corn now tried for the first time (Du Toit's corn).

"The ravages of the horse sickness early in the year caused a great addition to the trials of the farmers, many of whom lost every horse they possessed.

"The fall of lambs was unusually good; but, from the excessive succulence of the pasturage, very many young lambs have died.

"The growth of tobacco in the neighbourhood of Heidelberg, where it is cultivated to a considerable extent, and with success, becomes more and more stimulated as the article is better known, and consequently better appreciated. On more than one farm have plants, numbering from 150,000 and 200,000, been raised and planted out.

"Irrigation works have not progressed very much, though some few dams have been constructed on municipal land during the year; but these works were undertaken mainly with a view to finding work for the poor, who during the winter months were almost starving; indeed, so much distress prevailed at one period that the goodness of the prison fare induced very many to steal, as they admitted, expressly to get into prison, where they could procure better food, clothing, and lodging, than they could earn for themselves by honest labour. Crime of any more serious nature than wanton and wholesale stock-stealing has been unknown during the past year."

Of Swellendam in 1874 it was reported:—"The present season has been most favourable, and the farmers are fast recovering from the stagnation caused by the previous long drought. Comparing this division, however, with the Karroo, it cannot be said to be a very profitable sheep-walk; the animals are small, and never very fat, and the increase not more on the average than forty to fifty per cent. The quality of the wool, however, is superior to that of the Karroo, bringing higher average prices. There is, however, more scab here than in the Karroo sheep-runs, and the constitution of the sheep inferior.

"This district seems better adapted for cattle, horses, mules, and grain crops of all kinds, than as a profitable investment as a sheep-run, more particularly as the farms are too small for this description of stock, and the mountainous part of the division is sour, the general weight of a good animal here not being over forty pounds clean. . .

"*Dams.*—Little has been done in this division for the conservation of water, most of our streams running to waste. Much might be done by dams in central positions, and by leading out the streams to natural hollows, easily converted into dams, and it is certain that large tracts of country could be irrigated by damming up the Breede River in numerous suitable places along its winding bank, where there are natural stone ridges, easily converted into weirs, from which the water could be led out for miles, fertilising large tracts of country to the benefit of the Colony. Works, however, of this kind can never be carried out by private enterprise; such a scheme must be national, guided by large views as to permanent results for the country, as in parts of Piedmont and India. This is purely a Government measure, and would be a lasting glory to it, and add greatly to the country's wealth and happiness of its people.

"*Thefts of Stock.*—This crime is rife in this district, and mostly around the Zuurbraak and Slang River settlements, where numbers of the coloured race are together, with no check upon their actions whatever."

SECTION II.—*Robertson, Riversdale, and Mossel Bay.*

The district of Robertson lies to the east of Bredasdorp. It has an area of 2,500 square miles, and comprises a district known as Kannaland, or as the Little Karroo—"an elevated plain between the Langeberg and the Zwartberg range of mountains, about 2,000 miles above the level of the sea, which is the next step in the ascent to the interior. . . . The Gauritz River running in a deep, rocky channel, which the old travellers describe as one of the most awful scenes in Nature, bounds this region to the east."

The general features of the division of Robertson are similar to those of the division of Swellendam, but some portions are much more fertile. One district alone, that of Montagu, produced, in 1865, 106,235 gallons of wine and 127,770 lbs. of dried fruit.

In regard to irrigation, M. le Brun, the Civil Commissioner, reported in 1862:—"In reply to circular No. 16 of 1862, dated

7th May, submitting certain queries on the subject of irrigation I have the honour to state that I have submitted the circular to the consideration of the members of the divisional council, and as this subject has long been forced upon my notice by the severe circumstances which have lately visited this town and district, my personal knowledge of the farms in this division enables me to pronounce it most eminently adapted for an extensive system of irrigation; and I beg to offer thereon the following reply:—

"1. No general plan for irrigation works in this division has been systematically proposed; but some inquiries have been set on foot by me in regard to making the Breede River available for irrigation. The farmers along its valley were favourably disposed and willing to subscribe towards the project. This was proposed to be carried out by shares, by which, could I have seen my way clear by any professional advice and report, a capital of £10,000 was obtainable. This would have benefited an extensive tract of country along both sides of the Breede River, extending, in my district alone, from the Vinke River to the entrance of Cogmanskloof, at the place Zandvliet, —several thousands of morgen of the richest soil in the colony; besides watering this village largely. It will be readily understood that, in the absence of professional assistance, so large a project was left to await a favourable opportunity.

"Near the village attempts have been made by very expensive mills to lead out the water of the Breede River. A horse and a windmill have been fixed there, but from some cause or other they do not answer. In some farms behind the mountain dams have been constructed and extensive water-furrows, made by stray navvies, but no work that may be considered a plan can be reported.

"2. The mountain ranges which divide this district into sections, —Zonder End, watering Boschjesveld,—the Langeberg, this valley and the whole of before Cogmanskloof on one side, and behind Montagu on the other,—the Keizies and Koesberg, yielding streams which have rendered that fertile country so richly productive; these ranges have all kloofs favourable for the construction of dams: for instance, the Willem Nel's River, which has been the subject of an expensive litigation and loss to this village, might for a small sum be dammed up into a natural reservoir near its source, when running down into this valley, through a narrow gorge of some 500 yards. This river runs to waste half the year, and during the other half is the source of much trouble and litigation. The Breede River receives the overflow, and much is lost along the sandy bed.

Artificial furrows, leading actual rights from the main reservoir, would obviate these defects in our periodical supplies from it. The same can be said of the waters supplying farms laid on the Hoops River, Klaas Vogels Hoek River, and Vinke River. I have been assured by the residents on these farms that a greater advantage can be derived from their respective rights, if it were possible to carry out such works. The great fear expressed has been that the expense of constructing sufficiently strong dammed walls would be beyond their means. Mr Lourens Erasmus, a rich farmer, residing in the Boschjesveld, assures me that a properly constructed dam at his place would supply him with water sufficient to double the value of his farm. As this point has a wide meaning, I shall endeavour at a future opportunity to detail each farm in each field-cornetcy capable of such irrigation works therein alluded to. The sinking of wells is rather prevalent, and generally with good results.

"3. I am satisfied, and the council agrees with me, that the landed proprietors throughout this division would avail themselves of loans from Government for the purpose of executing works for improving the supply of water on their properties. The only question raised by them is a doubt whether any reasonable sum can do the work.

"In conclusion, I beg to assure his Excellency the Governor that no measure can benefit this district, or for which it has so great a need, as a general system of irrigation."

In his report for 1865, Mr Le Seur, the Civil Commissioner of Robertson, wrote:—"The effect of the continued drought afflicting the land is felt everywhere, and has here well-nigh paralysed every branch of industry. It has put an end to all trade; it has deterred the local bodies from making any attempt to undertake public improvement; it has forbidden all private enterprise, and thus deprived the labouring classes of all chance of gaining, without very unusual exertions, even their daily bread. Sluits, which in former years, at the driest season, were well supplied with water, contain now nothing but sand; so that even the extensive wine-farming, which used to make this village and Montagu independent of the crops beyond their vicinities, has this year received a considerable check. Outside the villages, matters are, if possible, worse. In and about the field-cornetcy of Touw there are, as travellers state, whole tracts of land where the colour green seems almost forgotten, and where only at long and rare intervals, in the immediate neighbourhood of the river from which the ward takes its name, a single thorn bush has contrived to outlive all other vegetation. In the Boschjesveld, which is the granary of the division, the wheat crop is so small that none but the wealthiest farmers intend to sell; nearly all the others have little more, many much less, than they require for their own use. For the coloured races there is in consequence next to no employment; and crime, principally goat and sheep-stealing, is much on the increase; and this more so in the neighbourhood of Montagu, where the scarcity and the dearness of bread and meat are greater than at the town of Robertson. . . .

"The village of Robertson is gradually recovering itself; and but for the continued drought would be taking many steps towards social advancement. The buildings are increasing, and are more ornamental; and a spirit of emulation has of late sprung up among many house owners.

"The great want is water. The prescriptive right proved by the occupier of part of a farm situated above the village, in the case lately tried by Mr van Reede van Oudtshoorn as special magistrate, to what is called *drink water*, has encouraged several of his neighbours to disregard the claims of the municipal sluits altogether, and the question is becoming so complicated that the formation of a company has been suggested for the purpose of buying these upper farms for the sake of their right of water, and re-letting the lands with only a portion of water; the rent to go for the payment of the interest borrowed, and, if necessary, the farms to be mortgaged to the shareholders as security for the capital.

"It is quite disheartening to see hundreds of acres of the most fertile soil along the Breede River lying waste and comparatively unproductive, owing to no other cause than the want of capital *to lead out* the Breede River, whereby the whole may be placed under cultivation."

In 1866 it was reported:—"No improvement has taken place in agriculture in this division during the past year; but a larger breadth of land was sown than during 1865.

"Nearly all the late-sown crops of wheat have suffered severely from rust; many farmers have turned their cattle into the grain, which was too much injured to be worth reaping. In the early stages of the bygone season barley was reported as being thin and backward; this, however, was not substantially correct, for barley is here an abundant crop; beans and pulse are also favourably spoken of. In some places mealies are suffering from insects.

"The vine disease appears to have become permanent in this division. Most of the farmers sulphur the plants shortly after the appearance of the fruit, and one or two successive sulphurings are generally sufficient to keep the *oidium* within bounds; the vintage otherwise promises well. Till the setting in of the present dry weather, there was abundance of rain, and in consequence the vines are full of good sound fruit; through injudicious trimming some farmers have had loss from the scorching of the grapes, but this is only in patches, and very trifling.

"An attempt is being made by one or two farmers to lead out the Breede River on their properties, by which five hundred acres of the best Karroo soil might be brought under cultivation."

I visited this division in August 1865, and again in 1866, and my own observations in a cursory journey through it were in accordance with these official reports.

On the occasion of my visit in 1865 arrangements were made by the Rev. Mr Macgregor, the minister of the Reformed Dutch Church, for conferences on subjects connected with agricultural operations being held by me in different central localities in his parish. These conferences were well attended, and they were conducted with spirit on the part of the farmers.

The country was at the time suffering from the drought. At one place there had been from the beginning of the year only two days' rain. Horses were suffering from a deficiency of forage; and I was told of hundreds of lambs having been killed so soon as born, lest the suckling of them should lead to the death of the ewes.

In these circumstances the importance of securing, if possible, a more continuous supply of water was brought under my consideration at all of the conferences which were held; and arrangements were made for my visiting, along with parties interested in the matter, kloofs, and basins adjacent to river beds, affording facilities for storing up water. The result was a conviction, shared apparently by all, that at comparatively little expense an abundance of water both for the supply of cattle and for irrigation might be secured in different localities.

To determine the height of the embankment which in each case it would be necessary to raise, and the quantity of water which might be thereby retained, would have required a survey and measurement and calculation, for which we had not with us the necessary appliances; and to estimate the expense which would be incurred in the execution of works would have required the professional aid of an hydraulic engineer; but data were obtained sufficient to indicate eligible spots for the location of such structures—such data as natural adaptability for the formation of reservoirs by embankments, materials existing on the spots for the execution of the works and facilities for the application of these, and facilities for important application of the water there stored.

I travelled from Worcester to Robertson by the road skirting the mountain range to the north, and in doing so passed through a range of productive farms moistened in some measure by the percolation of water under the surface of the soil from the higher lying land of the mountains, and by streamlets from these.

At a numerously attended conference in Robertson it was stated by more than one of the speakers that a copious supply of water was all that was required to render the whole of the land in the neighbourhood of the village most productive; that much had been done to secure this; that what had been done was as nothing compared with what might be done; but that a difference of opinion existed in regard to the practicability of measures which had been proposed.

To give me an opportunity of judging of these, arrangements were there and then made for a day being given to a survey of the locality, with a view to learning my opinion in regard to structures, relative to which a difference of opinion prevailed. On the day appointed, about twelve gentleman, including representatives of different portions of the community interested in the matter, met in the morning at Dassen Kloof and proceeded thence on foot to the top of Wolf Kloof, through which we wended our way back to the village in the evening. In Dassen Kloof are places well adapted by their outline and position for conversion into reservoirs, but which are covered with rich alluvial soil too valuable as garden ground to be so employed. But in Wolf Kloof there are several places where, by a contraction in the width of the kloof immediately below wide expanded spaces, a large quantity of water might be retained by a low wall or stone embankment; and where, for the formation of such a wall, abundance of material is at hand on the mountain precipices on either side, ready to be rolled down by the necessary application of the crow-bar. And, immediately above the spot where the water is at present led off for the irrigation of the village erfs, a spacious reservoir might be made to retain a great quantity of the water which at times rushes off in torrents to the Breede River as it passes through the district.

On a subsequent day I was taken in the morning to examine a

quantity of land belonging to the community irrigated from water retained by an artificial dam made at little expense. This showed what might be made by irrigation of a vast extent of land now lying waste. And I was afterwards taken to see Hope River, which showed what scenes of beauty might be created in mountain kloofs if the rain falling on the Colony were much more extensively retained and utilised for agricultural operations.

In returning on the following day to Robertson, by the lower drift from Bushman's River, whither I had been to hold a conference on subjects connected with agriculture, after holding one at Lady Grey, I saw the extensive water leading, on the southern bank of the Breede River, which was being constructed on ground belonging to Mr de Wet, whose water leading from the Hex River I have already mentioned in speaking of the water supply in the district of Worcester, and in regard to which water leading in this division, and the results obtained, I have there given some information.

Nearer Robertson my route on this occasion crossed a comparatively level ground on the northern bank of the same river, a valley debouching on the river bed, from which it had received at high water a quantity of water which was there retained, indicating the practicability of filling the whole ground when the river was full, and retaining the water by flood gates so long as it might be considered desireable to do so, securing thus some of the advantages which are gained from inundations of the Nile along with others connected with the system of agriculture generally adopted in the Colony.

From Robertson I proceeded to Montagu on an invitation from the Rev. Dr Hoffmeyer, minister of the Dutch Reformed Church, to hold conferences on subjects connected with agriculture in different parts of that district. My road crossed extensive beds of limestone, and I saw on either hand abundance of land apparently well suited for the growth of the vine if a sufficiency of water could be secured. At the time of my journey there was a manifest deficiency of this. But passing through Cogmanskloof, with its narrow defile between precipitous walls of rock, towering aloft at an almost perpendicular elevation, and its meandering stream crossing, or rather crossed by, the road again and again, I was informed, in reply to inquiries, that frequently the road was altogether impassable from the overflowing of that stream; proving that there was abundance of water to be obtained in regions beyond, which might be utilised if there existed the means of retaining it.

As at Robertson, so here. After a conference held at the village a party was formed to convoy me through districts where water might be stored. We first went through Dunkel Kloof, through which flows a stream of limpid water, by which a considerable portion of the village is supplied. Immediately above a mill, situated at the upper entrance of the kloof, is a wide expanse which might be made into a large reservoir by the construction of a short, low, retaining dam; and below this is a still wider expanse, in which at even less expense a reservoir to retain both the water coming from Dunkel Kloof and that coming from Bath Kloof might be retained to be given off as required for use.

On a subsequent day a party was formed to accompany me through this Kloof, and in passing through it it was seen that at no great distance from the mouth there existed great facilities for the formation of a reservoir; and that above this there might be formed a succession of reservoirs capable apparently of containing an unlimited supply of water.

It may be asked—Does a supply of water sufficient to fill so many reservoirs come down that Kloof? That remains to be seen. I had not the means of determining the fact. But while above I have adverted to the practicability of forming a succession of dams, I consider one dam only would be required; and of the different localities referred to the hydraulic engineer might make a selection, which might be determined in some measure by the object to be effected.

At the top of the kloof there is, it may be mentioned, a warm spring, which has been made to supply a small bath, and from this the kloof there has received its name; but the water is not offensive to the taste, though indications are given below of both sulphur and iron having entered into its composition.

The whole formation of the kloof is interesting. The admirer of the magnificent in Nature and the student of geology, as well as the agriculturist desiring to obtain a supply of water, either for his own use or for the benefit of others, meets in it with much to engage his attention. Though not in a continuous line with Cogman's Kloof it leaves on the mind an impression of its appearing to be a continuation of the same gorge through the mountain range in a diverging direction. In the one as in the other, the precipitous mountain sides rise to a great height; and in these, on both sides are seen dissevered continuations of waved strata, which have been twisted into tortuous courses. At one place after another, Dunkel Kloof widens into wide sky-roofed compartments, from the centre of which there appears neither entrance by which the explorer could have got thither, nor place of egress by which he may get thence. The entrances from the upper portion of the kloof and the lower continuation of this are narrow; the rocks on both sides of these entrances are so numerous and so disposed that it seems as if workmen with crowbars could with the greatest ease hurl them from their places, and thus begin to dam up the water-course which winds its way below. I write from memory, and I am quite aware that in so doing, after a long interval, fancy may combine with memory in producing the picture which rises up before me; but I feel certain memory has not played me false, and that the vision I see is in a great measure the counterpart of the reality upon which I have looked.

A curvature and vertical folding back upon themselves, such as memory pictures to me in strata in Cogman's Kloof, occur in the strata in the vicinity of St Abb's Head, on the east coast of Berwickshire; but in this case on a larger scale. There are there sixteen bendings in the course of six miles; and it is computed that these strata, extended and restored to a horizontal position, in place of being subtended as they now are by a horizontal base of six miles, they would be subtended by a horizontal base of about ten miles and a half. A representation of them, made by Sir James Hall, is given in the seventh volume of the "Edinburgh Philosophical Transactions." This occurs there in transition rocks. And Dr M'Culloch, in his "Geology of the Western Isles," has made two representations of strata of gneiss which occur in the island of Lewis, and which have shifted into a position which is highly inclined and incurvated, so much so that it is alleged that these strata, were they extended and restored to their original horizontal position, would be subtended by a horizontal base three times greater than that which now subtends them.

Of the hills on both sides of the Kariba, Dr Livingstone writes that they are much like those of the Kebrabasa, the strata being tilted and twisted in every direction, with no level ground. And again he writes:—"Some rocks in the water near the outlet of Kariba at a distance look like a fort; but such large masses, dislocated, bent, and even twisted to a remarkable degree, at once attest some tremenduous upheaving and convulsive action of Nature, which probably caused Kebrabasa, Kariba, and the Victoria Falls to assume their present forms; it took place after the formation of the coal, that mineral having been tilted up."

By Mr Charles Chalmers, of Merchison, it was submitted for consideration to his scientific friends, in a series of notes for inquiry, whether such curvature of strata might not be occasioned by the contraction of underlying strata by cooling during lengthened periods of intermission of solar incandescence. The suggestion is, I consider, one well deserving consideration; but I am persuaded an ordinary observer of the curved strata in Bath Kloof and Cogman's Kloof would find it easier to trace in them similarity to curvature such as might be occasioned at the angle of departure from the horizontal line, or at other places, by the upheaval of a soft or lubricated mass, through strata or layers, of plastic materials; and viewed thus they may be considered as supplying corroborative evidence in support of the hypothesis advanced in the treatise on Hydrology of Southern Africa, in the chapter relating to the primary and principle cause of the aridity of South Africa—the upheaval of the land and consequent escape and drainage of the water.

On another day a party was formed to go up the Kingnya River, to examine an existing dam, and discuss improvements upon this which had been proposed or suggested. On this occasion it was found that there was a succession of dams across the river bed at different places in its course, all of them of the simplest construction, and some of them at the time not in use; while rocks, cropping up at other places, offered facilities for dams being made also at these, and water being thence led on to the highest lying erven in the village, for effecting which there was an abundant supply of water if it were retained, and not allowed to run away to the sea. And before setting out on our return to the village my attention was directed to Lang Kloof, where, by running a mound of no very great elevation across a narrow outlet, there might be formed a lake, or, as it was styled by my informant, an inland sea, much more than a mile in extent.

Visiting afterwards the Kow, where there is a rich and productive soil, I was told by one of the farmers residing there, after a conference on agricultural subjects, that he and his brother had purchased land some six hours distant on the Karroo; that with a few weeks' labour they had formed a dam which had enabled them to keep there a large herd of cattle, though none had ever been kept there before; and I was informed that at some little distance from where we were there was a piece of Crown land upon which a similar dam might be made, and the value of the same thereby greatly increased.

It was proposed that I should also visit the Tradouw, and arrangements were made for a conference being held there also if this should be accomplished, but it was found to be impracticable to carry out the proposal.

Of the importance of securing a large supply of water for Montagu and the Karroo beyond, some idea may be suggested by the following evidence, given by the Hononrable J. Barry, M.L.C., before a committee of the House of Assembly, on the Cogman's Kloof, on the 18th July, 1862 :—

" 9. *Mr le Sueur.*] Are you aware that people are applying themselves to the breeding of woolled sheep in that district?—Yes, they have done that for some years past; because it is a most profitable article, and easy of transport.

" 10. And besides that they have the usual productions, such as wine, brandy, and raisins?—Yes, they make large quantities of raisins; no other part of the Colony produces finer raisins, or a larger quantity. The quantity of wine is also increasing very rapidly. Montagu, ten years ago, did not produce more than 20 or 25 leaguers of wine; and the year before last, before the disease affected the vines, it produced upwards of 500 leaguers in what is considered the municipality of Montagu; out of that, about 300 leaguers were distilled into spirits by us, because we found it impracticable to transport the wine to market, merely for want of conveyance.

" 11. Is not Swellendam generally the market for the wine?—Yes, the principal market, but I have transported a quantity of Montagu wine to England.

" 12. *Chairman.*] Is the wine of good quality?—Yes; and by a little attention it would be equal to any wine in the Colony. We were rather fortunate with our shipment of wine to England; whether it was the body of the wine or its peculiar flavour I do not know, but it realised £2 per pipe more than the ordinary Cape wine. It is a strong bodied wine, and the vines are more productive in the neighbourhood of Montagu than in any other part of the Colony, unless it is in the neighbourhood of Olifant's Hoek. In this part of the Colony the yield is, I believe, from one to one and a half leaguers per 1,000 vines; in Montagu three to five leaguers is the general return, and I have known many instances of even six leaguers. Of course, I am speaking of the time when there was no disease, as in the year before last; last year the whole district did not produce eighty leaguers of wine, in consequence of the disease. Some distance from Montagu along the range of mountains to the right and left of

Cogman's Kloof, it is Karroo country, very productive and very fertile, because they have water for irrigation,—and if some plan could be devised of preserving the waters in the Karroo, it would be equally productive. Opening the Tradouw would open up an equally fertile country."

Mr Woodifield, Assistant Commissioner of Roads, in giving evidence before a committee of the House of Assembly in June, 1861, in regard to a road through Cogman's Kloof to Seven Weeks' Poort—the route for which he had surveyed and recommended in preference to two other routes which had been proposed—in answer to query 49—"Are there good sites for dams?" replied—"Yes; many good sites with good retentive soil." And in a written report submitted by him there occurs the following statement:—"I now proceed to detail the duration and description of the springs met with along the line of the proposed road, only observing that in very few instances did it appear that any attempts had been made to augment the supply of water by artificial means. By the application of a very small amount of labour, guided by ordinary intelligence, I have no doubt that the present supply of water may be increased three-fold.

"Again starting from Montagu, which is plentifully supplied with water all the year round, we proceed to nine and a half miles, where are found two permanent springs, slightly brack. At twelve miles, one permanent, good, and strong spring. At thirteen miles, one weak spring of good water. At fourteen miles one permanent and strong spring of good water. The last three springs are some three or four hundred yards from the road, and are on Government ground. At twenty-one and a half miles one small permanent spring of very good water. At twenty-eight miles (Adder's Fontein), one mile from the road, one permanent strong spring, slight metallic taste. At thirty-one miles one permanent strong spring of excellent water; one weak spring in its vicinity, on Government ground, reported good. At thirty-two miles one permanent spring, metallic flavour. At thirty-three miles (Sand Fontein), at a distance of half a mile from the road, four fine permanent springs, slightly metallic flavour. At thirty-four miles one weak spring, good. At fifty-five miles the Touws River (Fisgat), permanent water. At sixty-three miles (Brewellskraal) one permanent fine spring, water of very good quality. At sixty-five miles three or four springs of good water, one permanent, and the water of very good quality. At sixty-nine miles, at half a mile from the road, a permanent water-leading; the water of good quality. At seventy-five and a half, Dwars River, permanent water. At seventy-nine miles (Ladysmith), permanent water. At eighty-four miles a dam, water not permanent. At eighty-six and a half miles permanent stream of good water.

"To this list I may add that at seventeen and a half miles, at thirty-eight and a half miles, and at forty-five and a half miles, good dams to store a large quantity of water may be constructed, if deemed necessary, at very moderate expense. It will be observed that the distance between Montagu and Ladysmith by the proposed trace is seventy-nine miles, and that the greatest distance between permanent waters is twenty-one miles."

From Robertson it was reported in 1869 :—"Throughout the year this division has been favoured with the most seasonable rains, and in consequence crops of every kind have been abundant. The vintage was excellent; but the producers received so little for their brandy and wine, partly owing to a monopoly existing here, that they were not much benefited by it. The approaching vintage (January, 1870) bids fair to be even better than the last. . . . Pasturage has been good all the year through. Stock of every kind has been in excellent condition; and, on the whole, the inhabitants have cause to be grateful for the blessings they have all enjoyed."

In 1874 it was reported :—"The year 1874, taken on the whole, has been a prosperous one; the grain crops have been good, although the wheat has in some parts of the district been slightly affected with rust, and the vines promise an abundant harvest.

"The rains have not been sufficiently severe to thoroughly saturate the ground, but they have been frequent, and consequently the veldt is looking fresh.

"Tree planting has been carried on to a limited extent in this village, by means of convict labour, and ostrich farming has been introduced into the division, the country of which appears suitable for the birds.

"The road through Cogman's Kloof is being carried on in a most workmanlike manner, but the labour question in this division, as in all other parts of the Colony, has become serious, and is evidenced by the comparatively slow progress of this work.

"Good roads are the greatest want in this district. The Cogman's Kloof is certainly the first consideration, and really ought to be taken in hand without further delay. It is well known that Montagu and the farms in its neighbourhood are unsurpassed as a grape and grain growing district; and yet this fertile country has no outlet for its produce, excepting along the bed of a mountain torrent, which after rains becomes so swollen that all communication with the outer world is cut off. Every year the Divisional Council spends a large sum of money upon what is by courtesy called a road; and every year the floods carry away the so-called road, and the construction or repairing process has again to be repeated.

"In July, 1868, the Civil Commissioner received from the Colonial Secretary a number of young white mulberry trees, which he planted in his garden. Nearly all of them rooted at once, and they have grown most luxuriantly, thus proving that this part of the Colony is well-adapted for seri-culture. One enterprising person (Mr Keet) reared on the wild mulberry, this last season, upwards of 20,000 silk-worms. The Civil Commissioner also reared a number on the leaf of the white mulberry, but there appeared to be no difference in the size or healthiness of the worms; neither was there any perceptible difference in the size or quality of the cocoons. Mr Keet has upwards of 20,000 cocoons for sale, which he intends to send to Capetown. Several persons have obtained cuttings from the Civil Commissioner, and in the course of a few years the white mulberry will probably be quite as common here as the wild or Cape mulberry."

The division of Riversdale lies between the Zwartebergen and the sea; bounded on the west by the Krombeck River, and on the east by the Gauritz. It has an area of 3,129 square miles, comprising part of the Little Karroo; but it is known by the name of Grasveld, from its abundant pasturage. The productions of the division are principally wool, horses, and corn.

Mr Rose Innes, then Civil Commissioner of the district, wrote in 1862:—"This division presents many facilities for irrigation in several parts. The Vette River, Kaffir-Kuils, and Gauritz Rivers, on the southern side of the Langeberg range of mountains, are all capable of being directed into channels for irrigation purposes; and the Brand, Groote, Buffels, and Hoeks Rivers, in the Karroo between the Langeberg and Zwarteberg mountain, present perhaps greater facilities for such works than those already enumerated."

The only plan for irrigation, said he, which has been proposed in this division is that of leading out the Vette River, with the view of obtaining a more certain and ample supply for the village; and in regard to this he refers to a memorial on the subject, addressed to the Government by the inhabitants of the village, and to correspondence attached thereto. This I have not seen.

In 1865 it was again reported by the Civil Commissioner:—"Parts of this division are well adapted for the construction of extensive dams, and the soil retains moisture for a long time where water is collected in any quantity; but enterprise to undertake such works is unhappily rare amongst the agricultural population. An idea seems to be prevalent that Providence looks upon the use of artificial means to develop the resources of the country as impiety, and such efforts are deprecated rather than encouraged by some farmers."

And he goes on to say:—"It is much to be regretted that these persons are so slow to believe in the injurious effects produced by burning the veld, which they do periodically and as a matter of course. Grass plants have almost been extirpated from this part of the Colony; and the inhabitants are daily using zealous efforts, not only to complete that work of destruction, but also to deprive the sources of the fine streams running through the division of all shelter from the evaporating influences of the sun's rays. Education can alone dispel the erroneous views which are held on this subject. The Forest and Herbage Preservation Act is virtually a dead letter, from the absence of any wish to bring offenders against it to justice."

And in the annual report it is stated:—"The sufferings which have been occasioned by the drought during the past year were unprecedented in the history of this division. The losses in live-stock, more particularly in the Karroo, have been extremely severe; and the failure of the wheat crops has necessitated the importation of meal and flour in large quantities into a division which generally sends considerable supplies into other parts of the Colony.

"The rust has affected the oat-hay crops to a great extent; and much difficulty is now experienced by people whose vocations render the possession of horses a necessity to them, in finding provender for these animals. In fact, food of all descriptions, both for man and for beast, is excessively dear; and this in a locality noted, in ordinary seasons, for plenty and cheapness.

"Fruit has not been abundant. Grapes have been rather more plentiful than any other fruit; and the *oïdium* does not appear to have been so extensive in its ravages as it had been for some years past. The other kinds of fruit are to some extent diseased, and have not in most cases ripened properly; the heat of the sun has baked the fruit before maturity.

"The worst is not yet; for, if the rains do not speedily fall, a famine must ensue. Already considerable distress prevails amongst the working classes, especially the coloured people at the missionary institutions of Amalienstein and Zoar."

In 1866 it was reported:—"The prospects of this division are widely different from what they were at the commencement of the year.

"The copious rains which fell during the months of April, May, and October, have furnished an abundant supply of water for irrigation and for stock purposes, besides having enabled the farmers to plough and sow more extensively than they had done for many years previously.

"The crops have yielded a fair average return, except that the wheat was more or less injured by rust, and in some cases entirely destroyed. Wheat has, however, already declined in price, and it is probable that ultimately breadstuffs will be cheaper in the course of the year than they have been for a considerable time past.

"Fruit is abundant and of good quality. The grapes are slightly affected by the *oïdium*, though not to such an extent as at one stage in the progress of the disease.

"What, with the signs of improvement just mentioned, and the high prices realised by wool at the sales held in October, the farmers have become more hopeful as to their future prosperity.

"The cultivation of tobacco is progressing. One farmer, who has met with great success in that department of agriculture, forwarded a sample of his tobacco for competition at the exhibition in Capetown. The soil is peculiarly adapted for the growth of this plant, and it is said that the finest colonial-grown tobacco is raised in this division.

"Some of the farmers are now beginning to perceive the advantages of improving their stock by the introduction of new and imported blood; but the belief in this principle spreads slowly, and there is ample scope for advancement in that respect.

"The epidemic called horse sickness committed great ravages at one period of the year, so much so as to necessitate the employment of mules instead of horses for agricultural purposes, to a greater extent than ever before. . . .

"In the month of August last, the magistrate, in conjunction with the Municipal Commissioners, selected a site for the construction of a reservoir to collect rain water. The spot chosen is on the south-west side of the village, on the slope of a neighbouring ridge above the level of the village. By the permission of His Excellency, the Governor, the prisoners under sentence of hard labour have since then been engaged in excavating the tank, and forming a very strong embankment on the lower side. The extent of the excavation, as far as at present carried on, is about 220 feet by 130 feet, the depth at the upper side being about ten feet; and, when finished, the dam will be of considerable benefit to the inhabitants by furnishing the dry erven with a permanent supply of water for domestic purposes, and more especially for the extinguishment of fires taking place at a distance from the watercourse. It is intended to convey the water through the village by means of iron pipes, but some time must elapse before the work is completed."

Of Riversdale, in 1869, it was reported :—"This division in the early part of the year was visited by a very severe drought, so that our agriculturists doubted whether they would not lose all opportunity of putting any cereals into the ground. The fine rains at the end of August and beginning of September induced them, although late, to sow their crops. The consequence will be that we shall have rather a late harvest. The unprecedented rains, in fact floods, which occurred in October, have completely changed the aspect of the country. Patches of emerald green are everywhere seen; pasturage is now abundant, and live-stock of all kinds are in most excellent condition; but ravages committed by the floods, and the damage thereby to many has been very great, especially in our Karroo. The Groot River rose to an extent which had hitherto not been witnessed by the oldest inhabitants. The farms which abut on the river suffered the most. A few days before where fine vineyards and flourishing crops of wheat and other grain were to be seen, were completely swept away, and nothing is now to be found but mounds of sand. The beautiful mimosa trees which lined both banks have all disappeared, giving the place a look of desolation and desertion. The river itself has completely changed its course, taking a snake-like bend through the different farms; the fords have become impassable, and new fords will have to be made. The roads have been fearfully cut up, and the loss in stock and produce very great. Two human bodies were seen floating down the Gauritz River, as also a number of oxen, horses, mules, goats, etc.; also a waggon and team of oxen. The mountains have been covered with snow to a late period of the year, without precedent in this division.

"The annual wool fair was held on the 26th October. The quantity offered for sale was much greater, being 295,150 lb. to last year's 127,860 lb. From the experience of last year, our farmers had determined upon bestowing more care and trouble to get up their clips. A number of the larger wool-growers went to the expense of erecting pumps in the rivers. They have now learnt that the more they improve their flocks, and the greater the care, the better prices they realise. The majority intended to exhibit their wools in bulk, but the unprecedented rains a few days before the market frustrated their wish and intention. . . . The breeding of Angora goats is now claiming more of the attention of the farmers, and thoroughbred Angora rams are in greater demand. Ostrich farming is also meeting with a fair share of consideration. . . .

"The increased cultivation of land on this side of the Platte Kloof, the demand for agricultural labour, and the cheapness of provisions, have tended very greatly to the diminution of crime.

"The grape crops promise as yet to be above the average, the *oïdium* having shown itself but in small patches. It is to be regretted that the wine farmer has not a better market for his wines; this would, perhaps, induce him to manufacture a better article than he does at present."

In 1874 it was reported:—"Although the crops in some parts of this division, in their early stage, suffered from drought, yet the recent rains were in time to save them, and the yield has been, in most places, much beyond the average. The heavy rains on some farms damaged the standing grain; but neither the loss from this cause, nor from rust, is material. The rains, however, while they have done so much good, have done a considerable amount of damage—roads cut up, ground washed away, and other indications, show how heavy the floods have been. Such drawbacks, it may be said, are fully compensated by the general good done; but to the farmer more consolation would have been afforded by the knowledge that he possessed adequate dams full of water to meet any drought that may be impending. Few farmers, however, possess any dams at all for the purposes of irrigation, and so the future must take care of itself. It seems a pity that, from time to time, a certain portion of convict labour is not made available for the construction of dams on private property. I think that arrangements between the Government and farmers wishing to have such labour under certain conditions might be made mutually advantageous. The veldt has not for years looked, at this season of the year, as it does now, and cattle have a good chance of getting through the summer in good condition.

" Ostrich farming is greatly on the increase, and up to the present time has been very remunerative. . . . The development, however, of this 'easy method' of farming will, in all probability, detrimentally affect other farming pursuits. In some cases less attention will be paid to the growth of wool than has been the case; in others, the seed lands will be turned into runs for the birds, and so the produce of grain will be lessened.

" As far as improvements in agriculture, properly so called, are concerned, I am sorry to say that I can report no advancement. If any means could be devised of establishing a model farm in each division, I think that many of our farmers would in the end see the benefit of a higher kind of farming than is now followed, and they would make greater efforts at improvement. We have, for instance, on most farms, little or no attempt at a succession of crops, but the same land is sown with the same grain year after year. The result of such a process, and the neglect to supply proper manure in anything like a proper quantity, cannot but impoverish the land, and cause the yield to be unremunerative and the land at last is left untilled for years. . . .

" During the close weather in the months of January, February, and March, and the great scarcity of water (owing to the severe drought), typhoid fever, slow fever, and dysentery prevailed throughout the district. There was also a good deal of ulcerated sore throat about the end of March. The health of the division was much better during the autumn and winter months; hooping-cough prevailed, of a mild type, but proved fatal in many neglected cases. There has been much illness during the last four months of the year, owing to the close weather."

Of 1875 it was reported:—" It may fairly be doubted whether this Division has seen a more prosperous year than that of 1875. The rains, coming from time to time at short intervals, have caused the Karroo as well as the grass veld to put on an appearance which they have not worn for many years. The result of this, of course, is that cattle and sheep, as well as other animals, have not only been kept in good condition, but there has been comparatively little mortality amongst them. In an equal degree, also, have the lands set apart for crops of different kinds been benefited. The heavy floods, it is true, have washed away some portions of the cultivated ground on some farms, but the greater returns on others have given more than an equal compensation. It is a matter of regret that more advantage has not been taken of the rains that have from time to time fallen than has been the case. One reason, no doubt, is the fact that the attention given to ostrich farming has interfered with other branches, both agricultural and grazing. Thus, while the favourable seasons have given us wool clips, fair and creditable, they might have given clips which would have been very excellent in quality and much larger in quantity; and, in like manner, had not ostrich farming interfered with agricultural pursuits, strictly so called, we should in all probability have increased our grain and other crops to a very large degree. That ostrich farming has become very general in this division the most cursory inspection will prove, and that most who have tried it have made money is a matter beyond dispute; whilst but few have been losers who have persevered in it and exercised the most ordinary care. Yet, although very many individuals have made money by this new branch of farming, I think it may be a question whether it has brought much real wealth into the country. At the present time buyers of birds are shy, and the seemingly steady fall in the price of feathers will make them more so; and if it should come to pass that prices should fall so low as to prove not to be remunerative, and ostrich farming have suddenly to be generally abandoned, there will be serious loss to many who have more recently given very high amounts for good birds, and who have thus drawn off their capital from other things.

"The breeding of horses and mules, I am glad to say, has again become more general, both in this and in adjacent divisions, than has been the case for some years, and unless some sickness break out in a virulent form, it will no doubt be steadily remunerative. Greater means of transport are being yearly required, and with the development of railways animals for transport purposes will be in greater demand."

The following is a tabulated statement of meteorological observations at Amalienstein, the mission station in this division; also a tabulated statement of number of observations of the wind at the several points :—

AMALIENSTEIN—SEVEN WEEKS' POORT. Lat. 21° 30′ S.; Long. 3° 0′ E., of Royal Observatory Capetown.

FROM DAILY OBSERVATIONS AT SUNRISE, NOON, AND SUNSET.	Jan.	Feb.	March.	April.	May.	June.	July.	Aug.	Sept.	Oct.	Nov.	Dec.	For the Year.
1867.	degs.	degs.	degs.	degs.	degs.	degs.	degs.	degs.	degs.	degs.	degs.	degs.	degs.
Mean maximum temperature of the air,	95·2	88·3	83·8	77·2	67·1	63·6	65·0	64·9	68·0	74·8	81·7	89·8	76·6
Mean minimum temperature of the air,	65·7	64·2	61·3	57·0	48·9	45·7	44·4	46·0	45·7	51·8	57·8	62.6	54·3
Approximate mean temperature of the air,	80·5	76·2	72·5	67·1	58·0	54·6	54·7	55·4	56·8	63·3	69·7	76·2	65·4
Mean range of temperature of the air,	29·5	24·1	22·5	20·2	18·2	17·9	20·6	18·9	22·3	23·0	23·9	27·2	22·3
Highest temperature,	107·4	100·6	93·9	92·7	74·8	77·0	74·8	72·5	90·5	91·6	97·3	101·8	107·4
Lowest temperature,	56·7	53·4	52·3	43·2	38·7	35·4	33·1	33·2	35·4	41·0	43·2	52·2	33·1
Extreme range of temperature on any one day,	42·7	36·0	31·2	31·4	27·0	30·0	36·5	32·6	40·5	40·0	36·0	37·0	42·7
Least range of temperature on any one day,	7·9	5·5	9·0	6·8	2·3	1·2	8·0	2·2	5·6	6·8	8·0	13·5	2·2
Quantity of rain in inches,	1·2	3·10	2·03	4·50	9·35	2·25	1·50	7·14	1·53	2·02	6·00	0·57	41·19
Greatest amount of rain on any one day in each month,	0·8	1·95	1·52	1·50	5·50	0·40	0·60	2·50	0·52	0·60	2·10	0·52	5·50
Number of days on which rain fell,	4	5	5	9	4	5	8	10	10	6	7	2	75
	degs.	degs.	degs.	degs.	degs.	degs.	degs.	degs.	degs.	degs.	degs.	degs.	degs.
Mean temperature of the air, from observations at sunrise, noon, and sunset,	80·1	75·4	72·5	67·9	59·3	55·4	55·9	56·2	57·3	63·1	69·5	75·3	65·7
1868.	degs.	degs.	degs.	degs.	degs.	degs.	degs.	degs.	degs.	degs.	degs.	degs.	degs.
Mean maximum temperature of the air,	91·8	85·1	81·5	73·6	69·6	65·9	63·5	67·5	72·7	74·5	81·9	87·7	76·3
Mean minimum temperature of the air,	63·7	62·4	60·1	51·8	47·8	41·2	43·0	47·5	50·7	53·0	58·3	60·1	53·3
Approximate mean temperature of the air,	77·7	73·7	70·8	62·7	58·7	53·5	53·2	57·5	61·7	63·7	70·1	73·9	64·8
Mean range of temperature of the air,	28·1	22·7	21·4	21·8	21·8	24·7	20·5	20·0	22·0	21·5	23·6	27·6	23·0
Highest temperature,	110·7	97·2	99·5	84·9	80·4	77·0	77·0	83·7	97·3	89·4	100·6	97·3	110·7
Lowest temperature,	59·0	56·8	52·3	38·8	35·5	27·5	27·5	36·5	36·5	41·0	45·5	50·0	27·5
Extreme range of temperature on any one day,	39·4	33·7	37·1	39·4	34·9	40·5	34·9	36·0	34·9	37·1	37·1	39·4	40·5
Least range of temperature on any one day,	9·0	6·7	9·0	6·8	13·5	10·1	2·3	4·5	4·5	4·5	11·3	12·4	2·3
Quantity of rain in inches,	1·37	3·85	2·48	1·04	0·96	0·52	5·08	1·56	0·37	1·87	1·83	0·26	21·19
Greatest amount of rain on any one day in each month,	0·50	2·50	1·25	0·51	0·26	0·25	2·75	0·39	0·21	0·51	1·02	0·17	2·50
Number of days on which rain fell,	6	6	9	6	6	8	5	9	6	9	5	3	73
	degs.	degs.	degs.	degs.	degs.	degs.	degs.	degs.	degs.	degs.	degs.	degs.	degs.
Mean temperature of the air, from observations at sunrise, noon, and sunset,	77·2	71·9	71·1	63·5	59·7	55·2	54·0	57·9	62·8	63·5	70·0	73·2	65·0

Number of Observations of the Wind at the several Points of the Compass during each Month of the Year 1868, made at AMALIENSTEIN.

POINTS.	Jan.	Feb.	Mar.	Apr.	May.	June	July.	Aug.	Sept.	Oct.	Nov.	Dec.	Totals.
1868.													
N., ...	1	2	6	1	3	7	8	2	-	1	1	1	28
N. by E., ..	-	-	-	-	-	-	-	-	-	-	-	-	-
N.N.E., ..	-	-	-	-	-	-	-	-	-	-	-	-	-
N.E. by N.,	-	-	-	-	-	-	-	-	-	-	-	-	-
N.E., ...	1	-	-	1	1	-	2	3	1	1	-	-	10
N.E. by E.,	-	-	-	-	-	-	-	-	-	-	-	-	-
E.N.E., ..	-	-	-	-	-	-	-	-	-	-	-	-	-
E. by N., ..	-	-	-	-	-	-	-	-	-	-	-	-	-
E., ...	-	5	5	7	5	7	8	-	-	11	1	4	48
E. by S., ..	-	-	-	-	-	-	-	-	-	-	-	-	-
E.S.E., ..	-	-	-	-	-	-	-	-	-	-	-	-	-
S.E. by E.,	-	-	-	-	-	-	-	-	-	-	-	-	-
S.E., ...	16	16	14	11	8	2	4	14	13	11	22	9	140
S.E. by S., ..	-	-	-	-	-	-	-	-	-	-	-	-	-
S.S.E., ..	-	-	-	-	-	-	-	-	-	-	-	-	-
S. by E., ..	-	-	-	-	-	-	-	-	-	-	-	-	-
S., ...	4	1	1	-	1	-	1	-	-	-	-	-	8
S. by W., ..	-	-	-	-	-	-	-	-	-	-	-	-	-
S.S.W., ..	-	-	-	-	-	-	-	-	-	-	-	-	-
S.W. by S.,	-	-	-	-	-	-	-	-	-	-	-	-	-
S.W., ...	2	-	-	7	4	1	1	-	3	4	4	3	29
S.W. by W.,	-	-	-	-	-	-	-	-	-	-	-	-	-
W.S.W., ..	-	-	-	-	-	-	-	-	-	-	-	-	-
W. by S., ..	-	-	-	-	-	-	-	-	-	-	-	1	1
W., ...	9	12	8	3	9	9	11	8	14	6	8	10	107
W. by N., ..	-	-	-	-	-	-	-	-	-	-	-	-	-
W.N.W., ..	-	-	-	-	-	-	-	-	-	-	-	-	-
N.W. by W.,	-	-	-	-	-	-	-	-	-	-	-	2	2
N.W., ...	1	-	3	-	1	2	4	3	1	2	2	1	20
N.W. by N.,	-	-	-	-	-	-	-	-	-	-	-	-	-
N.N.W., ..	-	-	-	-	-	-	-	-	-	-	-	-	-
N. by W., ..	-	-	-	-	-	-	-	-	-	-	-	-	-

The division of Mossel Bay, is designated sometimes the division of Aliwal South, from the name of the town, which is the seat of the magistracy, situated upon a rising ground on the north side of Cape St. Blaise, the west horn of the bay. It has an area of 560 miles.

Mr Marsh, Civil Commissioner, in answer to the queries issued in 1862, stated it as his opinion that all the irrigation works enumerated in the queries issued by the honourable the Colonial Secretary—viz., the damming up of rivers or other running waters, the diverting and raising of water from such sources, the construction of dams in favourable localities, the collection of periodical supplies of water, and the sinking of wells—are practicable in the division, and could be constructed with much benefit to the landed proprietors and to others. He also stated that no plans for irrigation works in the division had been officially proposed, although such works were much needed. But he added :—"A few of the landed proprietors have recently availed themselves of the presence of Mr Kearns at Mossel Bay to obtain his professional assistance in raising water from a running stream, to their homesteads and dwellings, by means of a force-pump with wind power, but at a heavy cost.

"The late Mr John Abraham Meyer, of Brandstock, incurred a great expense in an endeavour to lead out the Little Brak River, from a very considerable distance towards its source, for the purpose of conveying the stream to his property on the high road from hence to George; but, for the want of practical and *skilful* assistance, much expense has been incurred, which, had he had such assistance, would have been avoided; and this work, so far as it has been carried, has been very imperfectly performed, and remains yet to be completed, although some years have elapsed since it was commenced. These efforts evince on the part of landed proprietors a desire to undertake and accomplish such work, and which desire I believe to be pretty general."

In his report for 1865 the Civil Commissioner of Mossel Bay, after stating the distress occasioned in the district by the drought, proceeds :—"On these few farms where irrigation has been practicable, good crops have been raised, evidencing the incalculable benefit which could be derived from the storing of water during the rainy season. The divisional council and agricultural society, foreseeing this great desideratum, have for some time devoted all their energy to meet the difficulty. The division, however, is small, and the inhabitants not wealthy; and it has as yet been found impossible to raise local revenues for the purpose, either by shares, loans, or taxes. Attention, therefore, can only for the present be directed to the ventilation of the question, and the collection of data for future operations.

"Some farmers have been induced to sink wells, and raise water by machinery from the deep holes in the beds of rivers; but, for want of skilled workmen, the results have not been very encouraging.

"In the field-cornetcy of the Mouth of Gauritz River there is now scarcely water enough for man; but on the other side of the Gauritz there is a perpetual stream on Government land, which runs to waste (some 500,000 gallons per day) into the river. This stream could, for £200, be conducted in a pipe under the river bed, which is deep mud, to a large flat of rich soil, and the water thus made available, not only for stock but for irrigation. Another stream, nearer the river's mouth, not quite so strong as the other, could in the same manner be brought across some private property at a cost (it is stated) of £50. The adjoining farms are unfortunately occupied by a host of proprietors in undivided shares. In some instances the owners are compelled to plough in one spot. Some of them are beggars, but cannot sell their shares, except to the others, who are too poor to purchase. Combination for improvement is therefore hopeless; and farms remain barren for want of water, though there is abundance close at hand running to waste. Energy and enterprise are thus, as in many other instances, paralysed by the law of inheritance, though many are heard to say—'Why should I spend my hard-earned subsistence in planting and improving the farm, when it must be sold or portioned out at my death, to meet the legitimate portion of the children?' Whilst the children, on the other hand, continue sunk in apathy, consoled with the reflection that they can do as their forefathers before them, and *must* have their legitimate portion.

"One of the most productive farms in the division, with running water on it, is inherited by two brothers in undivided shares, to be again occupied in the same manner by their two eldest sons. They must plough and share in the same spot. The necessary consequence is, that where the original proprietor realised a little fortune, his heirs now produce barely sufficient to pay taxes and support life. In this locality large reservoirs for two years' supply could be formed by barriers of stone and clay across narrow kloofs.

"The division appears generally well adapted for the formation of reservoirs for surface water. The soil is good for cereals, and, in fact, for almost any produce, so that there be a supply of water. At Brak River the gums grow to a magnificent size; so would the fir in the mountain. Clay and limestone predominate towards the sea. Some stone is here and there to be found, with evident traces of copper. Some of the clays are fit for the best potteries, as they become hard and take a polish.

"Mr L. Vincent, surveyor, who, with Mr Melville, re-surveyed the division, thus reports on its capability for dams:—'The Great and Little Brak Rivers are, in my opinion, suitable localities where, with a comparatively small outlay of capital, some 500 acres of good land might be brought under cultivation. These two rivers yield an *abundant* and *constant* supply of water, which runs to waste. There are no great difficulties to be overcome in leading the water out of these streams. I should say that for £5,000 a very good supply might be obtained in these localities. Of course, this does not include the construction of large reservoirs for storing water; but the streams themselves are tolerably strong, and would, even without dams, yield a good supply. I believe the soil at the Brak River is adapted for tobacco; and the wild mulberry would grow luxuriantly, so that silk worm culture could be taken in hand. Potatoes, too, would thrive exceedingly well; so would mealies, pumpkins, and other vegetables and fruits. The Gauritz River might also be brought on the rich karroo soil of Melkhoutenbosch; but this would be a much more expensive and difficult undertaking than those at Brak River. Large dams for the collection of rain-water might be constructed in those localities; but I do not think there is a farm in this division where a larger or better extent of rich soil could be irrigated from a rain-water reservoir than on Voggelvalley, Gauritz River. The dam would have to be made very large, and would depend entirely on rains for its supply. There is a narrow and deep kloof, well adapted for such a purpose. The cost would be considerable. Extent of available land, at least 800 acres. Little or no manure would be required.'

"Along the main line of road from the western to the eastern boundary of the division there is now no water for transport riders. A dam could be made on Government land adjoining Welbedaght, near to which there is a small spring. Another dam at Kleinberg or Stink River, where there are small springs, would enable waggons to complete the passage of the division.

"Good cattle and horses, formerly very numerous, are fast decreasing. This year it might be difficult to collect a thousand in the whole division. The sheep have increased, but the breed is deteriorating from the land being overstocked, and the want of capital to purchase fresh stock. In the course of time farms are divided through the law of inheritance. Those which were formerly, as a whole, good for pastural purposes are almost useless in small portions. Every farmer will struggle for sheep, to the exclusion of other stock, because sheep require the least labour, with the greatest remuneration. A small division cannot by sheep alone support numerous landed proprietors without some other produce. Every additional flockmaster, who, by the law of inheritance, gets a slice of a farm or an additional individual share, must, under these circumstances, be a positive injury to the division. Those who have for a quarter of a century known the condition of the Boers cannot have failed to observe that the law of inheritance has had a tendency,—let prejudicial persons state what they like,—to render the majority, in the country at least, a pauperised nation of shepherds, without capital, energy, or hope beyond the necessaries of life. Its peculiar antiquated restrictions on the will of the owner might cause a sluggish people to thrive under a Dutch national government. It certainly does not seem adapted to overcome the *vis inertiae* of the boers, and to produce a provident, active body of intelligent free men. Agricultural societies have given a temporary impulse to improvement within small circles; but the mass of agriculturists are not effectively reached and moved. How can the proprietor of the undivided $\frac{1}{64}$th share of the farm, inherited from his progenitor, dream, in such a country as this, of earning capital to purchase improved machinery, or to experiment for new produce? Capital expended, however, in the tillage of one of these farms, upon the share in produce, would, over a term of three to five years, yield a more satisfactory result than 'winkling.' The farmers could give the labour, and the capitalist could wait six months for a remunerative market price.

"Dams for stock and irrigation, on a scale which Government alone can attempt, promise, in the opinion of most men, to be effective remedies against the evils of drought, disease, and the law of inheritance. The farms, too small for grazing purposes, or being overstocked with sheep, could be made to raise cereals and other produce. Model farms for experiment, and for the education of agriculturists, would soon follow. The expenditure of from twelve to fifteen thousand pounds in the erection of reservoirs throughout the division, sufficient to hold a year's supply, would yield greater satisfaction and quicker returns than one or two larger ones holding three year's supply. Those who have unencumbered property might make application through the Divisional Council or Civil Commissioner; inspection, and report on security and locality, would follow; the board at head-quarters could select the most suitable spots; and, upon a final inspection by some appointed engineer, the sum applied for might be granted; the performance of the work to be by contract, under competent supervision. If the position of dams must be arbitrary, all farms deriving benefit should be bound *in solidum* as security, each paying its annual rent according to quantity of water allowed; otherwise mortgages already existing would have to be bought up. The voluntary system would work the best.

"Mr. Vincent, who has lived in this division for many years, and has much advanced the development of its resources, thus speaks on this subject:—'That part of the division of Mossel Bay known under the name of the Droogveldt is much in need of irrigation. The undulating surface is not suitable for large works; but isolated works on each farm, or for two or three farms jointly, would be of immense benefit. To promote these it would be necessary that competent persons guided the operations, and that the so-much-needed funds be placed at the disposal of the farmers, upon easy and safe terms. The introduction of capital for this purpose, upon debentures, or under guarantee of Government, is likely to be one of the most useful and productive debts which ever has been or ever will be contracted by the Colony. The properties will give the best (and gradually increasing) security to Government, warranting the future repayment of the debt. A first mortgage in many cases could be obtained in this division; a second bond, with other collateral security, might in special cases be sufficient, and joint proprietors might *in solidum* become liable. The security need not, and could not well, be defined by law; it might differ according to circumstances; and an efficient general board could be made responsible. The importance of works for irrigation in this division can scarcely be overstated. Within ten miles of the seaport about five hundred acres of the most fertile land could be made to yield £5,000, at an outlay of, perhaps, £3,000. This is only one instance out of many, not of equal extent, but each of comparatively great value; and there is no doubt that both soil and climate are fully worth more skill and capital.'"

In 1866 it was reported:—"It appears, from the reports of the several field-cornets and others, that the brightening prospects of the farmers on the sudden fall of rain during September and November, after the long drought, have been sadly blighted. The rain came late, and the oxen were so poor as not to admit the usual extent of ploughing and sowing. Where irrigation was practicable, the Victoria, Golden Bull, and Klein Corn have thriven the best; but all the crops have been more or less affected by 'rust;' the barley the least. The Du Toit's corn, which had been in great repute for withstanding drought, has all perished. The mealies and pumpkins have also been attacked by the 'rust' and 'rupsen' (caterpillars). Some consider that the rust is a fungus produced by the sudden heat drawing out the superabundant sap from the soft stalks of quick growth. Others are of opinion that the rust is the effect of an insect like a minute worm, which comes out of the ground about November, and which, in a particular condition of the weather, is able to adhere to the young leaves and stalks, its course being traceable up to the ear, and visible under the microscope. The insect can be plainly seen in the mealies and pumpkins. Those on old land suffered most. Tolerable crops were in some places procured on new land, alongside of one totally destroyed by rust on old land. Sulphur as a remedy failed.

"The weather was hot and 'muggy' during November and December, with fogs and drizzling rain. The dew also remained on the crops late after sunrise.

"The quantity of wool shorn was less than last year, but the prices realised were, on the whole, higher.

"More than half the lambs died; and early in the year nearly all the horses at grass died of horse-sickness and *paapies*. Sulphur in bran mashes was found to be an excellent remedy for *paapies*, more effectual than sugar and rock salt, usually administered here. On opening a horse's stomach after death, the *paapies*, half an inch long, may be seen in great numbers holding on to the stomach by the nippers, and when they are pulled off holes inflamed round the edges, penetrating the inner coating, are visible. . . .

"The veld and stock are now in good order. So far, there is a good beginning; but heavy rain is yet required. The hopeful, patient temperament of the boers, under so many years' retrogression, is worthy of admiration. If they are not enterprising in running risks for improvement, they are nevertheless self-denying in adversity, and not perhaps so liable to insolvency, of which there have been fewer cases this year than might have been expected. With all the drawbacks or reputed indolence of the boers, there is no doubt of the division eventually recovering and shooting ahead again, provided there be two or three successive good seasons. In a retrospect of 25 years, it is said that there has been great progress since the introduction of merino sheep. And yet, whatever may be the apparently increased value put upon property by appraisements for purposes of taxation or loan, it is seen that there is a decline in the prices realised at public auction. Property seldom changes ownership save in insolvent estates, assignments, or bequests, by which the transfer duty is chiefly obtained.

"There was more extensive grass and bush burning this year than usual, and yet more rain fell during the last quarter than during some previous years, and on the ruggens rather than along the forest range. If this be correct,—and it seems well corroborated,—little practical reliance will be placed by the boers on the Colonial Botanist's theory;* it must remain to them an apparent contradic-

* I had alleged that the destruction of bush and herbage by veldt burning had promoted the desiccation of the country. My argument is given in detail in the volume on "Hydrology in South Africa; or, Details of the former hydrographic condition of the Cape of Good Hope, and of causes of its present aridity, with suggestions of appropriate remedies for this aridity"—(pp. 158-215).

tion, or a residual phenomenon, until further scientific explanation. Experience has at last convinced some that burning out woody kloofs will sensibly diminish, if not stop, the water from springs. The evaporation and desiccation have far more injurious effects than are produced by the trees and ferns. The rhenoster bush, since the introduction of sheep, has fearfully encroached on the pasturage, leaving no alternative but to burn it, especially as the sheep can, on some farms, feed on the young bushes for one or two years after the burning. Any attempt at arboriculture without water is hopeless.

"Water for irrigation, and, in some parts, for stock, is the great want of the division. Climate and soil are unsurpassed. According to the theory, therefore, that 'a bad climate and a bad soil are essential to civilisation,' this, as well as other important divisions, is eternally condemned, unless some legislative measure, even at great present sacrifice, induce such a general storage of rain-water throughout the Colony as to prove the fallacy of the doctrine. Premiums awarded to those landed proprietors who make reservoirs, according to a certain scale, might, after all, be found the quickest, if not the cheapest, method. The revenue will receive it back in the increased produce and consequent imports. Proprietors of land continue gradually to make small dams for the supply of a few months. They can be made at a small cost during the idle days of summer. Two years ago one was made by four labourers, at a cash outlay of only £4 10s. It does not leak, and now supplies 1,700 sheep and goats, 100 cattle, neighbours, and carriers. Were it not for this dam the owner would have to trek with all his stock. The embankment is only made of clay well tramped down, and is some 30 to 50 feet wide, and 60 long. This answers better here than facing the water-side with rough stone, for in the latter case the crabs harbour behind the stones and eventually cause leakage. It is hoped that the example will give great impetus in the construction of numerous dams.

"Each man waits too long to see his neighbour's success. Discovery and improvement are therefore too slow, and not equal to the requirements of the times; wherefore, model farms have been so strongly advocated by the Mossel Bay Agricultural Society for groups of divisions where experiments can be made upon scientific principles for the observation and guidance of all. There can be no doubt that the investment will be returned by the country, and the institution become self-supporting like any other well-conducted farm. The soil of the division is very productive under irrigation;

and it seems a national sin that it should, with others, be almost abandoned to the feeding of sheep and the growth of wild aloes. An epidemic similar to the one which raged in Australia might here attack the sheep. Inevitable ruin must follow, for the farmers have at present literally almost nothing else except sheep and goats to depend upon.

"The people require to be roused somewhat from the lethargic effects of the law of inheritance. Since this law has been so warmly discussed, it is surprising to see how rapidly the opinion is gaining ground, even amongst the Dutch, of the inexpediency of fettering parental affection by any artificial guide, which tends to paralysis rather than vigorous life. It cannot be denied that the effect is to cause excessive sub-divisions. If such sub-division were beneficial, the young owners would not continue to trek to the Free State in search of new land. If the land could be made immensely productive by irrigation, a different effect might occur. The unnatural restraint, however, would still be as much against true principle as any coercion in religion. . . .

"The active operation of the agricultural society has been somewhat retarded for want of funds. The members of these societies have done much good; and it is very unfortunate that their continued efforts should be arrested under the erroneous impression that the refusal of State aid is a necessary economy. The farmer might as well refuse to part with any of his wheat for future seed and keep it all for bread. Such institutions may not yield any immediate tangible revenue, but they are powerful agents in developing the resources of the Colony and reviving its prosperity. The expectation is futile that this division and a few others can, by agriculture and sheep farming, afford employment and support to a large population without manufactories for the conversion of raw material. Individuals will not generally follow up slight indications of latent wealth; but agricultural societies give official reports and collect data upon which reliance will be placed by companies. Last year this society put forward two schemes,—one for irrigation, and one for model farms. Though not entirely adopted, yet the vital principle of right of passage of water appears to have received the attention of Parliament, and resulted in a proposed Act by the Committee of the House. This society was among the first, in July, 1865, to urge the matter upon the press and the other societies."

He also reported:—"Mossel Bay is suffering severely from the prevailing drought. Few of the farmers have had any return from the grain sown in 1865, and stock is perishing for want both of water and grass. The greatest distress, approaching to a famine, is apprehended, by reason of the scarceness of provisions in the country. Many must surrender their estates if sued for debt or taxes."

In the report of the Colonial Botanist for 1866 it is stated that I had "publicly intimated, during the severe drought which prevailed in the beginning of the year, that I was prepared to lecture in any place in the Colony, however remote it might be, on the causes of the aridity of South Africa, and on the remedies which could be applied, and to hold conferences on the subject, either in the lecture room or on the veld, on condition that my travelling expenses should be met, either by several applicants for my services conjointly, or by one and another as I moved from place to place."

This measure was carried out, in whole or in part, in various parts of the Colony, and having enumerated these the report goes on to state:—"An offer was made to me of free conveyance throughout an extensive tour from Mossel Bay, if I could find my way thither. And on the occasion of a conference on irrigation being held at George a letter was addressed to the Colonial Office by the Civil Commissioner of Mossel Bay, in which that officer wrote as follows:—

"'By a resolution of the Mossel Bay Agricultural Society, I have been desired to solicit the official attendance of Dr Brown, the Colonial Botanist, if he would kindly come, at a meeting of delegates from other societies at George, invited for the purpose of discussing a scheme for irrigation and model farms which the Mossel Bay Society intends to offer for criticism and for improvement. This meeting will take place about the 7th March, during the combined agricultural show for George, Oudtshoorn, and Mossel Bay divisions.

"'Such a suitable opportunity rarely occurs for the hearing of that gentleman's scientific views and suggestions for the increase of products founded on the practical observations of others. Considering that many of the farmers from four important divisions will be congregated on the occasion, the committee trust that the general benefit to be derived will justify His Excellency the Governor in graciously acceding to the request, and in incurring the small expense which will be incurred thereby. The committee had hoped to be in a position to afford the expense, but unfortunately find that the show will necessitate again an extra subscription on the part of the members. The hospitality which will be gladly shown to Dr

Brown will limit his expenditure to the transport alone between Capetown and Mossel Bay.

"'In my official capacity I am of opinion that incalculable advantage to the central part of the Colony is likely to be derived from his *viva voce* lectures by the agriculturists, who are keenly alive to their present critical state, and anxious for some practical scheme whereby labour will be made remunerative in the production of better and other articles for export and home consumption.'

"It was intimated, in reply from the Colonial Office, that no provision was made for meeting such expense. And having thus an opportunity of showing what I have experienced in many other ways, that the salary and allowance of £400 per annum is insufficient to meet the expenses which are necessarily incurred in the discharge of the duties to which I have been called as Colonial Botanist, I presented to the Honourable the House of Assembly a petition, praying that adequate provision might be made for the discharge of these duties; but it was deemed more expedient, with a view to retrenchment, to abolish the office."

Of Mossel Bay, in 1869, it was reported :—"This division has been highly favoured during the last four months of the year with fine rains. The consequence has been a larger area of land brought under cultivation than usual, and the harvest one of the most abundant known for many years. There has been good grazing in places where grass was never before seen. Cattle and stock of all kinds have been in splendid condition. The increase in lambs has been good, and meat cheaper than during the previous twelve years.

"The clip of wool has been heavier than usual, and the quality better; but I cannot report any improvement in the 'getting up' of the article.

"There has been an importation of two threshing and two reaping-machines during the past year; they have given great satisfaction to their owners and the farmers who have witnessed the work performed by them, which is likely to lead to others following the good example.

"Several persons in this and the adjoining district of Oudtshoorn have purchased some of the Angora rams lately imported at Port Elizabeth from Asia Minor, and as these divisions, especially that of Oudtshoorn, are well adapted for the rearing of goats, we may soon hope for a new article of export.

"Ostrich-farming is being tried by several agriculturists in this division, and has so far answered very well, one of the farmers having sold one year's produce of twenty-five birds for nearly £200.

"The new pass over the Ruiterbosch, now named 'Robinson's Pass,' was formally opened in ———, but shortly after, the heavy rains washed away part of one of the retaining walls for a distance of twenty-five yards, causing a stoppage to traffic.

"Crime is now reduced to a minimum in this division, principally owing to the fruitful years of 1868 and 1869, although the stringent measures in force, especially with regard to spare diet, have doubtless had a very salutary effect.

"The health of the inhabitants has been good, which may to some extent be attributed to the better supply of water throughout the district."

At Mossel Bay, Mr Gamble, the Government Hydraulic Engineer, in November, 1875, by instructions from the Commissioner of Crown Lands and Public Works, examined certain questions in regard to the water supply of that place, and gave to the Chairman of the Municipal Commissioners a report in writing, a copy of which was sent to the Commissioner of Crown Lands and Public Works.

The following is a tabulated statement of daily observations made at Mossel Bay in the year 1868 :—

Mean height of the barometer at the temperature 32° Fahr. (not corrected for altitude above the mean level of the sea), for each month and for the year, from observations at 9 A.M., 1 P.M., and 5 P.M.												
Jan.	Feb.	Mar.	April.	May.	June.	July.	Aug.	Sept.	Oct.	Nov.	Dec.	For the Year.
ins.	ins.	ins.	ins.	ins.	ins.	ins.	ins.	ins.	ins.	ins.	ins.	ins.
29·762	29·805	29·813	29·893	29·866	29·983	29·974	29·948	29·868	29·928	29·863	29·797	29·875
(1) Maximum and (2) Minimum observed.												
29·989 29·510	30·205 29·496	30·294 29·481	30·414 29·536	30·068 29·604	30·430 29·664	30·380 29·517	30·419 29·609	30·169 29·604	30·264 29·498	30·157 29·506	30·047 29·430	30·430 29·430
(1) Quantity of rain in inches; (2) Greatest amount of rain on any one day in each month; (3) Number of days on which rain fell.												
0·304 0·190 5	7·733 5·548 9	1·368 0·608 4	0·532 0·266 3	1·520 0·684 5	0·228 0·170 2	1·824 1·140 4	1·216 0·304 6	0·689 0·190 8	2·370 0·342 10	0·658 0·380 8	0·722 0·418 3	19·164 5·548 66

The following is an abstract derived from meteorological observations made at Mossel Bay in the year 1862 :—

Abstract derived from Meteorological Observations made at Mossel Bay in the Year 1862.

Months.	Mean height of Barometer at Temperature 32°.	Temperature of the Air. Mean Temperature.	Mean Daily Range.	Mean of greatest range any one day of each month.	Mean of least range any one day of each month.	Mean of greatest range throughout each month.	Mean Temperature of Evaporation at 9 A.M. 1 P.M. 5 P.M.	Mean Humidity for each month.	Rain.	Clouded Sky.	Lightning. Recorded Number of days.
	inches.	degrees.	degrees.	degrees.	degrees.	degrees.	degrees.	per cent.	inches.	per cent. of visible hemisphere.	
January,......	29·580	70·85	18·18	28·0	12·0	33·0	65·97	68·22	0·08	31	1
February,....	29·531	70·48	18·55	30·0	9·0	40·0	66·04	70·91	0·57	41	...
March,........	29·602	68·67	17·45	30·0	10·0	43·0	63·37	67·56	1·19	40	...
April, ...	29·597	65·14	16·13	27·0	8·0	38·0	61·49	76·82	1·09	51	...
May,...........	29·670	63·66	18·57	29·0	8·0	42·0	59·10	75·57	0·80	36	...
June, ...	29·547	60·06	17·48	27·0	9·0	39·0	53·92	65·11	1·11	46	...
July,...........	29·625	56·74	16·84	28·0	5·0	39·0	52·09	70·52	2·52	39	...
August,.......	29·693	56·71	17·94	25·0	10·0	36·0	52·74	70·93	2·02	34	...
September, ..	29·644	59·18	17·70	29·0	10·0	43·0	55·15	72·42	0·77	46	...
October, ...	29·568	59·32	17·68	25·0	7·0	37·0	54·75	68·05	0·88	49	...
November,...	29·571	64·97	17·80	25·0	11·0	33·0	59·81	63·99	0·37	43	...
December, ...	29·519	71·02	19·00	26·0	9·0	36·0	65·96	63·57	0·15	34	1
Mean Annual Values.	29·596	63·90	17·78	27·42	9·00	38·25	59·20	69·47	11·55	41	2
	*+ ·012	...	...	...	...	...	...	...	...	...	...
	29·608	...	...	...	...	...	...	...	...	...	...

* ·012 inch is the reduction to the Observatory Standard Barometer.

Number of Observations of the Wind at the several Points of the Compass during each Month of the Years 1867 and 1868, made at MOSSEL BAY.

POINTS.	Jan.	Feb.	Mar.	Apr.	May.	June	July.	Aug.	Sept.	Oct.	Nov.	Dec.	Totals.
N., ...	-	1	-	1	2	-	-	1	-	-	-	-	5
N. by E., ..	-	-	-	-	-	-	-	-	-	-	-	-	-
N.N.E., ..	-	-	-	-	-	-	-	1	-	-	-	-	1
N.E. by N.,	-	-	-	-	-	-	-	-	-	-	-	-	-
N.E., ...	-	-	-	-	-	-	-	1	-	-	-	-	1
N.E. by E.,	-	-	-	-	-	-	-	-	-	-	-	-	-
E.N.E., ..	-	-	-	-	-	-	-	-	-	-	-	-	-
E. by N., ..	-	-	-	-	-	-	-	-	-	-	-	-	-
E., ...	1	-	-	-	-	-	1	2	-	-	-	1	5
E. by S., ..	-	-	-	-	-	-	-	-	-	-	-	-	-
E.S.E., ..	-	-	-	-	-	-	-	-	-	-	-	-	-
S.E. by E.,	-	-	-	-	-	-	-	-	-	-	-	-	-
S.E., ...	41	32	26	15	14	8	9	21	16	24	20	22	248
S.E. by S., ..	-	-	-	-	-	-	-	-	-	-	-	-	-
S.S.E., ..	4	2	-	-	-	-	2	9	8	3	-	-	28
S. by E., ..	-	-	-	-	-	-	-	-	-	-	-	-	-
S., ...	-	-	-	-	2	-	2	5	1	2	7	3	22
S. by W., ..	-	-	-	-	-	-	-	-	-	-	-	-	-
S.S.W., ..	-	-	2	-	-	-	-	-	1	-	-	-	3
S.W. by S.,	-	-	-	-	-	-	-	-	-	-	-	-	-
S.W., ...	38	23	23	27	11	17	21	22	36	30	27	36	311
S.W. by W.,	-	-	-	-	-	-	-	-	-	-	-	-	-
W.S.W., ..	-	-	-	-	-	-	-	-	-	-	-	-	-
W. by S., ..	-	-	-	-	-	-	-	-	-	-	-	-	-
W., ...	-	-	-	2	1	2	2	-	-	2	-	5	14
W. by N., ..	-	-	-	-	-	-	-	-	-	-	-	-	-
W.N.W., ..	-	-	1	-	-	-	-	1	-	-	-	-	2
N.W. by W.,	-	-	-	-	-	-	-	-	-	-	-	-	-
N.W., ...	2	5	18	18	19	16	26	18	20	17	14	10	183
N.W. by N.,	-	-	-	-	-	-	-	-	-	-	-	-	-
N.N..W, ..	-	-	-	-	-	-	1	-	-	-	-	-	1
N. by W., ..	-	-	-	-	-	-	-	-	-	-	-	-	-

SECTION III.—*Divisions of George, Knysna, aud Humansdorp.*

The division of George lies "along the south coast, between it and the first mountain range, with an area of 2,600 square miles. It is bounded on the north by Oudtshoorn division and Lange Kloof; south by the Indian Ocean; east by the Knysna; and west by the Gauritz River and Mossel Bay division. It is considered one of the most fertile divisions of the Colony, and is both well wooded and watered. Many torrents of inconsiderable size intersect the coast line, rendering travelling along it difficult. The high road from Capetown to the frontier runs along Lange Kloof, which is ascended from the coast by Montagu Pass in the Outeniqua Mountains."

This is "a difficult defile, formerly called Mostert's Hoek, in the second range of mountains which hem in the Breede River Valley in its upper course, through which an excellent road was formed in 1853 by the late Mr A. G. Bain, and, in connection or rather continuation of the line through Bain's Kloof *viâ* Wellington, forms one of the most important links of communication between Capetown and the interior."

Several lakes or lagoons occur along the coast between Georgetown and the Knysna, affording good fishing and pleasing scenery; and along the coast are dense forests of valuable trees.

In reply to the queries issued in 1862, Mr Davidson, afterwards Colonial Treasurer, then Civil Commissioner of the division, reported that he thought it probable, from the configuration of various parts of the country, that localities might be found for works of irrigation; but that he was not aware that any plans for irrigation works within the district had been proposed."

In regard to 1865 it was reported:—"No progress in agriculture has been made during the last year, owing to the impoverished state of the country, caused by the long and protracted drought. The farmers are considerably reduced in circumstances, in consequence of the loss sustained in stock and the failure of their crops. A few who have been fortunate in being able to irrigate have reaped a good harvest; and it is to be hoped that the lesson derived from the drought will be attended with good results hereafter, and that the farmers will see the necessity of husbanding the water in good seasons."

In 1866 it was reported:—"The past year has been one of great trial throughout this division. The drought was so severe that many people could scarcely procure the common necessaries of life, and numerous applications were made for Government relief, principally by coloured people, and those residing at Pacaltsdorp Institution. Among the woodcutters the distress at one time was also very severe; but much of this was brought about by their own laziness, in consequence of their neglecting to make gardens at times when every opportunity was afforded them; for there were many patches of ground that could be easily cultivated, even in drought. As the spring approached, the country was blessed with beautiful rains, which continued till the end of the year; and, although the rust has been very prevalent, nevertheless there is every prospect of a fair harvest. The late crops, such as beans and mealies, look very promising; and, if nothing unforeseen presents itself, the supply of these articles will be abundant. At one time of the year the labourers could scarcely find work, and many of them offered to work for the farmers for their food; and, in some instances, even then could procure no work; and no doubt several were forced to steal; but there were many idle and lazy vagabonds, who took advantage of the bad times, and lived upon plunder. The prospects of the country are at present more cheering, as there is plenty of work at small wages for the honest and industrious. The drought has effected some good, for it has roused energies hitherto dormant; and several pieces of ground have been cultivated that were previously lying waste, and dams and watercourses have been made on farms where such things were never known before.

"Crime has increased considerably, and the gaol has for several months been crowded; there are at present 26 prisoners awaiting trial for Circuit and the Magistrate's Court; and 174 criminal and 258 civil cases have been disposed of during the past year."

Of George, in 1874, it was reported:—"The prospects of the division during last year have been most favourable. The division has been favoured with seasonable rains, which have tended to the certainty of a good harvest. The crops have been most promising, and with the exception here and there of appearances of rust, a return beyond compare with that of 1873 is anticipated. It is therefore to be hoped that the year 1875 will show a reduction in the price of all cereals.

"Ostrich farming is carried on with renewed vigour, and as the prices of young birds and feathers rule high, a considerable augmentation to the incomes of those engaged in this very productive source of industry must have taken place.

"It is gratifying to be able to testify to the diminution of crime, a sure criterion of the prosperity of the Colony, and a guarantee that all who care to work can find employment."

The division of the Knysna includes a great extent of fine forest, extending between the sea and the Outeniqua Mountains, which is still the resort of the elephant and the buffalo, and produces abundance of excellent timber; and the lagoon of the Knysna is the best natural harbour in the Colony for coasting vessels. Further east than the River Knysna is Plettenberg Bay, in the neighbourhood of which is found excellent pasturage. The area of the division is 1050 square miles.

In 1862, Mr Fichet, Civil Commissioner, reported to the Colonial Office that no irrigation works had been proposed in that division, and that from inquiry which he had made there appeared to be no desire at that time on the part of the landed proprietors to avail themselves of any irrigation scheme."

In 1865 it was reported :—" This division has suffered much during the past year from want of rain, but not to the same extent as many others situated more inland. During the last three years there has been a visible decline in the rain-fall. The harvests have in consequence not been good. The part mostly affected by drought is the Plettenberg Bay ward, where the average yield of the crops has been about one fifth less, as compared with former years. Water has completely failed on some farms. It has been remarked that the rain-fall has been greatest in the neighbourhood of the forests. There have been no losses of stock hitherto; and, as a rule, the coast suffers less than places situated inland. In consequence of the distress prevalent in many of the latter, trade has been and is very dull. In wood the exports have fallen far short of the average of former years. The market appears to be overstocked, and there is little prospect, in the present state of the country, that there will be much (if any) demand for this product, at least for some time to come. The quantity of wood shipped at the port during the past year was 1,119 tons.

"The Knysna forests produce timber second to none in the world for usefulness and durability. Witness the ironwood piles driven by the Admiralty into the Saltwater Lake, upwards of forty-seven years ago (1818), now as sound as the first day they were placed there. But these forests, so worthy of preservation, will in a few years be denuded of their best timber, and completely destroyed, unless steps are at once taken to avert this by stopping the practice of felling timber without regard to the seasons of the year. An increased number of rangers is necessary, and it would be well worth the expense to organise a more extensive snpervision than exists at present."

In 1866 it was reported :—" Owing to the severe drought under which the division suffered during the early part of the past year, the hopes of a good harvest were blighted. The rains fell rather too late; and consequently agricultural operations were necessarily retarded.

"Rust has destroyed nearly all the wheat crops; and where the Du Toit's corn has been sown it was totally destroyed. Rust also injured the oat crops, but not to the same extent; mealies have also been affected by it. In the gardens, caterpillars, grubs, and other insects have done much damage. Still, there has been a very fair yield; and the potato crops have thriven very well.

"A landing and shipping wharf has been constructed by a company, and ships now lie alongside, and discharge and load directly from it. The former slow and unsatisfactory way of doing this by waggons and boats has been thus superseded; and this has certainly been a vast improvement to the port. Several vessels have visited both Knysna Harbour and Plettenberg's Bay. In December, H.M .steamship *Petrel* entered both ports; she came to take up a quantity of boat knees for the Admiralty, which have been supplied, and which have given much satisfaction to the naval authorities.

"The road into the Lange Kloof is so far completed, that waggons now come into Knysna along its whole length, instead of having, as heretofore, to go round through the Little Lange Kloof; and the horn of the post cart may now be heard echoing through the forest. This new route is certainly the right one; but no doubt the day is not far distant when not only the post cart, but the telegraph will run direct from Capetown through this division by the Keurboom River and Tzitzikamma Forests to the Frontier. There is no doubt that, both in a commercial and social point of view, this division is an important one. Hitherto, from the difficulty of approach to it, owing to the want of roads, its resources have not been developed to any extent; but now that this obstruction has been partly removed, it is to be hoped that brighter days are in store, and that a few years will see it improved in every respect."

Of the Knysna in 1869 it was reported:—"In the early part of the year the district suffered fearful losses in homesteads, farm stock, forest, crops, etc., from a conflagration, which raged with great fury for several days, rendering many families houseless and nearly destitute; some were made entirely so. The losses to property of all descriptions have been estimated at £22,500, which is rather within the mark. About 2,500 acres of forest, in itself of great value, was destroyed past recovery. The origin of this fire, or rather of these fires, for they were not confined to this district, but affected also those of Humansdorp, Uitenhage, and Mossel Bay, has not been traced, although efforts to that end have been made. Many of the sufferers were relieved from actual starvation by charitable donations of money entrusted to a committee which was formed here, but whose efforts were perforce contracted in consequence of the limited means at its disposal. About £750 were distributed in relieving the sufferers. As a counterpoise to this disaster, it is satisfactory to find that the harvest has been good indeed, and also that the trade of the port has steadily increased within the last year.

"Stock has been free from disease. . . . The health of the people has been good. . . . Crime has not increased. . . . Vegetables are plentiful. . . . There will be a failure in the fruit crop, owing to the terrible heat of the February fires, which must have affected the trees.

"It is gratifying to notice that machinery, by which manual labour may be economised, is being appreciated by the farmers, and the introduction of reaping and thrashing machines prove that we are gradually but surely emerging from a state of semi-barbarism, and discarding antiquated notions—ignoring the creed which some few years ago was a favourite one with the bulk of the rural population—'My father did so, and why should not I?'"

Of the extensive and destructive bush fires referred to, details have been given in the volume on the "Hydrology of South Africa: and causes of its present aridity" (pp. 178–193), and the subject will be found reported on also by the Civil Commission of Humansdorp.

In 1874 it was reported:—"Having had no opportunity of making a tour through the division, my knowledge of its general condition must be limited, and it thus becomes impossible fully to report on any improvement that might have been made in farming, or any other branch of industry. I can therefore only report from hearsay.

"The year 1874 has been more favourable than the preceding one; the crops having turned out satisfactorily, and there having been but little rust, there will be no famine in the land.

"The season has been favourable for sheep, and large numbers of lambs have been reared. Stock has been healthy, and realised high prices. Sheep and wool are on the increase, but still the butchers have to look to the back country for their supply. Wool realised very fair prices in the market.

"Ostrich farming is fast increasing, and the birds fetch high prices. I have credibly been informed that one enterprising individual, who had only commenced two years ago, cleared £700 last year in sale of chickens and feathers of two pairs of birds.

"The wood trade continues brisk, and monopolises all the labour."

In 1875:—"The general health of the division during the past year has been good. With the exception of a few cases of low fever there have been no epidemics or malignant diseases of any kind.

"Agriculture in general has been progressing. Ostrich-farming has been on the increase. Sheep-farming has not been neglected, though the butchers have to look to the back country for their supply of slaughter sheep.

"Trade seems to be in a healthy state, and the district increasing in wealth, although not so rapidly as the surrounding divisions. The enterprising firm of William Lloyd & Co. has imported a steam-engine of 12 horse-power, and erected in the Township a steam saw-mill, which is now in full operation.

"The wood trade is on the increase, and licenses to the amount of £1,078 have been taken out during the year.

"The crops have turned out satisfactorily, although it was feared that the heavy floods of October had caused irreparable damage to the lands.

"There has been more crime this year, but most offences were trivial. . . . No cases of cattle-lifting nor of thefts of sheep."

Of the district opened up by the new line of road from the Knysna to the Lange Kloof, I may state that Mr C. L. Stretch, while engaged in executing the preliminary arrangements necessary for the making of that road, forwarded a memorandum on the character and capabilities of that portion of the district through which it was contemplated to form the new road. And to this memorandum Mr Stretch, in reply to inquiries I addressed to him, referred me, as expressive of the views which he still held. In it, alluding to the plan of survey, then in the possession of the Road Commissioner, he says,—
"The section on the plan beginning at A. to Iron Neck embraces four miles of open country, and the intersection of the forest designated Sutherland's line.

"Within the first sub-division three or four good sheep-farms, or forty or fifty lots, not exceeding ten morgens each, could be advantageously disposed of to individuals employed as wood-cutters, and cultivated with some advantage.

"The Rev. Mr Squibb informs me that a number of people, in moderate circumstances, within his parish, are most anxious to avail themselves, by purchase or otherwise, of similar locations; but the country is particularly adapted for the depasturage of sheep.

"On Sutherland's line of the forest country, two lots only were disposed of to private parties, *vide* Ford's survey, and at least forty more could be sold on either side of the proposed line of road to Lange Kloof; but I am of opinion that, by alienating this description of Crown property, the interest of Government is injured, and thereby deprived of a steady increasing revenue, derived from the sums paid by the wood-cutters for their license to cut timber. With deference, therefore, I should recommend that all the forests of the Knysna be retained and controlled by Government, as at present, under the superintendence of the conservator and his officers; whereby, if advantage is taken of the intersection through the principal forest which I have recommended, and a system established of slip-paths branching from it into portions hitherto inaccessible to the wood-cutter, a considerable revenue will be received annually.

"From Iron Neck, in section B., to the division of waters of Kruis River and Klien River, the country extending west to the Gowna River and north to the Kruis River, on either side of the contemplated road, eighty or a hundred moderate sheep farms and agricultural lots could be profitably disposed of. The demand for such farms has been increasing for some time past, and I have no doubt they would be occupied by an industrious class of people. The country of the Gowna and Kruis Rivers is the most luxuriant pasture country in this district, and presents a most favourable opportunity, from its vicinity to the forests, to people of limited means acquiring a comfortable and productive location.

"Maize ranges over almost the entire district, although there is a considerable difference in the productive capacities of the different parts of it, but it excels in growth in the Gowna division. The grape, the peach, and the banana, fruits originally of warm climates, are in these parts splendid in growth. The country is copiously supplied with water issuing from the forests, and, therefore, the springs are not influenced, as in other divisions of the colony, by drought. Tobacco, and more particularly the finer and more aromatic qualities, has similar climatic affinities. Sweet potatoes, cane, and rice would also yield a productive return—particularly the latter grain, in the marshy lands lying by the forests.

"The valley of the Knysna has been so little explored, and its impenetrable woods and luxuriant vegetation throw so many impediments in the way of the botanist and geologist, that a long time must yet elapse ere we can hope for satisfactory intelligence. Beds of marl clays, of different colours, and porcelain clay, occur frequently, and sandstone, of which two churches have been partly built, and proved its utility here.

"A superficial view of the agriculture of the Knysna district argues a country of the most exuberant fertility, in which are not reared many products of the earth that might succeed, whilst the limited number produced are cultivated in a rude and slovenly manner.

"The wasteful manner in which the operations of procuring timber has hitherto been carried on is already beginning to attract the notice of the conservators; and it would be a great acquisition, and beneficial to the public interest, if foresters were encouraged by government to emigrate from Europe.

"The port, or harbour, is the exact miniature of Rio Janeiro, with a lake of ten miles in circumference, and the channel has depth for vessels drawing twelve feet of water, which can approach the Knysna town within pistol shot, and ride in safety at anchor.

"I make this statement upon the high authority of Admiral Sir Pultney Malcolm, Capt. Pole, of the *Maidstone*, Commodore Owen, and other naval officers, regarding its favourable situation and convenience; and it would be presumptuous in me to make any further allusion to this splendid and beautiful harbour of Knysna.

"The Admiralty still possess a considerable tract of ground adjacent to the township, for naval purposes, and three vessels have already been built, whilst the timber in this part has been pronounced almost inexhaustible for shipbuilding purposes."

Such is the account given by one spoken of in the Colony as well qualified to judge of the agricultural and horticultural capabilities of the district. In other parts there are extensive pasture land; and these, enriched by heavy dews, as they are, are said never to lose their verdure, even in seasons of the greatest drought. This is a more favourable account of these than I would be disposed to give; but I do not know a district in South Africa more favourably situated in regard to moisture than is this. The position of the mountain ranges in relation to the sea is conducive to securing a large deposit of moisture; the extent of the forest is calculated to regulate this deposit, and to prevent both its speedy evaporation and its rapid flow into the sea; and while want of water is the great complaint elsewhere there is little complaint of this here.

To the east of the division of Knysna is the division of Humansdorp, with an area of 2,630 square miles. It is a broken and mountainous region, lying between the coast and the Winter Hoek Mountains. There is valuable timber in the forests of Tzitzikamma. The resources are chiefly agriculture and rearing cattle; and a considerable quantity of wheat is grown.

To the queries issued in 1862, Mr Cox, Civil Commissioner, replied that in that district hitherto damming up or diverting the course of running water had sufficed to irrigate all farms lying low and contiguous to the stream; and that, except in a few and unimportant instances, all the farms in this division either lie in positions where water can easily be led to them, or where, as in the case of higher-lying lands, it would be almost impossible to lead to it.

Mr Alfred Kennedy, land surveyor, Humansdorp, supplied to me the following additional information :—

"With regard to your circular of 1st June, 1865, I fear I can afford you very little information. I am quite unacquainted with the karroo of these parts; but in the upper districts I have seen many places where the water issuing from narrow gorges—'poorts' they are called in this country—might have been collected in times of flood to any extent, and led over the country in beneficial irrigation in times of drought; and I have often wondered that advantage was never taken of such natural capabilities. When I have spoken of this to the inhabitants, the invariable reply has been,—'Oh, the ground is too porous, the water will escape.'

"The Crown lands in this division are, for the most part, very inaccessible, and form what are called the Congu hills, or rather mountains. The Congu river runs very deep, and carries annually a great body of water into the sea. I believe it is only on one farm that its water is used for purposes of irrigation. One enterprising young Boer of my acquaintance, at Nautenbach, whose farm is some two miles from the river, has conceived the project of raising the water to the top of the precipitous cliffs, which from its banks is, say 1,000 feet, by means of hydraulic appliances, and leading it thence over his farm. This is no doubt possible; but I fear the expense will be too much to be remunerative."

In regard to the division of Humansdorp, it was reported in 1865: —"The twelve months just passed have been the worst the division has known for many years. Drought, scarcity of money, shortness of credit, sickness in man and beast—everything, in a word, has been working injuriously against the farmer.

"The crops, where there are any, have been most indifferent. Prices consequently are very high. Meal sells at £2 10s., mealies at £1 10s., and potatoes at from 15s. to 18s. a muid.

"There has not been a great deal of lung sickness among the horned cattle; but horses have been, and still are, dying at such a rate, and those of them that are in health are in so low a condition that riding seems almost gone out of fashion. Nearly every horse in the Gamtoos River and Zuurbron wards that has been allowed to run has died. Nor has sickness been confined to the brute creation. Fever of all kinds, but especially gastric, has been very prevalent throughout the division—in many cases with fatal results.

"Land has decreased in value. The twenty lots of Crown lands put up for sale in March last realised very low prices. Applications to purchase Government land are not nearly so numerous as they were some two or three years ago, and leasing is greatly out of favour."

And yet not far from Humansdorp is the prosperous Missionary Institution of Hankey, where, by hydraulic works executed by the natives alone, under the sole direction of the missionary, carrying water of the Gamptoos by a tunnel through a mountain range, and again by pipe under the river at a lower level, extensive fields on both sides are irrigated and rendered abundantly and profitably productive!

In 1866 it was reported:—"The year just passed has been, on the whole, a bad one. The division has suffered much from the drought; the crops have been to a great extent a failure; the *nieuw ziekte* has carried off many of our horses, and lung-sickness among the horned cattle has also been prevalent.

"The applications to purchase Crown land have ceased; but, when the lots already applied for and surveyed are put up for sale or lease, fresh applications will no doubt pour in.

"The agricultural society is working well. Many of the farmers are giving much attention to the improvement of the breed and condition of their stock."

Of Humansdorp in, 1869, it was reported by the Civil Commissioner:—"The year just closed will long be remembered in this district. On the 9th February, a destructive bush-fire, urged on by a furious wind, swept like a fiery torrent over this and the adjacent districts, extending far and wide on either side, and leaving behind it many a homestead in ashes, and in some sad instances the charred remains of men, women, and children, and even of swift antelopes and hares, overtaken in their flight. Towards the close of the year, the antagonistic element, water, rushing along in foaming torrents, where but a few days before little rills had trickled, carried away many a promising crop, buried vineyards under sand, destroyed roads, and impeded postal communication.

"Not since the year 1847 has the Gamtoos River been seen so swollen. The ferryman was forced to desert his house, and to seek safety higher up the sloping banks. The punt, though well secured, could not cross, and even boats were with difficulty got over. At one spot, Diep River—where in ordinary weather there is not more water than a three-inch pipe could carry off—so deep and strong was the current at the time that the post could only be got over, after a delay of sixty-two hours, by making oxen swim over with it.

"For five days previously the rain had fallen almost incessantly, the wind coming from the south-east, or a little to the south of that point; and it is difficult to imagine what would have been the consequence of a continuance of the down-pour for twenty-four hours longer. . . .

"Various theories have been put forward to account for the extraordinary extent of country over which, on the 9th February, fires burst out and raged simultaneously. The intense heat on that day (the thermometer showed 139 in the sun, it is said), and the great distances apart at which fires broke out, led some to suppose that the sun's rays being concentrated into a focus by pieces of glass or other accidental lenses, situated in favourable positions, converted these into burning glasses, and thus originated fires wherever the combination of circumstances rendered it feasible.

"I believe the following to be a more reliable, if less ingenious explanation. A very wet season about the middle of 1868 had been succeeded by drought and great heat, which prevailed for some months before the fire occurred. The heat had prepared for instant ignition the grass, shrubs, and brushwood then unusually plentiful, in consequence of the heavy rains that had preceded the dry weather. Veld-burning—that barbarous system, or want of system, of agriculture, as it has well been termed in a recent publication—had been going on for some time, and on a calm night bright dots and streaks, and reflections in the sky marked the localities where it was being practised. The heat was most intense on the 9th February; a scorching hot wind from the north, blowing like a sirocco, withered and dried up all that came within its influence. Everything, therefore, combined to make ready for combustion the plentiful grass and bushes, so that where there fell cinders or sparks (from the fires kindled to burn the veld), which a strong gale of wind bore along to an almost incredible distance; a new centre of flame instantly burst into destructive activity, and this, in its turn, gave origin to many others.*

"The loss of life in this district alone amounted to twenty. Of these four were Europeans, and the rest natives. Of the Europeans three consisted of a mother with an infant in her arms and a child by her side, who fled before the fiery blast till they were overpowered, and sank victims to the flames.

"The damage to property was estimated at £16,364, exclusive of a portion of the Government forest, calculated at £5,000. Other districts have nobly come forward to assist in succouring distressed fellow-creatures, and as much as £1,389 in cash and £512 in kind, has been distributed to sufferers by this great calamity.

"The grain crops, watered by abundant and seasonable rains, have in most instances yielded a fair return, but rust has considerably blighted the prospects of several who, but for it, might have harvested a good crop.

"The rise in the price of wool (though somewhat affected by recent accounts) has revived the hopes of many farmers who were beginning to fear that it could not be produced here at such rates as would enable them to compete with more favoured countries.

"Thefts of stock have not been very rife, possibly because the scarcity of labour, in consequence of a partial exodus of Fingoes and Kafirs, has induced nearly all who could work to earn an honest living. Few cases have arisen that could not be dealt with by the Court here under the increased jurisdiction given by recent Acts.

"The cultivation of tobacco still continues to engage the attention of agriculturists, and some very creditable samples have been exhibited.

"Owing to the heavy rains that have fallen, pools of water (vleys) have been formed, and have been frequented by wild duck and other aquatic fowls, which had been almost unseen for several seasons.

"Generally speaking, the health of the inhabitants of this district has been good during the past year, and, notwithstanding the visitation of fire and flood, they have many reasons to be thankful."

In 1874 it was reported:—"The past year has been, in many respects, just the reverse of its predecessor. In 1873, the complaint throughout the division was the continued drought, which paralysed farming operations to such an extent that it was impossible to get a

* Of this conflagration detailed accounts are given in "Hydrology of South Africa"—(pp. 178–194).

plough into the ground, excepting in a few more favoured spots in the neighbourhood of the Gamtoos River, where the lands could be irrigated. In the early part of 1874, the fine rains that fell gave every promise of good crops, and restored the land in every part of the district for grazing purposes, but later in the year the rains continued to fall so heavily, washing away roads and bridges, and saturating the soil to such an extent that the old complaint of rust, particularly in the wheat crops, soon made its appearance, and continued to increase so rapidly that many of the crops had to be reaped before arriving at maturity."

And in 1875 :—"The year just closed has been a most favourable one for all agricultural pursuits, and as a pleasing result the crops have turned out so well that on many of the farms there were not sufficient labourers to have them properly harvested.

"In some parts of the district, especially along the banks of the Gamtoos River, the lands and crops suffered very much from an over-abundant supply of water, and at several localities where the river overflowed its banks it not only carried away the standing crops, but also the soil, damaging in many places the water furrows which had been constructed at considerable expense by the proprietors for the purposes of irrigation.

"The year has also been a favourable one for all kinds of stock. Horned cattle and horses were in healthy condition, and commanded high prices. Sheep farming also continued successful, and a large number of lambs were reared, and satisfactory prices were realised by the farmers for wool. . . .

"The roads, and particularly the drifts in the Kromme and Gamtoos Rivers, were much injured during the past year by tremendous floods, and this, with the damage already done to the roads during the previous year, arising from the same cause, obliged the Divisional Council to use very great exertions to keep the different sections in anything like a passable state of repair."

Such is the information obtained in regard to the districts along the south coast line of the Colony. I have visited most of the districts referred to, but have little to add.

Of a destructive flood at Hankey, in which the water rose in a gully to a height of seventy feet above its usual level, a detailed account is given in "*Reboisement* in France" (pp. 335–339), with a brief notice of the tunnel by which a large area of arable land is there irrigated.

In regard to George, Knysna, and Humansdorp, it may be stated that the extensive forests spread over these have secured in some quarters an excess of moisture beyond what the farmers can in some cases utilise. From the abundance of the supply the Tzitzikamma, it is alleged, has received its aboriginal name, which means, it is said, many waters. The enterprise displayed at Hankey shows what may be accomplished, and the encouragement that there is to improve the advantages of the district. And my observations at Montagu and Robertson show that in districts in which less has been done there are not awanting facilities for forming works of irrigation.

We are now brought to the eastern or south-eastern district of the Colony.

CHAPTER V.

Supply of Water, and Facilities for the Storage of it, in the South-Eastern Divisions of the Colony of the Cape of Good Hope.

Section I.—*Divisions of Port Elizabeth, Uitenhage, and Alexandria.*

The division of Port Elizabeth consists of a small tract of country extending from the sea to the neighbourhood of the town of Uitenhage. It has an area of 220 square miles, which is but poorly wooded and watered. It is less an agricultural district than one attached to the commercial port from which it obtains its name.

The division of Port Elizabeth was described by the Civil Commissioner, in 1862, as a division of very circumscribed limits, with no streams running through the country; and the inference was suggested that no formation of dams or water leadings were likely to be undertaken there in aid of agriculture.

But the spirit of enterprise was powerful in the energetic inhabitants of the district, and ere many years had elapsed the *Eastern Province Herald*, in an article on a scheme which had been initiated by the "Zwartkops Land Irrigation and Water-works Company," for damming up the Zwartkops River and leading it out for the purposes of irrigation, thus wrote:—"Mr Woodford Pilkington, the engineer and manager of the company, having, in the course of his professional engagements, observed that a large tract of land lying along and on both sides of the Zwartkops River, between Rawson Bridge and Uitenhage, was capable of being placed under irrigation, and, by position and other circumstances, was peculiarly adapted for irrigation works on an extensive scale, secured the right of purchase, and, associating with himself the gentlemen who form the provisional directors, started the scheme which bears the title which we have quoted above. To us there seems much that is sound and practical in the proposition, and we should be glad to see it fairly launched and successfully carried out. The land or tract of country upon which the Company propose to operate is that 'splendid property formerly in the possession of the late General Cuyler, known as Zwartkops Waggon Drift, and Fishwater Flats, in extent upwards of 12,000 acres, commencing about five miles from Port Elizabeth, and extending for nine miles along the banks of the Zwartkops River, on the line of the proposed Port Elizabeth and Uitenhage Railway.' The soil of this tract is said to be of excellent quality, and, when irrigated, capable of bearing the highest cultivation. It is proposed to place a considerable portion of this property under irrigation, and sell it for farming or building purposes. To do this it is proposed to construct six reservoirs on the Company's lands, each to contain 15,000,000 gallons, to receive the surplus water of floods, and the waste running into the sea. Mr Pilkington is at present occupied in gauging the volume of water which now flows down the Zwartkops River, and which he estimates at something like three millions of gallons daily; this, it must be borne in mind, being the *surplus* water after supplying the establishments along its banks. The reservoirs, when complete, containing 90,000,000 gallons of water, Mr Pilkington reckons would, in the case of an ordinary flood, fill in about ten or twelve days, and when full would afford a supply for a large portion of the year. Other reservoirs, in the direction of Port Elizabeth, are also proposed to be constructed, should it be deemed advisable to do so at a future period. But the scheme is a land scheme as well as an irrigation scheme, and the distinguished feature of this company is that it proposes at once to distribute a *land bonus* to each shareholder with each share taken. With this view one of the farms (Zwartkops Waggon Drift) has been sub-divided into 5,000 erven, each erf containing a tenth of a morgen, or sixty square roods—rather more than the fifth of an acre—ten of these erven forming a block containing one morgen. The shareholders will draw for choice of position, the holder of 50 shares and upwards drawing first, of 20 and upwards second, of 10 and upwards third, and the other smaller shareholders last; the holders of any two and not more than four contiguous blocks to have the right of inclosing them within one fence, and shutting up the public thoroughfare through them. These erven will be given out *in freehold*, so that there will be no annual quitrent or other charges upon them. We venture to think this feature in the proposed scheme will be an attractive one. Each shareholder will become the *bona fide* holder of an erf, which will afterwards be irrigated, and which he can at once use for any purpose he chooses, and this, irrespective of his participating in the contingent advantages to be derived from the other resources of the Company. We should mention also that the right to a large tract of commonage has also been reserved. After these erven have been laid out, and the commonage reserved, there will still remain the farm 'Fishwater Flats,' in extent 9,000 acres, capable, when watered, of being laid out in erven for villa residences along both banks of the Zwartkops River, which, under proper management, cannot fail to bring in a good revenue to the Company. Looking at the prospectus, calmly and dispassionately—for we have no personal interest whatever in the undertaking—we cannot but think that the scheme is one of a thoroughly practical character, and likely to prove remunerative to the investors. The names of the provisional directors are a sufficient guarantee of the *bona fide* character of the work. Such cautious men of business as Messrs Chase, Simpson, Fleming, Buchanan, and Hume, would not countenance or lend their names to any wild chimerical land speculation. They must see their way clear before they move in a matter of this kind, and have data before them which will warrant them in taking the subject up in earnest. With such provisional directors, and with the professional services of an engineer of the ability and standing of Mr Woodford Pilkington, we think the Company comes before the public under favourable auspices; and if the premises and calculations upon which they are founded be anything like correct, we shall be surprised if the project does not 'take' readily, even in the present depressed state of business. But there is one contingency which must, to a very considerable extent, affect the success of the object the Company has in view, and that is the construction of the railway to Uitenhage. Indeed, the prospectus before us not only does not hide this, but mentions it in express terms. It speaks of the proposition as 'by far the most lucrative investment now offering, particularly as a land one, *with the railway to Uitenhage a certainty.*' However valuable the land on the banks of the Zwartkops may be, and however desirable the site as a place of residence, it will be, comparatively speaking, unavailable to the public, without easy means of access, and this can only be obtained by a railway. Indeed, we consider the two projects to a great extent mutually dependant on each other."

The division, as intimated, consists of what, compared with some of the divisions of the Colony, is but a small tract of country, extending from Uitenhage to the shores of Algoa Bay. Interest centres in the commerce of Port Elizabeth; and the special requirements of the port have led to more attention being given to works of hydraulic engineering having the supply of these requirements for their object, than to those relating more immediately to agriculture. But a reference to what has been ascertained in regard to the supplies which may be obtained for this purpose may make more manifest what supplies of water are available for the promotion of agriculture in the Colony, if it should be deemed expedient to appropriate to that object what is now flowing into the sea.

When I first visited Port Elizabeth, which was about the year 1847, I found casks, jars, and vessels of diverse other names, forms, and of diverse dimensions, arranged in the court yards of the houses to collect and retain for use the rainfall on the premises. On revisiting the Bay in 1863 I found much more satisfactory arrangements had been made for the purpose—the water being collected and retained in capacious covered tanks, properly constructed and lined. Since then there has been brought into the town a supply from the Sharks River; and at one time it was proposed, as has been stated, to introduce a larger and better supply from the Zwartkops.

The editor of the *South African Advertiser and Mail* said of the account of that river which I have cited, that if it be correct to the extent of one half or even one fourth of the body of water in its flow, enough has been proved by experiments conducted with the greatest skill, and by experience on a vast scale, to satisfy us that 100,000 gallons of water may be conveyed in a three inch pipe from the neighbourhood of Uitenhage to Port Elizabeth daily. And in suggesting the importance of paying a competent surveyor to take the levels of a line between Port Elizabeth and Uitenhage, and between Port Elizabeth and Zwartkops River, at a point immediately below the late General Cuyler's farm, he wrote,—"Nothing precise can be said on this subject until it is ascertained at what height above the level of sea, which is nearly the level of Port Elizabeth, a sufficient quantity of water can be drawn either directly from the Zwartkops, or from the great spring on the left of its source, or from the reservoir formed somewhere above or below Uitenhage.

"If a reservoir can be formed on or near the bed of the Zwartkops, and fed by it, or rains, etc., at an elevation of 200 feet above the level of the sea, distant say 16 miles, a three-inch pipe will convey above 100,000 gallons of water in twenty-four hours. This gives a fall of about 12 feet per mile. The Croton Culvert, which supplies New York with 60,000,000 gallons of water every twenty-four hours, has a fall of only 13½ *inches* per mile. The projected supply from Loch Katrine for the City of Glasgow is to be conveyed by means of a culvert inclined at 1 in 6,336, with iron pipes over the valleys. The fall over the whole will be 1 in 1,000, or 5·28 feet per mile.

"With regard to rust or incrustation, this only takes effect when the water is checked in its current, and occasionally stagnant. The pipes which supply the city of Edinburgh with water from about a distance of nine miles have remained perfectly clean since they were laid down in 1819. The pipes in Capetown, through which the water flows *constantly* into the reservoirs, have been in action for more than forty years without the slighest incrustation. Where the piping is level, or flow of water very slow and occasionally checked, a considerable deposit takes place, which, however, is easily removed without injury to the pipe. All chance of inconvenience from this source will be obviated by discharging the water of Port Elizabeth into a reservoir, from which it will be drawn by shorter pipes for the service of the town and inhabitants."

It has also been proposed to introduce into Port Elizabeth water from the Van Staden's River, in connection with which it was suggested, in the beginning of 1866, that the flow of the river should be ascertained, the time being then most opportune. Mr Pinchin, civil engineer, in urging the matter, said,—"If there be sufficient water now, in the extremity of this long-protracted drought, I think we might depend upon that source as affording a permanent supply. In 1860, when Mr Archibald, the town engineer, first gauged the stream, he found 1,250,000 gallons per diem. In June, 1862 (a dry time), I found a quantity equal to 614,820 gallons per day, and taking the average of these figures at 933,410, or nearly a million, I gave evidence before the Parliamentary Committee at Grahamstown to the effect that we might reasonably calculate upon nearly a million gallons a day from that source; but the last three or four years have been excessively dry, and the whole country appears to be undergoing a gradual desiccation. In January last, 1865, just a year ago (then very dry), I gauged the water with great care, and found 431,723 gallons per diem. And here I may state that the quantities above given by Mr Archibald and myself are approximately correct for their respective dates (sufficiently near for practical purposes), but, in the gauging in January last, I erected a weir across the river, and caused all the water to flow through a feather edged triangular notch, carefully cut in the board, the result, giving 431,723 gallons, may be relied on as absolutely correct for that date.

"Thus we had in 1860, 1,250,000; in 1862, 614,823; and in 1865,

431,723 gallons—showing a gradual and rapid decrease. It would be advisable to ascertain whether this gradual decrease still continues, or whether the last gauging of 431,723 gallons may be taken as a minimum, for if decrease still goes on and dry weather continues, it may end in the failure of these springs altogether, and really I do not know of any other source from whence an adequate supply of water can be brought to town."

Such at that time was the quantity flowing towards the sea.

In "Hydrology of South Africa," (pp. 228–232), and again in "*Reboisement* in France," (p. 339–344), I have given details of damage done by a heavy storm of wind and rain which fell upon Port Elizabeth and the country behind, in November, 1867, when the rainfall, gathering in the kloofs near the town, overflowed the ordinary drains, and rushing down certain streets with irresistible force formed channels for itself, undermining houses, and occasioning great destruction to life and property.

From Port Elizabeth it was reported in 1869 :—"The year 1869 has been distinguished by several events, in which the division of Port Elizabeth has been more or less concerned. In February, after a season of unparalleled heat, extensive bush-fires ravaged the neighbouring divisions of Humansdorp and Uitenhage, causing the loss of several lives, and destroying much property. Though this division suffered from fire to a very trifling extent, it was the first to take action to relieve the sufferers in other parts of the Colony. . . .

"In October last heavy rains fell, and flooded the country, doing serious injury to the roads and bridges, and sweeping away crops and live-stock. Rawson Bridge, over the Zwartkop River, was so much injured as for a time to be considered unsafe, and it is now being reconstructed. . . .

"On the 19th September last, this port was visited by a south-east gale of unprecedented violence, which drove on shore eleven ships and every cargo-boat at anchor. Only two vessels out of thirteen rode out the storm. Eight men, including the captain and chief-officer of the Norwegian bark *Sea Snake*, were drowned, and a young man, named Leslie, was also drowned in endeavouring to render them assistance. The force of the gale may be estimated from the fact that the velocity of the wind several times during the day was registered at twenty-five miles per hour—nearly hurricane speed—giving a pressure of nearly thirty pounds upon the square foot. Considerable damage was also done to property on shore. Altogether, it was the wildest and most terrible storm with which this port has been visited for many years."

The following is a tabulated statement prepared from daily observations made at Port Elizabeth throughout the years 1868-1869 :—

Months.	1868.			1869.		
	Mean height of the Bar. at the tem. 32° at mean sea level.	Approximate mean tem. of the air.	Rainfall.	Mean height of the Bar. at the tem. 32° at mean sea level.	Approximate mean tem. of the air.	Rainfall.
	inches.	degrees.	inches.	inches.	degrees.	inches.
January, - - -	29·958	68·19	0·26	29·933	72·16	0·19
February, - - -	29·984	67·12	3·87	29·996	73·41	0·51
March, - - - -	30·062	65·79	1·59	30·024	66·18	0·69
April, - - - -	30·073	62·62	1·89	30·074	63·39	0·70
May, - - - - -	30·046	60·76	0·80	30·024	56·57	2·83
June, - - - - -	30·220	58·40	0·63	30·085	59·37	1·59
July, - - - - -	30·167	56·59	3·91	30·183	54·47	2·33
August, - - - -	30·177	58·72	1·23	30·218	57·63	3·20
September, - - -	30·055	59·83	0·38	30·180	59·54	3·60
October, - - - -	30·108	60·76	1·28	30·044	61·06	4·37
November, - - -	30·018	66·03	2·58	29·985	65·77	2·10
December, - - -	29·974	66·54	1·55	29·900	66·84	0·99
Mean annual values,	30·070	62·6°	19·97	30·05	63·03	23·00

(Signed) Charles Hammond, Meteorological Observer.

The rainfall has been heavier during the year than in 1868, but not nearly so great as in 1867.

The following are tabulated statements prepared from daily observations made at Port Elizabeth throughout the years 1867–1868; a tabulated statement of the number of observations of the wind at Port Elizabeth at the several points of the compass during each months of these years; and also a tabulated abstract derived from observations made at Port Elizabeth in the same years :—

Daily Observations made at Port Elizabeth in the Year 1867. Lat. 33° 57′ 43″, Long. 7° 8′ 32″ E. of Royal Observatory, Capetown.

From Daily Observations.	Jan.	Feb.	March.	April.	May.	June.	July.	Aug.	Sep.	Oct.	Nov.	Dec.	For the Year.
	ins.	ins.	ins.	ins.	ins.	ins.	ins.	ins.	ins.	ins.	ins.	ins.	ins.
Mean height of the barometer at the temperature 32° Fahr. for each month, and for the year, from observations at 9 A.M., 1 P.M., and 5 P.M., ..	29·785	29·812	29·802	29·832	29·860	29·945	29·974	29·966	29·988	29·868	29·894	29·797	28·876
Maximum observed, ...	30·107	30·131	30·137	30·141	30·085	30·330	30·436	30·344	30·420	30·135	30·180	30·101	30·436
Minimum observed, ...	29·451	29·469	29·372	29·488	29·652	29·626	29·448	29·559	29·584	29·605	29·603	29·452	29·372
	degs.	degs.	degs.	degs.	degs.	degs.	degs.	degs.	degs.	degs.	degs.	degs.	degs.
Mean temperature of the air for each month, and for the year (Fahr.), from observations at 9 A.M., 1 P.M., and 5 P.M., ...	73·43	71·38	71·40	66·92	63·64	61·21	60·09	60·77	60·83	63·51	67·00	70·25	65·869
	degs.	degs.	degs.	degs.	degs.	degs.	degs.	degs.	degs.	degs.	degs.	degs.	degs.
Mean temperature of evaporation for each month, and for the year (Fahr.), from observations, at 9 A.M., 1 P.M., and 5 P.M., ...	66·36	65·58	65·12	61.81	58·29	56·04	54·38	56·91	56·38	58·31	61·52	68·66	60·363
Mean humidity of the air for each month, and for the year (the integer number 100 implies saturation), derived from observations at 9 A.M., 1 P.M., and 5 P.M., ...	65·48	70·79	68·10	72·87	71·90	71·70	69·74	78·03	75·00	75·19	71·37	67·03	71·433
Maximum humidity, ...	85·0	99·0	90·0	98·0	95·0	96·0	96.0	97·0	97·0	99·0	98·0	93·0	99·0
Minimum humidity, ...	38·0	48·0	32·0	46·0	29·0	34·0	29·0	45·0	50·0	28·0	47·0	44·0	28·0
From the self-registering thermometers (Fahr.).	degs.	degs.	degs.	degs.	degs.	degs.	degs.	degs.	degs.	degs.	degs.	degs.	degs.
Mean maximum temperature of the air, ..	78·09	76·50	75·88	72·42	68·90	66·70	65·83	65·13	64·78	68·41	72·00	74·84	70·790
Mean minimum temperature of the air, ..	63·83	62·53	62·86	58·63	54·82	52·49	51·46	52·26	52·56	53·53	58·18	60·50	56·971
Approximate mean temperature of the air, ..	70·96	69·52	69·37	65·52	61·86	59·59	58·65	58·69	58·67	60·97	65·09	67·67	63·880
Mean range of temperature of the air, ...	14·26	13·96	13·02	13·79	14·08	14.21	14·37	12·87	12·22	14·88	13·82	14·34	13·819
Highest temperature, ...	92·0	91·0	82·3	85 0	86·7	81·5	79·5	75·5	80·0	75·3	80·0	84·7	92·0
Lowest temperature, ...	55·7	57·0	55·5	49·5	48·5	43·5	41·5	47·3	44·5	42·0	52·5	52·5	41·5
Extreme range of temperature on any one day,	23·0	27·7	22·6	22·5	32·2	29·8	27·5	23·5	18·3	22·3	22·6	23·5	32·2
Least range of temperature on any one day, ..	6·8	6·3	7·8	6·0	6·4	4·8	5·5	4·8	5·7	7·5	6·5	5·7	4·8
Quantity of rain in inches, ...	0·470	1·370	1·710	1·310	2·740	4·420	1·920	3·100	3·910	2·650	9·270	0·720	33·590
Greatest amount of rain on any one day in each month, ...	0·240	0·590	1·170	0·670	1·280	1·240	0·760	1·040	1·690	0·580	6·420	0·350	6·420
Number of days on which rain fell, ...	5	7	6	7	6	5	4	8	9	9	5	4	75

N.B.—Index error of Barometer, = 007, allowed for.

From Daily Observations made at Port Elizabeth in the Year 1868.

From Daily Observations.	Jan.	Feb.	March.	April.	May.	June.	July.	Aug.	Sep.	Oct.	Nov.	Dec.	For the Year.
	ins.	ins.	ins.	ins.	ins.	ins.	ins.	ins.	ins.	ins.	ins.	ins.	ins.
Mean height of the barometer at the temperature 32° Fahr. (not corrected for altitude above the mean level of the sea) for each month, and for the year, from observations at 9 A.M., 1 P.M., and 5 P.M., ...	29·766	29·792	29·863	29·896	29·866	30·041	28·987	29·998	29·878	29·829	29·839	29·794	29·879
Maximum observed, ...	30·044	30·189	30·147	30·431	30·227	30·450	30·400	30·486	30·171	30·385	30·090	30·034	30·486
Minimum observed, ...	29·464	29·389	29·470	29·524	29·641	29·620	29·565	29·658	29·553	29·557	29·471	29·494	29·389
	degs.	degs.	degs.	degs.	degs.	degs.	degs.	degs.	degs.	degs.	degs.	degs.	degs.
Mean temperature of the air, for each month, and for the year (Fahr.), from observations at 9 A.M., 1 P.M., and 5 P.M., ...	70·92	69·19	67·65	64·79	62·67	60·93	59·83	61·36	61·82	63·47	67·11	70·78	65·04
	degs.	degs.	degs.	degs.	degs.	degs.	degs.	degs.	degs.	degs.	degs.	degs.	degs.
Mean temperature of evaporation for each month, and for the year (Fahr.), from observations at 9 A.M., 1 P.M., and 5 P.M., ...	64·64	64·27	62·20	59·36	58·05	55·18	53·57	56·40	57·34	58·18	61·40	64·06	59·55
Mean humidity of the air for each month, and for the year (the integer number 100 implies saturation), derived from observations at 9 A.M., 1 P.M., and 5 P.M., ...	68·61	74·10	71·58	71·07	74·94	69·83	67·16	72·74	74·80	71·26	70·20	66·23	71·04
Maximum humidity, ...	92·0	96·0	95·0	95·0	94·0	93·0	93·0	94·0	96·0	94·0	98·0	91·0	98·0
Minimum humidity, ...	49·0	45·0	43·0	38·0	35·0	32·0	34·0	32·0	47·0	43·0	41·0	43·0	32·0
From the self-registering thermometers (Fahr.).	degs.	degs.	degs.	degs.	degs.	degs.	degs.	degs.	degs.	degs.	degs.	degs.	degs.
Mean maximum temperature of the air, ..	75·28	73·41	72·52	69·58	68·44	67·19	64·70	65·79	66·07	67·22	71·89	75·58	69·81
Mean minimum temperature of the air, ..	61·15	60·83	59·14	55·44	53·22	49·63	48·49	51·69	53·61	54·34	56·61	59·35	55·29
Approximate meam temperature of the air, ..	68·21	67·12	65·83	62·51	60·84	58·41	56·59	58·74	59·84	60·78	64·25	67·46	62·55
Mean range of temperature of the air, ...	14·13	12·58	13·38	14·14	15·21	17·56	16·21	14·10	12·46	12·88	15·28	16·23	14·51
Highest temperature, ...	79·7	81·5	80·5	85·0	89·5	88·0	79·0	83·3	72·5	82·5	78·5	88·3	89·5
Lowest temperature, ...	56·3	51·3	51·0	47·0	47·0	44·3	42·5	41·0	47·5	46·3	47·5	54·5	41·0
Extreme range of temperature on any one day,	20·2	22·0	22·2	25·0	31·7	31·5	30·0	30·0	21·7	28·0	26·5	30·8	31·7
Least range of temperature on any one day, ..	8·8	5·0	8·0	6·7	6·7	6·5	6·0	7·0	6·5	5·0	6·7	7·5	5·0
Quantity of rain in inches, ...	0·260	4·130	1·550	1·890	0·800	0·630	3·910	1·230	0·380	1·280	2·170	1·550	19·780
Greatest amount of rain on any one day in each month, ...	0·110	3·520	0·680	0·670	0·300	0·430	2·350	0·620	0·130	0·270	1·140	0·470	3·520
Number of days on which rain fell, ...	4	6	7	7	6	4	4	4	4	9	5	7	67

N.B.—Index error of Barometer, = ·007, allowed for.

Observations of Wind at the several Points of the Compass during each Month of the Years 1867–1868, at Port Elizabeth.

Points.	Jan.	Feb.	Mar.	Apr.	May.	June.	July.	Aug.	Sept.	Oct.	Nov.	Dec.	Total	Jan.	Feb.	Mar.	Apr.	May.	June.	July.	Aug.	Sept.	Oct.	Nov.	Dec.	Total
N.	-	-	-	1	4	-	6	3	-	-	-	-	14	-	-	1	-	5	1	3	-	2	-	-	-	12
N. by E.	1	-	-	1	4	6	5	1	1	-	-	-	19	-	-	-	1	4	4	5	-	-	-	-	1	15
N.N.E.	-	2	-	2	8	3	5	3	3	-	-	-	26	-	-	-	-	3	9	8	3	3	1	-	2	29
N.E. by N.	-	-	-	-	-	-	-	-	-	-	-	-	-	-	-	-	-	-	-	-	-	-	-	-	-	-
N.E.	1	3	6	6	2	8	5	9	4	2	-	1	47	-	1	1	9	6	8	6	7	3	1	-	-	42
N.E. by E.	-	-	-	-	-	-	-	-	-	-	-	-	-	-	-	-	-	-	-	-	-	-	-	-	-	-
E.N.E.	-	1	2	-	-	-	6	-	4	1	-	-	14	1	2	2	5	1	2	2	2	2	3	4	-	26
E. by N.	-	2	1	-	1	1	2	-	1	1	-	-	9	-	-	1	1	1	4	3	1	-	1	-	2	14
E.	2	1	2	6	1	-	6	3	-	1	4	1	27	2	1	2	6	7	6	4	7	3	3	3	6	50
E. by S.	-	1	2	2	2	5	4	-	3	-	1	1	21	-	-	-	-	3	2	-	2	2	7	1	2	19
E.S.E.	5	8	6	6	7	1	3	7	11	4	8	4	70	2	7	6	3	6	6	8	8	2	4	3	3	58
S.E. by E.	4	4	2	2	-	-	-	-	3	-	2	2	19	1	1	1	-	-	1	-	1	-	-	2	-	7
S.E.	19	12	12	13	3	2	3	5	8	12	24	22	135	30	20	8	4	3	5	4	13	9	21	12	15	144
S.E. by S.	10	3	5	2	2	1	-	1	1	3	4	4	36	5	7	5	2	-	-	1	-	-	3	7	2	32
S.S.E.	4	-	5	2	1	2	2	5	1	3	7	3	35	11	3	3	1	3	1	1	-	4	1	2	3	33
S. by E.	-	-	-	-	-	-	2	1	2	-	2	1	8	2	1	-	1	-	-	1	2	1	3	-	1	12
S.	2	3	-	-	-	-	1	4	-	4	1	3	18	-	1	1	2	3	2	-	1	2	-	3	-	15
S. by W.	3	1	-	-	-	-	-	-	1	1	1	2	9	-	-	1	-	-	1	-	-	-	1	-	1	4
S.S.W.	2	3	3	-	-	-	1	2	1	1	1	-	14	3	6	2	1	1	-	2	4	-	-	-	-	19
S.W. by S.	-	-	-	-	-	2	-	-	-	-	1	-	3	-	1	-	-	-	-	-	-	-	-	1	-	2
S.W.	4	5	5	5	-	2	2	3	4	5	9	8	52	-	5	4	7	1	1	5	3	6	7	5	11	55
S.W. by W.	-	1	-	-	1	1	-	1	1	-	1	1	7	-	-	-	1	-	-	-	-	-	2	1	1	5
W.S.W.	6	6	9	6	7	6	1	8	5	9	6	5	74	10	12	11	3	4	5	1	8	17	17	13	14	115
W. by S.	7	5	10	5	7	4	2	2	8	9	7	11	77	12	6	9	8	2	6	8	7	7	2	7	6	80
W.	11	14	9	10	9	7	6	6	9	12	6	16	115	2	7	18	11	11	10	7	8	14	8	12	15	123
W. by N.	5	4	4	5	9	9	2	6	3	3	1	2	53	5	2	7	10	14	2	9	3	5	2	7	5	71
W.N.W.	6	4	6	10	16	13	5	12	8	14	3	6	103	6	4	7	10	7	3	9	10	7	4	2	3	72
N.W. by W.	1	-	3	-	7	2	1	8	-	-	-	-	22	-	-	-	-	-	-	-	-	-	2	1	-	3
N.W.	-	1	-	4	-	11	12	-	7	7	-	-	42	1	-	3	3	5	7	2	2	-	-	3	-	26
N.W. by N.	-	-	-	-	-	1	-	-	-	-	-	-	1	-	-	-	-	-	-	-	-	-	-	1	-	1
N.N.W.	-	-	1	2	1	1	10	1	1	1	1	-	19	-	-	-	1	3	2	3	1	1	-	-	-	11
N. by W.	-	-	-	-	1	2	1	2	-	-	-	-	6	-	-	-	-	-	2	1	-	-	-	-	-	3

Abstract derived from Meteorological Observations made at PORT ELIZABETH in the Years 1867–1868.

MONTHS.	Mean height of Barometer at Temperature 32°.	TEMPERATURE OF THE AIR. Mean Temperature.	Mean Daily Range.	Mean of greatest range any one day of each month.	Mean of least range any one day of each month.	Mean of greatest range throughout each month.	Mean Temperature of Evaporation at 9 A.M. 1 P.M. 5 P.M.	Mean Humidity for each month.	Rain.	Clouded Sky.	LIGHTNING. Recorded Number of days.
	inches.	degrees.	degrees.	degrees.	degrees.	degrees.	degrees.	per cent.	inches.	per cent. of visible hemisphere.	
January,......	29·780	69·58	14·19	21·60	7·80	29·85	65·50	67·04	0·365	Not recorded.	Not recorded.
February,....	29·802	68·32	13·27	24·85	5·65	32·10	64·92	72·45	2·750		
March,........	29·833	67·60	13·20	22·40	7·90	28·15	63·66	69·84	1·630		
April,.........	29·864	64·02	13·97	23·75	6·35	36·75	60·59	71·97	1·600		
May,...........	29·863	61·35	14·64	31·95	6·55	40·35	58·17	73·42	1·770		
June,.........	29·993	59·00	15·89	30·65	5·65	38·35	55·61	70·77	2·525		
July,...........	29·980	57·62	15·29	28·75	5·75	37·25	53·97	68·45	2·915		
August,.......	29·983	58·71	13·48	26·75	5·90	35·25	56·66	75·38	2·165		
September,..	29·933	59·26	12·34	20·00	6·10	30·25	56·86	74·90	2·145		
October,......	29·848	60·83	13·83	25·15	6·25	34·75	58·24	73·22	1·965		
November,...	29·867	64·67	14·55	24·55	6·60	29·25	61·46	70·79	5·720		
December,...	29·796	67·56	15·28	27·15	6·60	33·00	63·86	66·63	1·135		
Mean Annual Values,........	29·878	63·21	14·16	25·63	6·43	33·77	59·96	71·24	26·685	...	...

N.B.—Index error of Barometer, = ·007, allowed for.

The division of Uitenhage extends from the division of Port Elizabeth on the south to the Zwarte Ruggen, a district shared by it and Graaff-Reinet, on the north, bounded on the west by Humansdorp, and on the east by Alexandria and Somerset. It has an area of 5,580 square miles. "It is traversed by a lofty range of mountains called the Winterhoek Mountains, rising in some places to a height of 7,000 feet, and affording well-known landmarks to vessels approaching Port Elizabeth. The country to the north, called the Zwarte Ruggens, consists of long low ranges of hills, covered with a prickly bush, and very arid and nearly waterless.

The division includes some fertile and well watered country, but it is chiefly pastoral. The town of Uitenhage is built on a very beautiful site near the Zwartkops River, about 220 feet above the level of the sea. There wool-washing is carried on extensively, and promises to contribute to the prosperity of the district. The division is watered also by the Sunday River, which is the boundary between it and Alexandria.

In regard to the district of Uitenhage, the Hon. Mr Chase, M.L.C., then Civil Commissioner there, replied to the enquiries of Government, in 1862, that no division could be more benefited by an enactment, such as seems then to have been contemplated, than that of Uitenhage, which is intersected by few permanently running streams, but in which great facilities exist for constructing dams to intercept the showers of rain falling, especially in the summer season, copiously; that already there were several such works in existence. Of these, the one surpassing all others perhaps in magnitude, importance, and manifestation of engineering skill and energy, is the tunnel and aqueduct at Hankey, a station of the London Missionary Society, to which I have already referred, in illustration of what has been done, and may be done, with only such appliance as a missionary can command, and which is not far from Uitenhage, though included in the division of Humansdorp.

I have had occasion to make mention of the water leading cut by the lepers under the superintendence of the Moravian Missionaries at Hemel-en-Aarde (ante p. 400). At Enon, an old mission station of the Moravians in a densly wooded country on the flanks of the Zuurberg and in the division of Uitenhage, which has been well described by Pringle and by Latrobe, the Brethren had also at that early period given their attention to similiar works. The following letter from Sister A. Schmitt tells at once of their success and of their disasters, more than fifty years back :—

"October 2nd, 1823.—My last to you was on the 5th of August, and Brother Schmitt, at the same time, wrote a long letter, and gave you a particular description of our new mill, which answers far above our expectations. Our river has not yet been full; but we have had plenty of water to drive the mill. If we should have a dry summer, there will be no want in this respect, as the Brethren find that the water is sufficient to drive two pumps when the mill does not grind. Brother Schmitt, in his last letter to you, requested to have iron pipes sent, and gave you the dimensions.

"The farmers often express great surprise at the construction of the mill and pump, a work which would be thought nothing of in England; but in this remote part of Africa I believe nothing of the kind was ever seen. Indeed, to have the water brought into our gardens and vineyard is a great advantage. The latter thrives well, since it gets water. It joins the garden, which is large, and in general supplies us with a good crop of vegetables. Bro. Lemmerz labours unweariedly, and has now much pleasure in it, since he finds his trouble so well rewarded.

"The fruit trees promise to bear plentifully; even those peach-trees, of which I planted the kernels not four years ago, bore fruit last summer. Orange-trees grow remarkably well. We have this year planted upwards of twenty. Those which we have reared from the seed are now more than a foot high. From this you will see that the soil is fruitful, when it has water enough. We hear the corn promises a better harvest than last year; in some places indeed the mildew makes its appearance, but not in our neighbourhood as yet. Many a prayer is offered up that the Lord may be pleased to give us this year something to reap, though it be less than in former years. The drought seems to keep the corn back in its growth; but we have good hopes that what we have sown near our house will yield much.

"November 2nd.—Little did I think when I was writing the foregoing, and pleasing myself with the thought how glad you would be to hear such good accounts from Enon, as, from the appearance of things, I was at that time able to give you, that, in a few days, our joy would be turned into mourning. On the 6th of October we had a fine rain, at which there was general rejoicing; but when it continued, with little intermission, for four days and nights, we began to be alarmed about the swelling of the river. Still none thought it would rise to such a height. Early on the morning of the 10th, Brother Halter, who had the charge of the mill, went to see whether the dam near the river was in order, so as to stand against any body of water which might be expected to come down; but there was no dam to be seen, and he was obliged to flee as quickly as possible, that the flood might not overtake him. He had scarcely got into our room before the river overflowed its banks, and we saw the mill in great danger, as the works were already under water.

"We now as quickly as possible, and during torrents of rain, brought our corn, flour, and all that was moveable, into the garret over the church. We could then do no more than pray to our Saviour to give us resignation to His will, and wait the event. Our poor Hottentots fled to us, many of their houses being washed down by the flood. All our eyes were fixed on the *mill*, which stood in the midst of a most fierce and rapid torrent. In about eight hours the water rose upwards of 16 feet, and the breadth of the stream in the smallest part was upwards of 600 feet. Our beautiful corn field was, I believe, more than five feet under water. There was not a bush or young tree to be seen. Large trees were torn up by the roots, and carried by the stream over our corn land. One of these struck the north-west corner of the *mill*, and, giving entrance to the water, soon brought it down to the ground. Some of the wall fell upon the water wheel, and broke it. The pump remained standing, and no material damage is done to the mill-work inside the building. The house would probably have stood against the torrent had not the tree struck against it. The water was now only four feet from our dwellings, and we were alarmed for our own safety; but about three o'clock in the afternoon it began to subside, so that we could spend the night at ease in our house. Great part of the Hottentots, whose houses had been damaged, lodged in our church. Our own buildings have suffered no material injury, nor our garden and vineyard; but the lower part of the Hottentots' gardens, which lie in a line with our corn-land, is all swept away. The poor people had been very diligent in planting, both here, and at the old place, and it was a pleasure to see how well all was growing; but now all is carried away, and in its place a bed of stones covers the ground. In some parts of the river the old bed is filled up, and the water has taken another course. Much as these misfortunes afflict us, still we have great reason to thank God that we have built just on this spot, for there is not as safe a place along the whole river in case of floods. Had we built on the old place, we must have fled to the hills, for nearly all, from one hill to the other, was under water.

"In the valley that bears your name dreadful havock is made,

The river has changed its course in several places, large trees are thrown down, and deep holes made, so that it is impossible for a waggon to pass that way. On the banks of the Sunday River, where are the best corn farms, large fields of corn are destroyed by the the flood. We have not heard that the mildew was so general as in former years, but this inundation has been equally destructive. Where the inhabitants will get bread the Lord only knows."

There are more than one hydraulic undertaking in this district upon a large scale. One of these is a reservoir constructed by the Messrs Parkes, of Wheatlands.

Of this undertaking one of my correspondents wrote to me:—"They are about the best farmers in the eastern province, and all is done by irrigation. They are now [since 1865] constructing a dam that will cost them, with the facilities of the ground, £1400, and that will, when once filled, keep a twelve inch pipe running full for three months."

Mr James Parkes, writing to me on another subject, in the name of himself and his brother, mentioned incidentally this undertaking, saying,—"We have now in course of construction a dam on rather a large scale by navvies. It will cover, when filled, from 150 to 180 acres of ground. This completed, we shall be able to do things on rather a larger scale."

Some particulars of this work I shall afterwards have occasion to communicate as these were communicated to me by Mr Hobson, a near neighbour of the Messrs Parkes, and a man of a like spirit, whose farm, like theirs, extensively irrigated, is situated within the district of Uitenhage, but nearer to Graaff-Reinet; and in connection with my statement relative to such works in that division I shall introduce his own account of what he himself attempted and accomplished.

Another structure of the same kind was undertaken by Mr E. Pullen, of Susannah Vale, in the ward of Van Staden's River, who wrote to me that he was constructing a dam 240 feet in length, 40 feet in width, and 16 feet in depth, in one of the vales on his adjoining farm, Augusta Park, which is fed by numerous perennial springs.

But all is not yet done which might be done to utilise the supply of water in the district.

One of the few permanently running streams referred to by Mr

Chase is the Swartkops River, of which I find the following description quoted in an old number of the *South African Commercial Advertiser and Mail*, as written by one believed to be well acquainted with the country :—"The Zwartkops River, which glides past the foot of the town, is a pure and constant stream, and has on its banks three wool-washing establishments, for which it is peculiarly well fitted from its copious supply and its proximity to the shipping port of Algoa Bay. The water used for irrigating the gardens has its source in the eastern extremity of the Winterhoek Mountains, distant from the town about six miles. In the driest seasons it never diminishes, and, at about 500 yards from its source, forms a rivulet, four feet broad and fifteen inches deep. It is supposed that *no more than half, or two-thirds of its supplies, reach the town,* being lost either by marshes or evaporation, but with this diminution, 2,512,632 (*Two millions, five hundred and twelve thousand, six hundred and thirty-two*) gallons are delivered every twenty-four hours at a height of above 100 feet above the level of the highest street, and 200 feet from that of the river. The sources of the splendid spring are well worth visiting, and on one of the trees, amid many other names, is found that of the good Sir Benjamin D'Urban who purchased the fountains for the use of the towns. On this stream are two water-mills, and very many more could be erected, and the town could be embellished with splendid fountains at a very trifling cost, with sets (jets) of full 100 feet high." And from this river it was proposed, as has been stated, to supply Port Elizabeth with water. But there are not awanting darker shades in the picture of the division !

Of the division of Uitenhage the Civil Commissioner reported in 1865 :—"Uitenhage during the past year, in common with other portions of the Eastern Province, has suffered most severely from a long and unprecedented drought.

"The loss of stock of all descriptions has been exceedingly great; and many farmers have been reduced from a position of comparative independence to one of abject poverty. Much distress prevails in many parts; and, in consequence of springs failing, many homesteads have been deserted, and all agricultural operations are at a complete standstill.

"The rust has all but destroyed the oat and other crops; and the prospects of the farmers are anything but cheering.

"Horse sickness, too, has been unusually virulent, and about two hundred horses have died in this town in the course of two months. There are at present no indications of any abatement of the disease. Several horses have been opened, and the small intestine has been found filled with 'bots;' and the general opinion is that death is caused by the presence of these in the small stomach.

"Two additional wool-washing establishments have been commenced during the year; and much activity prevails in this description of industry, which gives employment to a great number of the coloured people at remunerative wages. The town, from its abundant and constant supply of water, possesses capabilities for this department of labour equalled by no other in the Colony. Wools washed at these establishments command very high prices in the London market."

By others the same remark in regard to the cause of death in sheep and cattle has been made, And traced back, the production of *bots*, *flukes*, and allied parasites, may be found to be connected with the state of the water supply, and to require for prevention a system of drainage in connection with irrigation; and the existence elsewhere, if not here, of stagnant pools and watery soil indicates that there is in the Colony in many places a necessity for drainage as well as irrigation.

Of Uitenhage in 1869 it was reported:—" In the early part of the year 1869, and until the months of September and October, this division suffered very severely from drought, and the losses of stock in consequence were excessive, many flocks of sheep being diminished by hundreds, and most of the lambs destroyed to save the ewes; famers being compelled to move with their flocks from place to place in search of water for their stock.

" In October, the heaviest fall of rain which has been known for many years occurred, flooding all the rivers in the country to an unprecedented extent, causing much destruction to property on their banks. Since the month of September rains have fallen very regularly; and both pasturage and stock, in consequence, are in excellent condition, and the agricultural prospects of the division are improving, the harvest generally being a good one.

" Rust, which for several years has so much injured and, in many instances, destroyed the oat hay crops, appears to be gradually disappearing, although some crops are still affected. The wheat crops are good generally, but in some instances great losses have occurred by the flovers. Vines still suffer from *oïdium*, which appears only to be kept down by sulphur; fruits generally are neither good nor abundant; the supply of meal, Indian corn, barley, and oats, has been abundant, and prices moderate; vegetables in the town are also abundant, large quantities being taken to the market at Port Elizabeth. . . .

"Several lots of Angora goats have been imported during the year by some of the merchants of Port Elizabeth, one large flock by Messrs Blaine & Co.; but I regret to say that few, if any, have as yet found their way into the hands of breeders in this division, although there are many farmers who have large flocks of goats on their farms; there is no better pasture for goats in any part of the Colony than in the north and northwest portions of this division, where flocks are herded in lots of from twelve to fourteen hundred in one flock.

"The general health of the inhabitants of the division has been good. Amongst cattle, lung-sickness still prevails; but no cases of that scourge, horse-sickness, have occurred for the last year or two.

"Farmers complain of the want of farm labour, although numbers of Europeans are roaming about the country; but, after a few days' service, their general resort is the canteen."

In 1874 it was reported:—"During the first nine months of the year a severe drought prevailed in the district, which resulted in considerable loss to those engaged in pastoral and agricultural pursuits; but I am happy to state, that while this division has been blessed with a plentiful share of the late heavy rains, it has comparatively escaped those very heavy floods which have been so very disastrous in other parts of the Colony."

And in 1875:—"Stock of all kinds are in good condition and continue to realise the same high prices.

"The Sunday River pontoon, which was unfortunately washed away during a freshet in that river in August last, has been reconstructed under the supervision of the chief resident engineer, and is pronounced to be a far superior boat to the one lost, and much better adapted for the river. It is now in full work and a great convenience to the public."

In the reports from Civil Commissioners for 1874 is the following from Jansensville:—"The district of Jansensville is the north-western part of the division of Uitenhage, and includes a narrow strip of Somerset. It was formed into a district in September, 1874, and comprises twelve freehold farms, in extent 15,947 morgen and

355 roods; 118 quitrent, 297,563 morgen, and 563 roods; 31 pieces of land leased under Act 19, of 1864, in extent 23,827 morgen and 74 roods; and 29 pieces of Crown land, leased annually, in extent 97,370 morgen and 13 roods; making an area of about 1,500 square miles. The northern part is open country, intersected by ranges of low hills; the flats are covered with karroo and vygebosjes, in good seasons mixed with grass, towards Jansensville interspersed with norse and bush. The hills are clothed with spekboom, and other nourishing trees and shrubs. Sheep-farming is carried on with success in this part. The country towards the west, the Zwart Ruggens, is flat, slightly undulating, covered with the wild pomegranate and a valuable plant called Vinger Pol, and in good seasons, grass. This veld is eminently adapted for all kinds of small stock; Angora farming is doing well in this quarter. The southern boundary is the Zuurberg range, whose slopes are covered with bush, the top with long coarse grass. The country skirting this range is karroo and norse, and approaching Jansensville studded with norse and prickly pear, and is very rugged, and only fit for cattle and goats. The boundary on the east is the Riet and Sunday Rivers. This part may be considered norse veldt, though on some farms there are karroo and bush. It is a cattle country, but best adapted for the Cape goat.

"*The Norsedoorn* is, so to say, the grass of the country about Jansensville. There are four varieties,—the bok norse is used for small stock; the large kars norse for all kinds. In seasons of drought the norse is fired, the long grass between readily burns, the thorns are scorched off, and the plant partially roasted. It is then greedily devoured by goats and horned cattle, which not only live, but fatten on it, and this at a time when in a grass country cattle can barely exist, which gives the advantage to the Jansensville carrier in severe droughts.

"*The Vinger Pol* is a very succulent plant, with great fleshy fingers growing out of a crown a foot in diameter, which being chopped off by the carrier, provides a most nourishing food for his oxen. It abounds in that part which is free from norse.

"*Sheep Farming* is an enterprise to which Messrs Hobson, Biggs, and Nash have devoted considerable attention. Their flocks are large, noted for purity of breed, and freedom from disease. Their farms, all joining, make a most extensive sheep walk, and were originally the property of the late W. C. Hobson, and D. Hobson, who, by their remarkable enterprise, gave considerable impetus to his pursuit.

"*Angora Farming.*—A. C. Stewart and Co. have extensive tracts of land towards the Zuurberg, where angora as well as ostrich farming is conducted under the able management of Mr Featherstone. A large number of pure bred French merinos have been imported by them, and are now at Mount Stewart, which, in course of time, will become a model farm.

"*Cattle* are only bred for transport service, which is so extensive that it has proved exceedingly remunerative. Most of the farmers are carriers, and dairy farming has been entirely neglected.

"*Ostrich Farming* has been carried on very successfully for some years in this district, parts of which are eminently adapted for the rearing of this bird. Up to twelve years ago, large numbers ran wild in the Zwart Ruggens, where some are still to be found.

"*Agriculture.*—The banks of the Sunday River, along its whole course, are of beautiful alluvial soil; many hundreds of acres are under cultivation, and new lands are daily being made. The produce is remarkably abundant in wheat and forage, the soil being extremely deep and rich. The oat hay is very fine, and commands a high price. The mealie crop is very heavy, the produce on one farm being six hundred muids, whilst the crop of forage was thirty thousand bundles. The lands are irrigated from temporary dams, though there are a few permanent in the river. The flood which has just happened has proved destructive to the best agricultural farm, carrying away the water furrow, and forcing a new channel for itself through the lands. In several places crops of forage and wheat have been swept away, and many of the lands have been destroyed. The loss of stock is not worth mentioning. In former years, the vine was extensively cultivated, and wine and brandy made. Through the *oïdium*, and scarcity of labour, the vineyards have been allowed to fall into decay. This industry may be revived, for the yield is good, and the flavour of the grape delicious. There are a number of orchards along the river; but they have been neglected, and little fruit is dried for sale. The fruit garden at Ebenezer is one of the finest in the Colony, and was planted by Mr Hobson, who spared no expense in obtaining the best varieties.

"*Dams.*—When the late Mr Hobson settled in this district, thirty-eight years ago, he commenced a series of dams, upon the most approved principles. The one at Ebenezer is so large that an extensive piece of land is irrigated from it in the driest seasons. His praiseworthy example has been so generally followed that the arid karroo may now be described as 'the country of the lakes.' Mr

Featherstone has constructed seven, which hold water for his numerous flocks all the year round. Several of these inland lakes have been made for irrigation.

"Mr Hubsch deserves great credit for his steam mill on the Sunday River, at Waterford. The water is pumped out of the river by machinery, and irrigates a large extent of land. A soft wheat is grown by him, from which he makes flour.

"There is a mineral spring on the farm of Mr Gouws, which may be of advantage to the public, as it possesses curative properties.

"Limestone abounds in several parts, and very good lime is made. There is a quarry of splendid building stone, which has been opened for the bridge.

"I would call attention to the fact that the prickly pear is taking in considerable tracts of country to the south of Jansensville, and little is done to eradicate the same. On the north, Messrs Hobson deserve great praise for their strenuous and successful efforts.

"*The Village of Jansensville* is part of a farm called Vergenoeg; it was sold and laid out in erven in 1853. It is situated on the eastern bank of the Sunday River. There are forty-eight water, and thirty-six dry erven. The soil is remarkable for depth (say twenty feet) and richness. There are a number of fruit trees and vineyards. Owing to the depth of soil, a good crop of corn, forage, or mealies can be secured by only irrigating twice from the Sunday River. The water supply is, however, very uncertain. There are a couple of wells for drink-water, and two have been sunk by the magistrate for the use of the public. The village, though now of age, is in a very backward state, owing to several circumstances, the chief being the lawlessness of former times, the absence of permanent water, and the main road not passing through it. The first of these evils has been removed, the second is in the course of removal, and the third will be removed as soon as the bridge over the Sunday River, at the village, is completed; the main line will then go through it, and sixty waggons a week will pass and re-pass through its streets. The streets were made by the magistrate, and are being gravelled by the prisoners, and the water-furrows have been bridged. An application has been made for a piece of Crown land called 'Lapport's Kraal,' adjoining the village, to increase the commonage, which is very small and insufficient. The grant of this, it is hoped, will remove the only remaining obstacle to the village becoming a flourishing township."

In 1875 :—" The last year, owing to the seasonable rains, has been most favourable to the agriculturist, and the crops of cereals above the average.

" Sheep and Angora goat farming engage a good deal of attention. Ostrich farming is still carried on extensively, but whether it will much longer continue such a remunerative industry is doubtful. Probably the reports as to a fall in feathers will induce many to curtail their operations in this branch."

Of the works of Mr Hobson he wrote to me, in 1865, in answer to inquiries which I addressed to him :—

" Twenty-five years ago I came into this country (the Karroo), bringing the first flock of woolled sheep over Bruintzje's Hoek which had ever come in this direction to be depastured. I bought some farms, from which I chose to live on Ebenezer, then called Palmiet-fontein, although it had only one small fountain at that time. I saw the urgent necessity of securing water by dam making, and without delay made a commencement. Besides some small dams on the veld I began a large dam across the river (Klein Nel River), by building two dry stone walls across, forty feet from each other, and filling in between with dry ground, leaving an outlet over a rock about thirty feet wide. This plan was the old approved plan of the Colony and is yet much practised. I was the only European then resident any-where near this, and was laughed at for my stone breaking, dam making, and *vader-land-skaap* propensities. The river came down one day so high as to cover outlet, dam, and all the width of the valley, and wash the earth from beneath the walls, which fell in con-sequence. I was laughed at, of course, and often heard ' We said so,' ' It is in vain ; the karroo veld is not good for dam making, neither will it hold water like the grass veld whence you came from.' I judged differently, and I saw, if I lived here, I must make dams, and on another principle, and I accordingly began. Having a rock foun-dation, I packed the stones—the largest ones at the back, smaller ones in front, and gravel stones, then earth next the water, which I again covered with rubble for the water's wash or waves when the wind blew—all packed at a gradient of about 1 in 12, and forming an eliptic arc, convex side towards the water, and it answers well. The water, when the river comes strong, rolls over it *à la Niagara*. In conjuction with my son, Mr S. B. Hobson, I work at it always between more urgent jobs, such as sheaving, ploughing, etc., by which means we raise it higher and higher every season. It has now stood the battle for twenty years, and every year becomes more solid. It fills five other dams below it, and above the lands, by pipes laid through the embankment on each side. The small fountain has become stronger, winter and summer, for the last twelve years. We irrigate from it thirty acres of land, have a first-rate orchard, small vineyard, vegetable garden, and my flower garden. Twenty-five years ago there was not one dam within thirty miles of this farm, now nearly every farm has one or more, great or small, on it. Nearly all the numerous squatters have made dams or tanks on the Government ground; and for this they often 'strive,' like the naughty people of old. Some of them, after having found a spot which holds water well, dig a deep tank, eight or ten feet deep, and fence it round strongly and high, and then call it theirs, and strive for it; this they call their drink water *(plek)*; and I know some such places which have not been dry for several years. The largest dams in this neighbourhood are those of Mr John Bolleurs, of Galga Bosch, of Mr Jas. Leonard, of Beestjes Vontyn, and of Messrs Parkes, of Wheatlands—all my near neighbours. There are several others of considerable sizes, but not worth mentioning at this time; but I must say a few words about Parkes' dam, which is constructed upon a plan after my own heart,—*i.e.*, to turn out only a part of a river, which runs for months at a time, into the large basin which they have dammed up, and from this to irrigate about a hundred acres, and more if it answers. They already grow, on an average 500 muids of corn, and by this new plan they will in all probability double it at least. They will have a flood gate, or gates, fifteen feet wide, to be opened when the first muddy water has subsided, to prevent the dam from filling up. There are places where the river could be turned out higher up, and so to water the karroo flats. The Sunday River, also, ought be turned out, and, instead of filling up Algoa Bay with mud, its rich deposit should be led on to the vast flats near to it, and some of its waters detained in the innumerable hollows and kloofs on its margin. The water, once conducted out, should never be allowed to return, except as dew and rain."

A plan of the work was subjoined, representing the course of the Melk River; a large extent of Government ground, generally level rich karroo soil; the new furrow, with flood gate at the river; the present lands, vineyards, etc., watered from spring; a new dam then making; lands to be watered from new dam; houses, etc.; a fountain in the river, and the existing water course. And it was stated:— Wheatlands' dam—300 yards long, 60 feet thick, and 17 feet high—

has cost £750, when finished it will cost about £1200; made by hand, with wheel-barrows and picks."

This dam of Messrs Parkes, it was afterwards expected, would cost about £1400, and that it would store water sufficient to keep a twelve-inch pipe running full for three months.

In a subsequent letter, written in the beginning of 1866, Mr Hobson wrote to me:—"There is a dam at Stapleford, on the farm of Mr S. Robson, Member of the Divisional Council, about four hours below Graaff-Reinet, across the Sunday River, on a hard and convex blue sandstone reef, as ugly and unlikely as can be imagined to put a dam on. It has now stood the torrents, stones, and trees of the old river for about twelve years. It is made of upright one and a half-inch iron, with half-inch up stream stays, and the upside with deals placed one on the other, and fastened with common staples driven, from the lower side, into the deals round the upright iron: the iron being let into the rock from six to nine inches. The dam is about 80 yards long, 4 feet high, most of its length running to nothing at one end, and cost only £150. There are many places on the Sunday River where such dams could be made, and several where a furrow might be blasted for a short way through or along the bank, and the water would run out without any dam."

A plan of the work was given, representing upright irons let nine inches into the rock; deal, three-inch plank, pitch pine; staples; stay up stream, let four inches into the rock; rock; mud and sand backing up the dam, and which the river deposited there the first flood after the dam was made. And Mr Hobson says,—"I mention this dam because it is a novel mode of building dams, and it has answered well for so long a time, costing but little, and would be easily repaired. Lead need not be used to fasten the iron in the rock, the drill should be a very little smaller than the iron."

Such works, accomplished by great enterprise and energy within the Colony, combined with the preceding accounts of what has been done elsewhere, seem to me to supply illustrations of the Roman proverb to the effect that the gods help them that help themselves, and to warrant the hope that no insuperable difficulty will be found in the way of those who in earnest, with skill, labour, and capital at command, undertake works of irrigation in the Colony, whether upon a small or a large scale of operation.

A gentleman, writing in 1867 of these, and other undertakings in the district, wrote as follows:—"On leaving Graaff-Reinet I determined to abandon the main road to the coast, and pay a hasty visit to the far-famed Wheatlands, and the equally famed Loots Kloof farms. After having spent a pleasant evening, and slept soundly under the hospitable roof of Mr Thomas Parkes, I proceeded the following morning to inspect the lands and stock on the farm, and I feel unequal to a description of the admirable system upon which everything is done at Wheatlands. Mr Parkes informed me that he had 150 acres of land under cultivation; all these lands at the time of my visit were under crop, and looking exceedingly fine; all had been well ploughed, harrowed, and rolled upon the English system, and when ripe is cut down with one of Ransom and Sims' reaping machines. I was informed that last year the yield on this farm was 800 muids of wheat, 150 muids of barley, and 200 muids of Indian corn, besides a large quantity of oat hay. The grain when harvested is carefully thrashed and cleaned by machinery, and ground on the spot, the straw carted to the sheep kraals to be converted into manure. From the corn-fields I proceeded to inspect what was to me the principal object of attraction, namely, the dam, or, more properly speaking, the reservoir. In this, as in every other branch of farming, Messrs Parkes have surpassed every other competitor. A solid embankment, 2,100 feet long, 60 feet at base, and 30 feet high, had, at a great expense, been thrown across a valley on the flats, forming a veritable lake, that, notwithstanding all the late rains, has not yet been filled, by which means the Messrs Parkes will be able to double the quantity of land under cultivation. We then proceeded to the sheep kraal, and, after a careful examination of the thoroughbred flock, I am of opinion that finer sheep are not to be met within the Colony, and I would recommend persons desirous of improving their stock to visit Wheatlands before making a selection of rams.

"From Wheatlands I proceeded to Ebenezer, the hospitable residence of W. C. Hobson, Esq. Next to that at Wheatlands, the finest dam that I have seen in the Colony is to be seen on this farm, by which 30 acres of land are now irrigated, the whole of which is enclosed with a substantial stone wall. One of the best orchards in the Karroo is to be seen here, besides a large number of forest trees from Europe, Asia, and America.

"At Harefield, the property of W. Nash and W. C. Hobson, jun., extensive tracts of land have been brought under cultivation. On this farm is one of the finest farm buildings perhaps in the Colony, with a frontage of 90 feet, all of cut stone.

"At Wellfound, the property of Mr E. Biggs, extensive improvements have been made during the last few years, and large tracts of land brought under cultivation. Here also a monster dam, second only to that at Wheatlands, is in course of construction, which, when completed, will enable the proprietor to treble the amount of land now under cultivation. There is also a fine orchard abundantly supplied with the finest fruit trees of almost every description.

"At Careysbrook, the property of Mr Jonathan Hobson, two substantial dams have been constructed by building solid stone walls across a deep ravine, one 300 yards long and the other over 200 yards, which will enable Mr H. to bring about ten acres of land under cultivation.

"At Bonnindale, the property of Mr W. Berrington, a piece of land over 600 yards long by about 200 yards in breadth has recently been brought under the plough. In all these places the lands are enclosed with substantial stone walls, and the crops look exceedingly fine."

At a public meeting at which was discussed the best means of giving employment to numerous European navvies, who had been thrown idle by the completion of works for which they had been engaged, Mr M. R. Robinson, then the acting colonial engineer, said:—"There were works (such as the Van Stadens River Heights) on which these men might be employed; but if he might make a suggestion with reference to the employment of these men, he hardly thought a more beneficial employment than making reservoirs along the main roads could be found. He particularly noticed the necessity for these works in his annual report to the Government; it is the greatest difficulty the country labours under. It would be a very great assistance, not only to travellers, but also to the carriers and stock, who have now to go miles before they can get water for their oxen. It is just the work for these men to do, and precisely the work they are accustomed to. Unless people have travelled over the country and know the roads well, they could not realise the fearful difficulty and distress the want of water occasions. He thought these dams specially required on the Port Elizabeth and Graaff-Reinet line of road—the new line comparatively recently formed. He travelled that road with horses, and, though there was sufficient grass, he had to ride ten or twelve hours before he came to water. There was the same want between Grahamstown and Port Elizabeth. One reservoir at least was required in the lower country, about thirty-five miles from Port Elizabeth, when you get into the Zandveld. The cost of dams of such a nature, he thought, would be from £1,000 to £1,200 each."

The division of Alexandria, with an area of about 1250 square miles, lies between the Sunday's and the Bushman's Rivers, the latter being the boundary between it and Albany. The division is well wooded, and, in its production, is similar to Uitenhage and Humansdorp; but certain tracks, such as the Olifant's Hoek, are even more fertile.

Of this district, Mr Philpott, the Civil Commissioner, wrote in 1862 :—"As regards dams for irrigation this division is badly situated, being nearly destitute of rivers or any running waters, excepting in localities where they could scarcely be made available for irrigation to any extent;" but he intimated that dams for watering cattle could be constructed on almost every farm.

In 1865 it was reported :—"Alexandria has suffered very considerably from the long-protracted drought during the last and several previous years. The slight rains which fell about the ploughing season were very partial, and so slight that it was only in localities where the soil was light that farmers were enabled to plough at all. In very many parts not a furrow has been drawn, and it is only on a very few farms near the coast that any crops have been reaped; and these, in most instances, extremely scanty. On some farms, even close to the coast, there are instances of seed being sown, but not a grain reaped. The prospects of the farmer are at present very gloomy indeed; even the season for cultivating Indian corn and pumpkins has passed, and but very little planted.

"Rust has again this season appeared in the oat hay crop, but not to so great an extent as last year. Wheat is, however, free from this, as also the English oat; but the yield of wheat will not be by any means sufficient for the consumption of the division.

"For many years past the crops in this division have been so precarious that farmers have found it necessary to employ their waggons and oxen in the carrying trade, which has, during the last year, been extremely limited, owing to the drought, but more especially to the commercial difficulties in which the frontier has been involved.

"Cases of lung-ziekte *(pleuropneumonia)* amongst cattle have occurred on several farms, but not generally to any considerable extent.

"Horse sickness has also been raging, particularly during the months of November and December, and still continues. It commenced in the early part of the year; although not severely, till about the spring months. Many farmers do not now possess even a horse to ride.

"An attempt is being made by one or two farmers to cultivate tobacco in larger quantities; but the severe drought will prevent any crop, and it is feared a sufficient quantity will not be raised to induce farmers to compete for the prizes offered by the Government."

In 1866 it was reported:—"The year 1866 has been marked, like many previous ones, by want of rain. Crops of wheat, barley, and oats, are consequently not up to the average of former good seasons; and rust has also, as usual, made its appearance to mar the prosperity of the division.

"Lung-disease amongst horned cattle has not prevailed to any serious extent, and the prevalent opinion seems to be that it is gradually disappearing. Another of the ills that this country is heir to—namely, horse distemper—was severely felt during the first half of the year, so much so that many farmers lost every horse belonging to them. Persons engaged in the transport of goods have been put to great inconvenience and loss during the last few months of the year, by foot-rot and tongue disease in oxen; and as almost every farmer in the division is more or less engaged in this line of business, the aggregate loss thus occasioned has been very considerable."

Of Alexandria it was reported in 1869:—"This division in general is in a very prosperous condition. We suffered from drought up to May last. Since then we have been blessed with constant and copious rains, so that the country at present is in a magnificent state. The extent of land sown this season was more than usual, and the harvest, which is nearly all gathered in, is, with few exceptions, splendid—certainly better than any for the last five or six years. Very little rust made its appearance, and the oats in particular resemble the crops of former years.

"Threshing-machines are busy already; they are fast superseding the old and primitive style of tramping with cattle. . . .

"The health of the inhabitants of this division is, on the whole, very good. The climate at present is more like that of spring than summer.

"No locusts have made their appearance as yet, nor has horse-sickness. Lung-sickness is still prevalent here and there. The cattle everywhere are in splendid condition.

"Crime is decidedly on the decrease, and an air of contentment prevails among the people, who appear to be slowly recovering their former position."

And in 1875:—"The year 1875 has been on the whole a prosperous one for this district. The crops have been almost free from rust, and the yield of grain, etc., has been above the average. The lung sickness has also not committed such ravages amongst the cattle as usual. Sheep farming is carried on to a very limited extent in this district, the pasturage within forty miles of the coast not being suitable. Ostrich farming has been attempted by several farmers with good results. The fencing of farms with poles and wire is becoming more and more popular, so that, even without an Act of Parliament, I feel sanguine that before long a large portion of the farms will be enclosed in this manner.

"The great drawback in this district, as I believe in most others, is the scarcity of labour. It has, however, had one good effect—namely, in inducing farmers to rely more upon the use of agricultural machinery, and the saving of labour in that way. A large number of reaping and threshing machines have been introduced, and one gentleman is importing a steam plough.

"Only one case of stock stealing has been disposed of by me during the year; the other cases, consisting principally of assaults, breaches of the peace, petty thefts, and cases under the Masters and Servants' Act, numbering in all 172, mostly had their origin in a too free use of bad Cape brandy."

SECTION II.—*Divisions of Albany and Bathurst.*

Albany presents two divisions—the coast region, called the Zuurveld, or Lower Albany, principally adapted for cattle grazing and agricultural purposes; and Upper Albany, which is well adapted for sheep farming. The whole was settled by British immigrants in 1820, and many of the descendants of the old settlers still remain on their original location. The area is estimated at 1,730 square miles.

In regard to Albany, lying to the south of Alexandria, Mr Graham, then Civil Commissioner of the division, in reply to the circular issued from the Colonial office in 1862, wrote:—"1. I would remark, with regard to the first question of the circular under reply, that no plans or irrigation works within this division have been proposed.

"2. Without a careful survey of all the likely localities in the division this [the second] question cannot be particularly answered. Generally, I may state that all or the greater part of the country along the banks of the Fish, Bushman, and New Year River is capable of cultivation, requiring only irrigation to make it highly productive. Most of the banks being steep, and high above the usual run of the water, appliances for raising water would appear most requisite in these localities; but there are numerous farms on which ample supplies of water for irrigation purposes could be secured, were the means and inclinations of the farmers at all to keep pace with the necessities of the country. But it appears to me that stock-farming is the favourite pursuit of this district, and for some time to come I think it would be doubtful whether agriculture to any extent would be equally safe and remunerating, and this principally on account of the labour market.

"3. I am inclined to think that a considerable number of the landed proprietors in the division would, in course of time, be likely to avail themselves of a law empowering the Government to make them advances of money, upon reasonable terms, for the purpose of executing works for improving the supply of water on their properties, but not until agricultural labourers can be hired at a reasonable rate, and the farms become more sub-divided, so that the proprietors can no longer keep such extensive flocks of sheep and herds of cattle as to render them independent of any exertion in the shape of agricultural pursuits."

Mr Graham submitted the queries to the field cornets of the division, and forwarded replies which he had received. Mr Wilmot, field cornet of Lower Riebeek, wrote from Spring Vale,—"In answer to the first question in the circular, as to whether any irrigation works have been proposed in this district, I can only answer, none.

"To the second question, I would reply that in my opinion a great many works are practicable.

"In this district there are immense valleys along the New Year and Bushman Rivers, of the richest soil, that would produce anything were the water only got out, and I have no hesitation in saying that this could be done without any great expense, either by means of dams or machinery. There could also be great improvements made on a great many farms in the shape of dams, as the chief part of the country is well suited for such works; and at comparatively trifling expense the farmers could secure good crops by means of irrigation. By digging wells there is also, I am satisfied, much room for improvement.

"In answer to the third question in the circular, I am unable to state definitely to what extent proprietors would avail themselves of Government assistance, but there is not the smallest doubt that a great many would do so; and it is my own opinion, as well as that of a great many persons whom I have spoken to on the subject, that the proposed enactment would be a very useful and valuable one, and tend almost more than anything else to the real progress of the colony.

"In conclusion, I would beg to remark that I think that the services of a competent public officer should be secured, both to inspect localities, and who possesses likewise a knowledge of machinery.

"Could government enable parties requiring machinery to procure such as they require at cost price it would likewise be a great assistance to them, as the profits charged by the merchants are exorbitant, and, in fact, almost double the price of the machinery."

And he adds:—"I may mention as a fact which has come under my own observation, that on a farm where I happened to be the other day 300 to 400 head of cattle have been daily watered for the last three months, and six or eight of the neighbouring farms supplied with all their water, even for washing their linen, from a well about sixty feet deep, the water being drawn up by means of two half-aums attached to a rope, the one descending empty the other pulled up full. The name of the farm is Nanaga, the property of Mr Peter A. Crouse, senior."

Mr Leppan, Field-cornet of Upper Riebeek, wrote from Tea Fountain—that in that field-cornetcy the only practicable and beneficial dam to the public would be the embankment of a running stream above the village of Riebeek.

Mr Bowker, Field-cornet of North Fish River, wrote from Thorn Kloof:—"In reply, I beg leave to state that the whole of my field-cornetcy is purely a pastoral district, and generally so stony and the soil so shallow that only small patches here and there are capable of being cultivated. The rivers, the Great Fish and the Koonap, run so deep that it would require the outlay of considerable sums of money to bring the said small patches under irrigation. For my own part, after an experience of thirty years, I would not borrow money for the purpose of damming up either of the two rivers running through the field-cornetcy for the purpose of irrigation, being satisfied that it never would pay; but, at the same time, I am certain that there are many parts of the Colony where money might be laid out to advantage in constructing dams for the purposes of irrigation.

"Answer to question No. 1.—No plans for irrigation works have been proposed, because the inhabitants are well aware that, owing to the scarcity of labour and the bad quality when obtained, that the agriculturists, taking an average of years, only receive a day's food for a day's work throughout the colony.

"No. 2.—I am not aware that there are any places in my field-cornetcy where, by the construction of dams for the purpose of collecting the periodical supplies of water for irrigation works, which have of late years been very scanty, could be carried out with any prospect of remuneration

"No. 3.—The country not being adapted for agricultural purposes, I believe none would avail themselves of the offer of the Government, however liberal."

In his report for 1865 the Civil Commissioner of Albany states:—"Some dams have been made, but no rains have fallen as yet [14th February, 1866] to fill them. Other arrangements are in progress."

In 1866 it was reported:—"The want of rain is still keenly felt, and retards most seriously the progress of the Colony. The supply of water has seldom been known to be shorter than it is at the present moment.

"The rust has attacked all kinds of grain crops during the past season almost to a greater extent than on any previous occasion. From that cause, and the drought combined, many farmers who have sown large breadths of land have not been able to get even their seed back.

"The foot and tongue sickness has made great ravages among cattle; and between the losses from death, the loss of the services of the working oxen, and the poverty of the dairy cattle, all from this cause, many of the farmers have been much embarassed. Large numbers of cattle have also died within the last year from that scourge, the *lung ziękte*. Sheep and the smaller stock have, it appears, done better. In the beginning of the year horse-sickness was extremely prevalent, and great losses were sustained by some of the farmers among their stock of this description."

Of Albany in 1869 it was reported:—"The past year has been on the whole a favourable one for farming, though there were heavy losses of stock during the winter from drought. It is much to be regretted that the farmers do not adopt precautions to enable them to have food for their stock during the few months of the year in which it almost always is scarce. The growing of turnip or mangel-wurzel, or cutting hay, would apparently not be attended with any great difficulty. There is little doubt that a great number of farms are overstocked, the owners often preferring to keep more sheep than their farms are capable of sustaining in dry weather rather than sell them at a low rate. Abundant rains fell in the spring of the year, and agricultural produce of all descriptions is plentiful. There has been little or no rust in either wheat or forage. A great number of agricultural farmers have planted cotton this year; but its value can hardly be estimated at present. The principal difficulties appear to be the scarcity of labour and the grubs, which frequently destroy trees of a year's growth, and whole crops of young trees. Sheep-farmers are turning their attention to angora goats, and in a few instances to ostrich-farming, with an apparent prospect of success."

And in 1875 it was reported:—" During the past year this district has revived in great measure from the effects of the drought in the early part of the previous year, and the subsequent flood rains in December.

" On the whole this has been a prosperous year, both for town and country.

"In the country the ostrich farmers have been very successful. This branch of farming is gradually taking up the attention of all the farmers residing in the northern and eastern portions of this division. I regret that I have to report that sheep farming is being gradually neglected for the more paying speculation of ostrich farming, and the effect of this is gradually being felt, as the price of butchers' meat is increasing; almost every six months the price of this article of provisions is advancing. During this year the Fort Brown Bridge has been re-opened for traffic, and the Carlisle Bridge is fast approaching completion. The roads, which were very much damaged by the flood rains in December, 1874, have in a measure been restored."

In the course of the year 1865, Mr Kohl, the professional water-finder, visited Grahamstown. Of his success the following account was sent to me as one which had appeared in the *Journal*, one of the newspapers published in Grahamstown:—" Mr Kohl arrived in this city on Wednesday last, and his services in the peculiar avocation he follows have already been in requisition. Yesterday his professional aid was engaged by the Hon. S. Cawood, who drove him to his farm on the flats, where Mr Kohl pointed out five places at which water might be obtained readily; and, without any previous knowledge of the locality, indicated the sight of the well-known springs on Botha's Hill, three miles off, and also of the Grahamstown springs, within sight of the farm. On the day previous, while in conversation with Mr Cawood, he expressed his belief that a good spring of water could be obtained from Sugar Loaf Hill. This afternoon, it is understsod that a proposition will be submitted on his behalf to point out to the City Council, for the sum of £50, various sites on the municipal lands at which copious supplies might be obtained. We have been favoured with the sight of a letter addressed to Messrs Wood Brothers, by the Hon. H. Tucker, of Cradock, of which the following is an extract:—'Mr Kohl has pointed out several spots at Waai Plaatz where water may be found—at one of these spots we have worked and have found a valuable spring—going deeper, we expect to get a larger stream, as he told us. In the district there are already fourteen fresh fountains opened and available, owing to his directions. I went with him to Waai Plaatz, and am perfectly satisfied that those who consider him an impostor and a fool deserve the same titles for themselves. He goes by data as certain as if backed by eminent geologists; his practical knowledge is apparently more valuable than a vast amount of scientific theory.' Mr Kohl has already numerous applications for his services, and amongst them has accepted a liberal offer from Mr Wallace, of this city. It has been related to us that on his way from his farm at Middleburg, in a district in which there was no hotel, he claimed a night's hospitality from a Dutch farmer, and although he disclosed his name and offered to inspect the man's farm, his overtures were received so discourteously that he inspanned, and repaired to the house of the next proprietor. Here he was entertained so liberally that in the morning he pointed out a fountain, the knowledge of which increased the value of the farm materially. It was added—We will vouch for the credibility of the information, that not only did the fortunate farmer secure the full benefit of the obliging information, but his surly neighbour had his sole supply cut off by the diversion of the stream in consequence."

And in the issue of a subsequent day it was reported:—"Mr Kohl, the 'water-finder,' has been very busy of late. He has been with

Mr Blaine to Vale Krantz, near Riebeek, and indicated various springs there; with Mr George Wood, junr., to the farm near Aylesbury, and pointed out springs there; with the Hon. S. Cawood, to Thorn Park, and showed a place from which a valuable stream of water has been opened out; with the Hon. George Wood, to Cooper's Place, Blaauw Krantz, and marked a spot whence water enough for 400 muids of corn is to be obtained. On Monday he commenced his tour of the hills and kloofs around this city, for the purpose of indicating water sources, according to the term of his agreement with the Town Council. He was accompanied in his inspection by quite a little crowd. The Mayor, the Town Clerk, the Inspector of Works, Dr Atherstone, Mr James Roberts, and Councillors Watson, John Roberts, Wm. Roberts, John Webb, Rhodes, Page, and H. Roberts, left the town-house, with Mr Kohl, about ten o'clock. They proceeded at once to Goodwin's Kloof, above the Cape Corps Camp. Here Mr Kohl indicated two springs, and pronounced the lower one to be a good stream. Thence they crossed Allison Flat, and passed down in the next kloof (Graham's?), where a very strong supply was promised on condition of digging deep. The next kloof visited was Dyason's—the one through which the Cradock road passes—and here the amount of water to be obtained was pronounced to be scarcely worth the trouble of getting out. The company now adjourned to the flats, where they took tiffin. After partaking of the hospitality of the municipal cart, off they started again on the keen look-out for indications. They took a turn round on the flats themselves, but to little purpose. Turning back, they descended into Hottentot Kloof, where the local habitation of a small stream was pointed out in the upper part. Lower down, Kohl said there was a far more abundant supply. Passing over the rise, they next inspected the Mission Station Kloof, in which a spring was said to be, but not worth the expense of bringing out. Crossing the Kaffir location, they proceeded to the first kloof to the eastward, where an abundance of water was promised, not so much from the higher as from the lower part. Back again, through the location, to Pankhurst's place in the Hottentot quarter. There, said Kohl, would be obtained, by digging deeply, and opening a furrow of sufficient depth, a very large supply sufficient for half the town. Turning off to the new church, in the Kaffir location, it was found that two Kaffirs, by digging on their own behalf, had opened out a spring themselves. By this time it was quite five o'clock, and the company returned home, well satisfied, we are told, with the day's work."

Bathurst is a fiscal division, formed from the district of Albany, comprising what was formely called lower Albany; an agricultural and cattle producing country, with an area of about 650 square miles. It was in this part of Albany the agricultural settlers of 1820 were principally located. It lies on the Indian Ocean, having the Bushman River as its western boundary, Albany on the north, and the Great Fish River on the east.

The principal rivers, common to both Albany and Bathurst, are the Great Fish River, the Koonap, the Bushman, the Kariega, the Kasuga, and the Kowie.

The land is more suited for agriculture than is that of Albany, and there is a much larger extent of ground under cultivation. In situations where water can be depended upon the soil is very fertile, and agricultural and horticultural operations give a good return.

Of Bathurst, lying betwixt Albany and the sea, Mr Armstrong, Civil Commissioner, wrote :—" I am favourably disposed towards the construction of dams; and am of opinion that an immense body of water might be husbanded, if attention was paid to it, which is now wasted by running down rivulets into large streams and emptying itself into the sea. Many favourable localities present natural facilities for making dams, and there would be little or no difficulty in making them; this, together with the sinking of wells, would certainly be of much benefit." He stated that no plans of irrigation had been proposed in the district, but that he did not know anything more likely to give satisfaction than some such scheme of promoting irrigation as that upon which an opinion was then desired.

In 1865 it was reported of this division :—" The oat, barley, and wheat crops have, on the whole, yielded well this season, considerably in excess of the previous year. The oat crop has been less affected with rust than in 1864; and the yield has been a fair average, and of a good quality. The barley and wheat crops have been good; and large quantities have already been thrashed. The wheat has been altogether free from rust.

" Had rains fallen in October and November last, there is every reason for supposing that the harvest here would have been a rich and abundant one. There was, however, no rain whatever during the quarter ended 31st December, 1865; yet this important division has suffered but little in comparison with others, owing to the abundance of grass for all kinds of live-stock. Woolled sheep are doing well, and are fast increasing."

Of Bathurst in 1869 it was reported:—"This division, during the year 1869, has been favoured with abundant rain.

"There has been a good yield of wheat, oats, barley, maize, etc. The wheat and oat hay, which for the last few years have been subjected to rust, have escaped this virulent attack.

"The cultivation of the mulberrry plant, I regret to say, does not make much progress, although the soil is admirably adapted for its growth.

"A great deal of attention has been devoted to the culture of cotton and linseed. Numerous cotton fields may now be seen throughout the division, and are in a thriving condition. Several of the most thriving agriculturists have sown it largely, and a deep interest is taken in it. Cotton gins are expected, and there is every likelihood of its proving a great success. 391 lb. cotton, valued at £30, was exported *via* Port Alfred.

"The coffee plantation on the estate of the Hon. Mr Cock, at Port Alfred, continues to flourish, and will soon bear fruit."

SECTION III.—*Divisions of Victoria East, Peddie, King Williamstown, and East London.*

The electoral division of Victoria East comprises the section of the country lying between the Great Fish River and the Keiskamma and Chumie Rivers. It includes, as fiscal divisions, Victoria East and Peddie. The country, in its topographical features, has been described as "a terrace, flanked on each side with deep kloofs," on the north of which is Victoria East, and on the south Peddie. The soil is highly productive; the pasturage very healthy for sheep, and very salubrious. The principal rivers, common to Victoria East and Peddie, are the Chumie, Kat, Keiskamma, Beka, and Chalumna. It has an area of 300 square miles, is well watered, and includes in it the fertile valley of the Tyumie.

From Alice, the chief town of Victoria East, the Rev. Henry Calderwood, then Civil Commissioner of the district, wrote, in reply to the circular from the Colonial Office, under date of 28th May, 1862:—"I am informed by most of the field cornets in this division that no instances exist, as far as they are aware, in which landed proprietors would avail themselves of advances of money for the purpose of executing irrigation works. In some parts of this division, where it has been found practicable, water has been led out in furrows by the damming up of rivers, of which there are several instances. It appears to be the general opinion that in other parts, where the farms are situated at a distance from the rivers, it will be found quite impracticable to irrigate, on account of the hilly nature of the country and the consequent difficulty of retaining any quantity of water at a remunerative cost in those localities.

"I will write again as soon as I hear from the field cornets, who were written to immediately on receipt of your circular.

"If the terms offered by Government are such as could be met by the Fingoes, I should like to try if they could undertake something in this very important matter."

And again, under date of 4th June:—"Several dams have been proposed in favourable localities.

"Irrigation works in this division have chiefly been effected by the damming up of rivers and the leading out of water by means of furrows, of which the following is a description, in the different field cornetcies, namely:—

"*Field cornetcy of Alice.*—One water furrow, which irrigates about 250 acres of land.

"*Field cornetcies of Upper and Lower Tyumie.*—One furrow on W. Kilpatrick's farm; one on W. Key's farm; one in the Fingo location of Auckland; one on James Attwell's farm; and one on B. Woest's farm. These will irrigate about 360 acres of land. Two superior water-mills have been erected on the two last-named farm—one on each.

"*Field cornetcies of Willshire.*—One furrow on E. Webb's farm, which irrigates from 30 to 50 acres of land; one on W. Bezuidenhout's farm, which irrigates from 30 to 50 acres; one on C. Coester's farm, ditto, ditto.

"*Field cornetcy of Aberdeen.*—Jan Hartman has a furrow which will irrigate about 20 or 30 acres; M. Lotter & Son, ditto, ditto—thus making a total of from 740 to 820 acres which can be irrigated."

He adds:—"Since my letter of the 28th ultimo, the field cornets have reported that it is likely that landed proprietors in their different wards would avail themselves of loans from Government to the amount of about £3,000 in all."

Information was sent to me that Messrs Cock and Nicholson had built a dam in the Waterkloof River, for H. J. Louw, Esq., which, for simplicity, strength, and superior workmanship, was said not to be excelled, if equalled, in the Eastern Province. It is 80 feet wide; 11 feet deep, and contains a sheet of water upwards of 2 miles long.

In 1865 it was reported :—"Victoria East has been visited with a very long-continued drought, and has suffered considerably in consequence. Several farmers have been compelled to remove their flocks to British Kaffraria and the neighbourhood of the Windvogelberg, for grass and water. Very many Fingoes have removed their cattle to the Hogsback and beyond the Kei for the same reason. The crops in this division have also suffered much; indeed, if rain does not soon fall, they will be an entire failure."

Of Victoria East in 1869 it was reported :—"The crops in this district are more abundant than they have been for several years. Rust has not appeared in the wheat crops this season. Cattle and sheep have done well. A few Angora rams have been brought into the district during the year. There have been but few cases of thefts by natives residing in the district."

Of 1874 it was reported :—"Nothing particularly worthy of note occurred during the year, until the 6th of December, when the town of Alice was visited, after several days' rain, by an extraordinary inundation, consequent upon the Gaga and Chumie Rivers overflowing their banks. Happily, no lives were lost, but a vast amount of property was destroyed in the village and surrounding country, and some persons were for a time rendered helplessly destitute. The most pressing cases were, however, at once relieved by the generosity of the town and district, and as many of the sufferers set to work without delay to repair their losses, and to make the best of the unlooked for disaster. In a short time but little evidence remained of the calamity, which will never be forgotten by those who witnessed it. The farmers and Fingoes were heavy losers, especially the latter, none of whom have sheds for their sheep. Probably about three thousand sheep and goats were either swept away by the flood, or perished from cold.

"The public health has been good throughout the year, although the atmosphere has been unusually damp.

"The cost of all necessaries of life continues very high. Transport from the seaports is scarce and difficult to obtain, and it is no uncommon thing for a waggon to be a month and upwards on the road from Port Elizabeth, a distance of only 140 miles.

"The Kaffir Chief, Oba, who bought a farm, as reported last year, having paid the full amount of purchase money (£2,000) in cash, has entered into occupation, and is leading the life of a peaceful and well-conditioned citizen. He is desirous of having a school for Kaffir children established on his farm, and has promised to render it substantial aid. . . .

"The electoral roll has increased its numbers from three hundred and sixty-five to four hundred and eighty-two. About seventy of the newly-registered electors are Fingoes.

"There has been little crime during the year; only ninety-one persons were charged before the magistrate with offences, and, with the exception of eight convictions for stock thefts, and six cases which were referred to the Circuit Court, no crime of any seriousness was committed. There were thirteen convictions for drunkenness—eight of the delinquents were Europeans, and five were natives. It is a remarkable fact that not a single Fingo of this district was charged either with drunkenness or stock-stealing during the whole year. . . .

"The sum of £4,158, 1s. 10d. passed through the hands of the Civil Commissioner during the year; of that amount, £1,932, 11s. 9d. was actual revenue, towards which the Fingoes of the district contributed £716, 0s. 9d."

And in 1875 :—"The extraordinary floods which occurred in December, 1874, saturated the whole country to such a degree that the pasturage during the past year was unusually abundant, even in the driest parts of the districts; various grasses, shrubs, plants, and flowers, which were once common here, but which had disappeared for so long a time that they were almost forgotten, sprang up in every direction, providing food for live stock, and adding freshness and beauty to the landscape. The mealie crop was below the average, owing to the prevalence of a destructive caterpillar, but the Kaffir corn crop was not injured. Gardens were rather a failure, and for many months neither a potato nor any green vegetable was obtainable.

"The spring rains of 1875 were so late, and so much less copious than usual, that the natives are only now, in most parts of the district, beginning to plough and sow, and therefore a partial failure of the maize crop, or at any rate a less abundant harvest than ordinary, may fairly be anticipated; but this, although to be deplored for some reasons, may have, and probably will have, a beneficial result from another point of view, by compelling able-bodied natives, whose time is unoccupied, to seek employment on the railway works.

"Neither the natives nor the European farmers, as a rule, grow wheat to any extent; and, with the exception of the farmers in the Chumie Valley, who raise, in the aggregate, about 1,200 muids annually, the cultivation of this grain is neglected."

In the district of Peddie, the features of the country, the climate, and the products are very similar to those of Victoria East. It is, however, of much greater extent, and there is a very much larger quantity of land under cultivation; but this is mainly attributed to maize crops, sown by the very numerous Fingo population. It is a district lying along the coast, forming the southeast portion of what was fomerly called the Neutral Territory, lying between the Great Fish and Kenkamna Rivers. It has an area of 640 square miles, and a population of 18,706, the majority being Fingoes or Kaffirs. It is rather an elevated region, well-watered and fertile.

Of Peddie, Mr Eyde, Civil Commissioner, wrote:—"It appears to me there are localities favourable for dams, etc.; but the farms being generally small the proprietors do not seem disposed to construct them with borrowed capital." No plans for irrigation works had been proposed in the districts; but he had, at the date of his letter, received applications for £1,550 in loans, to be employed in the construction of dams, if this could be procured.

Of Peddie in 1865 it was reported:—"In the commencement of the spring there was a fine fall of rain in this division; and fair early crops of wheat, barley, and oat hay were raised; but the later crops, and the mealies and Kaffir corn, on which the natives almost entirely depend, having suffered severly from drought, and unless rain comes soon these will entirely fail; and great want will, it is feared, ensue. The rust in oat hay, though not so bad as last year, did considerable damage to the crops. The English oats, however, appear to be proof against this disease."

In 1866 it was reported:—"As was anticipated, the crops failed almost entirely last season, and the natives are suffering considerably from want. There has, however, been a succession of gentle rains during the present season, and fair crops are hoped for; but a heavy soaking rain would do an incalculable amount of good. The wheat and forage crops have been very much injured by rust. It is said that this disease has this season partially attacked the barley and English oat crops, which have hitherto been free. There can be little doubt that farmers must turn their attention to the raising of some other commodities, for they can hardly make a living under present circumstances. Some have already left off ploughing, and taken to transport riding and other employments, which are not liable to such wholesale losses as agriculture. At the same time it is probable that cereals will now maintain remunerative prices for those who can grow them, and this may be productive of good; hitherto, in favourable seasons, the farmers could not obtain any prices that would at all repay them. . . . .

"There has been a good deal of sickness amongst cattle during the year—lung-sickness *(pleuro pneumonia)*, which, however, does not appear to be now of so virulent a character as in former years; melt (?) sickness, which has carried off a considerable number; and tongue and *klaauw* sickness. The two latter, which almost invariably accompany each other, have been excessively severe; and as the diseases principally attack trek oxen, they are the source of very serious loss and inconvenience. The horse sickness has not made its appearance, but was very severe last season.

"Sheep have thriven moderately during the year, but do not do really well in most parts of the division, more especially near the sea. It is a fact worthy of remark that they generally do very well for the first year or two on a new place, and gradually fall."

Of Peddie in 1869 it was reported:—"Nearly all the farmers in this division have grown cotton, some as much as from 10 to 15 acres. Heavy rains have fallen, and the crops are looking well; there is every prospect of a good harvest."

Of Peddie in 1874 it was reported:—"This division continues to make progress, both in regard to agriculture and stock, and the condition of the inhabitants, European and native. Cotton was grown again this year, and a large crop gathered, but this article of produce will be discontinued, I believe, as not sufficiently remunerative to the grower or exporter.

"The extraordinary high floods in the beginning of December did great damage to crops, and some loss of stock, and also to some produce in transit in this division. The roads and drifts are much damaged; the principal outlet for traffic in this division, by Trumpetter's Drift, is destroyed, the bed of the drift being washed out, and a heavy expense of £60 or £100 will have to be incurred to make it passable. In the meantime the farmers and carriers of the division suffer great loss and inconvenience, many having to go forty or fifty miles round to arrive at Committee's Drift, often to find that also impassable. The company forming for the purpose of erecting a bridge at Trumpetter's, subscribed for shares to the amount of £5,000; Mr Newry, the resident engineer at Committee's Bridge, inspected the river and chose a site for the bridge, made a sketch drawing, and estimated the cost at £22,000. This was more than Peddie could raise, and consequently the idea of building without Government aid was abandoned. However, the recent floods have shown that had the intended bridge been already built, the money would have been uselessly expended, and that nothing, excepting a suspension bridge at a great altitude, would ever be of lasting service on this river, at least in these lower parts of it."

And in 1875:—"The last season was a very good one, and the prospect for a continuance very favourable for the present. Both Europeans and natives continue to make progress in agriculture and stock farming. The American stripper for wheat and other cereals has been tried and much approved for economy in every way, besides rendering farmers more independent of the precarious native hand labour. It is expected many such machines will be introduced as soon as they can be obtained.

"Cotton culture, I expect, will soon cease in this district. Ostrich farming is being tried with great success, and bids fair to extend during the present year, this district being well adapted, and known to have been well stocked with wild ostriches up to within the last 30 or 40 years.

"Crime in this district is not increasing."

When I made the tour of this region as Colonial Botanist, and visited Kaffraria, that land had not been incorporated with the Colony of the Cape of Good Hope, which it now is, being divided into two electoral divisions, that of King Williamstown and that of East London; aud no information has been obtained by me in regard to the water supply, and existing facilities for storing it, in this district. The division of King Williamstown has an area of 1,790 square miles —it is both well watered and wooded. The division of East London has a general resemblance to the coast region of the east frontier, being plentifully intersected by deep and wooded defiles, through which flow numerous rivers. It has a probable area of 2,000 square miles.

Of King Williamstown in 1866 it was reported:—"This division extends from the Chumie to the Kei River, probable distance about

80 miles, varying from 25 to 30 wide. The ranges of the Amatola and Buffalo Mountains come within the division.

"The sides of the above mountains adjacent, and other kloofs in the vicinity, abound with a variety of timber suitable for building purposes and for waggon making, as well as for erecting sawing and corn mills, also for cabinet and other works. Some of these woods are very beautiful.

"Since the occupation of the lands in this division by grantees and others, their present wealth in stock, etc., as well as comfortable homesteads, bear testimony that the lands are capable of rearing every description of stock, as well as of producing all kinds of grain and vegetables, in proof of which King Williamstown market is always well supplied with agricultural and other produce, attributable in a measure to the perseverance and industry in the culture of the land by the German immigrants. Notwithstanding the heavy losses by prevailing diseases, etc., among the cattle, as well as failure of crops through drought and other causes, the people in the main are thriving. This is proof positive that, with industry and perseverance, the lands in this division are quite capable of supporting a large population."

And of East London it was reported:—"East London is one of the two divisions of the recently annexed territory of British Kaffraria, and King Williamstown the other. East London is bounded east by the Kei River, west by the Keiskamma; its northern boundary is a line running from the Great Kei Drift, cutting the Berlin Hill, down to the Westleyville Drift of the Keiskamma River the line of coast between the two rivers is its southern boundary.

"In the year 1858 the Crown lands were parcelled out in farms, varying in size from 1,500 to 4,000 acres, and were granted to farmers under the Cathcart system of occupation. There are in this division 201 grantee farms. The unoccupied Crown lands are in extent about 200,000 acres.

"It is a magnificent pastoral country and its soil is also well adapted for cultivation; but, unfortunately, rust has, during the last two or three seasons, made sad havoc among the crops. The present harvest bids fair. The rust has again slightly attacked the oats, but the wheat and mealie crops promise fair."

CHAPTER VI.

SUPPLY OF WATER, AND EXISTING FACILITIES FOR THE STORAGE OF IT, IN INLAND EASTERN DIVISIONS OF THE COLONY OF THE CAPE OF GOOD HOPE.

SECTION I.—*Divisions of Fort Beaufort and Stockenstrom.*

To the west of Victoria East, and north of Albany, is the electoral division of Fort Beaufort, comprising the fiscal divisions of Fort Beaufort and Stockenstrom. The division is small in extent, but productive; and the pasturage is excellent. Wool is the chief staple. The great bulk of the inhabitants belong to the coloured classes, who, for their food, plant large quantities of maize; and this accounts principally for the fact that in 1865 nearly nine-tenths of the land in the division was under cultivation. From the town of Fort Beaufort, which is situated on the left bank of the Kat River, can be seen a long range of rugged and well-wooded hills, intersected by deep and precipitous ravines, and to which there attaches some historical interest on account of the locality having been the scene of several conflicts with the natives during some of the frontier wars. It has an area of 1,060 square miles. It is well-watered by the Kat, Koonap, and other streams rising in the Kat and Chumie Mountains. It is a country well adapted for sheep farming, and the hills are well wooded.

In regard to Fort Beaufort, Mr Stringfellow, then Civil Commissioner, stating that the observations he had to offer were founded upon an intimate knowledge of the physical formation of the division, and the character and circumstances of its inhabitants, wrote as follows in reply to the circular from the Colonial Office:—"At present no plans for irrigation in this district have been proposed, nor does it appear that the subject has engaged the attention of the inhabitants.

"Irrigation to a considerable extent is carried on at the sources of the Koonap and Kat Rivers, where the waters run near the surface, but as these streams advance in their course they become deeply seated, yet much land on their margins might be irrigated if the farmers were inclined to agricultural pursuits.

"It is believed that very few farmers in this division would avail themselves of the advances of money for the purpose of irrigation as the majority of them reside upon the banks of the rivers above named, which are permanent and copious streams, and used chiefly for sheep-washing. The occupants of these farms are sufficiently wealthy to effect any improvement of the nature referred to if they desired to cultivate any portion of their lands. There are a few farms, however, which have not the advantages of permanent streams upon them; but the owners have ample means of erecting dams should the necessity arise for such constructions."

In accordance with this is the following evidence, given before the Select Committtee of the Legislative Council on Irrigation, by Mr Upton, Member of the Legislative Assembly for the district:—

"182. *Chairman.*] Are you of opinion that in your part of the Colony much can be done in the way of irrigation; and will you be kind enough to favour the committee with any facts which your practical knowledge enables you to state?—I think a good deal might be done with irrigation in our part of the country. A good deal has been done already, but I think more might be accomplished if we had more immigrants to carry on such irrigation. There are a few rivers in our part of the country which have been more or less taken out and dams formed across them. The water is generally led over the land, but not to the extent the farmers might and would do if they had a better supply of labour to aid them in the work. Our part of the country is not alone an agricultural country, that is, a country requiring irrigation to develop it, but it is likewise a sheep country. A great portion of the farmers turn their attention to sheep together with agriculture, so that, the two things being combined, they do not carry on irrigation to the extent they might otherwise do. I think that prevents our farmers in a great measure from attending more to agriculture.

183. In your opinion, is it want of capital and want of energy, as well as want of labour, which prevents irrigation being more extensively carried out there?—In my part of the country it is not want of capital, for I think that there is no farmer there who is not in a position to irrigate his lands if he chose to do so. A far as knowledge goes, they have sufficient to carry on irrigation if they thought it would pay them sufficiently to do so; but, in many in-

stances, the farmers think they can do well enough without having recourse to irrigation, and, therefore, do not trouble themselves about it.

"184. *Mr Barry.*] Do you mean to say that they find pastoral occupation more profitable than agricultural?—Yes. . . .

"244. *Mr de Wet.*] In your district you do not require irrigation for the purpose of making kitchen gardens, I suppose?—Some farmers have no water on their places which they can lead out, but the greater portion have, and wherever it is practicable to get it they lead it out.

"245. *Chairman.*] You mean wherever it is practicable to do it without any large expense at the outset?—Yes.

"246. When, in fact, you might lead it out with a plough furrow? —Not exactly. Dams five and six feet high are thrown across a river, such as the Koonap; but anything above that I do not believe would stand. The Dutch Church at Adelaide, for instance, commenced a dam in the Koonap River, and failed twice. They built the dam each time seven feet high, and the water came down and took it away; but after inspecting the ground and going up a distance of three quarters of a mile higher, a better place was found, where there was not the necessity of building such a high dam. One was built five feet high, which stands well; but higher than that I have not seen them stand.

"247. Was the dam at Adelaide built by engineers?—No; it was contracted for and built the ordinary way that builders construct dams.

"248. Were the walls straight up?—Inclined both ways.

"249. *Mr von Maltitz.*] And built with lime?—Yes.

"250. *Mr de Wet.*] Could irrigation be practised in your district to such an extent so as to improve the pasturage, or would that cost too much in some places?—The pasturage is rich enough already.

"251. But in dry seasons?—It would not pay us to irrigate for cattle purposes.

"252. Why not?—I think we can do very well without it, and we have generally sufficient for our stock,—as much as farmers in general need.

"253. But supposing a person could not afford to buy stock, and was obliged, in consequence of straitened circumstances, to feed his cattle, oxen, and so on, for sale, would it not pay to irrigate his land? —I do not think it would pay.

"254. They get food enough without?—They generally do. It might pay on a small scale, but not to the extent of the profit already secured by farming.

"255. *Chairman.*] You suffer great losses in stock from time to time on account of droughts. Have you ever calculated whether,—or could you offer good reason for believing it impossible, by preserving the water in winter in different rivers, you would be able to use some part of it for irrigating pasture land, so as to enable stock to live through the dry season?—I think we could if we irrigated to that extent; but we irrigate to raise produce for the market, and not for stock. I do not think it would pay to irrigate to raise food for stock."

And the statements by Mr Stringfellow receive illustration from the following evidence given before the same Committee by Mr Zievogel:—

"96. Do you know the Kat and Koonap Rivers?—Somewhat.

"97. Do you think they contain much water?—Yes.

"98. Do you think the water could be led out?—I do not know whether there is a permanent run of water in them; I think I have gone over the Koonap dry-shod more than once; but, generally speaking, there is a flow of water which may be led out for irrigation.

"99. When are these rivers generally dry?—The Koonap and Kat rivers, I should say, must be seldom dry in winter, except in very dry seasons indeed, because the water comes from the high mountains northward of their sources, and a fall of snow in winter will produce a good supply of water; but in dry summers the supply would become less. I know the rains fall in summer in our part of the country, but then these rains generally come in large quantities and run off immediately. I do not think that thunder showers, though good for the pasturage, contribute much to the permanent running of the rivers; that depends more on the fall of snow in winter on the mountains from which the rivers issue."

At the close of 1865 the Civil Commissioner of Fort Beaufort reported:—"Very nearly seven miles of the new water furrow (by which a stream of water from the Kat River will be brought into Fort Beaufort) has been completed by the contracter, under the personal superintendence of one of the Municipal Board. Iron pipes have been ordered, and a large reservoir is to be built; when these works are completed Fort Beaufort will be one of the best watered towns on the eastern frontier." But he reported at the same time in regard to the district:—"The past year, particularly the last six months of it, has been one of unprecedented drought. Very little grain has, in consequence, been sown; and in many cases that little has been lost for want of rain or irrigation. The numerous natives on the commonage will have no crops this year; and those at Healdtown will not be much better off. In the Winterberg, the granary of the division, most of the crops have failed.

"Nor has the stock farmer fared much better. Thousands of sheep, and hundreds of cattle and horses, have been removed from some of the finest farms in the division, to seek sustenance in the highlands of Fort Fordyce, Bontebok Flats, etc.

"The Koonap River, one of the finest streams on the frontier, has ceased running; the village of Adelaide has consequently been deprived of water to irrigate its beautiful gardens, and the water in the river has become stagnant. The Kat River has fared very little better; irrigation has ceased almost along its entire length. Ainslie's fine water-mill has ceased to grind; and, although the river has not *entirely* ceased running at Fort Beaufort, the water is becoming unwholesome for domestic purposes."

In 1866 it was reported:—"During the first months of the year the drought continued unabated. In the commencement of the spring, a very fine rain fell, just in time to enable the agriculturists to plough and sow their lands, and soon changed the appearance of the country, from a burnt-up condition and no grass to the most delightful verdure and abundance of food for stock.

"This first rain was followed by refreshing showers almost every week, and farming stock rapidly improved in condition, while the crops of wheat, barley, oats, Indian and Kaffir corn, gave promise of an abundant harvest. The natives on the town commonage and the Fingoes at Healdtown and other places cultivated their gardens to a very large extent, and the most sanguine hopes were entertained that the harvest would be an extraordinarily good one.

"But these hopes have not been realised to so great an extent as was expected. About the month of November, when rains were much needed, they suddenly ceased, and since that period the showers have been infrequent, light, and very partial; and the crops of every description have suffered much in consequence, and will be much lighter than was expected. Rust has also again made its appearance, particularly in the oat crops; and this time it has, in several instances, which have been reported, attacked not only what is usually called the Cape oat, but the English side oat and the black oat as well. Notwithstanding these drawbacks, however, the reports which have come from different localities lead to the belief that the harvest will be a considerable improvement upon the last.

"The lambing season has been a good one, and there has been very little disease amongst the flocks.

"The reports with reference to horned cattle have not been so favourable. Lung-sickness has appeared in several localities, and the hoof-sickness and tongue-sickness have been very prevalent, and attended with considerable loss to farmers and carriers. . . .

"In the town of Fort Beaufort some progress has been made. Several new buildings have been completed; abundance of fresh water from the Kat River now flows through the town, iron pipes are daily expected, and a large wool-washing establishment is in full work. The water for the latter is lifted from the Kat River by means of an immense pump, imported by a private gentleman, which, in addition, supplies him with abundance of water for irrigating his lands. Nor must Prince Alfred's Grove be forgotten. The trees in the avenues have made wonderful progress, and many scores of choice plants have been introduced by the Civil Commissioner. Among these are the olive of commerce, the indigenous wild olive, the wild chesnut of our forests, and the *ailanthus glandulosa* (also called the 'tree of Heaven'), all of which are flourishing in perfection."

Of Fort Beaufort in 1869 it was reported :—"I very much regret that I cannot report progress during the past year.

"The failure of the Frontier Commercial and Agricultural Bank, the low prices for wool, drought during nearly three-quarters of the year, locusts, etc., have caused great depression in almost every industry. The loss of sheep has been enormous; few lambs have been saved, and full-grown sheep have died by thousands. Complaints are heard on every side.

"Stock thefts have increased. Owing to the high prices paid by dealers for goat skins for exportation, the natives steal and kill sheep for the skin, leaving the carcases untouched . . .

"Considerable improvements have been made in the repairs of the streets and water-furrows of the town. A reservoir is now in course of construction by military labour. When completed (probably by the end of March next), iron pipes will be laid down in the streets, and Fort Beaufort will be one of the best watered towns in the Colony."

In 1874 it was reported :—"The year 1874 has been marked by great floods. Much destruction has been caused to roads and bridges, and the loss in stock to farmers has been immense. The Kat River, near which Fort Beaufort is built, rose to an unprecedented height; the Victoria Bridge was completely covered with water, the parapets were carried away, and other serious damage done to it. It is now undergoing repairs for temporary traffic, but plans, specifications, and estimates have been forwarded to the Chief Inspector of Public Works, with a view to the permanent repairs and restoration of the bridge. The damage done by the rain and floods to private property can hardly be calculated. Houses in towns and villages have fallen in, or otherwise suffered severe damage. Some houses built near the banks of the river have been completely swept away. In the agricultural parts of the district, whole fields sown with wheat, barley, oats, etc., have been submerged, and the crops destroyed or swept away. Business of all descriptions has greatly suffered. In addition to the bridges which have been swept away, the roads have been so cut up by the rains that the transport of goods and produce has been rendered impossible for some time. So much damage has been done to the wheat crop that we can hardly look for a fall in the price of meal; it now rules at 60s. per muid, while, a few years ago, 20s. was considered a high price.

"Butcher meat is still quoted at a high figure, slaughter cattle selling at £15, and upwards, and sheep at 20s. each. Groceries, owing to the high rate of transport, are most expensive."

In 1875:—"Upon the whole, the year 1875 has been a prosperous one to this district, and production of all sorts have been cheap and plentiful."

The division of Stockenstrom comprises the old Kat River Hottentot settlement, with an area of 260 square miles. The country at the base of the Katbergen is wonderfully fertile and well watered, and calculated to support a much larger population than as yet inhabits it. The Katbergen are a range of lofty and precipitous mountains, which bound the Kat River region to the north, and rise to a height of about 5,000 feet. They are a continuation of the Great Winterberg range on the west, and join the Amatola and T'Chumie range on the east. Their coast face is well wooded and steep, but the northern or inland slope, like that of nearly all Cape mountains, is on an easy incline and devoid of timber. The outline is broken and picturesque in the extreme.

There are many very valuable sheep farms in the valleys between the offshots of the mountain range.

In regard to Stockenstrom, the reply of Mr Emmet, Civil Commissioner of this district, to the queries issued from the Colonial Office were as follows :—

"1. When this division was formed into a settlement for the Hottentots, dams to lead out water from the various streams flowing through it were made for the several blocks of erven, with a few exceptions. A plan to lead the water out of the Elands River to this village was proposed some time back, and the work has been completed by local efforts.

"2. The preceding answer to the first question will partly have answered the second. Great facilities exist throughout the whole division for leading out the water, excepting in a few places. There is, therefore, no necessity for sinking wells.

"3. I am not aware that any of the landed proprietors of this division would avail themselves of a law empowering the Government to make advances of money for improving the supply of water on their properties. One or two cases might arise, but generally speaking the facilities are such that water can be got out without the outlay of much money and by the voluntary labour and contributions of the erf-holders."

In regard to Stockenstrom it was reported in 1865 :—"The year 1865 has been one of the severest in every respect both for the mercantile and agricultural interests of the community; and the usual pressure, so widely felt in this Colony, has been largely participated in by the inhabitants of this division. The fall in the price of wool, the scarcity of money, and the extended drought, have produced a state of things under which many have succumbed. Estates have been surrendered; and many persons, who during previous years had been in comfortable circumstances, have suddenly found themselves very much reduced; whilst the low state of valuable commodities and stock, owing to the scarcity of money, has still further tended to complicate and embarrass their position.

"The wheat crop has been remarkably good; and the late mealie crops also gave a good return after the destructive locust, referred to in last year's report, had left; but only lands under irrigation produced these crops, for the drought had been so severe, that the crops on many lands have been completely dried up. During the year very little rain has fallen, the fountains became weaker and weaker, and the hitherto abundant supplies of water from the numerous streamlets of this favoured region were so diminished at the time when the water was most needed, that in many parts of the division the crops failed entirely; while in other parts the want of rain prevented the erf-holders from sowing any seed. The prices of provisions have, however, been moderate, in consequence of supplies having been introduced from various parts."

In 1866 it was reported :—" Stockenstrom has undergone a trying time, owing to the long continuance and depressing nature of the severe drought from which it, in common with other parts of the Colony, has suffered during the past and several previous years. On account of this, the agricultural society made no attempt to get up a show for 1866; to have done so would have been vain.

"An effort was made by the clergy and several other gentlemen charitably disposed to alleviate the distress amongst the native population. With this view a committee was formed, but, owing to the want of co-operation on the part of some, and the slender means of others, very little assistance could be rendered to the needy.

" Some of the natives who have erven have suffered mainly through their own indolence and extravagance; others, owing to drought and other causes, have not been able to do more than obtain a bare livelihood. Towards the close of the year, however, there were good rains, and the result is highly favourable. Notwithstanding rust here and there, the wheat crops look well, and there will be a tolerably good harvest. The oat crops have been a failure.

" The sheep farmers in this division have done well, notwithstanding the drought; and a larger quantity of wool has been sent away during the past than any previous year."

Of Stockenstrom in 1869 it was reported:—" The year 1869 has been remarkable for considerable pecuniary difficulties, depression of trade, and for severe losses of sheep, owing to protracted drought. On the other hand, fine rains in the spring have again fallen, grass has been abundant, and the cereals and other crops have been good. Agriculturists are not slow in cultivating their lands to the best advantage, and trying the best kinds of wheat. It is hoped that the difficulties men have had to pass through (many having been reduced to absolute want) will induce habits of frugality and industry which, perhaps, would not be attained without time, and many have resolutely put their shoulder to the wheel to overcome the difficulties of their position. The low price of agricultural produce has rendered the position of many still more difficult, and while there has been plenty of food, it is found difficult to turn produce into money, and until exportable articles are more generally taken to, this difficulty will continue. Some are beginning to pay more attention to the get up of wool for the European market, and until this is done generally, a low price will be given for an article which has almost glutted the market.

"Many erven have changed hands, and insolvencies have been not infrequent, though fewer than in the preceding years. Both white and coloured erf-holders have lost their erven through debt, and in some of the blocks there are only a few coloured erf-holders now.

"The usual amount of sickness has prevailed, but the paucity of deaths is a strong argument for the general salubrity of the South African climate. The close of the year has brought a summer remarkable for moderate heat and a more than usual amount of cloud, and dews at night. . . .

"Provisions of all kinds have been abundant and cheap; mutton and beef have decreased considerably in price, although the value of wool has lessened; sheep farming is still persevered in, but more attention is paid to the breeding of the Angora goat.

"Lung-sickness among cattle and horse-sickness have been less than in former years.

"This division has been nearly entirely exempted from locust ravages, but there are other plagues of a serious nature. The 'spring beestje' in the peach and other trees has been worse than ever it was before. The peach particularly has suffered by the uppermost branches withering up, and in that condition bears a little, but in most cases no fruit. It is remarkable that the St. Helena peach is hitherto free from the attack of this insect. It is likely, however, that to a great extent the want of continuous and proper dressing, the retention of old stocks, and the absence of a more scientific mode of culture, may be the principal cause of deterioration of a fruit which has been most abundant and excellent in these localities. Pears, although promising well by a superabundance of blossoms, have mostly failed, as have apricots and plums. Apples have borne well, and figs are particularly abundant, and fine. Grapes have in some kinds not been free from *oïdium*. Cotton growing has not been attempted on any scale, nor does the growth of silk appear to be progressing, but a preparatory work is going on by planting the mulberry. The *alanthus glandulosa*, a tree that grows readily, is also getting about. Linseed has been sown by a few persons, and it is to be hoped that the growth of this exportable article will be persevered in."

"In the early part of the summer an extraordinary quantity of blue-bottle flies appeared, their great increase being attributed to the numbers of dead sheep during the drought. Grubs and caterpillars have been more abundant than usual, and also a black and golden-coloured winged beetle called the 'tor-a-dor.'

"There has been sickness among the poultry, and fowls and turkeys have died almost everywhere. Extensive grass fires have occurred, extending in some parts to the Crown forests. During one of these conflagrations, on a very windy day, three coloured persons were burnt to death whilst endeavouring to make their escape from the devouring element. The miserable system of grass-burning, which neither legal enactment nor the voice of common sense appears to be able to prevent, is gradually denuding the country of herb, and sterility and barrenness must inevitably result from the continued destruction of forest, bush, and shrub. . . .

"Partial efforts have been made to extirpate the burr weed, which is increasing in some parts of the division, owing to the indifference of those who should, for their own interest, be foremost in its destruction."

It 1874 it was reported of Stockenstrom :—"The district was still suffering, in the early part of the year, from a very long prevalence of drought, which, however, happily broke up in time for a late mealie crop, and other winter cereals, and the fine rains brought forward the winter pasturage, which was the means of saving thousands of head of stock which must have otherwise perished.

"The very high prices of the necessaries of life have been maintained up to the present time, but our prospects for the year 1875 look brighter.

"Our winter was an unusually severe one, with several heavy falls of snow of from four to six inches in the village, but much heavier on and near the Katbeg rânge of mountains, but the losses of stock were not so severe as might have been expected, being sustained, no doubt, by the good pasturage. Sickness prevailed, however, to a considerable extent, more particularly amongst the children, and great hardship and suffering was endured by many, there being no district surgeon at present.

"Very extensive repairs were undertaken on the roads, drifts, etc., as well as the opening up of the Mitchell's Pass leading on to the Bontebok Flats, division of Queenstown, where a large number of farmers have recently settled, but I regret to say that the late serious floods have done this division very serious damage, by cutting up the roads, and causing heavy losses of stock of all kinds. The Mitchell's Pass works have sustained little if any damage, however, and they are now being rapidly repaired, and pushed forward, and will be of very great advantage to the Bontebok farmers, by opening up to them a good road thence to Fort Beaufort and other parts of the Colony.

"Our wheat and other crops are good, which I trust may be the means of lowering prices very considerably.

"Several houses fell down in the village, from the long-continued wet weather."

In 1875:—"The past year has been what may be termed a dry season; but the crops have nevertheless been a fair average, and stock generally has not suffered much. . . . .

"The forests, I regret to say, are not worked and conducted as they ought to be, and it is a matter that requires the serious attention of Government, and a little sharp practice in this and probably other divisions. It would be desirable that surveyors be employed to effectually and permanently lay out the forests, in order to their being properly worked and protected from plunder and destruction, as well as broad spaces being left as barriers against fire. The surveyors ought to survey the whole tract of forest, and delineate, with the aid of rangers, the wards or blocks into which it is to be divided for permanent culture, and the roads necessary for transport, fire, etc. This plan, with code of instructions, to be handed over to each forest officer. When a ward is cleared out the time of closing it ought to be carefully recorded, so that the date of renewal may be known.

"I now cull a few remarks from the *Edinburgh Review*, received in December last, so applicable that I trust no apology will be necessary for their insertion here:—'The foot-fall of man on newly discovered shores has been attended but too often by wanton destruction of that vegetable clothing of the earth, on the extermination of which, fertility in a measure ceases, and a garden becomes a desert.'

"'God Almighty,' said Bacon, 'first planted a garden. And on this key-note depends the tone of the entire treatment of the subject by the restorer, and the lover of modern science.'

"From 400 to upwards of 1,600 loads of timber per mile may be annually obtained, according to latitude and level, in perpetuity, by careful culture. The systematic care of self-sown seedlings, and the thinning of the young growth is the true method of providing for forest reproduction. But our thinning at present is carried on ruinously by the Kaffir tribes for the building of their huts.

"In France and Germany, forestry is one of the State professions of highest scientific character, and regularly filled from the better and well educated class, and the study of forestry or arboriculture ought to form part of the curriculum of the South African College.

"In France there exist 2,700,000 acres of State forest, in the conservation and management of which £500,000 annually is expended. The returns amounting to £1,740,000, other countries in proportion; of the plantations made in England in 1808, 40,000 acres were valued in 1840 at £1,087,777, and the value of the same when arrived at maturity was estimated at £10,000,000.

"It is strongly and sensibly urged that the State should control all forests, on the ground that private individuals could not be expected to let his wood grow long enough to furnish timber of a size fit for naval or other most important construction.

"British India has achieved a most brilliant triumph by the skill of the forester in the naturalisation of the precious Chinchona, and in 1873, 2,645,373 plants were luxuriously growing in the Government plantations in the Nilgiri Hills. But for this precaution, the Chinchona might have been exterminated in our own day.

"In the wholesale destruction and neglect of our forests the rainfall will become more irregular and diminished, and climatic law ought to be more studied.

"That a great system of rainfall, watershed, water supply, irrigation, drainage, and outfall, finds its cradle in woodland heights, and its outlet in the sea, has been much lost sight of.

"Enough rain falls along our Katberg mountain range here, if properly husbanded, and the natural advantages taken into scientific account in the construction of proper reservoirs, to supply the whole of the Eastern Province.

"Wherever planting has been systematically and largely carried out, the rainfall has increased in proportion.

"Near Valucia, in South America, there was a beautiful lake which was surrounded by forest. The trees were cut away, and the waters of the lake receded to the distance of four miles and a half. The trees were replaced by others, and the lake gradually returned to its original boundaries."

SECTION II.—*Divisions of Bedford and Somerset.*

To the west of the division of Fort Beaufort is the division of Bedford, which includes all the fine and fertile valley of the Kaga,

Great Winterberg, and Kromme ranges, well-wooded, and fit for agricultural purposes. Its area is 1,200 square miles.

In regard to the division of Bedford, Mr Liddle, then Civil Commissioner, wrote in reply to the Colonial Secretary circular:—" No plans for irrigation works within this division have been proposed.

" This division, unlike the Karroo, is usually well watered, possessing, as it does, the Fish River, Koonap, East Riet River, Kaga, Gaba, Cowie, Mancazana, Baviaan's River, and other small running streams. The chief, I might almost say the only, pursuit in this division is wool-growing, and, notwithstanding an unusually dry season, I have not heard any complaints of want of sufficient water for stock.

" I am not aware of any especial need for the construction of dams, etc., of the nature contemplated in the 2nd section of above circular, nor of any particular locality suitable for such purposes.

" It would, of course, be desirable in some instances to construct dams to lead water from some of the rivers I have named; but subject as they annually are to floods, which invariably carry all before them, the expense of sufficient dams, even with the assistance of money advanced by Government, would be so great as to render it beyond the means of individual persons, and I am satisfied no farmer would be disposed to incur the risk of having the result of his labour and means swept away in a single season."

In 1865 it was reported:—" During the year 1865 Bedford has been visited by one of the most severe droughts ever known in this country. For about eight months there has scarcely been any rainfall; and, during the greater part of the time, hot winds have prevailed to such an extent that at the present moment the country is almost destitute of vegetation. The heat, too, night and day, has been intense. During December, the thermometer for several consecutive days stood at 112° in the shade.

" Water has been so scarce in the village that wells have been sunk on private property, and excellent water has been obtained at a depth of fifty-seven feet. Neighbouring farmers, hearing of this, are now sinking wells upon their arid farms.

" The agriculturists have suffered heavy losses; and many would have been entirely ruined, had it not been for the grass and water afforded by the Crown land on the mountain tops at the Mancazana and Baviaan River. Numbers of farmers with immense herds of cattle and sheep have been permitted temporarily to occupy these invaluable lands, free of all charge.

"No garden or other crops were sown either by natives or Europeans at the proper season, and therefore there is but a dreary prospect to look forward to; in fact, a famine, similar to that of 1862, with all its attendant consequences, seems inevitable."

In 1866 it was reported:—"The past year has, on the whole, been one of general prosperity. Drought, the great bane of the country, has not visited the division to any serious extent; and, although horse and klaauw-ziekte have carried of large numbers of valuable animals, yet stock of all kinds has thriven and increased so rapidly, that the farmer, taking all things fairly into consideration, has no cause to be discontented.

"During the latter six months of the year copious rains have fallen everywhere, and large quantities of grain have been sown both by Europeans and natives. Unfortunately the rust, and a caterpillar hitherto uncommon, have made great havoc; but still a fair average crop is anticipated."

Of Bedford in 1869 it was reported:—"This district, in common with other portions of the province, has suffered very severely during the year from drought, and the loss in sheep, goats, and cattle has been greater than in any preceding year.

"In addition to the loss arising from want of pasturage, immense swarms of locusts passed over the district, and devoured the little pasture which had been carefully preserved for the subsistence of stock during the winter months, and a large quantity of stock of every description died from starvation. Many of the flock-masters were compelled to destroy the lambs to save the mothers, and in a large number of cases the drought has carried off no less than a half of their full-grown stock.

"This unprecedented loss of stock has reduced the yield of wool in the same proportion, and not a few of the farmers in this district have been compelled to manage upon half their usual farming income. In the Baviaan River Ward, however, the losses have been comparatively trifling, the mountain pasturage of the farms in this portion of the district proving sufficient for the maintenance of the stock.

"Owing to the scarcity of pasturage and want of grass, it was almost impossible to keep the sheep together, and being found straying in groups of fives and tens, great facility was thereby afforded for stealing, which the Kaffirs are not slow to avail themselves of, the absence of the Frontier Mounted Police in the Damara country giving them every chance of passing with the stolen property into Kaffirland without being detected. . . .

"To a portion of the district considerable distress was occasioned by a tremendous fire which broke out in the Kowie forest in the month of July last, destroying valuable and comfortable homesteads, the inmates barely escaping with their lives. Stacks of wheat, oats, barley, and other crops were completely burnt up, and a large quantity of mealies destroyed. In the upper part of the Kowie, where bush-workers and natives, with their families residing, the huts and lots were all burnt, and some lives lost. Relief was afforded to the sufferers out of the funds which had been raised towards the distress occasioned by the great fire in the Knysna, Humansdorp, and Uitenhage districts, and which, it appeared, was not likely to be all required.

"Much attention is given in this district to improvements for getting up wool in the best possible condition, and the Australian or warm-water process of wool-washing is coming into more general use.

"At an exhibition of wool, held at Port Elizabeth during the year, open to the whole Colony, the prize was carried off by Mr W. Dodds Pringle, of Baviaan River, his son (Mr Robert Pringle) being second.

"The rams introduced by Messrs Blaine & Co. have found several purchasers in this district, and no pains or expense are spared in improving the character of the wool and getting it up in the most respectable condition. The spirit evinced by the flock-masters of the Bedford district in this respect is highly creditable.

"The production of Angora hair is also receiving especial attention and with very favourable and encouraging results, some of the clips of goats, owned by farmers in this district, having realised as much as 2s. 6d. per lb. Several of the rams imported by the same firm, from Asia Minor, have been placed in this district. The wheat and other crops are a fair average and very free from rust."

In 1874 it was reported :—"The year 1874 will long be remembered in this division as one of the most disastrous on record. It began with a drought, and ended with a flood, and many of the farmers of small means who suffered severely from the drought, were nearly ruined by the flood. The circumstances attending the latter, and resulting from it, have been so thoroughly described in the public prints, that it is unnecessary to do more than make this allusion to them in my report. Since the very heavy rains which fell in the early part of December, frequent showers, particularly thunder storms, have fallen, more especially in the northern districts, the result of which has been that the Great Fish River has several times risen to its banks, and been thus rendered impassable for some time, thus suspending traffic between the coast and inland divisions.* Much damage has also been done to the drifts and main roads, after they had been repaired by the Divisional Councils. The pasturage is, as a matter of course, now exceedingly abundant, and the stock which escaped destruction is in excellent condition.

"The crops of mealies will probably be heavy, unless the caterpillar destroys a large proportion, of which there is, unhappily, some prospect. Owing to the heavy rains having fallen late in the season, the wheat crops will be scanty. Forage is much cheaper than it was some months ago. Fruit will be scarce, the country having been visited by severe hail-storms previously to the general rains.

"No agricultural show was held at the time appointed for the annual fair, the Society having found it quite impossible to get up a fair exhibition, owing to the low condition to which stock had been reduced by the drought of 1873-4. A show is to be held, however, on the second Wednesday of next March, when it is hoped the division will have somewhat recovered from the effects of the drought and flood of 1874."

And in 1875:—"The Bedford division has been tolerably prosperous during the year 1875, notwithstanding a drought which prevailed in some portions of the district during the latter part of the year.

"Stock has been comparatively free from disease, although fever prevailed as usual amongst sheep during the early months of the year, and lung sickness occasionally broke out amongst horned cattle.

"High prices have been obtained by farmers for their produce, and the wheat crops ripening in December have been very fine, giving hope to consumers that a material fall in the price of that cereal will take place in 1876. Prices of horned cattle have risen to a marvellous height, and the demand for slaughter stock has become so great that the supply has now nearly failed."

* In another volume I have had occasion to state in regard to a similar flood in 1863:—"I proceeded to Adelaide and Bedford *en route* for Somerset and Graaff-Reinet. But at Bedford I was again detained through the Fish River being again impassable. It had come down and had continued to flow for a great many days, I think ten days or a fortnight, and that with a current so impetuous that the ferrymen had refused a fare of £10 to carry across a medical practitioner, whose services were required on the further side; and there being no appearance of subsidence, I had to abandon my purpose of crossing it, and to change entirely my route in returning to the Western Province."

To the west of Bedford is the division of Somerset, with an area of 3,300 square miles. "It is intersected by several lofty mountain ranges; the most northern, called Zwagers Hoek Mountain, is a continuation of the Sneeuwbergen, connecting that range with the Great Winterberg. Another range parallel to it is the Boschberg (4,880 feet), running east and west, between the Sunday and the Great Fish Rivers, at the foot of which is situated the town of Somerset East, a well-watered village, with 2,000 inhabitants and several churches, banks, a college, and other public institutions. Somerset division is bounded on the west by that of Graaff-Reinet, on the north by Cradock, on the east by Bedford, and on the south by Uitenhage, the Zuurberg Mountains forming the division line. Owing to its hilly surface it presents some fine romantic scenery, and possesses some of the most valuable farms in the Colony, particularly those lying in the rich glen or valley called Zwagers Hoek. . . .

"There is much diversity in the soil of Somerset, caused principally by the occasional scantiness of the water supply. In certains parts of the division farmers are sometimes compelled to leave their farms on this account; but then again, after the rains, the pasturage is rich and more abundant. The grass of the division is particularly advantageous to the fattening of stock."

In regard to facilities for irrigation in the district of Somerset, Mr Maskew, then acting as Civil Commissioner in the district, wrote, under date of 16th May, 1862, to the Honourable the Colonial Secretary:—"In reply to your circular, No. 16, of 1862, received yesterday, requesting information on irrigation works within this division, I have the honour to report, for the information of his Excellency the Governor, on the following points therein named:—

"1. No fixed or systematic plans have, as far as I am aware or been able to learn, been proposed for irrigation works within this division, which is principally a pastoral one, and its inhabitants Dutch sheep-farmers.

"2. There are two or three farms in this division which are permanently, and five or six occasionally irrigated by diverting the water from the Little Fish River; and portions of several farms lying on its banks might be placed under cultivation by damming it up. Some eight or nine farms are also irrigated from the Vogel River, above Pearston. No water is led out of the Great Fish River and Brak River, the banks of the former being too steep, and the latter dependent on thunder-storms.

"Mr William Geer, of Craigie Burn, near Somerset, who has made this subject his study, informs me that to dam up our rivers is not generally advisable, but that the use of pumps for raising the water and conveying it by a furrow to a reservoir would place all the farms lying on its banks under cultivation at less than half the cost of dams, which are liable to removal by floods.

"There are many favourable localities in this division for the construction of dams. A considerable number have been made, but none of any importance, except one constructed by John Bolleurs, Esq., on his farm, Galgenbosch, Vogel River, which, owing to great natural facilities, has cost but a trifling outlay of time and means, and has secured so large a body of water as to enable him to cultivate hundreds of acres; and to supply, during the last drought, his neighbours, for three miles below, with water for their perishing flocks. Several, say seven, such artificial lakes might, I am assured, be formed in different parts of this division at a cost so small and with permanent results so great as to repay the enterprising individuals in four years.

"In the field-cornetcies of Vogel and Brak River, dams are essential to the prosperity of sheep-farming, for any continuous drought causes the farmer to trek with his stock, unless he has a never-failing fountain on his farm or a very large dam, either of which are advantages possessed by very few farms in those localities; and the streams from which those wards take their names being more or less dry during the greater part of the year.

"In the wards of Zwagers Hoek and Boschberg, where almost all the grain produced in this division is grown, the farms are generally blessed with one or two perennial springs, which have nearly always proved sufficient for the wants of stock in dry seasons; and in good years have, or would have, enabled them to grow more cereals than could be consumed in this division; but as the occupiers of these favoured spots are mostly well-to-do sheep-farmers, they have been so indifferent to the cultivation of the ground, that some few of them actually purchase their bread-stuffs in preference to selling wheat or oat hay at remunerative prices.

"As far as I am able to judge, and from reliable information, I should say that perhaps some twenty farmers in this division would avail themselves of a Government loan to aid them in making dams of such capacity and in such situations as would not only enable them to occupy their farms in severe droughts, but also induce them to cultivate winter fodder for their stock; and wheat, mealies, and oat hay for sale, as the soil is virgin and moderately fertile.

"If the Colonial Parliament is prepared to sanction an 'irrigation loan,' say £2,000 for this division, at such rate of interest as it would have to guarantee on debentures (six or seven per cent.), to be repaid in instalments, extending from three to six years, I have little doubt but that the whole of this amount would be applied for in sums varying from £100 to £500; and when once fairly in operation, a larger loan will be required.

"The success of such an investment cannot be questioned by any enlightened individual; and the providence of a Government which would, in these times of pecuniary difficulty, make cash advances for the development of the resources of a country, and for adding to its permanent prosperity at a season of commercial depression, deserves the deepest gratitude of those who will be benefited by such opportune enactment as is proposed in the annexure to your circular under reply.

"4. Although many years in this division, I have travelled few parts of it, and therefore deemed it expedient to request a friend, Mr Leonard, merchant of this town, to favour me with information on the subject of this report. This he has very kindly consented to do, and I therefore beg to annex copy of his letter, for the special information of his Excellency.

"I may, in conclusion, add that I know of no gentleman in this division more capable than Mr Leonard of satisfactorily answering the queries in your circular; and of affording reliable information and practical suggestions on the subject proposed, he having been brought up in this district, and travelled over every part of it; he is also intimately acquainted with the capabilities of the country as a farmer, and of the idiosyncrasies of its inhabitants as a general dealer, and capable of forming a correct estimate and judgment of both by the possession of natural talent and habits of observation. It is therefore needless to state that I unhesitatingly endorse the facts, conclusions, and suggestions contained in his communication."

Mr Leonard's letter is as follows :—"I have the honour to submit, in reply to your request for information upon the subject of irrigation, the following remarks :—

"1. From my extensive acquaintance with the inhabitants of this division, and my knowledge of all their public movements, I can confidently say that no scheme of irrigation has been even mooted among them in such a tangible form as that anything could be gathered therefrom to guide his Excellency or the Legislature in their deliberations upon the subject.

"2. The rivers which might be dammed up for the purposes of irrigation are the Great and Little Fish Rivers and the Vogel River. Upon the facilities that present in either of these streams for the formation of lakes to retain the whole body of water flowing down their valleys in rainy seasons, by blocking up the openings between the hills where the streams happen to flow through narrow gorges, I shall not speak, because I believe that such undertakings are entirely beyond the present means of the colony, and therefore for the present an impracticability.

"Neither do I think it desirable to direct attention to any measure in this division, at least for the present, for interference with the streams of the rivers named above, because no moderate expenditure would in my estimation benefit the farmers through whose several properties these rivers flow; nor, because even if the waters were led out for them, enabling them to irrigate the whole of their low land beside these streams, do I believe that any fair proportion of them would turn the same to account, since they are by a much easier method than agriculture in the receipt of more than adequate to supply all their wants, and the state of the labour market does not afford them strong enough inducement to embark upon the improvement of their property in this direction. An exceptional case might be met with, but I think I can confidently say that in the large majority of cases my opinion would hold good.

"3. The part of the country to which the attention of the Legislature might be most beneficially directed is that part lying between the rivers named, where extensive sheep-walks are solely dependent upon the rains which periodically fall for their water supply, where many natural facilities present favourable to the husbanding of water during times of rain, and where, to my knowledge, many persons would construct dams of an extensive character, but cannot, because they want the means.

"Dams in such localities would not only be of immense advantage in obviating the necessity that now periodically arises for the farmer to trek to some friend residing near the river for the purpose of securing water during a drought to save his stock from perishing, but could be turned to much more account in many cases, because frequently the situation of the intended dam could be easily chosen so as to command a considerable extent of fertile land below (as in the case of J. Bolleurs, Esq.), when oat hay, mangel wurzel, and other crops could be grown in abundance, whereby, in cases of drought, stock might be kept alive and in good condition, to say nothing of increased population that could be supported upon a farm where agriculture is combined with stock-farming.

"4. Local institutions professing to lend money to those in need of it have been multiplied to such an extent within the last five years that this may be called 'The Age of Banking;' but it is a notorious fact that the object of all these institutions, without a solitary exception, is antagonistic to the higher and nobler one of developing the resources of the country, since their sole aim is to turn to account the monetary need of persons engaged in trade and speculation, and by lending money at an enormous rate of interest, for very short dates, to secure to shareholders a return of from fifteen to twenty per cent. upon the amount of their investment. No bank lends money for a longer period than four months, and then the lowest rate in the Eastern Province is nine per cent., I believe indirectly it is often twelve. And this profitable mode of investment has drawn out all the surplus cash of those who, as private holders, might formerly have been willing to lend money to their more needy friends and acquaintances at five or six per cent., to enable them, by a little time, to gain their ends.

"5. Any measure, therefore, really calculated to secure to landholders of slender means sums varying from £100 to £500, at a moderate rate of interest, say five or six per cent., and allowing the borrower, under good guarantee, to repay the money in annual instalments, extending over, say four, five, six, or seven years, would in my estimation be of such real public advantage as to entitle the session of Parliament for 1862 to be ranked among the most useful that has ever been held in the history of the Colony up to this date; and I do not hesitate to say that if the measure were passed and the funds judiciously administered, so as really to be held sacred to the object for which the measure is contemplated, it would be largely turned to account by many in this division. I, for one, would be glad to show my neighbours the immense practical benefit it would be to the farmers in the neighbourhood where I am at present operating."

In a number of the *Somerset Courant*, issued some time before I left the Colony, it was reported :—"Mr J. L. Botha, (P. R.'s. son), who succeeds his father to the fine farm so long celebrated as the *Wooning van oude Philip smous*, and lying to the north of Pearston, is engaged upon a large dam, which is to drain a number of the secondary tributaries of the Vogel River. This piece of work is a specimen of what is needed, and may be produced, in every part of the country.

The embankment is already over twelve feet high, is about 300 yards long, and will, when completed, be fully 600 yards. It is to contain water enough to supply the stock and raise several hundred muids of grain, and is being so constructed as to contain water even during the progress of the work. Mr Botha is at the present time irrigating and watering his stock from the supply husbanded during the April rains. Such enterprise deserves notice, and we hope will soon find imitation."

And after my return to Europe, I received the following excerpt from a subsequent number of the same journal:—"The large dam which for some time back Mr John L. Botha, (P.R.'s son), has been making for the purpose of storing water for the use of his stock and for irrigation purposes has been finished. On Wednesday, the 29th of last month, in consequence of that happy event, he invited a few of his relations and friends to a rural feast, for the purpose of inaugurating the same. At eleven o'clock on that day a procession was formed, headed by the respected father of the enterprising young farmer (Mr P. R. Botha, sen., a gentleman nearly eighty years of age), which, after having passed the wall, named it 'Perseverance,' under a volley of fire-arms. This having been gone through, all sat down to a sumptuous dinner, laid out under a rural awning, the tables groaning with the good things of this world. Speeches were made. In one by the above-named party, he stated that he had commenced and finished this work without any other means than that which the generality of farmers have—viz., 'self labour.' It was not so much the want of money as the 'will;' and he trusted that many present who had farms would follow his example. The evening previous to the above, the heavens, as if to smile on the endeavours to 'advance,' opened, and copious showers of rain fell, which, had this work not been in existence, would have run to waste, instead of which the water which fell was husbanded. On the same day, a pleasing sight was seen—that of an old coloured servant (who had resided nearly forty years on the farm), aged 105 years, who had that morning walked nearly three miles to grace the proceedings with his presence; thus age and youth joined together in wishing prosperity to the young man who had carried out such an undertaking to a satisfactory end. Dimensions of dam: length, 600 yards; height, 27 feet; width at bottom, 25 feet; average width at top, 6 to 8 feet; which, when filled, will throw back a body of water nearly 3,000 yards. At the same time, while on the subject of storage for water, we might mention that Mr H. Coffie, of Pearston, has sunk two wells, in two different erven of his in that village, and obtained abundance of beautiful water. All honour to such men. Go thou, and do likewise!"

The Fish River bed has been described to me as being for hundreds of miles a deep ditch or natural canal; the waters of which could be drawn off into tanks and level plains of the richest soil, which could thus be covered with crops and vegetation of every description, where at present there is not a blade of grass nor a drop of water accessible for many months in the year. And it does seem desirable that this, if practicable, should be effected.

From the late Mr Pears, minister of the Dutch Reformed Church at Somerset, I received a letter some time before his death, in which he deplored the progressive desiccation of the district; attributed this to the pastoral farming of the inhabitants, to the burning of the veld, and to the destruction of trees; and earnestly advocated arboriculture and irrigation as remedies for the evils deplored. Let the following extract show his views:—

"I have been prevented by many things from answering your letter sooner, and even now can afford you but very little information, but I cannot delay longer lest you should think I am indifferent to the subjects mentioned.

"The great want in the Colony, as you justly remark, is water; and though there are many places where a greater supply could be procured, there are other places where it seems impossible to obtain it. I have repeatedly traversed the Colony from Hopetown to King Williamstown, and from Algoa Bay to Burgersdorp. I have often crossed most of the rivers, and some of them I may say weekly, as the Great and Little Fish Rivers cross my parish; and for years I was consulant of Queenstown, when I had at all seasons to cross the Fish River, Tarka, Koonap, Kieskamma, and Key, to say nothing of the smaller streams, such as Baviaan River, Kutkover, Little Fish, Klass Smots, Klop-plaats, etc. I have passed all of these sometimes almost dry; at other times, even the smallest, so swollen that, after remaining for days on the banks, I have been obliged to fasten my clothes on my horse, drive him before me, and swim through. When swollen, as they often are in the summer, by thunder showers, nothing can resist the volume and force of the rapid streams; and, in general, the bed is so deep and the banks so high that it seems little can be done to employ them in irrigation. Still in many, nay, I may say in all, there are places where, by skilful labour, and proper dams and cutting, the water might be led out, and in time of flood might in many places be drawn off to reservoirs, where it could afterwards be used for irrigation; and as the banks recede in many places, and leave large spaces of deep alluvial soil, that which now is covered with useless bush would be adorned with rich crops, and would in a short time abundantly remunerate the labour and expense incurred. Even in the extensive plains and karroos there falls sufficient rain, which, if collected in dams and reservoirs, might irrigate large tracts of country which now lie waste. There are very fine farms where, by dams, etc., sufficient water might be had to cultivate sufficient for the consumption of the Boers, and enough to sell to pay for their labour. But though many have made dams, it is only for water for their cattle. What is wanted is skilled labour and more capital. Were a few skilful persons employed in each district to point out the places where proper and efficient dams could be made, and to show the best and most economical way of making them, I feel assured that very many of the Boers would avail themselves of their knowledge, gladly remunerate them, and carry out their plans. The present mode of sheep-farming is fast ruining the country. Large flocks are kept, which are all brought to one kraal at night, and to one fountain or dam to drink; so that, in the ground, for almost a mile round the homestead, there is not a pile of grass. By the constant treading of at least 20,000 feet daily, the grass is fast disappearing, and karroo bush taking its place. When I came to the frontier, thirty-eight years ago, there was grass everywhere in abundance—in the plains, sweet, and on the mountains, sour—sometimes five or six feet high; now none, excepting near rivers or on the tops of mountains, are to be found. Formerly, also, the mountains were unoccupied, as no one chose to pay for them. The herbage was abundant and the moisture long detained, so that all the little streams continued to flow through the whole year. Now, these mountains are all occupied, and generally burned annually; and the consequence is that the water has failed. For instance, the mountain behind my house, which rises to the height of 1756 feet, was covered with high grass, and thousands of beautiful bulbous flowering plants and shrubs, and its whole face and off-shoots adorned with yellow wood and other valuable trees; now, these are all gone, not a yellow wood or other tree worth anything left, and only a useless growth of bushes occupying their place; and the consequence is, that a stream that supplied my garden and some others runs now only after rain. The whole face of the mountain, if planted with oaks, firs, and other useful timber, might not only be valuable, but again protect the water. But almost every year, by the idle and reckless, the mountain is fired and all is destroyed. It is now burning fiercely. In the kloofs there still stand the charred stumps of large yellow wood trees. I have strewed acorns and the seed of the gum tree in several places, but they are soon destroyed. Round my house are some large oaks and gum trees, and a few fir trees of my planting, one of the gum trees is more than ten feet round. At first I gave them water, but for years they have had none, and yet they continue to grow well, which proves that in the mountain, where there is always moisture, they would rapidly grow if protected. There is a range of mountainous country to the north, watered by the Little Fish River and its tributaries, called Zwagers Hoek, the many valleys of which are very fertile, and formerly produced a large quantity of grain; but now, in consequence of the multitudes of sheep pastured there, scarcely any corn is grown. In this large district, extending near to Cradock, dams might be made to almost any extent, and fine crops raised; but the Boers, who are all comfortable, and some of them wealthy, say that nothing pays but sheep, and therefore they cultivate little. Servants, indeed, are scarce and very idle, being mostly Kaffirs. The large plains that extend to the south, and are bounded by the Zuurberg, were formerly covered with sweet grass; they are now mostly covered with karroo bush, and are considered very fine sheep walks; but there is a scarcity of water, and though many dams are made they are in general small and shallow, so that in dry years they have not water even for their cattle. I am well acquainted with most of these, and know that there is not one farm but might, by skilled labour, be much improved. To the west for a considerable way the farms are rich and valuable; but after you cross Brootje's Hooghte you enter a very warm bush karroo country, which extends to Graaff-Reinet. It is very dry, but in many places there are fertile spots. All along the base of the mountain which stretches from Graaff-Reinet right east to the Amatola in Kaffirland, there are valuable farms, and those to the east of Somerset are considered some of the best in the Colony. The whole extent between the Northern range and the Zuurberg range is an undulating plain, which, if watered, would be very valuable. Vleys are found in several places where formerly there was water, where the wild beasts drank all the year round. These are now mostly dry, excepting after rain. The grasses have, as I have said, nearly disappeared, and the karroo bush has taken their place."

Of Somerset in 1866 it was reported:—"Stock has not suffered much from disease during the past year. In certain portions of the division farmers were obliged to leave their farms for want of water; but, on the whole, the lambing season has been good. Many dams have been made, and in some places additional lands have been brought under cultivation."

Of Somerset East in 1869 it was reported:—"One of the most noteworthy points is the recent importation of Angora goats from Asia Minor, many of these animals having been purchased by farmers in this district.

"The season, which in spring set in favourably, has latterly become very dry, and though sheep and goats are still in good condition, large stock is poor, and should rain not soon fall, our prospects for the winter will be gloomy.

"In October we had heavy, cold rains, and the losses of stock amounted to 25 per cent. at least on the whole number in this division, and in consequence of the previous drought, the lambing proved very unproductive, so that, even including the whole increase from this source, the number of stock is lower than it was last year.

"Stock thefts remain much as they were during the last year; but a new feature in these thefts is that goats are now stolen and killed simply to obtain the skins, which, if large, can be sold for two or three shillings each, the flesh of the animals thus destroyed being left to the vultures. Should this wanton destruction continue or extend, some severe and stringent measure will require to be enacted for its suppression.

"The want of labour still continues to be felt, and there appears no prospect of any improvement in this matter.

"During the past year a wool-washing establishment, on a large scale, has been commenced by Mr Robt. Hart in the neighbourhood of this town. Such establishments, though they should bring little profit to their proprietors, are a public benefit, as they enable the Colony to retain the money which was annually paid for the exportation to Europe of thousands of tons of filth."

Of Somerset in 1875 it was reported:—"Since the rains of December 1874, the seasons have been good and the pasturage has recovered to an extent which had been deemed almost impossible by those who had noted the gradual desiccation of the previous (I am informed) two years. Agriculture has consequently received unusual attention during the past year, and the district has perhaps never before yielded so much grain (principally wheat) in any one season. The price of bread must therefore become much reduced as the yield of this division is far in excess of its wants and consumption. There has also been an almost entire absence of rust. Cattle have been free from disease and increased rapidly, and horses, although receiving but small attention, are procurable at fair prices, and have also been free from disease. Angora goats are rapidly on the increase, and the coming mohair season will surpass the yield both in quality and quantity of any previous year. The farmers have also had an opportunity of improving their already excellent flocks by an importation of thorougbbred stock. The Angora is fast displacing the slaughter goat (Kapater) which are now more difficult to be got."

SECTION III.—*Divisions of Cradock and Queenstown.*

The division of Cradock lies to the "north of the Great Winterberg range, which separates it from Bedford and Fort Beaufort. It has Somerset on the south, and Graaff-Reinet and Middelburg on the west, Colesberg and Albert to the north, and Queenstown on the east. It has an area of 5,070 square miles. . . . It is one of the greatest wool-producing divisions in the Colony, but timber is scarce, although the mountain kloofs surrounding are admirably adapted for producing it.

"The division is for the most part flat country, about 3,000 feet above the level of the sea, and is surrounded by mountains, of which the Sneeuwbergen divides it from Graaff-Reinet in the west; the Rhenosterberg from Middleburg and Colesberg; the Zuurbergen on the north from Albert; and the Great Winterberg on the south from Somerset and Bedford."

Of Cradock, Mr Meurant, then Civil Commissioner, wrote in reply to the Colonial Secretary in 1862:—"No plans for irrigation works have been proposed to my knowledge for this division;" but he goes on to say:—

"I am of opinion that the productive resources of this division could be vastly increased by the damming up of rivers and the construction of dams in favourable localities. The principal river in this division is the Great Fish; and I have long held the opinion that, by throwing dams across this river in several spots, a large quantity of water might be collected during the season of thunder-storms, which, by the aid of pumps, or other means, might be raised and employed in irrigating the fertile soil on the banks on either side. In proof of this opinion, I need only instance the village of Cradock, where a large number of gardens are irrigated by means of a dam thrown across the river by the municipality of the village.

"I am of opinion that the landed proprietors of this division would be likely to avail themselves of a law empowering the Government to make them advances of money upon reasonable terms, with convenient periods for repayment, for the express purpose of executing works of this description. A little difficulty might be felt at first, but I think that the example of a few enterprising men would be followed by the same happy results as was the introduction of merino sheep about thirty years ago. I can mention several instances in this division where 'dry' unproductive land has been converted into fertile arable land and gardens, by the formation of dams; and I am quite sure that the same thing might be done in this district to a very large extent. My own opinion is, that money could not possibly be laid out better, and would yield a large return."

In this division also Mr Kohl had opportunities of showing that subterranean supplies of water were at command.

The following is an extract from a private letter from Tarkastad, (formerly the military post of the Zwarte Kei, a few miles north of the Great Winterberg Mountains):—"Kohl, the great water-finder, is here. He has, I understand, made £500 within the last three weeks. The drought continues, and Kohl will at least do some good, whether he really has knowledge or not, as he has set the farmers to work digging for water, and in many instances with great success. I think Government ought to take the matter up, and get experienced people, from England or the Continent, who thoroughly understand the laws relating to subterranean waters."

In 1866 it was reported:—"The division of Cradock is principally pastoral; a great quantity of grain, however, is grown, mostly wheat and oats of very superior description (the oat hay is unequalled by any in the Colony), but all by irrigation only; dams are constructed on almost all the farms, but generally small, and, with few exceptions, by no means well made.

"There are many places where very large reservoirs could be made (at no great expense, but yet beyond the means of private enterprise), by which very large tracts of country could be brought under cultivation. If this subject were more generally attended to, Cradock would be able to supply the greater portion of the eastern divisions with cereals, for the spoil is most productive.

"The crops generally this season have been very good, and on many farms large quantities of wheat have been grown. Rust, so destructive to the wheat and oats in the lower division, is scarcely known here; the winter crop oat hay alone suffering by it.

"Fruit of all descriptions, and of superior quality, is most abundant on all farms where there is irrigation.

"The general aspect of the division is extremely sterile, but yet for grazing it is unequalled. Large flocks of merino sheep are depastured on the stony hills, which, to a casual observer, appear as if an animal could scarcely find subsistence on them; yet the stock of all descriptions is in excellent condition, although a large portion of the division is still suffering from drought, and the rains have been so very partial."

Of Cradock in 1869 it was reported :—"The division of Cradock has suffered very much from drought during the greater part of this year; the rains which have fallen in such abundance in many parts of the Colony have not visited this district, and in consequence the Karroo bushes have been all but dried up, and the stock of the district is in a very low condition. Water is scarce, and fountains are drying up. The effects of the present drought are quite as great as those experienced in 1861; but the drought of that year was more severe, because it was more general.

"Locusts have appeared in great numbers, but have moved off to better pasturage, without doing much injury to crops. The harvest of the present season is a fair average one."

To the east of the division of Cradock is the division of Queenstown, "once the country of the Tambookie Kaffirs, from which they were driven in the war of 1852. It is bounded North by the Stormberg range, South by the Amatola, West by the Cradock Division, and East by the Indwe River to its junction with the Great Kei, an area of 4,000 square miles, with a population of 44,500 souls, it being one of the most populous divisions in the Colony. The country is well watered, and the rivers flow in a manner favourable for irrigation purposes, not being sunk in the deep channels in which nearly all our colonial streams flow."

It is divided from Aliwal North by the Stormbergen range on the

North; Kaffirland and British Kaffraria separates it from the Sea, on the east and the south east, while from Victoria East and Peddie it is separated by the chain of high mountains comprising the Katberg, Gaikas Kop, and Amatola ranges. The country is for the most part a highly fertile and well watered region. A gentleman well acquainted with the division writes:—"The inhabitants of this part are daily attaining more information and experience in the storage and conserving of water, and slowly, but surely, beginning to understand that it is their duty to intercept and retain all water endeavouring in the course of nature to regain the level of the sea, and for such storage the country is very well adapted, under moderate irrigation wheat can be grown to great perfection and abundance, rust seldom making even a partial appearance."

The circular addressed to the Civil Commissioner of the district in 1862, was, in his absence on duty, submitted to the Divisional Council, the Secretary of which was instructed to reply, to the effect that up to that time no specified plans had been proposed for irrigation works within the district; but that the district presented great capabilities for carrying out all the kinds of irrigation works referred to in the circular.

In regard to the division it was reported in 1865:—"Queenstown has suffered but little from the effects of the late drought. The spring rains, having set in very early, were the means of enabling the European and native agriculturists to harvest a plentiful supply of wheat and oat hay. The effects of the drought will be felt, however, with regard to the winter crops, of which it is feared none will be raised.

"During the year a great many Fingoes from the Kamastone and Oxkraal locations have moved into the Transkeian Territory.

"No improvements in agriculture, nor introduction of any machinery for agricultural or other purposes, are reported.

"The matter of leasing Crown land in this division, under Act No. 19 of 1864, has already been brought to the notice of Government by the Divisional Council."

In 1866 it was reported:—"The severity of the drought during part of 1866 has been very severely felt in this division, more especially amongst the natives. The crops of 1866 have been far below the average. In some instances the farmers have barely recovered the seed they have sown, especially of wheat and oats. During the past year mealies and Kaffir corn have been at famine prices; and when the present sowing season first set in there was every prospect of a famine amongst the natives of this division and on our border. They were unable to sow any seed for want of rain, and there was every appearance of the sowing season passing over without it. Fortunately, before it was too late, copious rains had fallen, and now there is reasonable prospect of a most abundant harvest; there is scarcely a spare acre of land in the Tambookie or Fingo locations that has not been sown with either mealies or Kaffir corn.

"With regard to live-stock of all sorts, the past season has been a most satisfactory one, and has in a great measure compensated the farmer for the badness of his crops. The increase in sheep and wool during the last few years has been very great.

"The Queenstown agricultural society is not in so flourishing a condition as it was a few years ago. The native agricultural society held its second show at Kamastone on the 28th March last, and proved a complete success, notwithstanding the drought, and the exodus to the Transkeian territory."

Of Queenstown in 1875 it was reported:—"On returning to the division of Queenstown, after an absence of seven years and a half, I find many changes. In the town are four or five steam engines, the power being applied to various purposes, principally wool-washing, grinding, and sawing. In the country, the area of land under cultivation has very greatly increased; better buildings have been erected on the farms, and the rural population numbers three times what it did some years ago.

"The farmers are fully alive to the advantages of fencing, and many of them have enclosed large portions of their farms with stone walls; others are proceeding with this work. Many are turning their attention to ostrich farming, which promises to be a most profitable undertaking in this division.

"The inhabitants have carefully protected the herbage and bush, the consequence of which is that great tracts of country which were formerly bare are now clothed with a luxuriant growth of mimosa, which not only affords protection to the stock in bad weather but a plentiful supply of fuel, which was rather a scarce article some time ago.

"I have not heard of any serious losses of stock, which, as far as I am able to judge, is improving steadily in breed through the importation of new blood, and in condition and health through the splendid condition of the pasturage. We have had most copious rains, the advantages of which have been felt not only in this district but in all the surrounding country.

"The natives in this division are improving steadily; they appreciate the advantage of civilisation in which they are progressing; they are gradually adopting European customs and appliances; they are peaceable and orderly, and when it is considered that there are 36,500 of them in this division I believe they will bear favourable comparison with an equal number of inhabitants of any other division in the Colony. Crime for so large a population is pretty nearly at its minimum.

"A Divisional Police, under Act 8 of 1873, having been established during 1875, has done a great deal towards the detection and suppression of crime. The force is most efficiently worked by a very zealous inspector, and, I think, gives universal satisfaction."

CHAPTER VII.

Supply of Water, and Existing Facilities for the Storage of it, in Midland Divisions of the Colony of the Cape of Good Hope.

Divisions of Graaff-Reinet, Middleburg, Murraysburg, and Richmond.

The division of Graaff-Reinet has an area of 5,350 square miles. A considerable portion of the surface is mountainous, the Great Sneeuwberg Mountains rising in some places to a height of 7,000 feet or more (Compass Berg q.v.), and the valleys are well watered. Along the foot of the mountains extends a Karroo region, called the Camdeboo, excellent for sheep. On the mountain tops the country is called the Koude (Cold), or Winter Veld. It is badly wooded, and very cold in winter, lying high. The principal stream is the Sunday River and its affluents.

"The division of Graaff-Reinet is generally a pastoral region, especially celebrated for its valuable sheep farms, and the excellence and quality of the wool produced thereon; and, although suffering, in common with the surrounding divisions, occasionally from drought, is eminently prosperous, and would become much more so were greater efforts made to construct dams in places where the rainfall is extremely uncertain and small."

In regard to this division, Mr Berange, Civil Commissioner, wrote in 1862:—"The construction of dams or reservoirs for the use of cattle has of late years become general throughout the division, and similar works will have to be undertaken and executed by the purchasers of Crown lands (from time to time to be disposed off), for these lands, being devoid of springs, would otherwise be comparatively useless."

The works for purposes of irrigation at present undertaken are described by him as being of a primitive character; but he intimates the possibility of works of a better, and eventually cheaper, construction being devised.

Within the electoral division of Uitenhage, but territorially in the district of Graaff-Reinet, are the large dams of Messrs Parkes, at Wheatland, and of Mr W. C. Hobson, at Ebenezer. The latter was described to me as the last farm on the northern extremity of the division of Uitenhage and on the commencement of the Karroo Flats, and I have given details of these in that connection. It was in correspondence relative to arboriculture in different parts of the Colony that I was brought into communication with Mr Hobson. Having answered certain queries on that subject, he added in a postscript:— "After all, water is the great thing needful in this the Karroo veld. The fountains are generally small, but permanent. Limestone and tufa abound; and every drop of spring water is impregnated with lime more or less. I believe there is not 3,000 morgen of Government ground here, but a dam might be made on it which would always contain water for stock and irrigation on a greater or less scale. Hitherto the Government land has been measured by those who wished to purchase it, to suit themselves (and no wonder), and with not the least thought about the future welfare of the country. There are yet remaining fine suitable spots for dams; and there should, I think, be a general survey of the waste lands before another inch is sold. I say this contrary to our own interest as farmers, but it is my opinion nevertheless."

The following evidence was given by Mr Ziervogel before the Select Committee of the Legislative Council on Irrigation. It relates to the Karroo generally, but may be considered as relating more immediately to this division:—

"41. *Chairman.*] Can you give the committee some idea of the value of good Karroo or other land, when not capable of irrigation, and when the same land can be abundantly irrigated for growing artificial crops; or what would the difference be in the quantity of grain or other crops grown in such soil?—Well, it is difficult, almost impossible, to state what the relative or proportionate value of land capable of irrigation and land incapable of cultivation may be, but that the difference is enormous there can be no doubt. In the interior parts of the Colony, where you have Karroo lands, and where you can irrigate, those lands are of the richest and most productive description; but the question is, whether if you raise produce to any very great amount you can dispose of it until your population increases to a much greater extent than it is now. Certainly, however, where these Karroo lands can be irrigated they are most productive. In that respect I may just mention that there are a great many agricultural farms in the part of the country where I reside, where irrigation must necessarily be had recourse to in order to raise any crops at all; but it is generally unnecessary to manure the land. When the farmers use manure they must use it sparingly, otherwise the crops become too luxuriant; and during the time that the rust prevailed the crops of manured land were found more subject to it.

"42. And is there much of such soil in the Graaff-Reinet district, and to your knowledge, in the Colony generally?—Yes; it is to be found not only in the Graaff-Reinet district, but all the way coming from there to this (Capetown) through the Karroo.

"43. Are there facilities for obtaining water to irrigate such land, provided capital and labour could be found?—In very many localities there are great facilities. I know that as long as thirty years ago the people in the Burghersdorp district began to make dams, not only for collecting water for the use of their stock, but for irrigation. One was made on a farm where a man could not raise, except in favourable seasons, grain enough for the use of his own household. He made a dam with a pipe to lead out the water, which enabled him to cultivate not only more than necessary for his own consumption, but he could send a deal to market. He made a dam where two ridges came close at their points, and, by blocking these up, secured a large body of water which, even if it were not to rain for one or two years, would not dry up altogether. That dam was made about thirty years ago by an old colonist. Now there are several others of a similar nature in that part of the country. About midway between Graaff-Reinet and Somerset there is a farm formerly belonging to a man who only used it for depasturing his stock after rain, when the water lodged in small pools; but as soon as that supply was exhausted the party was obliged to take his stock away, although the pasturage remained good. The man died, and one of his step-sons obtained the farm. He made a small dam also where two ridges came close together, with an embankment of only about five or six feet high, and this immediately gave him a good deal of water. The first rain the dam was filled, and he could reside constantly on the spot with his flock. Since that, another person, Mr Bolleurs, purchased the farm and raised the embankment much higher, so that there is enough rain-water to last for two years, if no rain fell. Mr Bolleurs has a good fruit garden and plenty of arable land irrigated from that time; and some squatters down below on Government ground, when the water at the pools where they reside is exhausted, apply to Mr Bolleurs, who, for a slight consideration, allows the dam water to run down into their natural pools. In our part of the country the people make their dams, not only in spots where two ridges come together, but even on almost level land, scoop out the soil and raise embankments, thus collecting water that will last for a very long time, where, otherwise, they would have nothing; and I know very many spots, some on private propery and some on Government ground, where you might make dams such as Bolleurs. Even in the Karroo, along the main road, I have seen several spots where such dams can be made; but then you must not construct such playthings as have been made by Government in the Karroo.

"44. What has been the price on an average of wheat per bushel in your district during the last four or five years?—I have not the means of ascertaining that here; but I should think about nine or ten shillings.

"45. What has been the highest price?—Meal has been as high as upwards of £2 for the three bushels,—between £2 and £3, and very seldom comes below £1 10s.

"46. Then I suppose there is very little cultivation in Graaff-Reinet, except with irrigation?—Except with irrigation, scarcely any. There are only a few spots on very high mountain lands in localities where water is apt to lodge, where there are 'vlei lands.'

"47. Is there any Government land in your district, or in any of the neighbouring divisions, which would be greatly enhanced in value by the construction of dams for watering cattle?—A good deal. Of course for watering stock you would have dams of a smaller description; and wherever there is undulating ground you do not want two ridges exactly running with their points towards each other, for between the ridges then there is ground you may dam up. . . .

"51. *Mr de Wet.*] Do you know of any lands in your district or neighbourhood let at high rates because they have irrigating facilities, and could you give us some facts showing the value of such land compared to dry land?—There are not very many leases of land in our part of the country.

"52. Do you know of any?—The municipality of Graaff-Reinet have lately let a farm which was placed under their management by the Government, but that is under peculiar circumstances. There is a mineral spring which the parties had to open and take care of, provide bath-houses, and so on; but I do not think that is a case which would explain anything of the sort the committee requires.

"53. But do you know of any private bargains of land leased?—

Yes; there are some cases in which farms have been let; and, of course, wherever cultivation can be carried on on a farm, it fetches a higher rent than where you have no cultivation.

"54. Have you any idea of the value of the rent per acre or per morgen?—It is impossible to make any calculations of that kind.

"55. Then you do not know of any bargains of parties letting or hiring at so much for a certain extent?—Not at present.

"56. Can you give us any information with regard to the value of dry and water erven of the same extent in villages?—Dry erven of two acres in extent are not worth a third or fourth of the value of those of the same size on which there is a supply of water for cultivation.

"57. What quantity of water?—Sufficient to cultivate a piece of ground fairly. A few years ago a village was established between Graaff-Reinet and Somerset, at Pearston. There were a great number of dry erven, and a number of water erven. The water erven were larger than the dry erven, but they fetched, I believe, some three or four times the price of the dry erven; and such dry erven are scarcely saleable, except in particular situations, near the church, near the market, for stores and shops. In such situations they fetch a decent price, but otherwise it is hard to dispose of them.

"58. *Chairman.*] Then, of course, as these villages become more and more peopled, the water erven increase in value in a greater ratio?—Certainly. I had an erf in Graaff-Reinet, the old Drostdy, which I purchased from Government. In the water distribution much more water was reserved to the Drostdy than to any other private erf,—I believe the quantity in proportion to the size was double. I disposed of a good deal of the land, which has since then been built upon, and consequently not so much water is required as formerly. A good deal of the ground is covered with buildings; and some time ago I sold two hours' supply of water twice a week from the stream for £200. That stream supplying this ground is generally strong enough to turn a small mill.

"59. Have you no means at present of ascertaining the actual quantity of water supplied by hour—has no one in the town ever attempted to measure the quantity?—No. The water which supplies Graaff-Reinet is, at the entrance to the town, divided into five different streams. In average seasons each of these streams can turn a small mill; in very dry seasons the quantity becomes somewhat less, but it is always sufficient for the requirements of the village, and for the irrigation of all the water erven.

"60. What, in your opinion, is the principal reason why municipalities and proprietors of erven in villages do not undertake more extensive works for irrigating their lots in dry seasons?—I should almost say it is merely want of energy.

"61. Not want of capital?—I can scarcely think so. I mentioned Pearston just now. Some persons connected with the Dutch Reformed Church wished to establish a village. They bought from the owner a homestead with thirty-six thousand morgen surrounding it, bad Karroo land, but with a good supply of water. They bought the land at the rate of £1 an acre, and at the first sale of erven they sold erven to the amount of about £11,000. Where people can pay such prices for the erven, surely they must have some capital to enable them to make improvements.

"62. *Mr de Wet.*] But does not the sale in such instances as you have mentioned take place on the part of an individual, who, of course, thinks himself fortunate in pocketing the advantage?—So far true; but when a number of people can afford to give such advanced prices for the ground, where they can give £11,000 in this way, I should suppose that these people amongst themselves must have capital to make improvements.

"63. *Chairman.*] Is there not often an artificial competition to establish a village, with a church in it, on a particular spot—do not people pay more than they otherwise would to secure the advantage of having a church near their residences, and to save themselves the expense of going a greater distance from home?—Still they must remember that they have to pay.

"64. But do you not think that this fact enters into the calculation they make, that it is better to have a church near their doors than to ride a distance of perhaps three or four hours or more?—That does come into the computation, but still they must be able to afford to pay. They must have some means, although I am aware that a great many poor people buy such erven too.

"65. Do you think that, if a company in England were to offer loans on good security, six per cent. and more would be given by landed proprietors in your division for the purpose of dam-making and similar improvements?—Certainly; they would get fully that and more.

"66. Are there not many rivers in your district, and in the Colony, the super-abundant waters of which might be turned into a different course during the rainy seasons, soaking a large extent of country, so as to improve the pasture, and counteract the effects of a subsequent season of drought to a considerable extent?—As to irrigating the pasture lands, I do not think that can be done to a very considerable extent. In some districts with which I am slightly acquainted, something of the kind can be done with propriety and good success; but in those portions of the Colony with which I am chiefly acquainted, I am afraid that not much can be done for irrigating pasture lands. The expensive works that would be required for the purpose would be so costly that it would not pay.

"75. *Mr de Wet.*] It is only in particular seasons that the Karroo requires irrigation for pasturage?—It all depends upon the rain. For long periods no irrigation may be required, but in other seasons irrigation would be useful. But, as I said before, to irrigate for the purpose of pasturage would not be a paying thing. It might at some distant period, but not now.

"76. I believe you say that some parts of the Karroo would admit of irrigation; what may be the obstacles to the irrigation of the other part?—I am speaking of irrigation for pasturage; you may irrigate at all seasons for cultivation.

"77. What is your opinion of the practicability of establishing in the Karroo large reservoirs of water, to facilitate driving of cattle?—That might be done, and very usefully, and at no very great expense. I have seen a dam made by the Road Department, in the Worcester district, between Driekoppen and Zoute Kloof; but in comparison with the dams made by the farmers in the interior, that piece of Government work is a mere plaything. In that neighbourhood, however, and in other spaces along the Karroo, much larger dams could be made that would contain large quantities of water.

"78. If such facilities existed in the Karroo for the purpose of driving cattle through it, do not you think it would influence considerably the price of butcher meat in Capetown?—Very much so.

"79. And of course the meat would be of a very superior quality?—Of course; but a good deal would also depend on the pasturage. If the pasturage is moderately good, but there be no water, the cattle will suffer from want of water; on the other hand, if there were a large dam they might have water, while, if the pasturage was insufficient, they would suffer for want of food.

"80. In your opinion, do you think that the system of collecting water could be carried to such an extent as to facilitate the breeding of cattle in the Karroo?—No doubt with water you might have grazing farms in any part of the Karroo; but I doubt whether you can make such dams as to irrigate the pasture lands.

"81. What is the nature of the country?—Some parts are a little hilly, but a great deal is almost level.

"82. If almost level, what is to prevent irrigation?—You would require immense quantities of water to irrigate the Karroo; even on grass lands, consider for a moment a farm of six thousand acres to be irrigated, what enormous quantity of water and expenditure it would require,—there is not the slightest doubt that an immense deal of the country now occupied for short periods during the year for grazing purposes, could, by making dams to provide a sufficient quantity of water for drinking, be occupied continually. Of course, even if you had water, when you get such dry seasons as we have had lately, there would be water for the stock but no food for them. A great many years ago, in the Karroo, there was a dry farm where there was a permanent spring with plenty of water; the sheep died of starvation, and, when cut open, there was nothing in their paunches but water.

"83. *Mr Barry.*] I believe a great many dams were made by the Dutch Government; are they still kept up, or allowed to fall into disuse?—I never saw any of them, and I do not know where or what they are.

"84. *Mr Wood.*] From your experience, as a practical man, can you state the value of land with water and without water, the quality being the same?—Of course you convert a mere grazing farm into an agricultural farm if you have an adequate supply of water.

"85. And what would be the difference in value—say, between a farm worth £2,000 well-watered and a farm without water, of the same size, the land being of equal quality?—At least fifty per cent.

"86. Do you, as a practical man, view a standing fountain as of equal value to a river running through your land?—If the permanent fountain or spring is of sufficient strength, I would almost prefer that to a river, unless the river has always a good body of water running, and there are great facilities for turning off the water; because if you have a spring you would require less labour and expense for leading it out than you would require for damming up and leading out a river; and a river dammed is so frequently liable to be injured by floods that the expense of maintaining such works must be considerable; whereas, with a spring there is scarcely any expense.

"87. Will you please to state the size, I believe it is the custom of the farmers to designate it by inches, of what you would consider a moderate spring for a farm?—I cannot describe the quantity of water by the designation of so many inches. I know that many farmers are in the habit of doing that, but they talk at random. When a farmer speaks of so many inches of water, a scientific man cannot rely upon what he says as anything approaching to a basis for calculation.

"88. Would you consider a volume of water, the size of this ink-bottle (1¾ inch diameter) enough for a farm?—Not for all purposes; it depends on circumstances, whether it flowed slowly or under pressure; but, of course, such a stream would considerably increase the value of a dry farm, because you would at all events have constant clean drink-water, and you could have a small garden.

"89. I may tell you that a leaden pipe of half-inch bore would throw out about eighty gallons per hour, and the size I mention would give four or five times that, do you think that such a farm would be increased twenty-five per cent. in value by such a stream?—I think it would.

"90. And do you think, from your practical experience, that, as a rule, the land in the district where you live would be increased in value to that extent by such means?—Of course the supply of water increases the value of land. . . . .

"100. Do you think the farmers in your district would be willing to burden their farms under hypothecation, if they could ensure irrigation?—Many farmers who have not the means themselves of constructing dams would, no doubt, be willing to borrow money for the purpose, and mortgage their lands; but they would strongly object to mortgaging all the lands in the district generally, for the purpose of works of irrigation being constructed.

"101. But those who did would benefit?—Yes, individually, for their own lands. I may state that the practice of making dams is becoming very general in our part of the country by people who have not a good supply of water on their farms.

"102. Has there not been, during the last few years, a large quantity of Government land, formerly said to be useless, taken up and rendered valuable, simply by making dams upon it?—Not simply from the knowledge the people have obtained of the way of making dams. This of course alters the value of the land very much; but the increased value of that land is not attributable to that cause alone; the better farms not being to be had, men must content themselves with inferior land.

"103. Upon these lands large quantities of stock are depastured, are there not?—Yes.

"104. And if means could be devised of cultivating roots and other sustenential matter for the cattle, would that not enhance the value of land—is that practicable?—Of course it is practicable, but I doubt whether it would pay, on a large scale, for a long time to come. I will give you a reason:—Feeding stock in this artificial manner would be very costly, and people would find that stock fed in that way would cost more than stock fed upon good farms in the ordinary manner. There are a great many farms where stock thrive very well throughout the year without artificial feeding; consequently, where artificial feeding is had recourse to, the price of cattle must come to more than the price of stock not artificially fed. . . .

"107. *Mr Barry.*] Do you know the value of a farm on the banks of the Olifant's River?—I have been in that country, but am not acquainted with it sufficiently to give an opinion.

"108. There are localities in which farms with frontages to the river are valued at £5,000, while at the back of them a larger extent of the same kind of land is valued at £1,500. Have you observed any difference between a farm cultivated with trees and one bare of trees; or do you consider the cultivation of trees upon a farm calculated to induce moisture?—I do not think that that would apply very much to the low lands; but I am certainly of opinion that if our mountains could be planted with forest trees we would have a much better supply of water. So much am I of that opinion, that I have endeavoured, on a mountain farm which I possess, to rear firs, oaks, and such trees on the mountain.

"109. And have you observed any difference?—I have not been successful. At one time there was a pretty good number of young firs and oaks growing, but some party at a distance set the grass on fire, and they were all burnt down; and that is the reason why I agitated some time ago for the passing of that law to prevent setting grass on fire. . . .

"113. *Chairman.*] If you have anything else that you wish to add the committee would be glad to be favoured with it, either now or afterwards in writing?—Having spoken of dams, I may mention another circumstance. It has been found that where there are weak springs, when a dam is scooped out just above the springs, it is somehow strengthened and the supply increased, whilst at the same time there is no perceptible diminution in the dam itself; this may arise either from some of the water percolating through, or from the ground near being kept cool. There are many instances of this in our district."

Many of the queries which I have omitted elicited opinions deserving of great consideration. This I say though not a few of them relate to matters of economy and policy on which I have come to opinions opposed to those at which Mr Ziervogel has arrived; and it is with regret that I have tried to confine my citations to testimony in regard to the water supply of the locality, facilities for the storage of this, and probabilities of inhabitants of the district availing themselves of these.

Mr Kohl visited Graaff-Reinet when in this district, and he found no lack of indications of subterranean supplies. While he was in the locality, there was sent to me, by some friend unknown, a Graaff-Reinet newspaper, in which it was reported:—"The water-finder, Mr Kohl, has been to Aberdeen and the neighbourhood, where he has indicated to the farmers many places where they may get water, and has bagged a good sum in cash. He was in negotiation with the Municipal Commissioners to show them how to double their present supply to the town. Meanwhile the Kerkraad got hold of him, and for £30 obtained his advice. A few of the inhabitants, on behalf of the municipality, then seized Mr Kohl, and for £15 more he was induced to point out the same spot to them; so he got £45 easily enough. The rights of the Kerkraad over the town lands and water is a matter of dispute, and counsel's opinion is to be taken as to the limits of authority on the part of the municipality and the church-wardens. Mr Kohl offered to enter into a bond of £1,000 to deliver to the town a stream of four inches of water if they follow his directions; or he was willing to do all the work and give the water for the sum of £300."

In 1865 it was reported by the Civil Commissioner in regard to Graaff-Reinet:—"The favourable anticipations which were expressed in the report at the end of 1864 have unfortunately not been realised during the past year. Trade has continued depressed, caused, in a measure, by the severe drought from which this division has suffered in common with others. The loss in stock has been very great; but the principal sufferers have been those who squat on Government lands (and there are numerous persons possessed of a great number of sheep and cattle who do so), and who are dependent on pools of rain water for their stock. The yield of the grain crop has, for the same reason, not been the half of what had been expected; and there is reason to fear that the late crops of pumpkins, mealies, etc., will be a total failure. . . .

"On several farms a great deal has been done to collect rain water by artificial means. Dams have been constructed; but, in many instances, from want of means, they are not of sufficient size and depth to retain water when the rains fail, as they have done during the last eight months, except an occasional shower here and there.

"The necessity of making provision for the periodical droughts is now so generally felt, that if assistance could be afforded, by the loan of money at a reasonable rate of interest, there would be no want of energy on the part of the farmers to construct the necessary works. Several wells have been sunk on different farms, and water has generally been found at a depth of between twenty and forty feet. The renowned Mr Kohl, commonly designated as the 'water-finder,' has, in this division, with more or less success, pointed out places on several farms where he thought water could be found by digging; and the facts of large sums of money having been paid to him merely for indicating the spots, when the chance of success was still uncertain, is another proof that the farmers will spare no expense or labour to procure this indispensable element."

In 1866 it was reported:—"The prospects of the farmers in this division are much better at the end of this year than they have been for several years before. Copious rains have fallen, and have brought the country to such a state that every kind of stock is thriving. A few such years in succession would bring this division back to its former state of prosperity.

"If during the drought the loss in sheep has been great, and still further losses have been suffered after the first rains by *geil ziekte*, the lambing season just past has been so good that with most of the farmers the numbers now even exceed what they had been before. Some small farmers, who after the drought had not sufficient stock left to start with, have not been able to recover lost ground, and with them their is still poverty and suffering.

"There has not been any unusual disease among cattle. The hoof and tongue sicknesses come periodically, and are now prevailing so much as to interrupt the transport of goods between this and Port Elizabeth.

"The grain crops generally have turned out remarkably well, and this has already influenced the market by bringing down the prices of wheat and meal twenty-five per cent. There is also every prospect of good crops of mealies and pumpkins, which will so reduce the prices as to bring them within the means of the poorer classes.

"We may reasonably hope that in future years agriculture will be carried on in this division to a much greater extent than is done at present. Formerly the farmers were content to construct dams merely for stock to drink at, which would retain water only for six months or so; but, stimulated, no doubt, by what has been done at Wheatlands and other places, dams are now being made at a cost varying from £300 to £1,200, which will contain sufficient water to irrigate great extents of land.

"The following description of the dam at Wheatlands may not be uninteresting:—

"Estimated contents of dam, when filled to within three feet from the top of the embankment, 36,000,000 cubic feet, or 1,000,000 tons. Allowing one-fifth for soakage and evaporation, there would be enough to irrigate 230 acres, at a rate of supply equal to a rainfall of two-thirds of an inch per week for one year, supposing the dam to be filled only once a year. The furrow leading from the river to the dam is 12 feet wide at the bottom, with a batter that could give a stream of water of an average of $13\frac{1}{2}$ feet wide. When 3 feet deep in the furrow, at a velocity of 1 foot per second, it would take ten days to fill the dam.

"Length of embankment 850 yards, 300 yards of which is 18 feet high by 72 feet at the base, the two ends rising or tapering from 18 to nothing.

"The face or water side of the dam is being faced with stone, to prevent the waves from washing the bank, or cutting into it.

"Contents of embankment, 30,000 cubic yards of ground.

"Estimated cost, including furrow, flood-gates, lime-stone, dam across the river to turn water into furrow, £1,400, more or less, including also two sets of pipes, valves, etc."

In the *Graaff-Reinet Herald*, January, 1866, there appeared a statement that "the stream of water which Mr F. T. Ochse has dug out on his farm, 'Quaggasvley,' has proved most valuable to him, and, we are told, is equal to a flow of two inches constantly. The work undertaken at Aberdeen by the Municipal Commissioners is finished, and a stream of water equal to two and a half inches in diameter has been obtained. This is a most valuable addition to the water supply of the village. Mr Kohl's instructions were followed exactly, and the water was found at a depth of nine feet from the surface—four feet of earth, and five through solid rock. We are assured that in Murraysburg division the farmers have derived great advantage from Mr Kohl's talent in discovering water for them."

Of Graaff-Reinet in 1869 it was reported:—"During the former part of the year this division suffered very severely from drought, many of the dams in the farms, especially in the Zwarte Ruggens and along the Kariega, dried up entirely, and the farmers had to remove their stock to places where water was to be obtained, in some cases even by purchase. The consequence was that many thousands of sheep died, and some farmers were very much impoverished.

"When the drought broke up, the division was visited for a time with very heavy rains, such as had not been known for years before, and many thousands of the sheep which had survived the drought were washed away by the floods caused by the rains. Goats did not suffer so much from drought as sheep, but during the floods many farmers sustained very heavy losses of these animals.

"In succession to this, the division was visited with an unusually heavy snowstorm, which left the snow in some places, especially in the Sneeuwbergen, six and eight feet deep, smothering upon many farms great numbers of sheep.

"After this, and it would seem as a crown to all these calamities, the locusts made their appearance in great numbers, in the shape of 'voetgangers,' and committed great havoc to the growing crops in places where the farmers could not muster sufficient hands to divert their course.

"The year 1869 has been a very trying one to the farmers, especially of this division, owing to what I have advanced, and it is almost surprising that they have maintained their position as they appear to have done.

"The Dutch Boer, being accustomed to these trials, seems to bear them with greater resignation than most of the English, and manage to plod on in the way of their forefathers, hoping that better times will some day come.

"I have not heard of much sickness among stock this year, beyond the usual trifling ailments, which prevail more or less every year. Lung-sickness is very little heard of now, and appears to be gradually dying out.

"An enterprising merchant, resident here, has introduced for sale a number of pure Angora goats from the lot recently imported by Messrs Blaine & Co., and as this division is more especially adapted for goats, beneficial results may fairly be expected to follow in the course of a few years. Mohair is, without doubt, superseding merino wool, and will most likely, in the course of a few years, become one of the principal staples of this part of the Colony. Farmers begin to see this, and are consequently turning their attention to the improvement of their breed of goats much more than they did in former years. Fortunately, the rains fell just in time to enable farmers to get in their crops, and the yield of grain has been tolerably fair, perhaps not quite so much as during the former year. . . .

"Crime is slightly on the decrease, the promptitude with which punishment follows offences under the recent Acts extending the magistrate's jurisdiction is one reason of this happy result, and it is to be hoped that this state of things may continue. It is a fact worthy of note that at certain periods of the year the coloured population are more disposed to commission of crimes and offences than at other times. Whether this state of thing arises from want of employment or from a liking to pilfer, it is impossible to say. . . .

"An enterprising farmer, Mr J. H. Booysen, living within ten miles of Graaff-Reinet, has for a few years past turned his attention to ostrich-farming. He has an enclosure of pretty considerable extent, within which he keeps from thirty to forty full-grown birds, and has been hitherto successful in breeding yearly sufficient to afford him a good income. His efforts in this branch of farming receive very fair patronage from others who purchase with readiness any number of young ones he may have to dispose of."

In 1874 it was reported :—"For many months there was a very scanty fall of rain, but at the end of the year it rained continuously and so heavily that the destruction of stock and grain was considerable. Houses and dams were also washed away, and for some time the Sunday and other rivers in the division became impassable, causing interruption to postal and other communication. . . . Locusts have made their appearance in some parts of the district, and a good many small swarms have been effectually destroyed."

And in 1875 :—"A very large quantity of grain has been raised during the year. . . . The last year has been favourable for farming generally; the gleaning season has been good, and on the whole there has not been any great mortality among stock owing to disease."

When the Hydraulic Engineer was at Graaff-Reinet, on the tour of observations, in accordance with instructions from the Commissioner of Crown Lands and Public Works, he gave advice to the Municipal Commissioners, as to their water supply, in a letter to the chairman of the Municipal Commissioners, and he reported to the Government :—" While I was at Graaff-Reinet I visited a place called Booysen's Poort, where Mr Pilkington, C.E., some years ago, proposed to make a large reservoir. I believe that Mr Pilkington made a survey and took levels, but the only account I could obtain of his scheme was from a letter in a newspaper. Mr Pilkington said that the reservoir would have a water-surface of ten square miles and an average depth of fifteen feet; he estimated purchase of land and compensation at £20,000, and the cost of the embankment with culvert at £32,000.

"This is certainly a good situation; two streams, Moordenaar's River and Zwart River join above the poort, there is rock in the bed of the river, and the poort is narrower than Salt River's Poort. I have no means, however, of checking the estimate, as I have seen no survey or levels. I believe no Crown lands would be affected."

The following is an abstract of daily observations made in 1868:—

From Daily Observations.	Jan.	Feb.	Mar.	Apr.	For four months.
1868.	ins.	ins.	ins.	ins.	ins.
Mean height of the barometer at the temperature 32° Fahr. (not corrected for altitude above the mean level of the sea), for each month, and for four months, from observations at 9 A.M., 1 P.M., and 5 P.M., ...	27·351	27·388	27·443	27·468	27·412
Maximum observed, ...	27·549	27·563	27·693	27·750	27·750
Minimum observed, ...	27·125	27·126	27·235	27·273	27·125
	degs.	degs.	degs.	degs.	degs.
Mean temperature of the air for each month, and for four months (Fahr.), from observations at 9 A.M., 1 P.M., and 5 P.M., ...	83·0	74·1	71·0	71·9	75·0
	degs.	degs.	degs.	degs.	degs.
Mean temperature of evaporation for each month, and for four months (Fahr.), from observations at 9 A.M., 1 P.M., and 5 P.M., ...	67·3	65·9	61·9	62·9	64·5
Mean humidity of the air for each month, and for four months (the integer number 100 implies saturation), derived from observations at 9 A.M., 1 P.M., and 5 P.M., ...	44·0	55·0	59·0	61·0	55·0
Maximum humidity, ...	69·0	88·0	88·0	94·0	94·0
Minimum humidity, ...	23·0	27·0	31·0	36·0	23·0
From the self-registering thermometers (Fahr.).	degs.	degs.	degs.	degs.	degs.
Mean maximum temperature of the air, ..	93·3	86·5	78·1	77·4	83·8
Mean minimum temperature of the air, ..	64·7	61·9	58·5	62·1	61·8
Approximate mean temperature of the air, ..	79·0	74·2	68·3	69·8	72·8
Mean range of temperature of the air, ..	28·6	24·6	19·6	15·3	22·0
Highest temperature, ...	104·0	95·0	92·0	83·2	104·0
Lowest temperature, ...	53·0	53·0	48·0	49·0	48·0
Extreme range of temperature on any one day,	41·0	32·0	32·0	26·0	41·0
Least range of temperature on any one day,	21·0	13·8	11·0	6·0	6·0
Quantity of rain in inches, ...	0·360	3·970	3·990	2·130	10·450
Greatest amount of rain on any one day in each mounth, ...	0·125	1·345	0·970	0·560	1·345
Number of days on which rain fell, ...	5	10	12	10	37

Abstract derived from Meteorological Observations made at Graaff-Reinet in the Years 1863, 1864, and 1865.

Months.	Mean height of Barometer at Temperature 32°.	Temperature of the Air. Mean Temperature.	Mean Daily Range.	Mean of greatest range any one day of each month.	Mean of least range any one day of each month.	Mean of greatest range throughout each month.	Mean Temperature of Evaporation at 9 A.M. 1 P.M. 5 P.M.	Mean Humidity for each month.	Rain.	Clouded Sky.	Lightning. Recorded Number of days.
	inches.	degrees.	degrees.	degrees.	degrees.	degrees.	degrees.	per cent.	inches.	per cent. of visible hemisphere.	
January,......	27·395	75·66	26·20	39·2	13·8	48·3	68·64	55·0	1·453	33	8
February,....	27·416	73·18	21·86	34·2	9·0	43·7	66·99	63·9	2·664	52	9
March,........	27·414	69·08	24·42	39·2	14·0	49·1	63·51	55·9	0·970	34	10
April,.........	27·480	62·18	23·49	34·5	13·4	48·3	56·94	57·1	1·129	33	8
May,...........	27·555	58·23	20·17	30·1	9·0	44·8	52·74	60·3	0·810	34	4
June,..........	27·616	54·49	21·19	35·3	9·7	42·6	48·32	56·9	0·307	30	0
July,...........	27·602	53·42	21·21	32·3	9·5	43·6	48·39	60·9	0·510	34	3
August,.......	27·572	57·79	29·94	41·3	10·4	51·7	51·17	51·5	0·292	32	3
September,..	27·524	60·67	25·11	39·3	12·0	50·4	54·63	55·7	0·824	42	4
October,......	27·422	66·51	26·89	44·0	10·9	55·0	59·95	53·8	1·563	42	12
November,...	27·409	67·29	29·09	44·8	14·0	56·1	60·03	49·7	1·567	37	5
December,...	27·352	74·44	28·64	40·3	13·3	51·3	66·73	51·1	1·107	33	3
Mean Annual Values.	27·480	64·41	24·52	37·88	11·58	48·74	58·17	55·98	13·196	36	23
	* + ·028	...	...	...	...	...	...	...	...	...	...
	27·508	...	...	47·0 1864. Mar. 8	2·8 1863. Oct. 6	58·5 1864. Nov.	...	...	...	...	...

* ·028 inch is the reduction to the Observatory Standard Barometer.

The following is a tabulated statement of observations on the rainfall, made in this division in 1869, the register at Goliad's Kraal being kept by Mr Brook, and that at Wheatlands by the Messrs Parkes :—

Months.	Graaff-Reinet.	Goliad's Kraal.	Wheatlands.
January,	·390	·570	·210
February,	2·480	2·080	1·815
March,	1·860	2·800	·820
April,	·470	·000	·640
May,	1·480	1·260	·935
June,	·360	·000	·160
July,	·090	·160	·200
August,	3·420	4·140	1·365
September,	1·290	·820	·165
October,	1·505	1·190	·950
November,	2·910	2·100	1·580
December,	1·950	2·000	·580
Totals,	18·205	17·060	9·420

From which it appears that the rainfall at Wheatlands was only half the quantity falling at Graaff-Reinet; and yet by husbanding it so much was accomplished.

The division of Middleburg is an excellent pastoral country, lying to the east of Graaff-Reinet, with an area of 2,180 square miles. It is watered by some of the head branches of the Great Fish River. It is a fertile and well-watered tract of country. The town is situated about the centre of a circle that would enclose the towns of Colesberg, Richmond, Graaff-Reinet, and Cradock; and the division partakes of the character of all of the divisions of which these are the seats of the magistracy.

Middleburg abuts on Cradock, on the north-west of this district.

Mr Boyes, Civil Commissioner, replied to the inquiry addressed to him in 1862 :—" No plans for irrigation works have been proposed within this division.

" There are several localities favourable for the construction of dams in this division, and in many instances the means thus afforded for husbanding the periodical supplies of water have been adopted. As a general rule, however, the farms in this district are well supplied with water from permanent and never-failing springs, and irrigation to a much greater extent than at present prevails might be carried into effect, were the farmers more energetic and alive to their own interests, as the lands under cultivation might be considerably enlarged and the yield of this district quadrupled."

In his report for 1865 the Civil Commissioner of Middleburg reported :—" With the exception of the fearful epidemic among horses, prevalent in the months of March and April last, by which it has been estimated that at least 2,000 died in this division alone, agricultural and pastoral pursuits during 1865 have met with fair success.

" The winter was exceedingly mild and favourable, so that sheep and other stock were healthy and in good condition, and well prepared to meet the severe drought of the last five months.

" The crops of wheat and other cereals, which are here raised by irrigation, were above the average, and were reaped in good condition.

" There is promise of abundance of fruits, which, however, it is feared, will be deficient in size and flavour, owing to the drought. . .

" Mr Karl Kohl, known as the 'water-finder,' has created some excitement in this division, and the farmers have very generally availed themselves of his services, but, in the majority of cases, without success; and the opinion now is that the ability to which Mr Kohl lays claim has been over-rated. In a few instances, however, Mr Kohl has been the means of developing a good supply of water; and there is reason to believe that he does possess a certain amount of practical geological knowledge, which, under favourable circumstances, may prove of great service to the farmers of the Colony.

" During the past year, many farmers have gone to considerable expense in the construction of dams, as well as in digging for water; and the late drought has, in consequence, been less felt in this than in the adjoining division."

In 1866 it was reported :—" The past year has, on the whole, been a prosperous one, and the inhabitants generally have not had much cause for grumbling.

" Genial rains, a favourable and plentiful harvest of every description of grain, and the absence of any extensive epidemic among livestock during the year, have been the means of adding greatly to the stability and wealth of the division. A desire has also been manifested amongst the farmers to improve their dwellings; and more extended efforts have been made in the opening of fountains, the construction of dams, and the sowing of a greater breadth of land."

Of Middelburg in 1874 it was reported:—"The monotony of a protracted drought in the earlier part of the year, broken up by heavy rains coming in with the commencement of summer or end of spring, has been varied this year by a flood of such magnitude as has never before been experienced in this division within the memory of any of its inhabitants. This event, which occurred in the middle of November, added an incalculable amount to the losses of stock previously occasioned by the drought, destroyed and damaged many buildings, and swept away great quantities of grain from the farms along the rivers and water-courses. The rain, which lasted three days, was accompanied by a strong south-east wind; all the mountains were capped with snow, and the cold was so intense that many thousands of sheep, and hundreds of cattle and horses, in addition to those swept away by the waters, succumbed to it, whilst hundreds of springboks were killed on the open flats by the extreme severity of the weather. The losses have been so general, and the difficulty of obtaining complete and authentic information is so great, that it is impossible to form anything approaching a correct estimate of the value of property destroyed, but it will take several prosperous years to counterbalance the disastrous effects of this unprecedented flood, which will be remembered for many a long year in these parts as the 'great flood of 1874.'

"Since the rains, young locusts are appearing in vast numbers, fortunately, however, too late to do damage to grain crops, and the pasturage is so luxuriant that it is hoped they will not be able to destroy it to any great extent.

"Crime has increased considerably during the past year, as the following statistics will show:—

	1873.	1874.
Number of cattle thefts,..................	8	20
Number of petty thefts,..................	18	27
Number of cases—masters and servants,	31	32
Minor offences,..............................	146	192
Total criminal cases,...........	203	271

"A mounted police force, under the provisions of the Act No. 8, of 1873, is in course of organisation, which it is hoped will prove effective for the detection of crime, and for the apprehension of criminals."

And in 1875:—"The effects of the drought and flood of last year are still painfully manifest in the greatly reduced numbers of the flocks, the small quantity of wool, and the high prices of provisions. The year 1875 has been an average one, in regard to the increase of stock, but a very good season for crops. The health of stock has been very good, no disease of any kind prevalent. Had we had earlier rains the veld would have been in splendid order, stock would have been in better condition, and the increase of lambs greater.

"The absence of locusts and rust has secured to us good crops. Trifling damage has been done on a few farms by hailstones and late frosts. Much larger breadths of land have been cultivated, owing no doubt to high prices. As the harvest is so abundant prices of bread-stuffs and forage will come down, without however reducing the income of the farmers from that source. The district is capable of very great agricultural progress, the soil is fairly fertile, water abundant, and could be greatly increased. The principal drawbacks being want of energy and enterprise, and the absence of steady and cheap labour.

"The show of fruit was considerable, but has been greatly diminished by late frosts.

"There is a marked absence of the growth of timber. Poplar and blue-gum would thrive in most parts with little trouble, and afford timber for building, shelter to stock during inclement seasons, and fire-wood, which is scarce and very dear.

"Great credit is due to the municipality for their successful efforts in tree-planting, whereby the town has been greatly improved and beautified.

"Ostrich farming has been progressing, but fortunately there has been no mania for it.

"Horse breeding would answer nearly as well as in the adjoining Hantam district, but no efforts have been made to introduce good blood; in fact, the breed has gone back, whilst there has been sufficient encouragement in the prices for good animals to revive this branch of farming.

"This district has been celebrated for its sheep-walks, but owing to its being over-stocked, and successive droughts, the capabilities of the pasturage have been reduced. Formerly two sheep could be kept on a morgen, now there is scarcely food enough for one. Enclosing would, to a large extent, improve the veld by allowing the sheep to sleep out. Though timber is so scarce, stone is abundant and handy; all that is wanted is energy and labour. From the good water supply and the number of vleys, artificial pasturage in large paddocks or meadows would be successful. Enterprising Scotch and English farmers with some capital are wanted for this purpose, and the ever increasing price of slaughter stock and dairy produce would at least warrant the investment. Some improvement of this kind, and the extended use of agricultural machinery, is absolutely necessary to increase our food supply, which is not keeping pace with the rate of population.

"Crime has considerably decreased towards the close of the year, by the prevention system: all idlers and vagabonds being made to enter service whenever by law compelled; and those for whom there is no vagrant law are well looked after by the police, and their wrong doings well punished. But there is still quite enough of theft, laziness, and insubordination of servants to require a vigorous policy and police surveillance. The police have been reduced in number as crime has diminished, and a saving of £300 a year effected.

"The health of the inhabitants has been good, especially since the rains."

To the west of Middleburg is the division of Murraysburg, a high upland region lying between the Rondeveld and the Sneeuwberg ranges, about 4,000 feet above the sea; it is traversed by what is called the Upper Road to Beaufort West, and has an area of 2,146 square miles. The village is about sixty miles north-east of Graaff-Reinet, and a hundred east of Beaufort West; is situated in a well-watered though badly timbered region, very cold in winter, but with a fine healthy summer climate.

It is the northern and most elevated portion of the electoral division of Graaff-Reinet. The highest mountain in the Colony, the Compassberg,* is in Murraysburg, and, flanked on the east and west by the Great Sneeuwbergen range, towers up between this division and Cradock and the Karroo plains of Graaff-Reinet. On account of its great elevation, Murraysburg is subject to many changes of

* Compassberg is 8,500 feet high, presenting a steep face to the coast; from its flanks flow in all directions the waters of the Great Fish, Zeekoe, Sunday, and Buffels River; hence its name. It is easily ascended from the back or north side, which is quite a gentle slope. Modern observations have reduced its height from 10,000 to 8,500 feet.

temperature, and some diseases of the respiratory organs are thought to be rather more common than in less elevated regions. In its general capabilities for production the country is very similar to Graaff-Reinet; but the farmers in the vicinity of the Sneeuwbergen have to contend with very severe winters, and thunder-showers are frequent in summer."

Of the district of Murraysburg, Mr Henderson, Civil Commissioner, wrote:—"No general or combined plan of irrigation works has been proposed within my division.

"The most favourable localities for damming up rivers or other running waters, or for the collection of periodical supplies of water within my division, have already been taken advantage of. There are now dams existing in my division constructed at a cost of about £5,400, irrigating about 824 acres of land, upon which is raised annually about 5,095 muids of grain and 79,400 lbs. of oat hay, and maintaining a great quantity of stock at localities which it would otherwise be impracticable to depasture; besides yielding an ample supply of vegetables and fruit of considerable variety for the consumption of the inhabitants, and turning sixteen mills. These works might, however, be considerably improved and extended.

"If such a law as that indicated in your letter now under reply were enacted, from the information I have gleaned on the subject, the landed proprietors in my division, I believe, would gladly avail themselves of it, to the extent, I should think, of about £4,000."

In the end of 1865 the Civil Commissioner of Murraysburg reported:—"During the first three months of the year immense swarms of locusts passed through the division; but whether owing to the ligneous nature of the herbage, or the fact that they were frequently harrassed by large flights of locust birds, they seldom settled anywhere long enough to do much damage.

"The horse sickness raged with extreme virulence during the months of March and April, and swept off fully a third of the horses in the division.

"Mr Kohl, the famous water-finder, made a tour through the division during the past year, and was employed by a considerable number of the farmers, who have since then bestowed a good deal of labour, and expended a moderate amount of money, in opening up the places indicated by him; but hitherto, unfortunately, with very slender results.

"An excellent inn has been established on a public outspan, about half way between here and Graaff-Reinet; and a large dam is in the course of construction there by the divisional council. These will be a source of very great convenience to travellers and transport riders between Port Elizabeth and this, and the divisions beyond. . . .

"At the commencement of the year, some fine showers fell occasionally, and in July it rained copiously, so that the springs increased in strength; and as cultivation is prosecuted here only when water can be procured in sufficient quantity for irrigation, the farmers are not so dependent on rain as they would otherwise be; hence, notwithstanding the existing drought, the produce of the crops just reaped will doubtless be fully an average. With regard to stock, however, the drought has been severely felt during the last two months of 1865; and, unless rain fall soon, there will be great distress."

In 1866 it was reported:—"The drought which prevailed at the close of 1865 continued until the 12th of January, 1866, being the day which His Excellency the Governor had appointed to be set apart, and which was observed, as one of humiliation and prayer for its removal. It commenced raining copiously then, and continued to do so, at short intervals, for several weeks, and seasonable showers followed during the remainder of the year, so that stock of every description has thriven in an uncommon degree. The wool is longer in staple and better conditioned than it was last year, and the clip will be improved in quality, and considerably increased in quantity.

"The crops, notwithstanding their being nipped by late frosts, and somewhat checked by rust, promise to yield rather more than an average return.

"The vines have grown vigorously, and are bearing abundance of fruit; but, unfortunately, the *oïdium* has made its appearance to a considerable extent within the last two or three weeks. It is believed, however, that its progress has been checked in some places by the plentiful use of sulphur and by thinning the foliage, so as to admit a free circulation of air about the grapes. The crop of other fruits, which promised to be abundant at the time of blossoming, has been greatly diminished by late frosts.

"Considerable efforts are still being made to increase the supply of water for stock and irrigation, by the formation and enlargement of dams, and the opening of springs.

"Large swarms of locusts have passed through some parts of the division, but without doing much damage."

Of Murraysburg in 1869 it was reported :—"This division suffered a good deal from drought during the first two thirds of the past year, which so reduced the condition of the stock that in August, when a great quantity of snow fell, the farmers occupying the more elevated portions of the country lost heavily, the animals being unable to withstand the severe cold. Two months later heavy rains fell, which caused another severe loss in stock.

"Although sheds would not altogether prevent loss on such occasions, they would greatly diminish the calamity; yet I do not believe there is a single erection of the kind in the division, notwithstanding the reported losses the farmers sustain by such periodical visitations.

"Some remarkably fine Angora goats, selected from those brought to Port Elizabeth a few months ago, have been introduced into this division. They will doubtless effect a marked improvement in the flocks in the course of a few years. Several she goats were among the number, so that there is a prospect of the purity of the breed being preserved. . . .

"Crime has decreased, especially with regard to thefts of stock. This, I believe, is to be ascribed principally to the operation of Act No. 17 of 1867, together with the salutary effects of the infliction of spare diet. . . .

"The wheat, barley, and oat crops have been unusually good.

"The summer, so far as it has gone, has been very cool."

In 1874 it was reported :—"In the beginning of the year rain fell in abundance for several months, so that the crops of mealies and pumpkins were unusually plentiful. A heavy fall of snow occurred in the beginning of July, and lay three inches thick in the streets; on the higher parts of the country it lay for several days to the depth of about three feet. It did not, however, produce the beneficial effects generally ascribed to snow, such as greatly increasing the strength of fountains, and promoting the growth of vegetation on the approach of spring; but perhaps that result may be attributed to the fact that no rain fell thereafter for months, and the drought which followed was becoming very perplexing, when on the 23rd of November rain fell, and continued at one part or another of the division for twelve days; latterly it is reported to have continued without intermission for upwards of seventy hours, in the Sneeuwbergen, accompanied by snow, and consequently extreme cold. The result was severe losses of stock to the farmers in that locality. Since the breaking up of the drought, the aspect of the country has been entirely changed; the pasture is in prime condition, and the stock much improved. The prices obtained for farm produce have been very much in advance of what used to rule a few years ago, for some articles certainly treble the amount formerly realised.

"Ostrich farming has proved so remunerative to those who have made the experiment, that several others have been induced recently to embark largely in the same pursuit.

"A tannery is being established in the village. It will doubtless answer well, and prove a convenience.

"The cereal crop is about the average; that of fruit, very fair."

And in 1875:—"Since the tempestuous breaking up of the drought towards the close of last year this division has been favoured with a succession of very seasonable rains throughout the year 1875; consequently, in the absence of any calamity, such as disease among stock, or damage to crops, everything connected with agricultural and pastoral pursuits has been uncommonly prosperous, while the price of every description of produce has continued to rule very high; hence, though the producer has profited much, the consumer has not benefited by the flourishing state of the country, having had to pay most exorbitant prices for the common necessaries of life.

"Ostrich farming has been considerably extended with great success. . . .

"The Commissioners of the Municipality have completed a large dam in the Commonage, which, besides being of great advantage to the cattle and horses belonging to proprietors of erven, adds much to the beauty of the locality.

"Drunkenness has somewhat abated, which is doubtless to be ascribed to the vice being checked by a rigid administration of the law regarding it."

The division of Richmond consists of the northern region of the Winterveld around the Brakke River. "It is bounded on the north by Hopetown, south and east by Graaff-Reinet, and west by Victoria West. It has an area of 2,950 square miles. It is an entirely pastoral country, depending for water on the formation of dams, and at the cold upland region badly supplied with fuel. Its principal village and seat of magistracy is Richmond. It is situated in the Cold Upland Region, behind the Sneeuwbergen, called the 'Winterveld,' and owes its importance to its being the depôt for the wool of the surrounding sheep farms. It lies about half-way between Graaff-Reinet Town and Colesberg."

With regard to facilities for irrigation in this district or division no information has been obtained by me. The only reply of Mr Blake to the circular issued from the Colonial Office appears to have been the following:—"With reference to the third query contained in your circular letter, No. 16, 1862, I have the honour to state, for the information of his Excellency the Governor, that there is no doubt that landed proprietors in this division would avail themselves of such a law, but to what extent would depend mainly upon the nature of the security that would be required for such loans. I may state that the farmers generally are averse to mortgaging their properties."

At the close of 1865, the Civil Commissioner of Richmond wrote:—"The past year has been one of average success, both to the farmers and to the other inhabitants of this division. Drought is, however, beginning to be felt by the former class. No rain has fallen since the commencement of the spring; and unless relief speedily come, in the shape of a copious down-pour of this very necessary element, serious losses in stock will be sustained. There are no means of watering the flock and herds, after the entire absorption of the water collected in the dams and reservoirs (such as they are); and this supply is nearly at an end.

"Mr Kohl, the so-called water-finder, has been the cause of several new springs of water having been discovered; but it is stated, on the authority of old and experienced farmers in this part of the country, that water can always be obtained at from ten to twenty feet of the surface in the immense Karroo flats which exist here; and this perhaps may be the cause why the stunted Karroo bush is so well able to withstand long and severe droughts. This, therefore, suggests matter for serious inquiry, whether, in opening fountains and digging wells, farmers will not be draining the country of that moisture which now sustains the invaluable Karroo bush; and whether it would not be far better (looking to the future) if the rain which falls were collected into great sheets of water. If water for drinking purposes could be secured for the flocks in this part of the country droughts would be little or never felt. This subject is one of vital importance to a very valuable portion of the Colony.

"Owing, perhaps, to the absence of rain, the locusts have not made their appearance; and the crops have, in consequence, been a very fair average as compared with those of former years.

"The wool clip has also been an ordinary one."

And in 1866 it was reported:—"During the past year this division has been visited by copious and seasonable showers, which maintained the pasturage in an excellent condition. This may account for the prosperous lambing season, which some of the oldest inhabitants are not aware to have been surpassed." And in a local newspaper it was reported:—"Mr W. Mellersh, of Leopard's Vley, Richmond, has lately opened out a new fountain on his farm, as his supply of water for irrigation was diminished by the drought. He first dug vertically where he suspected there might be water, and, on finding a good supply, commenced a furrow about 250 yards long to lead out the stream. He finished the work in five days with twenty men, at a cost somewhat over £10—a sum so small as to be of no value in comparison to the worth of a constant stream of water led on to cultivated land."

Of Richmond in 1869 it was reported:—"The season during the past year, although not amounting positively to a drought, cannot be called favourable, on account of the scarcity of rain, and the wheat crop has also materially suffered from late frosts and rust.

"Attention has been more directly turned to the breeding of Angoras. The first attempt was made during the year 1858, and stock was obtained from the Somerset and other divisions, but only one farmer appears to have started with imported stock. During the year 1863, those who had undertaken the enterprise began to be discouraged on account of the small demand for mohair, and the small price offered, viz., 9d. per lb., and while many entirely gave up the undertaking others allowed their stock to deteriorate. During the last year increased demand for mohair induced merchants to import goats from Asia Minor, the first shipment of which arrived, per *Grace Darling*, of which it is to be regretted but a small number found their way into this division; but the last importation by Messrs Blaine & Co., of Port Elizabeth, caused considerable excitement, and, considering the low price and increased production of wool, together with the success attending the efforts of those who had perseveringly adhered to the breeding of Angoras, caused the farmers of this district to follow the example of those other divisions; consequently, almost every enterprising farmer has now on his farm a number of Angoras, so that no doubt, in a few years, the yield of mohair in this division will form a considerable item in its products. Some of our farmers who cannot realise the opinions expressed concerning Angoras and the production of mohair are determined to improve their flocks of merino sheep. The removal of the Act prohibiting the introduction of horned cattle caused sheep to be introduced from Australia, and many supplied the want of imported blood from these animals; but a fear is entertained by some of the most experienced farmers that the cross of the Australian with the merino in this Colony will not tend to improve the latter in this division. That considered best in this district is the Saxony merino, which are eagerly sought after. A few have been introduced during the past year, and arrangements are being made for obtaining a larger number, so as to supply the want of imported stock so much felt, consequent on the Act preventing the introduction of imported stock. . . .

"The production of cereals has markedly increased, and the division now fully supplies its wants at remunerative, yet reasonable, prices; and very little, if any, grain from other divisions now finds its way into the Richmond division.

"The breeding of horses, it is to be regretted, has been entirely abandoned, although the country offers every facility. The argument against the undertaking is that the tracts of land required for that purpose can be more profitably employed for sheep-farming."

In 1875 it was reported :—"The past year has been a very favourable one for this division. The harvest has also been exceptionally good and plentiful.

"A steam wool-washing establishment, the first in the district, is in course of construction, on a farm four miles from the town, which it is anticipated will be the means of effecting a considerable reduction in the cost of transporting wool to the seaport."

CHAPTER VIII.

Supply of Water, and Facilities for the Storage of it, in North-Eastern Divisions of the Colony of the Cape of Good Hope.

Divisions of Hopetown, Colesberg, Albert, Aliwal North, and District of Wodehouse.

The division of Hopetown is bounded on the south by Richmond, on the west by Victoria West and Bushmanland, on the north-east by the Orange River, and on the east by Colesberg, The village of Hopetown is situated close to the Orange River. The division has an area of 4,500 square miles. "There is but little agricultural farming in the division, owing to the great scarcity of water; and it very frequently happens that the farmers find it necessary to reserve the rain-water collected in dams for the sustenance of their stock. These difficulties, however, do not prevent a large and increasing wool trade, for sheep thrive well and multiply fast."

From the reply of the Civil Commissioner to the circular issued from the office of the Colonial Secretary it appears that that circular gave rise to some discussion in regard to the best means of raising the money necessary for the construction of irrigation works; and from this it seems probable that facilities for the forming of such works, upon a scale more or less extensive, must exist in the district; but of these no particulars have come to my knowledge.

It was some facts recorded by Mr Wylie, late Geological Surveyor to the Colony, when he was in this district, which first directed my attention to the importance of irrigation works as a means of moderating the extremes of temperature observed in some districts of the Colony. In his notes of a journey in two directions across the Colony, made, as Geological Surveyor of the Colony, in the years 1857–58, with a view to determine the character and order of the various geological formations, writing of the Narrows on the Orange River about twenty miles from Colesberg, he says:—"The night spent here, on the banks of the Orange River, was one of the coldest I have experienced in the Colony. Inside of a tent, in which three persons slept, bowls of hot coffee, and a compound liquid of still more potent character, were converted into a solid mass of ice.

"While in the neighbourhood of Hopetown, a gallon of water in a tin bucket was sometimes frozen quite solid during the night, and yet the ice on the dams was never more than a quarter of an inch thick. This discrepancy evidently arises from the water of the latter [and the earth] receiving so much heat during the day as to prevent its freezing at night, except to a very limited extent; whereas the water in a small vessel, exposed on all sides to the cold wind, is soon reduced to atmospheric temperature, or *even below it.* The freezing power of the air is greatly assisted by its dryness, which causes rapid evaporation to take place even in solid ice itself."

The more arid the air the greater the rapidity of evaporation, and the greater the rapidity of evaporation the more rapid and the greater the reduction of temperature; the moisture passing into the atmosphere from an extensively irrigated field cannot cool down the mass of ground to a very great degree, and it tends to produce an equalisation of the temperature of day and night, and of summer and winter, by impeding the radiation of heat from the sun to the ground and from the earth into space. And the observations of Wylie are cited as indicative of the degree of aridity to which the air must have attained when such phenomena occurred.

At the close of 1865, the Civil Commissioner in his report, writing of the want of rain and the consequent necessity of trekking about for water experienced by the farmers, goes on to say:—"During the early part of 1865 there were some heavy thunder-storms in this division, which supplied the dams with as much water as they could hold, and in some cases broke away the walls. The winter was a very mild one, and the spring was ushered in with some good heavy rains. The pasturage, therefore, throughout the year has been good and sufficient, and stock of every description is at present in beautiful condition.

"For the last two months, however, the heat has been very excessive, and there have been heavy, scorching winds. The want of rain has been greatly felt, and most of the farmers have been trekking about for water. At the same time the country, so far as the pasturage is concerned, is still in good condition, and the farmers have only themselves to blame for running short of water. It is a great pity that they do not see the necessity of exerting themselves a little in making larger and deeper dams, and more substantial walls to them, for they could secure sufficient rain-water for all their wants.

"There are, however, a few exceptional cases amongst the farmers, which prove that the want of energy in the others is the only cause of their want of water; for while the latter show what their stock can suffer on account of their own indolence, the former, on the other hand, exemplify what benefits are to be derived from industry by being enabled to remain on their farms with their stock, while others are roaming about the country begging for water. Though the majority of the farmers in this division have for the last two months been complaining of the want of water, and moving with their stock in search of it, yet there are some who, though their farms were quite as dry as those of their neighbours, have simply, by a little exertion, made large and deep dams on their farms, and dug out fountains, so that they have sufficient water for all their wants, and can even supply other people.

"The lambing season has been particularly good. Stock of every description has been healthy. In fact, the year has been a very good one for the division, though a slothful class of people can reap but little benefit from their advantages.

Very little was seen of the locusts during 1865. Though swarms passed backwards and forwards at some places they did no harm. . .

"This division is pronounced, by men who are good judges, as one of the finest stock divisions in the Colony. The want of water is the only drawback; but that is attributable to the people's indolence, for abundance of water could be collected in proper dams. There is also abundance of water underground, which only requires a little hard work to bring out. At some of the driest farms water has been found at no great depth, and good running fountains have been opened. If some of the stock farmers nearer the seacoast, who have to work hard year after year, and yet find that they can make nothing at farming, and whose stock is subject to all sorts of diseases, could but know what sort of a stock country this is, they would be glad of the opportunity of leasing Crown lands here; and they would find that a few years' residence, with very easy work compared to what they have to undergo along the seacoast, and with little disease amongst their stock, and good increase every year, would amply repay them, and perhaps induce them to remain here altogether."

In 1866 it was reported :—"During the first nine months of 1866 there was no lack of rain in this division. The farmers had sufficient water, and the pasturage was very good. But during the last three months the weather was very dry; there was no rain, and its want is now being very much felt. The locusts have also been doing a great deal of damage to the pasturage. Notwithstanding this, the year has been a good one for the farmers; their stock has been in good health and condition; the lambing season has been a favourable one; and more wool has been brought by the farmers to the market during 1866 than during any other year since the drought of 1862. And although the want of rain has more recently been greatly felt, it does not appear that any farmers have been obliged to trek about in search of water, though a good many of them have had hard work in pumping water from wells for their stock. And though the grass has during the last three months suffered from want of rain and from destruction by locusts, the bush pasturage (Karroo bushes and others) is still good.

"There is not much cultivation carried on in this division in consequence of the scarcity of water. Whatever rain is collected in dams is reserved for drinking purposes for stock. The very small crops of wheat and oat hay that were raised during 1866 were nearly all destroyed by severe hailstorms."

From Hopetown in 1875 it was reported :—" There is but little to report this year, except the prevalence of a severe drought, which has caused much loss to the farmers. Late frosts have injured the gardens. Cases of stock lifting have been comparatively but few in number. Public attention has lately been directed to the desirability of irrigation in this division, by leading out the water of the Orange River in suitable places.

East of Hopetown, and like it, bounded on the north by the Orange River, is the district of Colesberg. "It is the most northern division of the Eastern Province; is bounded—north by the Orange River, east by Albert, west by Hopetown, and south by Middleburg and Graaff-Reinet divisions. It has an area of 6,080 square miles, and presents generally a vast elevated plain devoid of wood, studded over with numerous kopjies, or little hills, spitz or tafelbergs, according as they are of the pointed or flat-topped shape. These plains were formerly tenanted by immense herds of the larger game, and were a favourite resort of the now scarce gemsbock. It is an excellent sheep and ostrich farming country."

The Orange River winds along its whole extent on the north and north-east. There is but little agriculture in the division; but for successful sheep-farming, and the quantity of wool produced, it stands unrivalled in the Colony. The rivers in the division are the Zeekoe, Ongars, Brakke, and Oorlogs Poort, all delivering themselves into the Orange River, one of the largest and most remarkable in South Africa. When swollen by rains in the interior it is a magnificent object, in some places about 1,000 feet broad, flowing with a full rapid current, justifying the name of the *Groote River;* at other seasons it is easily forded, being shallow at the ordinary drifts."

In an elaborate report submitted by the Civil Commissioner of Colesberg at the close of 1865 it is stated :—" In the census returns for 1855-56 the returns of stock for Middleburg and Colesberg were included together, and Hopetown, which was included with them in the population returns, was omitted. The following is the result of a comparison of the two periods :—

	Mules and Asses.	Horses.	Horned Cattle.	Sheep.	Goats.
1865 Middleburg,........	113	7,166	12,817	284,948	33,651
,, Colesberg,...........	399	8,682	16,057	549,681	41,149
Total,...........	512	15,848	28,874	834,629	74,800
1855–56 ,, ...	117	12,421	32,734	618,100	15,888
Increase,.......	395	3,427	...	216,529	58,912
Decrease,......	...	...	3,960	...	...

"The increase in the number of sheep and goats—3·5 per cent. of the one and 3·73 per cent. of the other—is very satisfactory, more especially as the increase of the former has been entirely in woolled sheep (the Cape have actually diminished by 4,717), and of the latter to the extent of 10,782 in Angoras, yielding also a profitable article of export. Horned cattle show a falling off of 3,960 in number, which is not surprising when the great discouragement of the farmer on the frequent decimation of his herds by lung sickness is considered. This fatal malady seems quite domiciled among us, and this division is never entirely free from it.

"The last census was taken in March, 1865. During the following month this part of the Colony was visited by that fearful epidemic generally known as the horse-sickness. The extent of its ravages has been recently ascertained by Mr David Arnot, who was employed by the divisional council to appraise all the fixed property of the division, in the course of which undertaking he visited every farm, and carefully collected information on the subject. The following is the number of horses which died and which survived in each of the rural field-cornetcies in this division :—

	Horses died.	Horses remaining.
Hantam,	1,757	1,979
Lower Sea Cow River,	2,065	1,909
Upper ,, ,,	790	727
South Middenveld,	478	760
	5,090	5,375

"The loss in the Middleburg division was probably in the same proportion.

"The agricultural operations of Colesberg are upon a very limited scale in proportion to its extent, arising from the impossibility of growing anything without irrigation, and from the scarcity of springs of sufficient strength. On a few farms artificial reservoirs or dams for collecting rain-water have been constructed; but, in general, dams are exclusively used for the watering of stock.

"The following table exhibits the breadth of land under cultivation, the number of springs by which it is irrigated, the number of dams in the division, and the quantity of grain grown in 1864 in each field-cornetcy :—

Field-Cornetcy.	Land under irrigation.		Springs of irrigation.	Dams	Crop of 1864.			
	Morgen	Roods.			Wheat, Bushel.	Barley, Bushel.	Mealies, Bushel.	Oatsheaves, No.
Hantam,	484	96	122	131	6,627	1,155	672	43,340
Lower Sea Cow River,	423	134	125	84	8,136	1,725	426	55,370
Upper ,, ,,	201	186	52	45	3,960	1,068	267	16,900
South Middenveld,	102	564	45	55	540	84	43	6,000
Total,	1211	381	344	315	19,263	4,032	1,408	121,610

"It thus appears that 1,211 morgen and 381 roods of ground yielded only 24,703 bushels of grain; and, averaging the oatsheaves at five lbs. each, 271 tons of oat hay—a very insignificant return for the extent under cultivation. The crop was below the average, for the late frosts were unusually severe; but in good seasons the return is but small in proportion to the average, owing to the havoc produced by the ravages of locusts, hail storms, and late frosts, from which some portion of the division invariably suffers, if not from drought. These impediments, and the necessity of irrigation, oppose an insuperable bar to agricultural operations on anything like an extensive scale. In favourable seasons sufficient grain will be grown for home consumption, but not more; and, even with a railway open to the coast, it would not pay to grow corn for exportation. To exemplify this it may be stated that for several years past wheat has never been lower than 12s. 6d. per bushel on the Colesberg market until this season, when, in consequence of the 1865 crop having been considerably above the average, it has fallen to 9s. The farmers, however, say that this price does not pay them; and those who can afford to hold will not bring it in if it does not rise to the old price, and will sow less this year.

"It will be seen, upon reference to the table, that there are 315 dams to 311 farms, or about one to each farm. Some of them are of considerable capacity; but all are gradually yet surely silting up with the washings of clay, etc., brought into them with the rain water. It is presumed that dams in other parts of the Colony are liable to the same drawbacks, which is one of the most important points connected with the topic of the day—irrigation, or rather with the intercepting of the flood water for that purpose. The civil commissioner of Colesberg has twice had an opportunity of noticing these deposits, at the flooding of his dwelling-house during violent storms, when, although the flood water had only possession for a few minutes, it left behind a deposit of mud, at least six inches in depth. These storms usually occur after an interval of drought, when, in consequence of the bareness of the pasturage, the surface soil is much exposed, and trampled into dust by the constant roaming about of stock in search of food. When the rain falls, the dust, converted into mud, is borne along with the flood water, and is deposited wherever a check to the impetuosity of the current affords an opportunity. If this grand difficulty with ordinary dams cannot be overcome, the only way left of improving the water resources of the Colony seems to be to lead out the rivers into large reservoirs after the first flood water with its muddy accompaniment has run of; but to lead out the stream of a river, very deep and expensive cuttings will be necessary, for their beds are generally much lower than their banks or the surrounding country. Dams can, of course, be cleared of their accumulated deposits, but farmers assert that the expense of doing so would be as great as the construction of a new dam."

And the following statements, though not connected immediately with the question of water supply, are indirectly, if not directly, connected therewith :—" Farmers, and other persons who employ unskilled labour, cannot complain of a scarcity of servants, but they make many complaints of their negligence. This arises in a great measure from the little inducement their low wages offer them to attend more diligently to the interests of their employer. It is an invariable rule that a good master has good servants. If a Kaffir is paid fair wages and is well fed, he appreciates it, and serves his employer well; but many farmers give their servants only a heifer for a year's wages, equal to five shillings monthly, no clothing, feed them very badly into the bargain, and expect them to be diligent in their service. It is not only that the wages promised the servant are small; but he learns, by report, when too late to draw back, that there is little chance of his getting anything without litigation, for his master will endeavour to pick a quarrel with him shortly before his term of service expires, with the view, if possible, of evading payment altogether.

"The criminal roll in the circuit and magistrates' court of this division has been remarkably light during the last two years compared with 1863, as the following return shows :—Statement of convictions for theft in the district of Colesberg, in which sentences of three months and upwards have been passed during the years 1863, 1864, and 1865.

Year.	Magistrates' Court, No. of convictions.	Circuit Court, No. of convictions.	Total.	REMARKS.
1863	69	10	79	
1864	13	12	25	
1865	17	7	24	None of the convictions in the Magistrates' Court were for theft of stock.

"1862 and 1863 were both years of drought, which broke up before the close of the latter. It is a well-established fact that a severe and protracted drought produces a heavy criminal roll, showing that the people who commit offences (thefts) are not naturally vicious, but are driven to such courses through want. During the years mentioned, when thefts of sheep and cattle-slaughtering for food prevailed, it was ascertained by inquiry that, in general, the class of persons by whom such offences were committed were Bushmen and Hottentots, discharged from the service of their masters because they had no food to give them,—at least, such was the reason assigned. Thefts at that time became so numerous that the farmers took the alarm, and held a meeting with the view of procuring funds to pay for the raising and equipping of an additional police force. The civil commissioner advised them to re-employ their discharged servants instead. In the meantime the drought broke up, servants became in demand, and crime ceased.

"A considerable revenue could be derived from leasing the Crown lands in the division, which would be a much more advantageous way of disposing of them than by sale, while the present scarcity of money prevails. It is a fact that there are not more than half a dozen men just now in Colesberg who have the means of purchasing, if so inclined, so that there would be little or no competition. On the other hand, there are numbers of young men, farmers' sons, who have flocks of sheep for which they are in want of pasturage, and who are able and willing to pay a good rent for land, but who cannot afford to buy. If a fair allowance, at the termination of the lease, were made for all improvements, the tenant would set to work with a good will, and the lands would treble their present value, even after deducting the allowance for improvements.

"Colesberg has been fortunate in escaping the drought. It was, however, upon the verge of much suffering from that cause when the first rains fell in December. Nearly all the dams were dry, and the springs were weaker than had been known for more than twenty years. The veld is now in good order, and another fall of rain before the frosts set in (usually in April) will give abundance of pasturage for the winter."

The relation of crime to the varying water supply in different years, to which attention is called, may have been remarked in many of the reports in regard to other divisions which have been cited.

Of the success of Mr Kohl in this district, the following notice was sent to me:—"We were last week, says the *Colesberg Advocate*, at Middlewater, the farm of Peter Maltitz, Esq., and had much pleasure in looking at the various improvements which Mr Maltitz is introducing. The ostriches are doing well. They certainly are most voracious birds. Bones, stones, anything goes down with an ostrich. The other day one of them swallowed a gimlet. We also saw hundreds of young trees, which have lately been planted, and the whole farm bears the mark of progress. Nothing, however, pleased us more than to find a fine stream of water where a short time ago no water was, or at least no water appeared. Mr Maltitz consulted the water wizard, Mr Kohl, and commenced digging at a spot pointed out by him. Having penetrated the superincumbent soil he came to a stratum of rock. Boring through this, there poured up a stream of delicious water. Six openings have been made in the rock, and there is now an abundant supply. The value of the farm is of course increased, as agriculture can be carried on now to a considerable extent."

Of Colesberg in 1869 it was reported:—"During the last two months of the past year the district has suffered from a severe drought, and many farmers have consequently sustained a considerable loss of live stock, but the drought does not appear to have been much more severe than is usual at that season of the year, that is during the months of November and December, and it fortunately broke up on the 2nd January, 1870, when partial but heavy rains fell. . . .

"Property throughout the district is heavily mortgaged, but a few land-owners are gradually clearing themselves, and very few farms are encumbered with a second mortgage.

"The price of land still continues low, and town property is almost valueless.

"Land-owners are at last beginning to feel the necessity for exertion, and a number of dams, wool-washing establishments, sheep sheds, and other improvements are being carried out. Unfortunately, in most cases, capital is now wanting to second the efforts of the land-owners, and in such cases, until the land passes out of the hands of those so circumstanced, and is possessed by comparatively wealthy men, little improvement can be looked for.

"Ostrich farming is still pursued with considerable profit by a few farmers in this district, although it is doubtful whether it will ever become a very general branch of farming, nor does it appear desirable that it should do so, as the loss, owing to the mortality among the birds in times of drought, is very great, and is far more severely felt than losses of sheep and goats, owing to their great value. Domesticated ostriches are found to be subject to a variety of fatal diseases, but it is to be hoped that, as the nature of the bird is more fully understood, losses from these sources will partly be guarded against. There are five ostrich farms in this district.

"Sheep stealing is still very prevalent; in fact, it appears to increase monthly, and farmers are continually reporting losses of sheep—tens, twenties, and even fifties, disappearing at once, and neither sheep nor theives, in most cases, to be traced. This arises from the facility with which cattle-stealers can sweep off a lot of cattle or sheep and drive them over the Orange River. It is to be feared that the herds in most cases connive with the thieves, as the grossest carelessness can scarcely account for there losing large lots of sheep of which there is seldom any trace discovered. If the country was very hilly and bushy these losses might be put down to carelessness, but on the wide open flats of this district it would appear almost impossible that sheep could be lost; notwithstanding which, servants are daily charged with the loss of sheep of which they have the care.

"Most of the sheep and cattle-stealers who are apprehended and indicted for the theft are Bushmen and Bushmen Hottentots; they usually plead guilty, and hear the sentence passed upon them with the greatest nonchalance. In two separate cases since October last the prisoners have been sentenced to be flogged, and the sentence having been carried out they were released. These men were again apprended, on a similar charge, a few days after their release, and again pleaded guilty with the utmost indifference.

"The only punishment which these men appear to dread is spare diet and solitary confinement, and several prisoners are now undergoing this punishment in the gaol here. Corporal punishment altogether fails from deterring the natives from thieving. In fact, no punishment in any sensible degree lessens the evil. Until farmers learn to treat them more justly and liberally, and do not try to take every mean advantage of them, they will always be robbed by servants, and the few who pursue a more enlightened policy must suffer for the faults of the uneducated mass. At present a native does not believe in kindness from a white person, and invariably suspects some self-interested motive as the cause of any act of the sort.

In 1874 it was reported:—"The past year was ushered in with copious rains. The Orange River rose to an extraordinary height on the 21st and 22nd of February, inundating large portions of the farms along its banks. At Onverwacht much damage was done. On the farm there are six buildings, the residences of several families. The water rose six feet in these houses, every one of which fell in. On the farm Riet Bult, the property of Mr Roos, the lands, garden, and some of the buildings were completely submerged, and much injured. When the flood was at its highest, the water extended three thousand paces from the banks of the river, on the colonial side, and aged persons, who have lived in this locality nearly all their lifetime, declare that they never experienced such a flood.

"The lambing season has not been a good one, on account of the poverty of the sheep. Winter was unusually cold; heavy falls of snow occurred in June, and violent wind-storms followed for several months.

"The drought which prevailed during the latter portion of the year was completely broken up on the 19th November, when the rains fell, and replenished vleis and dams. Fearful showers have since visited this district, and were repeated on almost every alternate day, and caused much destruction of property. Mr Theunissen, of Oorlog Poort, late M.L.A., was a severe loser. The water rose to such a height on his farm as to wash away the stone wall round the ploughed lands, and carried off some five thousand oat sheaves, and all the standing crops. He also lost one thousand and fifty sheep, two hundred and twenty-five goats, eighty-five head of cattle, four ostriches, and several horses. In his orchard a number of valuable trees were destroyed, and about one hundred springboks perished on the farm. Mr Theunissen's losses are estimated at £1,200. Several other farmers have also been great sufferers. Heavy falls of hail occurred in this town lately, and in the direction of Phillip's Town and Hanover, and caused much damage to the standing crops, fruit trees, and buildings. The veldt is now in good order, and a little more rain will give abundance of pasturage for the ensuing winter.

"The drought may be judged by the price of meat, which is selling at 8d. per pound, and meal £4 the muid. Trade was completely paralyzed. Carriage from Port Elizabeth to this was as high as 28s. per hundred pounds in October.

"Ostrich-farming is being pursued with great success by about fifteen farmers. Several large stone wall enclosures exist for these birds, and some are now in course of erection.

"Locusts, in countless myriads, in the shape of *voetgangers*, made their appearance soon after the breaking up of the drought; fortunately most of the wheat, barley, and other grain, had been reaped before the locusts appeared in great numbers. I am happy to state they are now taking wing, and it is hoped that ere long the district will be cleared of this scourge. . . .

"Colesberg has for a long time been noted for the healthiness of its climate, several parties from Europe suffering from consumption having been restored to perfect health after a short residence here.

On medical authority, I can safely state that no better place for a sanatorium could be found, either in the Colony or in the neighbouring States."

And in 1875 :—" During the year this district has been favoured with splendid rains, and in consequence crops have been above the average. The soil remained well saturated up to October. The late rains set in very partially, and did great damage on some farms in the field-cornetcy of Hanover, destroying dams, land-walls, and submerging the crops. At other farms towards the northern side of the field-cornetcy hardly any rains have fallen, and the loss of stock has been considerable.

"The lambing season has not been a very good one, on account of the severe winter. There has been but little disease among cattle; lung-sickness has appeared, but not so extensive as in former years.

"Locusts made their appearance, but have done little damage, owing to the luxuriance of the vegetation. They were soon pursued by large flocks of various kinds of locusts birds.

"When sheep were first introduced into this district the pasturage consisted chiefly of grass and sweet bush, but now the 'bitter bush' and intoxicating melicæ, or 'dronk-grass,' are spreading in every direction, the former being of a nauseous character enjoys immunity and is only eaten by sheep when in blossom, or in cases of dire necessity. The valuable schaap-bosch and other sweet shrubs are gradually disappearing in some parts.

"The division might be greatly improved by the construction of a greater number of dams of large dimensions. The physical features of the country are well adapted for the conservation of water; the soil is very fertile, yielding from fifty to sixty fold. The selection of sites, especially for large dams, should, I think, be left to an hydraulic engineer, and his advice should be followed in every respect. Small dams are only disappointing.

"Ostrich-farming has been carried on very successfully for several years. The continued decline in the prices of ostrich feathers in the English market has greatly discouraged many of the farmers, who begin to fear the effects of over-production; there appears, however, to be still a wide margin. Some are again turning their attention to sheep-farming, and consider it more safe.

"High winds and late frosts have greatly injured the fruit this season. The heavy rains have done much damage to the roads, which are still in many places in bad repair; but the Divisional Council are using means to have them put in a tolerable state. I anticipate that within the present year the roads generally will be vastly improved. . . .

"The district police are giving great satisfaction. Many theives, chiefly Bushmen squatters, preferring other parts of the Colony where facilities are not provided for their detection and apprehension, have left this district. There is consequently a marked decrease in the number of criminal cases for theft of stock.

"Horse-breeding is still a favourite pursuit in the Hantam ward. During the past year several thoroughbred horses have been introduced by Messrs Van Zyl from some of the principal studs in the Western Province.

"The town of Colesberg is progressing. . . .

"A public garden has been laid out on the Vaalbank flats, near this town, and a number of trees have been planted therein which are thriving well.

"Two dams have been constructed on the commonage here, one at the back of Coleskop, which now contains a large body of water, and the other above the public garden above referred to.

"At Hanover an aqueduct has been constructed by the Municipality, for a distance of 1,000 yards, from the fountain which supplies the village with water for drinking and other purposes, and is a valuable acquisition. It consists of a rectangular covered drain, 14 inches by 9 inches, built of stone and mortar. This drain is erected partly on a wall of massive stone and partly under ground, at an average cost of 22s. 6d. per yard. The fountain is protected by a 'water-house.' The water used formerly to be led through an open furrow. The increased strength of the stream, which does not lose now by evaporation or cattle drinking in its course, has enhanced the value of the water-erven considerably, and gardening is carried on to a much greater extent in the village."

In reply to queries issued in 1862, Mr Green, Civil Commissioner, at Colesberg, wrote:—"1. As to whether any plans for irrigation works have been proposed in this district?—My reply to this query is in the negative. I may at the same time state that almost every farmer has one or more artificial reservoirs for damming up and collecting rain water in for watering his stock, and that these are occasionally large enough to permit of part of their contents being applied to the irrigation of a few acres of corn-land; but no extensive works for irrigation have hitherto been undertaken or projected.

"With regard to your query No. 2, as to what irrigation works appear practicable in my division, I have to state that the Sea Cow River, the principal one next to the Orange River, is led out in one place by a cutting of three thousand yards in length for irrigation purposes; but the land of a proper level to which the stream can be applied is of very limited extent. There are other places where the same thing might be done, but there is no land as near the banks upon which the water could be used; much longer and very expensive cuttings would therefore be necessary for the purpose. As the river ceases to run during a drought, much benefit would not accrue from such works after their execution without dams were constructed across the stream to restrain the flood-waters during a rainfall, and carry them, by means of a suitable cutting, into reservoirs, which the natural features of the country frequently permit of being constructed of great extent, and at a comparatively trifling expense, by filling up a gap in a chain of hills with masonry. A great drawback to the collecting of flood-waters in this manner is the large quantity of sand and mud carried by them in solution, which would necessitate the occasional cleansing of the reservoir—an expensive process.

"The chief difficulty in the Government carrying out irrigation works on an extensive scale seems to me to lie in the fact of nearly the whole of the land in the district being in the hands of private parties, and that such works as diverting the channel of a river would not benefit the proprietor through whose ground the heaviest cuttings are made, but rather the reverse, inasmuch as they would be dangerous to stock, and that the proprietors whose land was available for irrigation by the collected waters might not be sufficiently alive to the advantages placed within their reach. These difficulties might, of course, be provided against by legislative enactment, which should give the Executive the power of damming up streams, making cuttings and reservoirs wherever necessary, and resuming such lands as may be required for irrigation purposes at a valuation, provided that no equitable arrangement could be made with the neighbouring proprietors for the use of the water after its collection in reservoirs.

"In projecting plans for giving an impulse to agricultural operations on a large scale in a district so remote from the coast as this, and with no navigable river or other cheap mode of inland transport, it must be borne in mind that although a great scarcity of all the cereals undoubtedly exists here at the present moment from the long-continued drought, and that prices during the most favourable seasons are far too high, the consumption on the spot is so limited that, by doubling or trebling the production, prices would fall to such a degree that cultivation by irrigation would not pay. I am therefore of opinion that, until the advantages of the railway system are extended to this quarter, the measure most advantageous to the division would be that which is suggested by your query No. 3,—namely, the Government making advances to landed proprietors upon reasonable terms for the purpose of their executing works for improving the supply of water on their properties.

"I find, upon looking over what I have written, that I have not replied to that part of your query No. 2 which refers to the practicability of raising water from rivers or other running streams, or of sinking wells; and I may briefly state, in reply, that there are many places along the Orange and Sea Cow Rivers where deep holes exist, sometimes of a couple of miles in length, from which a considerable quantity of water can be obtained; but, as the banks of the rivers are high, rather expensive machinery is required to raise it to their level. This is now being obtained by private enterprise—the proprietors of two farms on the Sea Cow River having imported force-pumps, at a cost of £150 each, for the purpose, and I am told that they are found to answer well for irrigation; but, not yet having had an opportunity of visiting the spots where they are in use, I cannot speak from actual observation. Wells may be sunk in most places with some advantage, but the supply of water obtained from them is far too limited to be used in irrigation."

And again:—"In reply to the last question proposed in your circular of the 7th instant, as to the extent landed proprietors in this division would be likely to avail themselves of a law empowering the Government to make them advances of money on reasonable terms, with convenient periods for repayment, for the purpose of executing works for improving the supply of water on their properties, I have the honour to state that, in my opinion, if the interest on the advance did not exceed six per cent. per annum, many proprietors would gladly avail themselves of such a privilege; and that you might reckon upon £10,000 being applied for during the first year following the passing of such a law."

To the east of the division of Colesberg is the division of Albert, on the banks of the Orange River, lying between the river and the division of Cradock. It includes the region forming the northern slope of the Zuurberg and Bamboesberg Mountains, extending to the Orange River, between the Stormberg Spruit on the east and the

Oorlog Poort River on the west, an area of about 3,400 square miles, which was formerly called the New Hantam district. It is well watered, although its rivers are rather torrents with a short and impetuous course. The climate is healthy, and sheep thrive, although, from its high elevation, it is very cold in winter. Its principal town is Burghersdorp, which is situated on the Stormberg Spruit, about twenty miles south of the Orange River.

The Hantam is a district consisting of high-lying mountain tops, considered healthy for horses in time of horse sickness. The same name is given to a similar sort of country in the division of Calvinia, to which the Boers send their horses when that epidemic prevails. It is a large mountain, with a flat top.

"The country in the division of Albert, both north and south, lies high, and has a cold climate in winter. In its physical formation it has been described as similar to the adjacent division of Colesberg; and sheep, horses, and cattle of most descriptions are bred with success. Albert is second only to Colesberg and Richmond for the quantity of wool produced."

In reply to queries issued in 1862, Mr Cromar, Civil Commissioner, wrote:—"I have the honour to acknowledge receipt of your circular No. 16 of 1862, dated 7th May, 1862, and to report—First, no plans for irrigation works within this division have ever been proposed; second, there are dams formed on every farm in this district that I have visited for the collection either of spring water or rain water, or both.

"Some of these dams are very large and others very small; but generally speaking they only contain water sufficient for the stock on the place, and not for irrigation.

"In only a few cases, where the springs are strong, or the dams situated in naturally favourable situations for the collection of rain, is there sufficient water collected to be employed for irrigation to any considerable extent. The largest extent of the land of any farms in this division under irrigation which I have seen is about fifty acres.

"With very few exceptions there is a fountain on every farm in this division, on some so weak as to supply only water sufficient for household purposes, and in most sufficient for stock as well as household purposes, and in only a few sufficient also for irrigation purposes.

"With the exception of the Orange River, of which the banks are so steep that I believe it hardly possible to lead water out of it, there are only two rivers—namely, the Zuurberg Spruit, which separates this division from that of Colesberg, of which I do not know whether it is applied to irrigation purposes, and the Stormberg Spruit, which runs throughout the entire division. I am aware that most of the farmers living along this river, where it is possible, lead off its waters to irrigate their lands.

"The drought for the last two years has been in many parts of this division so severe that this river has been stagnant for at least one third of each of these years, and in some parts of the division there has not been a drop of water in many dams during this period. I am not aware of any wells being sunk in any part of this division, except in the town of Burghersdorp, the inhabitants of which are wholly supplied from sunk wells of the depth of from twenty-five to thirty feet, there being no fountain or running water in or close to the town. I do not consider that sufficient water for irrigation could be produced by sinking wells, unless very expensive machinery should be employed for pumping the water.

"Third, I consider that an advance of money, on the part of the Government, on reasonable terms, for the purpose of enabling the landed proprietors of this division to execute works for the supply of water on their properties, would induce several of them, I can hardly say the majority, to avail themselves of the offer, and thus improve the division."

Four years prior it was stated, in the *Albert Times*, "vast dams are in the course of construction." It was in reference to a dam in this locality that Mr Ziervogel stated, in evidence given before the Select Committee on Irrigation, which I have already quoted, that "in very many localities there are great facilities. I know that as long as thirty years ago, the people in the Burghersdorp district began to make dams, not only for collecting water for the use of their stock, but for irrigation. One was made on a farm where a man could not raise, except in favourable seasons, grain enough for the use of his own household. He made a dam with a pipe to lead out the water, which enabled him to cultivate not only more than necessary for his own consumption, but he could send a deal to market. He made a dam where two ridges came close at their points, and, by blocking these up, secured a large body of water which, even if it were not to rain for one or two years, would not dry up altogether. That dam was made about thirty years ago by an old colonist. Now there are several others of a similar nature in that part of the country."

And the following evidence, given by that gentleman, shows that a scheme of still greater magnitude had been discussed, in view of the facilities afforded by the physical character of the district, for rendering the water passing by it and from it available for agricultural operations.

"Do you happen to know the Orange River?—Yes.

"Do you know its relative height as compared with Somerset?—I do not know its relative height, but I am aware that a great many years ago some people spoke of leading out the Orange River, and bringing it, by a cutting at Kniehalter's Nek, into the Great Fish River; but I cannot speak as to the practicability of the plan.

"You are aware that there is a very large body of water running generally in the Orange River?—Yes.

"And do you think, if that water could be taken out of the river and distributed over the country, the land would be increased in value to any extent?—If it could be done; but we must not think of such a thing as leading that river into the Fish River,—the cost would be tremendous.

"But if it could be done, do you think the farmers receiving the water, with such a pressure as the Orange River at the back of it, would be willing to pay a rental for it?—They would, but not such a rental as would be sufficient for such a work. Though, if it could be done, I think it would be a piece of work which is not yet equalled in Europe. I know that in other countries rivers are led out; but then they have comparatively level land; while the land on this side of the Orange River rises to a great extent.

"Are you aware that Sir George Grey submitted a plan of a canal from the Orange River down to the Colony?—I am not aware that Sir George Grey actually entertained such an idea; but I am confident that the expense of doing that would be so immense that we cannot entertain it.

"Then it is only a matter of expense, in your opinion?—It is a matter of expense, but it would be so enormous that we could not entertain it. The lowest piece of land where a cutting might be made, Kniehalter's Nek, between Cradock and Somerset, is a considerable distance from the Orange River, and at a considerable elevation above the river. But the Orange River might be led out to irrigate an immense quantity of land lying along its borders."

In 1866 it was reported:—"Albert enjoyed during the year 1866 abundant rains; and the crops throughout the division have been very good."

Of Albert in 1869 it was reported:—"This division is suffering severely from drought, so much so that stock is dying in consequence. But little rain has fallen this year, and the showers have been very partial. Locusts are in immense quantities.

"Twenty-six Angoras from the flock lately imported by Messrs Blaine & Co. have been introduced, and some small flocks of colonial-bred Angoras have been purchased.

"There are many very fine dams in the division, in many cases rendering the farmers comparatively independent of rains for a length of time, but this year some of them have become dry in consequence of the long-continued drought."

In 1874 it was reported:—"My report for the year 1874 must be very much the same as that for the previous year. Carriage has maintained the same high rates, and the necessaries of life have not diminished in price.

"The division suffered from severe drought up to the first week in December, when heavy rains set in from the south-east. The cold increased for several days, until the result was a very heavy fall of snow along the Stormbergen, and between Burghersdorp and the Orange River, causing the death of large numbers of horses, cattle, and sheep. Since then the country has been deluged with rain and hail, causing much damage to the roads, and heavy losses in sheep and farm produce. Fortunately, the division has not suffered anything like the damage suffered by many other divisions, the benefits accruing from the rains more than counterbalancing the damage."

To the east of the division of Albert is the division of Aliwal North, including the country between the Stormberg range and the Orange River; it has an area of 4,450 square miles, and is a region well watered by the Kraai, Stormberg Spruits, and other streams running into the Orange River. The country lies high, and is cold in winter. The chief town, Aliwal North, is built close to the Orange River. The physical features of the country are similar to those of the division of Albert.

Mr Burnet, Civil Commissioner, in reply to the circular of 1862, submitted the following report:—"1. No general plans for irrigation works have been proposed within this division, but upon almost every farm one or more dams have been constructed, as well for purposes of irrigation of corn lands and gardens as for the watering of the flocks and herds. In general, these dams are not very large, although on some farms they are found of such magnitude, and in situations so well chosen, that they never fail unless in very extreme droughts.

"2. The Grey River, commonly called the Kraai River, is the chief stream of the division of Aliwal North. It rises in the Drakensberg, on the southerly border of the district, receives several tributaries, and flows northward into the Great Orange River, about a mile east of the town of Aliwal North, bounding the town lands on the eastward.

"It appears by the eye practicable to raise the water of the Grey River at several points, and to irrigate a number of farms along its course, as also to lead water from it over the town lands, or through the town of Aliwal North.

"The town is supplied for irrigation purposes by the 'Mineral Government Springs,' at Buffels Vallei, about two miles distant. The water for family purposes is now carried from the Orange River, on the south bank of which the town is situated.

"There are a number of smaller streams in the district, the waters of many of which might be raised and diverted, so as to be more generally useful, by the construction of dams, whilst numerous situations present themselves in kloofs and valleys, where large additional supplies of water could be collected by the same means; but no correct estimate of such works can be made without an inspection of the division. Water would, I believe, be found at no great depth by sinking wells.

"3. As the best means of ascertaining the views of the farmers of the division on this point, I called a public meeting for the 14th inst., and took much pains, by numerous circulars sent through the field-cornets, to explain the object which his Excellency has in view. Only one field-cornetcy of the division has taken the matter up with any spirit, the ward of Kraai River, which held a ward-meeting, and sent an intelligent deputation to the divisional meeting.

"The views of the farmers of this ward unanimously and substantially are these:—That the proposed system is one of the best things Government could set in operation; and that their own district and the whole Colony would derive great benefit from it, and become a flourishing land. It is estimated that from £4,000 to £5,000 would speedily be taken up for irrigation purposes in this ward.

"Several farmers from the other field-cornetcies stated that the farmers were much divided in opinion, many did not understand it, and supposed it might end in a 'tax.' No interest was generally felt in the matter at present, and the general impression was, that such a law, to be thoroughly understood and appreciated, must be in active operation, and actually affording the assistance proffered to all who choose to avail themselves of it.

"It appears to me that many farmers would take advantage of such a law who now think altogether unfavourably of it, were it in practical operation."

In 1865 it was reported in regard to Aliwal North:—"Although comparatively much less rain has fallen than during 1864, the past year has been favourable for both stock and crops. The lambing season has generally been very good—fully an average one; and in numerous instances more lambs in proportion have been reared than in some previous years. A number of fine showers fell in August, so that the fields were well covered with grass before the summer heats set in; and, although we have had little or no rain for nearly four months, no general scarcity of either grass or water has been felt. On the whole, cattle have not improved in condition; but no losses of any consequence have taken place. Heavy rains appear to have fallen in Basutoland, at the sources of the Orange River and its tributaries."

In 1866 it was reported:—"During the first months and winter of 1866 little rain fell. The winter months were remarkably cold, but bracing and healthy. No sicknesses of a serious character have prevailed, beyond cold and influenza in some localities. During the summer and autumn months fine and successive rains fell, and the crops generally are a full average of good years."

Of Aliwal North in 1869 it was reported:—"At the commencement of the year there was great depression in the trade of this town, owing to the unsettled state of affairs between the Orange Free State and the Basutos, but since the articles of convention were entered into between His Excellency, Sir P. E. Wodehouse, and the deputation from the Orange Free State, there has been a gradual improvement; but a great drawback now seems to be the want of a bridge across the Orange River. The only step taken during the present year towards gaining this desirable end was the securing certain concessions from the Orange Free State Government. The cost of the bridge is estimated at £30,000, and if one were to be constructed by a company, a return of at least 10 per cent. might be expected. Smithfield, Rouxville, Wipenaar, and Lady Brand are towns in the Orange Free State which would avail themselves of the bridge, and it would be advantageous to the trade of the Colony, the Orange Free State, Basutoland, and Transvaal.

"Fortunately, wheat, Kaffir corn, mealies, and other articles of consumption have been very cheap during the year, owing to the abundant crops reaped in the Wittebergen Native Reserve. Prices have been as follows: Meal, from 10s. to £1 per muid."

In 1874 it was reported:—"The commencement of the year 1874 was noted for the unusually heavy rains, and on 19th, 20th, and 21st February, 1874, the Orange River overflowed its bank on the Orange Free State side, and the water is calculated to have risen twelve feet higher than it ever was before. The breadth of the river is about three hundred yards, and, considering that the water spread beyond the bank to the hill on the opposite side, a distance of three hundred or four hundred yards, twelve feet additional height of water will enable the persons erecting the bridge here to make allowance for a similar flood on a future occasion. The highest water mark has been taken, and can be pointed out to the engineer when required. The large mill erected on the bank of the Orange River, and also Mr Tudhope's wool-washing establishment on the Kraai River, near its junction with the Orange River, and about three miles from Aliwal North, suffered considerable damage. The latter has been repaired, and wool-washing carried on for some time, but the former had the two upper stories—which were of brick—fall in. Fortunately the roof settled on the lower portion of the building, and saved the machinery from floating away. . . .

"The last winter is said to have been much more severe than any previous winter, as known even to the oldest inhabitants. Cabbages were killed by the frost, and many blue-gum trees almost completely bared of leaves. Yet there was very little sickness, excepting the measles, which prevailed throughout the district, but in a very mild form.

"The farmers had considerable losses of stock, owing to the severe winter and the very late rains, which only commenced towards the end of November. Lunk-sickness is prevalent amongst the cattle, and one farmer, Mr P. de Wet, informed me that he had lost about seventy head. The sheep are dying off very much in some parts of the district, from the "fluke," for which the farmers appear to have no remedy. Field-cornet Penny, of New England, lost, in ten months, over one thousand sheep.

"The crops of wheat and other cereals have been below the average, owing partly to the late rains, and partly to the heavy rains and hail-storms at the reaping season. The increased demand caused the prices to be higher than in former years. The wheat, mealies, and Kaffir corn in the district of Herschel, which is the native reserve, and in Basutoland, could not be brought here in consequence of the fearfully boggy state of the roads, and the ruling price during the year, for mealies and Kaffir corn, has been from 20s. to 30s. Meal, from 50s. to 60s. per muid."

The following is a tabulated statement of meteorological observations at Aliwal North in the year 1867 :—

Tabulated Abstract of Meteorological Observations made at ALIWAL NORTH in the Year 1867.

FROM DAILY OBSERVATIONS.	Jan.	Feb.	Mar.	April.	May.	June.	July.	Aug.	Sept.	Oct.	Nov.	Dec.	For the Year.
1867.	degs.	degs.	degs.	degs.	degs.	degs.	degs.	degs.	degs.	degs.	degs.	degs.	degs.
From the self-registering thermometers (Fahr.)													
Mean maximum temperature of the air,	82·4	79·9	76·4	70·8	67·1	56·0	55·6	61·1	70·4	74·6	76·1	83·4	71·1
Mean minimum temperature of the air,	61·8	59·5	58·9	51·3	37·0	29·2	28·7	38·1	44·4	49·3	53·1	59·5	47·6
Approximate mean temperature of the air,	72·1	69·7	67·6	61·0	52·0	42·1	42·2	49·6	57·4	61·9	64·6	71·4	59·3
Mean range of temperature of the air,	20·6	20·4	17·5	19·5	30·1	26·8	26·9	23·0	26·0	25·3	23·0	23·9	23·5
Highest temperature,	91·0	87·0	86·0	78·0	70·0	68·0	63·0	76·0	82·0	88·0	87·0	93·0	93·0
Lowest temperature,	50·0	45·0	43·0	35·0	27·0	20·0	21·0	28·0	31·0	35·0	38·0	45·0	20·0
Extreme range of temperature any one day,	32·0	31·0	30·0	26·0	38·0	35·0	37·0	32·0	37·0	44·0	39·0	34·0	44·0
Least range of temperature any one day,	12·0	12·0	7·0	10·0	12·0	13·0	15·0	11·0	12·0	11·0	11·0	15·0	7·0
Quantity of rain in inches,	1.55	2·50	2·31	1·06	0·62	1·35	·00	0·65	0·65	1·35	2·38	4·49	18·21
Greatest amount of rain on any one day in each month,	·75	·62	·75	·40	·42	·75	·00	·53	·53	·57	1·07	1·79	1·79
Number of days on which rain fell,	8	7	12	4	4	3	0	6	6	7	9	12	78
Aneroid barometer :—													
Mean height of the barometer at the temperature 32° Fahr. (not corrected for altitude above the mean level of the sea), for each month, and the year, from observations at 9 A.M., 1 P.M., and 5 P.M.,	25·37	25·36	25·36	25·41	25·43	25·45	25·49	25·44	25·36	25·31	25·37	25·36	25·39
Maximum observed,	25·50	25·52	25·58	25·62	25·58	25·56	25·70	25·67	25·66	25·47	25·56	25·49	25·67
Minimum observed,	25·12	25·32	25·14	25·10	25·22	25·26	25·16	25·20	25·14	25·00	25·18	25·20	25·00

Wodehouse is a district lately formed, in 1871, of the east part of the district of Aliwal North, in which are located numerous bodies of Fingoes and other natives. It comprises the slopes of Witterbergen and Orange River valleys. The principal place is Dortdrecht. The climate, as this region lies high, is very severe in winter.

Of Wodehouse in 1874 it was reported that the district had suffered somewhat from drought and very late frosts; that the harvest had not been good, and the crops of the Tambookie location had not been as abundant as in former years; that there had been but little sickness amongst horses, cattle, and sheep, and the losses from the heavy rains had not been so severe in this division as in many other parts of the Colony; that the natives, generally, had willingly paid up the taxes demanded of them, and many had large flocks and herds; and that cases of stock-lifting had been comparatively few in number.

And in 1875:—"This division continues to make progress, both in regard to agriculture and stock, and in the condition of the inhabitants, European and native. . . . In some parts of the division the disease among sheep, known as the 'fluke,' has been prevalent, and some farmers have been heavy losers.

"Crime has not been on the increase; the revival of the system of patrolling by the F.A.M. Police has been attended with good results. Stock-stealing has not been of such frequent occurrence, the fact of the police being on the look out for stock-lifters acts as a great check to this particular species of crime.

"The crops of wheat and other cereals have been below the average, and prices have ranged high. The crops of the natives in the location attached to this division have, on the whole, been good; but the great drawback to the improvement of the location is the want of a road to Dordrecht; at present there is only a footpath, and that unfit, even for travelling on horseback. Thus, in the greater portion of the location, the natives have no outlet whatever for their produce."

PART III.

Supply of Water, and Facilities for Storage, in Colonised Lands adjacent to the Colony of the Cape of Good Hope.

Chapter I.—Basutoland.

To the north of Aliwal North is Basutoland, separated from the Colony by the Orange River, and bounded on the west by the Orange River Free State. In this territory springs and streams are scarce. It is a mountainous region; but rains, it is said by some, seldom fall, excepting on the Quathlamba range, by which it is separated from Kaffirland and from Natal. There, also, Baldwin considers that wells if sunk would be a priceless boon to the natives.

The country measures about 150 miles long by 50 miles wide, and contains about 7,000 square miles. Though the rainfall be reported as small, the country is described as well watered. Its principal rivers, as stated in the "Hand-book of South Africa," issued by Messrs Silver & Co., are "the Orange, Caledon, Cornet Spruit, Klein Caledon, Putiatsana, Tlotse, and Sengunyana. The chief mountains are the Drakensberg, Maluti, Queme, Masiti, Berea, Tsikwane, Leribe, and Langeberg. The climate is cool and temperate in summer, with abundant rains; and dry and bracing in winter, when snow occasionally falls. There is but little wood found in the country, though grass is plentiful. The mountain scenery is very picturesque, several spots being of great beauty. The average height of the plains above the level of the sea is about 5,000 feet. The Drakensberg and Maluti mountains average here about 9,000 feet above the sea, and the other mountains about 6,000 or 7,000 feet."

At the extreme northern point of Basutoland is the highest point of the Drakenbergen, 10,000 feet above the level of the sea, called by the natives "Pofung" (the name given by them to the elan, *antelope coruna),* but a mountain known to Europeans by the name "*Mont aux Sources,*" a name given to it apparently by Messrs Arbousset and Daumas, agents to the Paris Missionary Society, on an exploratory tour made by them in 1836, undertaken mainly with a view to publish the Gospel to the tribes then inhabiting the Maluti mountains, but partly with a view to trace the upper streams of some of the principal rivers of South Africa.

M. Arbousset, in his narrative,* says :—" No European traveller had yet penetrated to the sources of these rivers, and consequently their exact situation had not been ascertained. We have had the satisfaction to explore some of the sources ourselves; and with regard to the others, we have gathered so many particulars in our interviews with natives as to warrant us in offering to geographical science a small tribute of information calculated to settle some questions, which, till now, have remained unsolved. Short and imperfect as our investigation may have been, we have satisfied ourselves that the rivers of which we are about to speak, the Caledon, the Orange River, the Namagari, and some other streams of less note, take their rise in a mountain which the natives call Pofung,† because there they have frequent elan hunts; but which we have designated in our map by the name of "*Mont aux Sources.*" This mountain, which runs from east to west, is situated near the 29th degree of south latitude, and about the 30th degree of east longtitude, at the extreme north of the chain of the Blue Mountains, of which it constitutes one of the culminating points. It is about fourteen hundred feet above the level of the surrounding soil, and may be about twenty miles in circumference at the base; as to its height above the level of the sea we were not able to take the exact measurement, but it cannot be less than ten thousand English feet. The summit of this mountain is in the form of a plateau, or table land, and is clothed with the richest verdure. On the west side of the mountain rises the river which the Europeans call Caledon, from the name of Lord Caledon, formerly Governor of the Cape, but which the natives name Mogokare,‡ because it flows through the country of the Mantetis and the Basutos. The Caledon, which near its source is a stream of a few feet of water, enlarges considerably in proportion as it approaches the secondary valleys, and it is dangerous to cross it when swollen by the fall of rains, and the melting of snow. At Merabing it receives the Tlotse, which rises, like it, on the west of the Blue Mountains, and now a river sixty

* "Narrative of an Exploratory Tour to the North-east of the Colony of the Cape of Good Hope, by the Revs. T. Arbousset and F. Daumas, of the Paris Missionary Society." Translated, from the French of the Rev. T. Arbousset, by John Croumbie Brown. Capetown: A. S. Robertson, Heerengracht; Saul Solomon & Co., S. George's Street. 1846. Second Edition—Aberdeen: George King & Co. 1852.

† The elan. ‡ By the middle.

feet wide, it carries with it, as does also the Caledon, a blue gravel, speckled with particles of mica. Opposite Thaba Bosio, the waters of the Caledon are augmented by those of the Saule, which rises in the Blue Mountains, and flows nearly due east. The Caledon is everywhere confined and rapid; in the summer, its waters being prodigiously swollen, flow nearly on a level with the banks, and have a depth of not less than twenty feet, with a width of nearly three hundred. Notwithstanding its depth, and the steepness of its banks, the Caledon might, in certain places, be led and employed to irrigate and fertilise the soil.

"Both rivers are bordered by willows, which generally rise to twenty-eight or thirty feet in height; and which, besides the shade they afford to the traveller, furnish the natives and missionaries with good fire wood and timber, very valuable in a country where there would be great difficulty in procuring any from other quarters. Rafters, joists, and beams may be got, some of which are not less than twenty feet long, and two feet in diameter; also planks, door and window frames, boards, etc. The fresh leaves of this willow furnish the cattle with a food, of which they are very fond. The Caledon,—which, throughout its course, can be forded in many places during the dry season,—cannot be crossed without difficulty, and even danger, when it has been swollen by the rains. At the fords, which a little before might have been crossed with a firm footing, nothing can be found but a deep quick-sand, on which it is dangerous to venture. When the ford is no longer passable, the river is crossed on the trunk of a tree, which the hardy swimmer propels to the opposite bank, or on a triangular raft, hastily formed of some pieces of wood rough-hewn, clumsily put together, and covered with underwood and briers. It is to that frail machine, dragged over by rope and dint of labour, that the traveller successively confides himself, his family, his waggon, and his baggage. The bed of the Caledon, wherever it is not sandy, is formed of masses of stones composed of clay, sand, mica, and oxide of iron, mixed in different proportions in different places. This stone, which the English call *iron-stone*, is common in the Malutis and the adjacent countries. The colour varies from a dull yellow to a very deep brown; but near streams it acquires, by the action of the water and the sun, a tint of glossy black. In all the sources of the Caledon may also be found plenty of onyx and sardine stone. Agates in the greatest variety of form and colour, plain or striated, clouded with a bright yellow or a deep brown, are also abundant here, as elsewhere are quartz and silex.

Here also may be found variolettes, a species of agates speckled with white and glossy grains, and so strongly embedded in the surrounding mass that they cannot be separated from it without difficulty. The more precious stones, such as opals, calcedonies, cornelians, although less common, are not entirely awanting in the Caledon, and some very regular crystals are occasionally found in its bed.

"The Orange River, the principal stream in South Africa, rises on the south side of the Mont aux Sources. The water comes gushing from the earth, and soon contracts a dark colour, which has obtained for this river the Sechuana name of *noca unchu* and *black river*. To begin at its source:—It flows for a distance of about forty leagues through a valley formed by the two chains of the Blue Mountains, the direction being from north-east to south-west. In that part of its course its waters are increased by a number of small tributaries which descend from both chains of the Malutis. One of these tributaries, named the Makaling, or the river of Aloes, rises in the neighbourhood of Mariah, and flows from north to south, following the western bend of the Malutis.

"After crossing the thirtieth parallel of south latitude, the Orange River issues by a narrow pass from the valley in which it has hitherto flowed, and takes a westerly direction, but gradually turns again towards the north. This part of its course, between the end of the valley, and the place where it receives the waters of the Caledon, has received from the natives the name of *Sinku*, which is also the name of the largest shield of the Matebeles. The Dutch farmers named it the Great River, and Colonel Gordon gave it the name of the Orange River, from the colour of its waters. The willows, the mimosas, and the olive tree, which shade the banks of this river, the transparency and delicious freshness of its waters, the masses of brilliant rock which border its course, the cascades, and the verdant islets scattered upon its surface, present the most agreeable contrast to the arid, rough, and uncultivated aspect of the neighbouring plains, which are desert, sandy, and rugged, with here and there rocky eminences of a yellowish colour. The bed of the river, very much confined towards its source, gradually widens as it proceeds; it is similar to that of the Caledon, and at certain places there is collected a sand spangled with mica, of the colour of gold and of silver. Its width, in the neighbourhood of Bethulie, is nine hundred and thirty feet, its depth, about two feet and a half, and the height of its banks, from twenty-four to twenty-five feet.

"The Orange River, like the Caledon, is subject to periodical risings.

These occur three or tour times between the end of November and the middle of April; the first rising generally lasts from ten to twelve days, the others continue five or six weeks. These risings frequently delay travellers who require to cross the river, and sometimes they overtake those who are attempting to ford it.

"Without stopping to notice the pretended medicinal virtues which the Dutch settlers and the Griquas ascribe to the waters of this river, we may briefly mention that a fine hot spring on the right bank of the stream, in the Buffalo Vley, flows into the Orange River, and that this spring is not the only one to be found in the neighbourhood of the river. The banks of the Orange were once frequented by the buffalo and the hippopotamus; but the numerous hunts of the colonists and of the natives have driven these animals away, and they have sought a quieter and safer retreat towards the Black River.

"The river Namagari, or the Fal, the most considerable tributary of the Orange, also rises in the same mountain, but on the north side. Four miles from its source, it is about eight feet wide, and brings down a bluish gravel. Its direction at first is towards the north; it then turns, with an immense sweep, towards the west; and, completing its course in a direction from north-east to south-west, it comes and empties itself into the Orange River, to the south of Campbell's Dorp, not far from the 29th parallel of south latitude, and the 24th degree of east longitude.

"Among the streams which descend from the east side of the Malutis and fall into the Indian ocean, the Letuele is one of the most considerable. Rising in the eastern prolongation of the Mont aux Sources, it flows to the north east, passes not far from Mococutlufe, and forming a *delta,* is lost in the Indian Ocean. The Matebeles name it *molampo o mokolu*, the *Great River,* and agree in representing it as being as large as the Sinku. The native travellers tell us that it winds, with difficulty and many meanderings, across a country of high table lands. The bed of this river is deep and confined; the current rapid; the banks fringed with willows and mimosas. It shelters hippopatami and crocodiles, some of which are from ten to eleven feet long.

"To complete what we have to communicate in regard to the rivers that rise in the Mont aux Sources, we must mention also the Monuenu, flowing parallel to the Letuele, about forty or fifty miles further to the south, and, like it, losing itself in the Indian Ocean."

The Putiatsana—called sometimes by the natives *noka-e-bogale,* the "wicked river"—seems to be a name given to a tributary of the Klein

Caledon, known also as the Saule. Of this river M. Arbousset writes :—" By some, the Saule is called the Little Caledon, from the name of the river to which it is a tributary. It has two sources; the principal one is about thirty miles to the north-east of Thaba Bosio,—the other is about twelve or fourteen miles to the south-east of that capital of the Basutos. The waters of the Saule are pure and limpid, and in some places deep. Where we crossed it, it rolls over a bed of very hard black grit, or ironstone. About a hundred paces distant the view is bounded by an eliptical semi-circle of rocks, terminating in a narrow passage which the Saule has opened for herself between the two hills of which we have spoken. There she stealthily disappears amongst the rocks, which seem determined, if possible, to prevent her escape. A little further on, finding a channel less encumbered, she flows on peacefully for about five miles, and then mingles her waters with those of the Caledon. We never tired looking on the scene; everything around us was grand, magnificent, and life-like, contrasting strongly with many a gloomy monotonous scene which had wearied the eye on other journeys we had taken. That circus, girt about with rocks, the rough and rugged sides of which may be seen reflected in the waters below;—that little wood, whose refreshing green does every thing possible to soften whatever may be too harsh in the outline of the landscape; every thing, in short, concurs to excite in the soul emotions of inexpressible delight. Never had we experienced a sweeter, or more extatic, joy than we did, when, with the Bible in our hands and with prayers on our lips, we turned towards the outlet from that mountain temple to gaze on the work of God, and the magnificent vestment with which he had there enrobed creation. Never before had the contemplation of His works excited such transports of grateful feeling, or such deep-toned devotion; not that the Lord had at any time, or in any circumstances, left us without an assurance of His presence; but we have found that those scenes which elevate the soul, like those which soften the heart, deepen the feeling that the Lord is near, and render manifest to the eye, as it were, the wonders of His goodness and His power.

" Before taking leave of the Saule, I must say a few words in regard to the inhabitants of its waters. The form of two species of these arrested my attention. One of them was a fish about sixteen inches long; the head and body were flat, the colour a uniform grey, the flesh white, and very delicate. The other was a little larger, with a cylindrical body, of a deep brown colour, and an elongated head. It is a species of barbel, having long barbels at the mouth.

I shall afterwards have occasion to speak of the way in which the natives catch these fish.

"If we may believe the natives, the Saule must also be inhabited by crocodiles, and these of two different species; one which the Basutos call the *Kuené*, and another which they call *Léfitué.* The former is said to be about the size of a young calf; the latter, so large that its body forms a kind of bank when it lays itself across the stream. The Basutos have a great and somewhat superstitious dread of both. We may relate a few of their traditions, leaving our readers to attach to them what importance they may think proper.

"'When a man,' say they, 'walks on the banks of the river, woe to him if he happens to be seen by a *Kuené*, or a *Lefitué!* The crocodile has the power of seizing the shadow of a man passing by, and by it dragging him into the river, where it will certainly kill him, though it will not eat a morsel of his flesh.' 'How do you know that?' we one day asked some of them who were speaking on the subject; 'they take cattle in the same way,' was their reply. 'One of them, in this way, caused a cow to fall into the river; when she was pulled out again she was found to be dead, but in no way mangled; they opened her, but they found neither blood in her heart, nor marrow in her bones; without a doubt the crocodile had sucked away both?' Another story which we had from the natives was this:—'The *kuené* seized a man who was passing near the Saule, and dragging him into his dwelling, which was dug obliquely upward from under the water, left him there, and went to call the *Lefitué.* While it was away, a ray of light coming through a chink in the earth, the prisoner saw lying just at his hand some copper bracelets, some strings of beads, and a pickaxe. He took them, cut an opening with the axe, and escaped. These things, say they, may have belonged to some former victim of the crocodile.' The Basutos have not the least doubt of the truth of this story. They believe also that the glance of the *Kuené* or *Lefitué* may prove fatal. 'A year ago,' said some of them to us one day, 'two young children were playing together on the banks of the river; they amused themselves by dipping their heads like ducks. In the evening, their parents being alarmed at not seeing them return, went to seek them, and found them drowned in a little shallow basin. A crocodile, cried they at once, has killed them by its glance.'"

From this may have been derived the character given to the river by the natives. The Putiatsana may be considered as the first stage of the missionaries on their journey northwards from their station of

Thaba-Bosio; another of their resting places was Kuening. Of this place M. Arbousset writes:—"Directing our course at first to the south of Merabing, we rode on horseback for six hours across high hills, over elevated flats, and through fissures and fens which are always found about the approach to mountains; and before sun-set we arrived at the foot of the chain, where we made our first encampment. Kuening, for that is the name of the place, was formerly inhabited by tribes of Bechuana-Bakuenas. In the neighbourhood there is a number of deserted kraals, and everywhere around the ground is covered with human bones, and skulls, and broken pots, and such like remains. 'Look at the work of the Matebeles,' said Monaile, 'they kill the Bechuanas as we would kill dogs. It is well that you men of peace have arrived in the country; but for you we had been all dead men; this is what was being done with the black nation.'

"Kuening, both from its fertility and its agreeable and commanding situation, is admirably adapted for a missionary station. The winter, indeed, is very severe, but this is of little consequence there. The former inhabitants will return with joy to cultivate these fertile valleys, from which war has driven them for a time. They will return with eagerness to localities to which they have given names expressive of the abundance in which they lived. One of their rivers they have named Atana *(where the cattle multiply)*, and another Khomokuanu *(where the cows love to feed)*; these two small rivers are among the number of streams which feed the Caledon.

"The first object which attracted our attention at Kuening was a rock about a hundred feet high, and four hundred and fifty long. It is composed of a soft but fine sandstone, the colour of which is a dull yellow. It is disposed in regular, horizontal beds. The front of this giant of the desert overshadows a little wood, through which winds the limpid Atana. At its base is an excavation in the form of a grotto, which served us for a lodging, as it usually does to the native travellers who go from Thaba-Bosio to Dingaan's, and to the king of the Basutos himself when he goes to hunt antelopes in those quarters."

Their next stage was the Tlotse; of this he writes:—"In pursuing our route beyond Kuening, we remarked numerous indications of the soil containing porphyry. We should not be surprised if there were, in the neighbourhood of Kuening, a quarry of marble. The fine-grained blue granite is very common there. We picked up a small lump of porphyry, containing beautiful rock crystals, and bits of quartz of all colours.

"After having gone thirteen leagues in a northerly direction, we arrived at Macosane, a town under the government of Mota, younger brother of Sekoniela. This town is composed of different groups of houses on a wide spread elevated table land; the view is bounded in the distance by a chain of mountains arranged in the form of an amphitheatre. At the foot of the table land flows the Tlotse, there a rapid stream, which may be regarded as the second source of the Caledon. On all sides, the eye rests on rich and fertile valleys, fields of millet and maize, and numerous herds of oxen and sheep which are tended by young shepherds. It would be an excellent place for a missionary station. Its population is concentrated, and the environs of the town alone present a sufficient field for the activity of two or three missionaries."

They next reached the Namagari, whence they advanced to the Malutis, of which he writes:—"The range of mountains which we then visited is called by the natives Malutis, or Peaks, many of their summits presenting this form, in which respect they are unlike the other mountains in South Africa, which, in general, assume the flat or tabular shape. In going from the flats, the ground traversed, before coming to what geographers denominate the *rameaux* of the second order, presents to the eye a succession of undulations gradually rising in elevation as they approach the lower range of hills, which are covered with fertility, and bordered with valleys of the richest green. The higher valleys are crossed in all directions by eminences,—sometimes isolated, sometimes in groups, but generally of a gentle declivity, and terminating, at no great height, in a level plateau. These elevations are composed of a sandstone of a yellowish colour, disposed in horizontal strata, with a rugged surface covered with black mould, consisting principally of decayed vegetable matter mixed on the surface with a small quantity of sand; but, at the depth of two or three feet, we find all kinds of loam and marl, heaps of pebbles and gravel, intermixed with clay. Some geological characteristics of this alluvial soil intimate the probable existence of iron ore in the vicinity. The long grass and herbage which cover these valleys, and which suggest the idea of a sea of pasturage, shelter and support a multitude of animals, such as the springbok, the grimme, the rietbok, the klipspringer, and the canna, surpassing all the others in beauty, and moving amongst them like a king. These animals multiply there, and they never migrate, although they are constantly exposed to the snares of the natives, and the attacks of lions, hyænas, and panthers. The woods supply shelter to innumerable tribes of the smaller birds; while the eagle, the vulture, the kite, and the sparrowhawk hover about in the higher regions. The climate of these valleys is salubrious, although keen and even cold in some parts; the streams are limpid and abundant; mineral or brackish springs are rare. In all our journey we remarked but two hot sulphurous springs.

"This country, the appearance of which more than once recalled to me that of the *Basses-Cévennes*, contains a population much inferior to what it might support. The tribes of Bechuanas, which have by turns occupied it, have never ceased there to keep up an exterminating war. There are now to be found but a few weak remains of these tribes, while the number of destroyed or abandoned kraals greatly surpasses those which are inhabited. From Mokoto to the Orange River, that is, in a zone of three degrees, the western district of the Malutis scarcely numbers forty-five thousand inhabitants (1836). The middle region of the chain, into which we enter on quitting the lower valleys, is of a different character. The rocks present themselves in close continuous groups of a steep acclivity, and almost inaccessible. At this height we found fine sandstone, flint, crystals, and some appearances of marble. The temperature is lower; vegetation, although still beautiful, becomes less vigorous. In the highest region granite still constitutes the foundation of the ravines; but the top of the mountains is formed of a rough-grained stone, coarse, brittle, and of a deep-grey colour, which gives to this ridge that bluish tint which has secured for it the name of Blaw-Bergen or Blue Mountains. Although this name applies particularly to the northern part, and is less general than that of the White Mountains, it has been preserved in the map accompanying the original. In this part of the chain the beds of stone assume a much more decided inclination than in the valleys at the bottom of the mountains; this inclination in some places approaches the vertical, and we no longer find those crumbled stones,—those remains which cover the sides of the lower valleys. Here are to be seen only one mass of smaller links heaped one upon the other, or closely associated together, carpeted with a strong perennial grass of two or three feet in height, with a few shrubs here and there interspersed. In the southern part of this range vegetation is richer and more varied. The eastern side of the Blue Mountains presents a multitude of terraces which, gradually descending, terminate in the shores of the Indian Ocean.

"During four months of the year—viz., from May till August—the summit of the Malutis is covered with snow; from the beginning of October to the end of March, it is deluged with rains; and during the two following months it is exposed to violent winds, or to fearful waterspouts, which render it uninhabitable. In winter the climate is sometimes so rigorous, that cattle, and even shepherds, have been known to perish with cold. At Moriah we have every year a little snow, and ice of about a quarter of an inch in thickness; and in the neighbourhood it has been found even three or four inches thick. The rain falls in torrents, and the hailstones are occasionally so large, and fall with such violence, that lambs have been killed in the fields by them. Sometimes, in less than five minutes, all the windows of our dwelling have been broken; often, too, has the hail come through the glass, leaving behind it a hole as neat and round as that made by a musket ball.

"The eastern aspect of the ridge, from its exposure, enjoys a climate more mild, and a vegetation more vigorous and varied. We find, however, to the west of the crest of the ridge some magnificent trees, among which we may mention, as one of the most common, a variety of the *Cunonia Capensis*, the trunk of which is no less than from twenty-two to thirty feet in height; it is a white beech, and has a very loose fibre. It is found within the colony of the Cape also. There, however, it scarcely attains the height of ten or twelve feet. On both sides of the chain of the Malutis flourish the *Olea Capensis*, the *Euclea racemosa*, a variety of the *Quercus Africana*, differing from the latter and from other kinds belonging to temperate climates, in its colour being white."

The Namagari takes a circuituous course to the north and the west from the part where the travellers touched on it at this part of their journey. And at a later stage he writes:—"We had gone on horseback to examine the confluence of the Namagari and the Lekua, two rivers of about the same magnitude, and whose united waters form the Yellow River, otherwise called the Fal *(from the Dutch Vaal)*.

"The Lekua comes from the east, and throws itself into the Namagari, at the base of a ridge of mountains previously unknown, which we called *Fransche Bergen* or French Mountains, for lack of a better name. They form a regular chain, which runs at first from the south-west to the north-east, and stretches very far into the interior, according to the uniform testimony of the natives who have observed them, as has been already seen amongst the Baperis, and further. These mountains, at the spot where they commence, may be upwards of two thousand feet above the surrounding soil, and about five thousand feet above the level of the sea. The natives assure us, moreover, that they increase greatly in magnitude as they advance to the north. They are covered with snow in the month of August, and three months later there falls at that place abundance of rain. Every thing then assumes a lovely appearance of verdure, and antelopes swarm there, as do also lions and hyenas. At all times there is found there great numbers of baboons, of coneys, called by others rock badgers, of panthers, of jackals, of wild dogs, and of wild cats. Olive trees and acacias are also very common there, with many other trees of a great height."

"I might have remarked, in speaking of the Fal, that its banks are very bare where I have seen them, only a few small willows show themselves here and there, while after its junction with the Lekua it is covered with firewood and even with timber.

"Along the river in the driest spots there grows a prickly bush, scarcely a foot in height, which bears a fruit pretty like the jujube in form, and having a juice which is rather sweet. Monaile and I supported ourselves on these almost the whole day of the 16th of April. We had left the waggon at an early hour without any other provision than a large piece of the flesh of the gnu, which was as tough as leather, and which hunger alone could lead us to consider eatable. It was in the midst of the fields, on the banks of a small lake, where we lapped a little water to complete our repast. Then, remounting our horses, we held on towards Mauliri, a small stream running into the Namagari. Its bed is almost even with the ground, and its water as clear as crystal."

On their return journey they crossed the Enta, of which he writes:—"As for the Enta, the upper sources of which we had crossed beyond Motlomo, it is at this place about twenty feet wide, and flows gently towards the west, away to the Yellow River, of which it is one of the most lovely tributaries. Thick mimosas shade its banks, along with willows of considerable height, and wild olive trees, etc. Many hippopotami frequent its waters, which are generally deep, and consequently preferred by the amphibia.

"On the 27th, following still a southerly direction, we arrived at Matloang, a charming little Lighoya village, built on a stream whose name it has appropriated (a source of the Tikue)."

M. Arbousset made a careful record of meteorological observations throughout this journey on the Malutis; but they embrace only those of the temperature and the clouds, made daily at morning, noon, and night.

The territory of Basutoland, which had long been under the sovereignty of the great chief, Moshesh, was proclaimed British territory, with the consent of its people, 12th March, 1868; and was annexed to the Colony by the Act No. 12 of 1871.

In the report by the Governor's agent at Maseru, dated 31st Jan., 1874, it was stated:—"I am afraid that the yield of grain this season will not equal that of the previous one, as the crops have suffered a good deal from want of rain. Nevertheless, larger areas of land are brought under cultivation each year, and consequently the demand for ploughs continues unabated, and the production of grain is annually increasing. Last year the number of ploughs purchased by the Basutos from the traders in Basutoland was six hundred; this has been carefully ascertained from actual returns furnished by the traders themselves."

CHAPTER II.

THE ORANGE RIVER FREE STATE.

To the west of Basutoland, and to the north of Colesberg, lying between the Orange River and the Vaal, is the Orange River Free State.

Of the liability of this land to sudden and destructive inundations I have given an illustration in "Hydrology of South Africa" (p. 251), citing the reports given in the journals of the day of the disasters to which Bloomfontein was subjected through a storm of rain in November, 1874. There, where in a previous drought I was told of emaciated horses having been seen hanging about the doors of stores while the floors of these were being swept, returning again and again when driven away, and afterwards tearing and eating old gunny bags and sheets of paper swept to the street, bridges were now washed away by the river, and houses toppled and fell as if built of cards or of clay.

The state is oval in shape, containing an area of about 70,000 square miles. On the eastern side of the state are the Drakenberg and Maluti ranges of mountains. "From this river," it is stated in the hand-book issued by Messrs Silver & Co., "the average height of which is about 6,000 feet, the Free State descends, in more or less gentle slopes and vast plains, towards the Vaal and the Orange—the general level being about 5,000 feet above the sea. The 'flats' of the Free State are characteristic features of the country. Wide stretches of grass land appear to be without bound, but the distant horizon; occasionally there are undulations, and, in some parts, conical hills, the sides of which are covered with large and rounded stones. Very little wood or bush appears anywhere, except along the winding river lines. Great herds of deer graze upon the unfenced lands, and are, at certain parts of the year, to be seen close to the road.

"The air of the Free State is remarkable for its dryness. This arises from the surroundings of the country. On the side towards the Indian Ocean the Drakenberg hems it in, and receives on its lofty eastern slopes the moisture which the easterly winds would otherwise bear to it. The Stormberg range, in the Cape Colony, acts in the same way on the southern side of the plateau; while on the north-west, and on the west, are the arid and sun-burnt wastes of the Kalahari. One result of this environment is the frequency of drought. . . . In winter the cold is sometimes severe, and in summer heavy thunder-storms occasionally occur.

"The Free State is almost surrounded by rivers. The Klip, a little tributary of the Vaal, the Vaal itself, the Orange, and the Caledon, form quite five-sixths of its boundary line. With the exception of the Caledon, which flows into the Orange, the streams which traverse the state run from the Drakenberg and the Malutis on its eastern side into the Vaal on the west and north-west. These are the Wilge, the Valsh, the Sand, the Vet, the Modder, and the Reit. They are not navigable, and at present are but little used in irrigation. In the dry season all of them can be forded, and some of them have then but little water. During the rainy season they are deep and rapid, making travelling difficult, as there is not at present a bridge over any one of them. Reservoirs for saving water are common features of the homesteads, and without them farming operations could not well be carried on. . . .

"The chief articles of production are wool and agricultural produce. The Free State is at present a grazing country, and sheep are herded in flocks on every farm. Each homestead has its orchard and vegetable garden, in which peaches, grapes, melons, oranges, and figs are grown; and enclosed ploughed land, for grain and forage, are common. Cattle are kept for draught, and in the neighbourhood of Harrismith especially horses are bred with success. The territory about the Caledon, conquered from the Basutos in the last war, is good for agriculture."

CHAPTER III.

Griqualand West.

To the west of the Orange River Free State is Griqualand West, including the diamond fields,—a small territory, measuring about 150 miles from east to west, and 100 miles from north to south, bounded on the west by the Langeberg range of mountains, which overlook the southern part of the Kalahari Desert, but bounded on the south by the Orange River from Ramah to Keis, and traversed by the Vaal and its affluents, the Hart and the Modder Rivers.

The district, however, is more remarkable for its valued mineral production—diamonds and other precious stones—than it is for its agricultural capabilities; and it is in view of these, and the development of these, that we are at present engaged in the consideration of the water supply of the country.

CHAPTER IV.

TRANSVAAL TERRITORY.

To the north of Griqualand West, the Orange River Free State, and Natal, and conterminous in part with all of these is the Transvaal Territory, lying to the north of the Vaal River, from which it takes its name, and which, with the Buffels, constitutes the boundary between it and the Orange River Free State. Its northern boundary is the Limpopo, which forms also part of its western frontier line, which again is completed by the Nuoguari River, which runs northward into the Limpopo, and the Hart, which flows southward into the Vaal, and a line between the sources of these rivers, skirting the Kalahari Desert. The area of the Transvaal is estimated to be 120,000 square miles.

"The elevation of the country is from 5,000 to 7,000 feet above the level of the sea, and on this account the climate is very healthy, although the temperature, in some parts especially, is favourable to the production of tropical plants; and the northern districts are above Capricorn. Being well watered with numerous tributaries of the Vaal, the Limpopo, and the sources of the Maputa and Umvolosi, and above the region of frequent droughts, this country is the most beautiful of the interior lands of South Africa. In some parts it is well wooded, almost everywhere it is green, and its mountain forms are picturesque. It is also remarkable for its fruitfulness, and not less so for its mineral wealth.

"No scientific observations have been as yet kept relative to the rainfall. The wet season begins in September and lasts till April, during which months a plentiful supply of rain falls. From May to October, during the winter months, there is little or no rain. . . .

"The mineral wealth of the Transvaal is remarkable. . . .

"The Transvaal is in some parts well wooded, particularly in the districts of Wakkerstrom and Utrecht, also to the north of the Magaliesberg and Drakenberg ranges, and from the 25th degree of south latitude. . . .

"As a grazing country the Transvaal is unsurpassed in South Africa. Sheep, cattle, and horses thrive generally, and certain districts are especially suited to one or other class of live stock. The Cape Colony receives considerable contribution to its cattle market from this favoured country.

"The Transvaal is considered to be the granary of the interior, being rich in corn land. It is also favourable to sugar, coffee, and cotton, as also to almost all tropical fruits."

A descriptive narrative of the productive character of the various districts of the Transvaal was issued some years since by authority of the Transvaal Society for Promoting Agriculture, Stock Breeding, and Industry—a society of which the President of the State was patron. From this the following information has been obtained :—

"The district of Potchefstrom—which name is a strange compound of the names Potgieter, Sherf, and Stockenstrom, the names of three men held in high estimation by the Boers—possesses a soil eminently suited to the growth of all cereals, of which two crops may be obtained in the year, the culture of tobacco, indigo, vegetables of all kinds, vine, fruit, and forest trees that are grown in a temperate and semi-tropical climate; besides which, the extensive pasture lands of good quality offer the best facilities for the breeeding of horned cattle, sheep, and Angora goats; most of the farms are likewise well adapted for ostrich farming and silk culture.

"The beautiful Mooi River takes its source about forty miles from the town, where several mills are already erected. This, and the fact that eight more are situated at the north entrance to the town that are either already at work or in course of construction, proves that the water power the town possesses in this magnificent stream can be turned to very great advantage by its enterprising citizens.

"The suitability of the soil of the Pretoria district for agricultural purposes and stock farming is very similar to that of Potchefstrom, with this advantage, however, that the northern part of this district, by enjoying a warmer climate, can produce, besides the different cereals, tobacco, indigo, and the orange tree, also coffee, the sugar cane, cotton, and the different kinds of tropical and semi-tropical products; added to which, various minerals have been found in the Wit-water-strand, and very good coal on the high table-lands."

Of the district of Rustenburg it is said :—"This district, having its chief town of the same name, may in truth be called 'The Garden of the Republic;' besides all the products of a temperate climate, those of a tropical nature can be cultivated to very great advantage in some parts of this district. The breeding of horned cattle answers admirably, but all parts of the district are not suitable for the breeding of small stock, excepting goats; Angora goats are most likely to answer well. . . .

"The district of Lydenburg [the site of the gold fields] has ever been acknowledged to be one of the best districts for the growth of wheat and for the breeding both of horned cattle and of small stock. . . . The country to the north, and the north-east of the gold fields, is eminently adapted for the growth of coffee, the sugar-cane, cotton, and other products of the same nature.

"Besides gold, this district is rich in other minerals; but it is especially noted for its coal beds."

The district of Marico is spoken of "as well suited for agriculture and the breeding of stock, for the culture of tobacco and indigo, all kinds of vegetables, the vine, orange and fruit trees, wheat, mealies, oats, barley, and other cereals; the breeding of large and small stock likewise holds out strong inducement, as being very profitable."

Christiana, Heidelberg, Wakkerstroom, and Utrecht are said to be "well suited for the breeding of cattle and sheep, the last three districts producing likewise large quantities of grain. Wakkerstrom is especially suited for the breeding of horses; and supplies likewise the different markets of the Republic with building timber; at Christiana several diamonds have been found. Coal is plentiful in the Wakkerstrom and Utrecht districts.

"Waterbergen and Zoutspansbergen, besides being large grain producing districts, are fitted for the growth of a variety of other produce of various kinds, such as tobacco, indigo, fruit, and vegetables, and in several parts coffee. The sugar cane and cotton also thrive remarkably well. . . .

"These districts," it is added, "are, without doubt, rich in minerals and various kinds of valuable woods, and the farms there situated, being on the whole well provided with timber, and having an ample supply of water, have a fine future before them, and offer, to the man of small means endowed with energy and industry, a guarantee of success, whether he be an agriculturist or stock farmer.

"The district of Nazareth formerly formed part of the district of Lydenberg, and possesses most of the properties of that district in regard to agriculture and stock farming, as well as soil."

Of the appearance of the country generally the following account was given some years ago in the *Transvaal Argus*:—"The scenery of the Transvaal Republic is generally of a very high description, especially from Marico and Rustenburg up to Waterberg, Lydenberg, and Zoutpansberg. It is a succession of undulating valleys and high mountains, watered by beautiful rivers, fountains, and rills, thickly interspersed with forests of thorn, mimosa, and acacia trees. The white houses of the Boers peeping through the trees; the herds of cattle, sheep, and goats peacefully grazing on the green lawn; the waving cornfields, the luxuriant grass, with the clear blue sky above, forms a panorama of great beauty; while the groups of sprinkbucks and antelopes bounding past, and immense herds of zebras and gnus, delight the huntsman, who can range here at will unchecked by game-laws."

And along with other information there was supplied the following:—"The territory of this State is situated between Lat. 22° 30′ and 28° S., and Long 24° and 29° 30′ E. The high range of the Drakensberg mountains divides it on the east from the Colony of Natal and the different Zulu Kaffir tribes to the north of that Colony; the Vaal River forms its southern boundary, separating it from the Orange River Free State; while the Limpopo, with its several tributaries, and the Harts River close it in on the north and west, dividing it from the countries at present occupied by the powerful tribes of Moselekatze, Sechomo, Sicheli, and Mahura. Towards the east, north, and west the Transvaal Government acknowledges no particular boundary, as in taking possession of the country the Boers have reserved their right to the vast territories situated in the directions mentioned. They have lately taken possession of a great part of land on the south-west taken from Mahura and other Kaffir chiefs, which is not included in the boundary line marked down in Hall's Map of South Africa. Besides this, the Transvaal Government is negotiating with Cetchwayo and Panda about the purchase of a strip of country to the north of Natal, with a seaport at the mouth of the Umhlatoozi River.

"The extent of the country might be estimated at about 80,000 square miles, which would be about half the size of the Cape Colony, and more than four times that of Natal.

"The soil of the Transvaal Republic is exceedingly fertile, consisting of loam, clay, and sand, and is abundantly watered by the numerous rivers and streams, which traverse the country in all directions. On every side the eye meets extensive pastures and meadows diversified by gaily coloured flowers, brushwood and groups of mimosa trees, animated by the various kinds of game common to South Africa, and by birds of the brightest plumage. Passing along the road, large herds of cattle, sheep, and goats attract the eye, attended by Kaffir herdsmen; now and then groups of plain, straw-roofed Boer houses and Kaffir huts or kraals present themselves, surrounded by small but luxuriant cornfields, and by hedges of pomegranate, figs, or cactus; while before the door, surrounded by groups of playing children, poultry, dogs, etc., are seated the old-fashioned, patriarchal owners of the farm, who are always ready to offer a hearty welcome and the shelter of their humble roof to the stranger.

"The climate of the Transvaal Republic is similar to that of the south of Europe. On account of the height of the country (about 5,000 feet above the level of the sea), the heat is much less oppressive than in most parts of Natal and in some districts of the Cape Colony. The summer is the rainy season, the winter dry and cold at night. The number of old people, and the great increase of population proves the climate to be remarkably healthy, and especially suitable for persons afflicted with pulmonary affections. Since the year 1852, when a great number of people died of dysentery and fever, there have been no epidemic diseases; but in those parts of the country which are most thinly inhabited people often die for want of proper medical attendance."

While I was in the Colony there appeared in the *Eastern Province Herald* an account of the Republic written by Mr Charles Jones, of the firm of Messrs Dunell Ebden & Co., Port Elizabeth, in which he says:—"The districts I visited were Potchefstroom, on the Mooi River, and Rustenburg, under the Magaliesberg. In these two districts sheep have not thriven well during the last few years, heavy rains having two or three years since carried off a large proportion of the stock, and farmers pay more attention to the breeding of horned cattle, for which the veld is better adapted. In the immediate neighbourhood of the Vaal River, and in the Utrecht and Wakkerstroom districts, the veld is, I believe, more suited for sheep-walks, and if the sheep-farming company recently formed in Glasgow can arrange with the Transvaal Government for sufficient land to carry on operations on an extensive scale, we shall soon hear of sheep-farming being carried on in that fine country on Australian principles, and of the introduction of many Scotch families of immigrants to work the farms, which will be the first grand step towards the development of the vast resources of the Republic.

"The great want of the country is exportable produce to a much larger extent than is produced at present. Sheep, as I have stated, do not thrive in all the districts. Corn, from the great facilities for irrigation, can be produced in great abundance, but the farmers (indolent as they are in this Colony) cultivate their lands to a comparatively small extent only. . . .

"I have rarely if ever seen any part of this Colony, or of the Orange Free State, more thickly studded with farm home-steads than some parts immediately under the Magaliesberg, which is a proof of the fertility and productiveness of the country."

The territory of the Transvaal is described as a land abounding in springs and running streams; and in the neighbourhood of Magaliesberg, in the centre of the state, abundance of rain falls. There, according to Baldwin, wells might be sunk with advantage.

In a preceding chapter much information has been supplied in regard to subterranean rivers in limestone strata [ante pp. 135–160], and in connection therewith mention is made of a subterranean river flowing into the Mooi River from beyond Wonder-Fontein, through which it passes [ante p. 134].

In my report as Colonial Botanist for 1866 it is stated :—"When the recommendation of the Select Committee on Retrenchment, appointed by the Honourable the House of Assembly, that the office I hold should be abolished, was adopted by the House, I was engaged in correspondence relative to a journey towards the Limpopo, to be undertaken in like manner, at my own expense, in the beginning of the year, provided His Excellency the Governor should approve my going so far beyond the boundary of the Colony. This journey would have taken me through the Karroo, the country intermediate between that and the Orange River, the Orange River Free State, the State of the Transvaal Republic to the advanced northern limit in lat. 22° S., and either through the Colony of Natal or through the country of Moselekatze and the district of the Kuruman on my return to the Colony.

"The proposed object of this journey was to obtain verifications or corrections of observations and conclusions embodied in the memoranda prepared on the Hydrology of Southern Africa, on the forest and forest lands of Southern Africa, and on the natural history of South African trees, before these should be published, and at the same time to secure a more extensive observation of South African flora. But the resolution of the Assembly to abolish, at the close of the year, the office of Colonial Botanist, rendered it necessary for me to abandon at once all thoughts of prosecuting this enterprise."

In the discussion of the primary causes of the desiccation of South Africa, in my volume on the "Hydrology of South Africa," it is stated in connection with a reference to the subterranean current from beyond Wonder-Fontein in the Gartzrand to the Mooi River:—"All fountains, indeed, may be considered outlets of subterraneous water-courses, of greater or lesser size, carrying away water from a higher level, and some of them must be of considerable magnitude. Such is the Kuruman, described by Mr Backhouse, at its source, as 'a stream, which would turn a mill, issuing from under sand rocks.' It is a clear permanent stream, but one which, unlike most, is largest at its source, and ceases about ten miles to the north-west, disappearing and reappearing at intervals, and at length terminating in a dry water-course, which is only filled in rainy weather. It is thus described by Dr Moffat:—

"'It issues from caverns in a little hill, which is composed of blue and grey limestone, mixed with considerable quantities of flint, but not in nodules as found in beds of chalk. From the appearance of the caves, and the irregularity of the strata, one might be led to suppose they have been the results of internal convulsions. The water, which is pure and wholesome, is rather calcareous. It is evident that its source must be at a very great distance, as all the rains which fall on the hills and plains for forty miles round, in one year, could not possibly supply such a stream for one month. Although there are no sandstone formations nearer than thirty miles, great quantities of exceedingly fine sand come from it; it appears to boil up out of the smaller springs in front of the larger, and it is to be found in deposit in the bed of the river for miles distant. The substratum of the whole of the country, as far as the Orange River, is compact limestone, which in some of the Hamhana Hills rises considerably above the neighbouring plain; but these only form the basis of argillaceous hills and iron-schist, on the top of which the compass moves at random, or according to the position in which it is placed. The strata of these schistose formations are often found to bend and curve into all shapes, frequently exhibiting an appearance of golden asbestos, but extremely hard. The common blue asbestos is to be found at Gamaperi, in the neighbourhood, the same as that found near the Orange River. The limestone extends to Old Lithako, where there are hills of basalt and primitive limestone; among which masses of serpentine rock of various colours, usually called pipe-stone, are to be met with. Beyond the Batlapi dominions, towards the Molapo, there is abundance of granite, green stone, etc., while the limestone foundation, towards the west, terminates among the sandy wilds of the southern Sahara. Fountains, throughout the whole extent of the limestone basin, are precarious, and independent of causes described in a preceding chapter; nor does that of the Kuruman continue to send forth the torrents it once did.'

"The mention of limestone, and the mention of its being so abundant and extensively diffused over the district, is in accordance with the supposition that there may be there numerous and extensive lines of caves, and in accordance with the fact of there being there this subterranean river, which only from its issue is called the Kuruman; and the Kuruman is only one of such.

"There is, I am informed, a large plateau in the latitude of Delagoa Bay, 26° S., and extending from 28° to 30° E., long., from which the river systems of the Limpopo to the north, and of the Orange River on the south, take their rise. It may be about 7,000 feet above the level of the sea. On this plateau, if I have understood my informant correctly, there are what can scarcely be called springs—they surpass so much in magnitude what are generally so designated—but streams, issuing like Minerva from the head of her father, in the full perfection of strength; not springs giving birth to rills and streamlets, spending their infancy and childhood in play amongst the hills, glancing or glinting in the sunshine, tumbling over stones, and playing such like childish freaks,—but streams in all the vigour of early manhood, like that of the Kuruman which has just been described. There are more than thirty such earth-born streams to the north of this plateau, and upwards of twenty to the south, but they do not all die in their bed like the Kuruman, some of them commingling themselves, and unitedly draining the valleys of the Limpopo and Gariep; and besides these there are many of a smaller size.

"I am further informed that there are here and there scattered over the plateau hollows with somewhat precipitous sides, it may be 100 feet deep and from 100 to 200 feet across, which may have been formed by a swirl of water drawn off into a subterranean channel when the plateau was a water-bed, or may have been formed by a subsidence of superficial strata roofing a cavern, which I consider more likely to have been the case, but which, in either case, are indicative of subterranean hollows, which might serve as reservoirs or water-courses.

"I may not ignore the existence of subterranean cavities of great extent into which water may escape by percolation or by rents in the rock, but I attach no importance to these as by themselves affecting the withdrawal of water from the surface of the land. I look upon them as bearing the same relation to subterranean rivers which lakes do to the rivers which flow through them, and estimate the drainage by the flow at the exit, not by the contents of the reservoir.

"Besides these, the numerous hot springs—some of them such as those at Brand Vley near Worcester, at Caledon, at Barmoral near Uitenhage, and elsewhere, of considerable magnitude—and the water found, as it used to be found, by the divining rod [?], all tell of the draining off of water from higher levels by subterranean water-courses."

Beside the study of these, I had another object in view. In the volume cited it is stated:—"I was informed by the Rev. Chas. Murray that, in travelling from the Cape Colony to Potcherfstrom, or Mooi-Rivers-Dorp, he observed many trees of Olivienhout *(Olea verrucosa*, Link), growing in remarkable quadrangular patches on the eastern slopes of hills, and on rising grounds. The trees in some of these he estimated to be of about 12 feet in girth. They were large and old trees. He remarked that in general the trees of each patch were apparently of uniform magnitude, but that the magnitude of the trees in different patches varied considerably. The statement may appear vague; I give it as I received it; it was the statement of an intelligent traveller through the district, observing what he saw, unbiased by any foregone conclusion; and it was these observations, communicated to me by Mr Murray, which first suggested to me the views I have just advanced. These views subsequently received confirmation from my learning that numerous observations, similar to those mentioned to me by Mr Murray, had been made by the late Mr James Chapman, in journeys made by him far beyond the Colony, and noted by him, though he found nothing significant in the fact until we had together talked over the matter; so that he also was unbiased in the observations made by him.

"My conjecture is that the trees observed by Mr Murray commenced their growth when the slopes and rising grounds on which they stand were a little above the level of the waters covering the ground below—the seed having germinated before the land on which they grow had become anything like so arid as it now is. And supposing such a position and condition of soil to have been characteristic of the land on which patches differing in magnitude are found, at different periods, separated it may be by centuries, it seems to follow that the time which has elapsed between these different periods and the present might be ascertained, or determined approximately, by counting the number of concentric rings in the trunks of the trees in different patches; while the comparative thickness of rings in the same, or in trees felled in different patches, and the order in which these appear, might supply data for the solution of other questions connected with the progress of desiccation; and this, even though it should turn out that these trees are only remains, and at the same time the only remains, of more extensive forests of which the other trees have perished by fire or otherwise.

"Mr Chapman had travelled extensively in the interior of the country beyond the Colony; he had travelled from Natal to the Zambesi, from the Zambesi to Walvisch Bay, and hither and thither in many directions in the country between. From him I received much information on subjects embraced by Natural History which was afterwards verified by the observations of others; while I never had information communicated by him subsequently disproved.

"By him I was informed that to the west of Lake Gnami are beds of limestone, on which grow a great many Motjcharra *(Combretum ?)* and Omboroomboongo *(Acacia)* trees, the former in clumps, and the latter around vleies or other spots where there is a moist soil. On intermediate ground is found in abundance the Haakdorn *(Acacia detinens)* of a comparatively later growth. On elevations which appear to have been at one time islands are clumps of eight or ten baobab trees *(Adansonia digitata)* with trunks averaging 60 feet in circumference, with bark nearly a foot in thickness, as is seen when, in accordance with the usuage of the nation, from time to time immemorable, vertical stripes have been cut off for fibres to be employed in the manufacture of bags and baskets; and over the whole district are scattered thickly trees of the Sweet Gum Acacia.

"To the westward of Twass is a sandy wilderness covered with trees of no great height, but amongst these appear here and there fine groups or small forests of Kameel doorn *(Acacia giraffaea)*. Near Elephant's Fountain these forests of Kameel doorn are very extensive, and the trees are of considerable age. And the Karroo doorn *(Acacia horida)* is conspicuous in the valleys and in the vicinity of vleies. To the west of Elephant's Fountain are here and there large trees of the same kind, and forests of smaller ones varying greatly in bulk and apparently in age; but most or all of the trees in each clump, or patch, or forest are apparently all of the same age, suggestive of either a succession of periods of drainage, in accordance with the views now advanced, or the recurrence at lengthened intervals of seasons favourable to the germination of seed—there being neither young trees nor solitary trees of intermediate ages.

"In the course of his journey Mr Chapman travelled some way up the banks of the Shua, a large periodic stream flowing westward from the country of Moselekatzi. In the country through which that river flows he found large baobabs *(Adansonia digitata)* growing in the vicinity of springs and limestone rocks, and mopane *(Bauhinia)* forests growing on level ground which looked like hardened mud containing a good deal of limestone, but near the river were only grassy plains, intersected by what appeared to have been river beds, and these were covered with an efflorescence of salt.

"It is conjectured that these baobabs grew on islands of limestone rocks rising above the waters of the lake, while these as yet filled the basin; and that at a later period the mopane trees sprung up and grew while the waters were receding, and covered only the space now appearing as grassy plains; and that these stood above the level of the waters while they still filled, and afterwards only partially filled, the river beds; and that at a later period still did the waters left in them by the disturbance of level evaporate and leave the efflorescence of salt.

"By sections of these baobabs and mopane trees interesting chronological data might be obtained. Nor is it the times of the more marked eras alone which might thus be determined; for, according to Mr Chapman's observations, 'The mopane trees are small in the lower portion of the Shua valley,'—which is that part which would be longest under water—'but they are longer and stronger the higher that valley is ascended,'—in proportion, that is, as the soil has been longer free from the covering of water—'and very much longer in the vicinity of the River Naté, which comes in a direction from the town of Moselekatzi, where they have attained a considerable magnitude;' and sections of trees at different elevations might reveal approximately the respective periods at which the spot became fitted, by the draining off of the waters, for the production of these trees.

"Of the Madénisana and Kaleghari deserts, he said—'There are here continuous forests, but more frequently the trees are in clumps or patches, consisting sometimes of one kind of tree and sometimes of several different kinds of these, with occasionally solitary trees of gigantic growth towering above the others or standing in solitary grandeur. Some kinds are found covering mountain slopes, others are found on sandy plains, others near to rivers or river beds, and others in the neighbourhood of leeghties of greater or less extent; some in all of these situations, *but decreasing in number or in magnitude, and perhaps increasing in one of these particulars while decreasing in another, according as they grow more near to one or other of the situations referred to.*' These may supply in a similar way chronological data which might prove important in such an investigation.

"'To the south of Tsamasechie,' he said, 'there are found springs and fountains in the sand, and in some places immense ripple-like parallel elevations thirty or forty feet high. On the tops of these are generally large trees growing, the most prevalent being the Kushé *(Milletia Caffra)* and the Shashanga, while in the level plateau of sand between them, measuring three, four, or five miles in breadth, are forests of the Magonane *(Grewia cana?)* and in the valleys and depressions, in the soil of which water is found near the surface, grows the Makow.

"'At Mottomoganyani, to the west of this, are beds of limestone like consolidated mud. To the east is Moselekatzi's country. It is very level, and abounds in salt-pans and salt-vleies, especially to the south, where they seem to represent the beds of ancient lakes. In the north are forests of mopane trees, but the most characteristic arborescent feature of the country is baobabs growing on islands in these salt-pans, and around the fountains, at a distance of about two hundred feet, and slightly above the water's level.'

"And in connection with the same thing he mentioned, 'that to the north-eastward of this are incipient forests of very young mopane trees. No large trees, however, are there to be found.'

"These mopane trees in other districts are found of a great magnitude. The youth of the forests found here I attribute to the comparatively recent time at which the ground had become suitable for the germination of the seed and the growth of the trees; and in the drying up of the vley, within the memory of the inhabitants of the district, I see an indication of the natural operation whereby this preparation of the soil was effected. A section of the trunk of one of these mopane trees would reveal, by the number of concentric rings in its substance, how long it has been growing there; and from similar data much might be collected in regard to the hydrography of the district.

"If this be admitted, as I doubt not it will, and that at once, a ponderous volume of hydrologic records is at our command."

And I hoped, had I been allowed to visit the district in question—which would have cost the Colony nothing, beyond the amount of my small salary for the time occupied in the journey—to have had obtained, by personal observation, along with a more extended acquaintance with the flora of South Africa, and of the forests and forest lands of that region, much valuable information in regard to the past and the present hydrographic condition of the country.

Of the aridity beyond the colonised portion of South Africa, illustrations supplied by accounts given by Dr Livingstone of his experience at Kolobeng, by the sufferings of Mr Helmore and his family to the north of Kamakama, and by accounts given by Mr Mackenzie of the same region, are given in the volume on "Hydrology of South Africa" (pp. 216–223); and illustrations of the water supply in the region beyond the colonised portion of South Africa are given in the same volume (pp. 223–226) in accounts of the country given by Bakwains to Livingstone, Oswell, and Murray, in notices of the Congo, of the plains of the Lobele, and of the plains of Lobee.

From Dr Livingstone we learn that even in well watered lands in the interior droughts occur, and that with something like regular periodicity. In regard to such periodical droughts, he writes:—"It would not be fair, while giving the results of one's inquiries, to keep out of view one serious drawback, which we believe is characteristic of every part of Central Africa. Periodical droughts must be expected. If a rainy zone exists under the equator, that is the only exception known. These droughts are always partial, but may prevail over areas of from one to three hundred miles in extent. Our inquiries led us to believe that from 10° to 15° south they may be looked for once in every ten or fifteen years, and from 15° to 20° south once in every five years. What the cause of these may be, we cannot tell; but lack of vegetation cannot be assigned as any reason either for their occurrence or greater frequency now than at any former period. The hills are covered with trees and grass to their summits. The valleys are often encumbered with profuse and rank vegetation; but suddenly, and without any warning, the years of plenty are succeeded by one in which there is neither earing nor harvest. A shower has fallen on one spot a mile square; there the grass has sprung up, but has died off again. The rest of the country is parched and burnt; the grass of the preceding year, which may have escaped the annual fires, is discoloured, and crumbles into powder in the hand; and the leaves of the trees, though alive, look withered. One who had seen the landscape in all its glorious freshness and verdure after rains, could scarcely believe that the brown and dusty world before him was ever green."

It may be remarked that such periodical occurrence of a drought does not prove a diminished supply or transport of watery treasure from the sea any more than does the alternation of seasons of sunshine with seasons of rain in the same year; and on this point he supplies the following interesting information, in regard to rainfall in the district of Lake Nyassa:—"When the sun is vertical over any part in the tropics on his way south the first rains begin to fall, and the effect of these, though copious, is usually only to fill the bogs and pools. When on his way north he again comes over the same part we have the great rains of the year, and the pools and bogs, being already filled, over-flow and produce the great floods which mark the Zambesi, and probably in the same manner cause the inundations of the Nile. The luxuriant vegetation, which the partial desiccation of many of these rivers annually allows to grow, protects their bottom and banks from abrasion, and hence the comparative clearness of their water in the greater floods.

"The first or lesser rains take place in this region in November, when the sun is vertical, going south. The greater rains fall in January, February, and March, when he is on his way back to the equator. Supposing our observations of the cause of inundating floods in south inter-tropical Africa to be applicable to the north inter-tropical district," all this can be accounted for satisfactorily, in accordance with the meteorological observations embodied in Part I of this volume; and the droughts spoken of by him may have followed a slight variation in the altitude of either the north-going or of the south-going atmospheric currents, in whatever way this may have been produced. That such disturbances do occur may be conjectured, if not proved, by meteorological observations in Europe.

CHAPTER V.

Zululand and Inter-Colonial Region of Nomansland.

To the east of the Transvaal Territory, and the north-east of the Colony of Natal, is the land of the Amazulus and other Kaffir tribes, in which, in latitude 26°, is Delegoa Bay and the Portuguese settlement of Lorenzo Marques, from which a railway to the Transvaal, passing through the Lydenberg gold fields, has been projected.

Of a region called Nomansland, which lies in the southern portion of this district, adjacent to the Colony of Natal, a correspondent of *The Colonies* gives the following account:—

"A desolate region without a single human inhabitant, black or white—a solitude given over to wild beasts, and birds, and serpents, where the waters team with crocodiles and hippopotami, and the hillsides resound nightly with the roar of the lion, the prolonged howl of the jackal, and the fiendish laugh of the hyena—such a region may fairly be designated as a howling wilderness; and such a region exists in the neutral ground, or "No Man's Land," that divides the British Colony of Natal from the territory of Cetewaye, the King of the Amazulu Kaffirs. The northern boundary of Natal is the Tugala River, one of the most important of the many streams that have their rise in the western mountains, and, running for the most part parallel to each other, empty themselves into the Indian Ocean. Between the Tugala and the southern limit of Zululand, which is marked by a stream bearing the euphonious name of the Umslatuzana, lies the howling wilderness I allude to, a belt of uninhabited country about seventy miles wide, which it takes on horseback at least two good days to cross, as there are neither roads nor bridges, nothing but the rude tracks of the hunters' waggons that here and there confuse the traveller by appearing and disappearing in a most perplexing manner. Across this rude country I once journeyed on horseback, experiencing similar sensation to those so beautifully depicted by Pringle, the South African poet, in his lines commencing—

'Afar in the desert I love to ride,
With the silent bush-boy alone by my side;'

the only difference being that my 'silent bush-boy,' instead of being an almond-eyed, flat-nosed, high-cheek-boned Hottentot, was an intelligent young Englishman, who spoke both the Dutch and Kaffir tongues—was a very Nimrod indeed, and certainly not silent. The sun had set over the western mountains, looking red as blood, as seen through the smoky atmosphere caused by distant fires; and darkness came upon us long before we reached the banks of the broad Tugala, where we intended to pass the night at a village, or 'kraal,' belonging to a once powerful chief, named Gudu, who, having displeased the king, fled with his people from the Zulu country, and became a refugee, under British protection, on the Natal side of the river. In the short African twilight we pushed on as fast as we could, and shortly entered a long narrow glen, through which we groped our way amidst rocks and stones. The herbage on either side was sparkling with myriads of glow-worms in every direction, that gave the ravine a fairy-like appearance as they lit up the foliage with their tiny pale green lamps. 'A fit spot for Oberon and Titania,' said I. 'A likely place to meet a lion,' rejoined my companion. At last we reached Gudu's village, made manifest in the darkness by the barking of a multitude of dogs that rushed out to meet us, followed by all the inhabitants, men, women, and children. Gudu, a Kaffir of Herculean proportions, whose right arm had been bitten off by a crocodile, received us with cordial hospitality; the women bringing us 'hemazi,' or thick curdled milk, which we ate with enormous wooden spoons about three feet long. Whilst thus engaged, the people crowded around us, gazing at us by the light of ignited straws, which they held in their hands. We passed the night in a hut exactly resembling a flattened bee-hive, the only orifice being an arched doorway about a foot high, into which we had to wriggle the best way we could on our hands and knees.

"Next day we rested at the Tugala, and Gudu took us to visit another refugee chief, whose name was Mankazana, who resided at a neighbouring village. This Mankazana had, like Gudu, given offence to the king, and fled over the border. He had been sent as a spy into Natal, and on his return he reported the English as being very strong, for which 'falsehood' he was condemned to be buried alive; but not relishing such a termination to his career, he managed to make his escape, two only of his sixty wives being successful in following him. On our return we skirted the river side, where magnificent baobab trees spread their shade around. We observed many crocodiles basking on the sand-banks in the river, looking, at a distance, like motionless logs of drift timber, till, on hearing us shout, they rolled lazily into the water. Gudu pointed out to us the spot where he lost his arm. He was swimming the river, at a point where the flood was deep and the current strong, when a huge crocodile rose and seized him by the arm. He grappled with his foe, and, after a desperate struggle, succeeded in making his escape, leaving his limb in the maw of the crocodile, whilst the eddying waters around were red with his blood. In the evening a grand carousal was held in the village, copious libations of 'outchualla,' or beer made from millet, being handed round in honour of our visit. That night, the deafening noise of the crickets singing in the thatch of our hut, combined with the unpleasant sensation of a clammy gecko lizard falling at intervals on our faces, just as dreams of home were stealing over us, rendered our repose anything but refreshing.

"Early next morning we took leave of Gudu, and attempted to cross the river at a ford below the village. Getting too much to the left, our horses suddenly went down, and we were nearly lost in a dangerous and dismal quicksand. The sensation was horrible; a couple more steps forward and we should have disappeared altogether. Somehow we fortunately managed to back out of it, and regained a more solid footing. Here we remained, not knowing which way to proceed, until some Kaffir women, who were at work in an adjoining plantation near the river side, good-naturedly tucked up their leather skirts and piloted us safely over, making an angular course to the opposite side. We were now in the Neutral Ground, which we hoped to cross in a couple of days. Our prospects were anything but agreeable, as the sky became overcast, and the rain commenced to descend in torrents, driven by a strong wind from the Indian Ocean. Presently we descried a solitary human form rapidly approaching in the distance. Who could it be that had thus crossed the lonely wilderness, and on foot? It proved to be a fine, stout young women, with a little boy on her shoulders, running as fast as her legs could carry her towards the Natal frontier. She, too, was escaping from the vengeance of the king, to whom she had been denounced as a sorceress, and had braved the dangers of the wild beasts at night, and the crocodiles in crossing the streams, to save her otherwise forfeited life. We gave her some food, and bade her push on to Gudu's village, where she might rest calmly and defy her pursuers.

"As the day wore on, the rain, fortunately for us, abated, and the sky again became clear and blue. We had been anticipating a dismal night in the bush should the rain continue, with the prospect of its extinguishing the fires we must necessarily make round our halting-place to keep off the wild beasts. By-and-bye we saw a herd of elands quietly grazing at a distance on the open undulating downs over which we were now journeying. These enormous cream-coloured antelopes, with their horns bending backwards over their shoulders, present a magnificent appearance when seen together in a flock in their native freedom. Guinea-fowl and bustards were numerous, and my companion shot one of the latter, of a kind called 'koran' by the hunters, which he slung to his saddle and carried on for supper. Entering a belt of thickets near a ravine, we were startled by seeing several small antelopes rushing swiftly and frantically past us—when lo! bounding after them was a noble lion, which, with tail erect in the air, pursued the terrified 'riet-boks,' without deigning to notice us as we paused, motionless with astonishment at so sudden an introduction to the 'king of beasts.' Within a few moments we saw three more, two of them apparently lionesses, bounding leisurely away to the brow of a neighbouring hill, over which they soon disappeared in the bushes beyond. I began to think that we had now entered the region of lions in earnest, and that there were too many of them to be either safe or pleasant; however, my friend re-assured me by saying that there was no danger of their attacking us while they were in pursuit of the antelopes.

"Pushing on as fast as we could, we arrived about sun-down at a deserted hut on the side of a hill, formerly built by elephant hunters, but now in a state of ruin and decay. Finding it full of fleas, we soon beat a retreat, and encamped under a tree outside; tying up our horses to the poles of the hut as soon as it was quite dark, it not being safe to let them graze any longer for fear of the wild beasts that would soon be prowling about. Having meanwhile collected all the firewood we could procure, we lit our fires, taking it in turns to keep them up during the night. We had filled our tin pannikins with water from an adjoining brook, and soon made ourselves some tea. We plucked the 'koran,' and cooked it, the ramrod of a gun doing service as a spit; and after supper I took the first watch, my comrade soon falling off to sleep. Tired as I was, I could not sleep when my turn came; the constant roaring of the lions, reverberating like distant thunder amongst the lonely hills, combined with the melancholy howling of the hyænas and jackals, some of which came very close to us, kept me wide awake. The first pale yellow streak in the eastern horizon was a welcome signal that the day was approaching, and when it was fairly dawn, and the wild beasts had retired to their hiding-places, we turned out our horses to graze, and prepared a hasty breakfast. I went down to the creek to fill the pannikins with water, at a clear spot where there was a wide margin of smooth sand. On looking down I saw the fresh footprints of the lions that had been drinking there only a few hours previously. Breakfast over, we again made a start, and had soon left our mournful hut and smouldering fires far behind, hoping by nightfall to reach Mathlapi's kraal on the Zulu boundary. All day it blew a hurricane, so that we could scarcely keep our saddles. About noon we arrived at the Umlalazi river, a black-looking, sluggish stream, about 120 yards wide; fortunately the tide was out, so that we were able to wade our horses across, the water only reaching to the saddle-flaps. Had the tide been up, we should have been compelled to swim the horses, a dangerous alternative, as the river swarms with crocodiles, several of which we could see lying on the uncovered sand-banks lower down. As it was, the shallow muddy waters were driven into short rolling waves by the violence of the wind, and we got pretty well drenched with the spray before we reached the opposite side of the river, where we had to struggle up the steep muddy bank and force our way through a belt of tangled canes, some twenty feet high, which were completely festooned with a large blue convolvulus, now in full blossom.

"On ascending a hill beyond we obtained a grand view of the surrounding country. To our right lay the mouth of the Umlalazi, with the blue Indian Ocean rolling in majestic waves upon the beach beyond. Northwards, far as the eye could reach, extended vast open plains, belted at intervals with patches of forest. Further inland rose the abrupt chain of the Engooi-Mountains, and to the north-west extended a chaos of hills and peaks displaying the most fantastic and picturesque outline imaginable. Rounding the border of a sedgy lake, we came upon the footprints of hippopotami in all directions in the muddy clay, and from the next hill we looked over the lake and there saw the huge beasts themselves, some eight or ten in number, playing about, and rising with their ponderous jaws above the surface. Graceful palm-trees grew along the borders of the water-courses, and clumps of the *Strelitzia*, looking like gigantic bananas, were scattered about on the sides of the hills.

"Late in the afternoon we reached the banks of the Umslatuzana, and soon descried a small Zulu village on the opposite shore. As we drew nearer we heard the welcome sound of the lowing of cattle being driven homewards, and saw dusky forms moving about the huts. Here we obtained some milk and millet cakes, and then passed on to another village, where we were met by a tall gentlemanly-looking Kaffir, who invited us to partake of his hospitality for the night. It being too late to go on to Mathlapi's kraal, whither we were bound, we accepted the invitation, and soon discovered, from the din and laughter we heard going on inside, that beer-drinking and merriment were the order of the evening. Tired as we were after our journey across the 'howling wilderness,' there was no help for it but to join in the festivities, which were kept up till a late hour of the night. Next day we reached Mathlapi's kraal."

CHAPTER VI.

NATAL.

To the south of Nomansland, by which it is separated from the country of the Amazulus, is the Colony of Natal—a diamond-shaped territory, with a sea-board of about 150 miles and an area of 17,000 square miles (11,000,000 acres). In the "Handbook of South Africa" it is described "as bounded on the south-east by the Indian Ocean; on the north-east by the rivers Buffels and Tugala, which divide it from the Transvaal Republic and Zululand; on the south-west by the Drakensberg Mountains, the upper waters of the Umzimkulu and the Umtamfuna, which separates it from the Basutoland and Nomansland districts of the Cape Colony and Pandoland; and on the north-west by the Drakensberg, which divides it from the Free State.

"The Colony lies wholly between the eastern rim of the great interior tableland of South Africa and the Indian Ocean. At the Drakenbergen the vast plateau is arrested abruptly, and by a descent, at first almost precipitous, then in rugged steps and now in gentle slopes, the land inclines to the ocean shore. Between the rim and the strand there is an average distance of 130 miles. Travelling inland from the coast, the edge of the tableland has the appearance of a lofty mountain range, the highest point being no less than 10,000 feet high. This is the Mont aux Sources, so called because it is a most prolific watershed. Another point on the ledge, as the Drakenbergen really is, known as the Champagne Castle, is 9,500, while the Giant's Castle is 9,000 feet above the level of the sea. The aspects of the great precipice along its whole length are grand and romantic; and as the land at its foot does not subside to the sea by easy levels but by broken steps, tumbled hills, and sweeping undulations, Natal is everywhere picturesque in its land forms. The region on the right of the road from Durban to Maritzburg, after Pinetown has been passed, is remarkable for its fantastic assemblage of sugarloaf hills—sugarloaves with their tops cut off. The midland districts have in many parts the look of downs—rolling sweeps of grass. The coast lands are singularly beautiful, with their rounded bosses, rich in bush and glade; while the shore presents a bold outline with projecting bluffs thickly covered with jungle, and long stretches of land broken by rocky floors and reefs, on which the majestic surf of the Indian Ocean perpetually breaks. Amongst the rivers which skirt or pass through the Colony are the Tugela, the Umvoti, the Umgeni, the Ilovo, the Umkumanzi, the Umzimkulu, and the Umtamfuna. Not one of these is navigable. Some of them, however, especially in the rainy season, are considerable sfreams, and all of them have their tributaries, so that the land abounds in water-courses. Cataracts are numerous, and the Umgeni Falls, ten or twelve miles north of Maritzburg, are famous for their beauty. Granite, trap, and sandstone underlie the beds of shale and vegetable soil which form the land surfaces; and here and there those rocks, especially trap, show themselves in bare and eccentric forms. Table-mountains frequently appear, and one of the finest objects to be seen from Maritzburg is Taffel Berg, a splendid specimen of its class, about sixteen miles from the city.

"The climate of Natal is one of the boasts of its inhabitants. Nearer the tropics than the Cape, its mean annual temperature is but little in excess of that of the more southerly Colony. At Maritzburg it is 64° to 71°, while on the coast the general range of the thermometer is from 53° to 90°. The winter is bright and comparatively cool, and the summer heat is softened by a clouded sky and frequent rains. The spring and autumn are agreeable periods of the year. Thunderstorms are of frequent occurrence during the summer months, and it is tot hese that the moderated heat of November, December, January, and February are due. Occasionally a hot wind from the land side blows over the Colony, but seldom lasts two or three days in succession. The following tables give an abstract of meteorological observations for the year 1875, taken at the Natal Botanic Gardens, Durban; approximate position, lat. S. 29° 56′, long. E. 31° 5′; height above sea-level, 128 feet. The instruments are all by Negretti & Zambra, numbered as at head of columns. The anemometer is Robinson's, improved by Negretti & Zambra, and observed for twelve minutes, morning and afternoon. The ozonometer is Sir John Clark's. To the barometer readings one-tenth of an inch is added for the elevation, and the temperatures are corrected according to the certificates with each instrument. Observations at 9 A.M. and 3 P.M. daily :—

ABSTRACT OF METEOROLOGICAL OBSERVATIONS FOR 1875.

1875.	Jan.	Feb.	March.	April.	May.	June.	July.	Aug.	Sept.	Oct.	Nov.	Dec.	Yearly Abstract
Barometer :—													
Highest Reading,	30·185	30·245	30·385	30·445	30·315	30·525	30·595	30·665	30·485	30·605	30·345	30·285	30·665
Lowest Reading,	29·865	29·905	29·785	29·745	29·355	29·995	30·055	29·775	29·785	29·775	29·825	29·895	29·355
Range of Month,	·320	·340	·600	·700	1·160	·530	·540	·890	·780	·830	·520	·390	1·310
Mean Daily Pressure,	30·070	30·070	30·106	30·147	30·216	30·234	30·326	30·217	30·194	30·163	30·116	30·086	30·162
Thermometer :—													
Highest Reading,	89·	89·	87·	94·	83·	78·	77·	82·	99·*	82·	87·	87·	99·*
Lowest Reading,	63·	63·	61·	55·	52·	50·	47·	45·	46·	50·	56·	57·	45·
Range,	26·	26·	26·	39·	31·	28·	30·	37·	53·	32·	31·	30·	54·
Mean Temperature of Month,	75·3	72·75	74·075	70·000	67·55	63·11	61·74	64·193	64·800	668·38	70·883	74·274	69·042
Geatest Daily Range,	23·	20·	23·	34·	22·	24·	25·	25·	34·	25·	27·	23·	34·
Least Daily Range,	10·	8·	12·	11·	15·	11·	8·	16·	9·	9·	7·	11·	7·
Mean Daily Range,	16·3	15·5	16·35	17·23	18·5	18·3	18·83	20·19	16·86	17·09	16·30	16·76	17·35
Approx. Mean Daily Temp.—													
Highest,	78·5	81·	79·	77·	73·5	67·5	66·5	72·5	82·	69·50	76·50	77·5	82·
Lowest,	70·	69·	67·5	64·5	61·	56·	56·	56·5	56·50	59·	62·50	68·5	56·
Dry Bulb, Mean,	78·245	78·5	77·1	67·9	73·5	66·27	66·58	66·54	69·53	70·	73·90	76·93	72·
Wet Bulb, Mean,	73·95	74·	73·3	69·3	68·2	63·39	63·87	62·03	65·60	65·83	71·16	73·70	69·52
Cloud, Daily Mean,	5·45	4·655	4·073	4·00	3·85	4·08	4·06	2·72	6·41	6·33	6·96	5·19	4·815
Rainfall, Days of Fall,	11	10	14	10	6	7	6	2·5	15	16	19	15	131
Inches Fallen,	3·64	3·80	6·06	2·11	0·22	2·05	5·57	0·27	5·56	3·35	16·62	5·53	54·78
Greatest Fall in 24 hours,	1·20	1·33	1·89	0·67	0·06	0·96	3·58	0·19	2·85	1·23	8·57	1·83	8·57
Evaporation :—													
Inches of Water Evaported,	5·	4·46	3·31	2·18	2·	1·74	1·35	1·93	2·13	3·13	2·80	4·44	34·47

* 21st Sept.—Hot Wind at 12 noon, N. [Anemometer 1·65 miles in 12 minutes]; Force, 5·00 estimated; Cloud, 0; Barometer, 29·805; Attached Thermometer, 87·; Maximum in Shade, 99·; ditto in Sun, 150·; Dry Bulb, 99·; Wet Bulb, 78.

Abstract of Meteorological Observations for 1875—*Continued.*

1875.	Jan.	Feb.	March.	April.	May.	June.	July.	Aug.	Sep.	Oct.	Nov.	Dec.	Yearly Abstract.
Wind—Direction :—													
N., 9 A.M.,	0	0	0	0	0	0	0	0	0	0	0	0	0
N., 3 P.M.,	0	0	0	0	0	0	0	0	0	0	0	0	0
N.E., 9 A.M.,	13	13	13	12	10	15	11	9	8	13	12	10	139
N.E., 3 P.M.,	14	10	13	15	19	16	11	13	9	12	7	13	152
E., 9 A.M.,	1	0	0	1	7	3	4	6	1	0	2	3	28
E., 3 P.M.,	1	4	2	3	1	0	0	4	3	3	3	2	26
S.E., 9 A.M.,	4	1	0	2	12	9	7	9	10	1	3	4	62
S.E., 3 P.M.,	6	2	1	4	9	11	12	11	9	7	9	6	87
S., 9 A.M.,	1	0	0	0	0	0	1	1	1	1	4	1	10
S., 3 P.M.,	0	0	0	0	1	1	3	0	0	1	1	2	9
S.W., 9 A.M.,	11	14	14	9	1	3	8	5	10	16	8	13	112
S.W., 3 P.M.,	9	12	15	7	1	2	5	3	9	8	10	8	89
W., 9 A.M.,	0	0	0	0	0	0	0	0	0	0	0	0	0
W., 3 P.M.,	0	0	0	0	0	0	0	0	0	0	0	0	0
N.W., 9 A.M.,	0	0	2	4	1	0	0	1	0	0	0	0	8
N.W., 3 P.M.,	1	0	0	0	0	0	0	0	0	0	0	0	1
Calm, 9 A.M.,	1	0	3	2	0	0	0	0	0	0	1	0	7
Calm, 3 P.M.,	0	0	0	1	0	0	0	0	0	0	0	0	1
Force—Mean Force, 9 A.M.,	3·09	2·28	1·8	1·6	1·45	1·63	1·67	2·22	2·16	2·74	2·06	2·54	2·14
" " 3 P.M.,	3·06	3·00	3·25	2·4	1·70	1·9	3·06	3·41	3·63	4·35	2·63	3·58	2·99
Ozone—Mean of Month, 9 A.M.,	5·4	5·28	5·6	5·5	5·35	6·4	6·64	6·29	6·40	6·27	5·30	4·83	5·77
" " 3 P.M.,	2·38	2·40	3·3	2·6	3·12	3·6	4·09	4·03	4·16	4·41	3·53	3·61	3·43
Daily Mean,	3·89	3·84	4·45	4·08	4·23	5·03	5·36	5·16	5·28	5·34	4·41	4·22	4·60
Storms :—													
Thunder and Lightning,	3	5	7	2	2	6	4	1	3	2	6	5	46
Thunder without Lightning,	0	2	0	3	1	0	0	0	4	4	2	0	16
Lightning without Thunder,	4	4	1	1	0	1	5	3	4	2	5	7	37
Lunar Haloes,	0	1	0	1	2	0	0	1	0	0	0	0	5
Hail,	0	0	1	0	0	0	0	0	0	0	0	1	2

Comparative Rainfall at Merebank for Nine Years.

From observations taken by Mr G. McLeod in the second fortnight of January, 1876, about the hotest period of the year, at Byrne, nearly 3,000 feet above the sea-level, boiling water 206° verified. The maximum in sun is taken in *vacuo*, and reads higher than an ordinary thermometer.

Months.	1866.		1867.		1868.		1869.		1870.		1871.		1872.		1873.		1874.	
	No. of days Rain fell.	Inches.	No. of days Rain fell.	Inches.	No. of days Rain fell.	Inches.	No. of days Rain fell.	Inches.	No. of days Rain fell.	Inches.	No. of days Rain fell.	Inches.	No. of days Rain fell.	Inches.	No. of days Rain fell.	Inches.	No. of days Rain fell.	Inches.
January,	5	1·01	5	1·11	11	3·26	8	1·96	11	4·35	11	2·29	4	1·90	7	3·42	12	4·01
February,	11	2·87	8	5·41	14	44·70	3	3·83	9	8·38	12	4·97	7	0.76	7	2·51	15	4·41
March,	9	20·19	9	3·66	12	5·36	13	7·77	8	4·86	11	3·42	17	3·34	11	5·85	12	7·86
April,	2	1·20	6	6·89	5	2·49	7	1·98	4	5·12	8	8·09	13	17·48	7	1·97	6	4·31
May,	4	2·41	1	0·70	—	—	5	0·72	1	0·17	6	1·92	10	3·57	3	1·10	5	0·38
June,	—	—	3	1·26	2	0·23	1	0·53	2	0·80	2	0·30	5	3·86	2	0·15	1	2·35
July,	3	1·21	2	0·16	6	2·38	1	0·15	—	—	2	0·51	5	4·21	4	0·54	1	0·48
August,	4	2·13	4	0·43	5	17·15	3	3·62	2	2·28	4	2·37	8	0·30	2	1·93	1	0·30
September,	8	3·98	6	2·21	7	2·14	2	1·02	7	1·49	2	0·45	11	3·90	5	2·73	6	1·82
October,	6	4·37	10	4·24	10	6·69	13	6·80	11	5·74	15	6·22	8	2·90	12	3·42	10	1·21
November,	7	5·76	10	5·49	12	6·10	10	6·51	14	6·81	6	2·34	13	7·06	16	4·57	15	4·52
December,	6	3·61	7	1·52	6	2·42	10	2·67	7	2·90	19	8·04	5	3·11	15	8·42	16	11.12
Total,	65	48·74	71	33·08	90	62·92	76	37·56	76	42·90	98	40·92	106	52·39	91	36·61	100	42·77

The preceding table represents the comparative rainfall at Merebank for nine years, taken by Mr Lamport. Merebank is situated eight miles S.W. of Durban, at a distance of two miles from the sea, and 30 feet above its level. The following were observations made in 1876 :—

January 1876.	Rain in 100ths of inch entered, including previous day.	Max. Temp. in Shade.	Min. Temp. in Shade.	Max. Temp. in Sun in Vacuo.	Min. Temp. in Night on Grass.	Wind at Seven A.M.	Wind at Seven P.M.	Standard Barometer at Seven P.M.	Attached Ther.	Standard Barometer at Two P.M.	Attached Ther.	Dry Bulb, Two P.M.	Wet Bulb, Two P.M.
17	·13	70	54	141	50	S.W.	E.	26·674	64	26·690	71	69	65
18	·06	79	54	144	50	E.	S.W.	·640	62	·586	73	77	71
19	—	83	61	150	56	N.	S.E.	·480	68	·444	76	80	72
20	1·96	64	60	—	64	E.	S.	·500	66	·470	66	61	61
21	·92	59	54	—	60	E.	E.	·594	62	·580	64	54	54
22	1·48	61	49	—	—	S.	E.	·710	58	·680	64	61	58
23	·03	76	50	141	46	S.	E.	·602	60	·544	69	76	69
24	—	79	54	142	54	S.E.	S. by E.	·584	65	·604	73	79	71
25	—	85	51	144	47	N.	N.	·582	64	·560	75	84	67
26	—	83	59	145	—	N.E.	E.	·608	67	·642	76	81	72
27	—	87	59	149	62	N.	N.	·660	70	·654	77	85	69
28	—	86	56	147	51	N.	E. by N.	·556	69	·500	80	84	70
29	·08	65	62	—	—	E. by S.	E.	·582	68	·632	70	65	60
30	·11	71	57	139	—	S.E.	N.	·784	64	·752	70	71	66
31	·01	88	54	147	50	N.	N.	·630	64	·566	80	88	73
2d ½ m 1st ½ ,,	4·78 2·40												
Total	7·18												

"*Observations.*—January 17, fine rain, 3 P.M.; 18, fine rain, distant thunder; 19, heavy hail, 3 P.M.; 20, rain all day; 21, rain all day; 22, slight rain; 23, fine; 24, fine; 25, fine; 26, fine; 27, fine; 28, fine, very hot; 29, rain, 7 P.M.,; 30, slight rain; 31, fine. Nine fine days, six wet days.

"1866.—Great flood of February—February 28 to March 3, 15·37 inches.

"1868.—Flood of February—February 19 to 22, 8·17 inches.

"1868.—Great flood of August—August 28 to 31, 17·11 inches (Maritzburg, 12·47 inches). During this disastrous flood 15·60 inches fell in forty-eight hours.

"1872.—Flood of April, 12·57 inches. From 1 P.M. on the 11th to 2 P.M. on the 13th, 8·23 inches fell.

"1873.—Flood of March. On the night of the 4th, 4·40 inches fell.

"1874.—Flood of December. From the night of the 7th to the morning of the 9th, 5·85 inches fell. Though not ranking among the 'great floods,' the bed of the river Umhlatuzan was scoured to an additional depth of 50 feet, and the weir, if not destroyed, was covered with sand, which still remains."

Of a flood at Natal, destroying property estimated from £100,000 to £150,000 in value, I have given details in the volume on the "Hydrology of South Africa" (pp. 235–238).

In a work entitled "The Colony of Natal," compiled and edited by Dr Mann, and published in 1859, there are given accounts of several thunder-storms in the vicinity of the capital of the Colony, which are confirmatory of the views which have been advanced, in a preceding chapter, in regard to the origin and natural history of such storms at the Cape of Good Hope, (ante pp. 52–53; 55–56); and of such thunder-storms, details are given in the volume cited (pp. 246-249).

The following is a tabulated statement of meteorological observations made at Pietermaritzburg, in Natal, in the year 1865, and for eight years preceding :—

A Tabulated Statement of Observations made at PIETERMARITZBURG, at Natal, in the Year 1865, and for Eight Years Preceding.

FROM DAILY OBSERVATIONS.	Jan.	Feb.	Mar.	April.	May.	June.	July.	Aug.	Sept.	Oct.	Nov.	Dec.	For the year or years
	degs.	degs.	degs.	degs.	degs.	degs.	degs.	degs.	degs.	degs.	degs.	degs.	degs.
Mean temperature of the air for each month, and for the year (Fahr.)	72·4	71·8	72·7	67·0	59·2	53·9	59·1	62·8	64·2	69·1	67·4	70·6	65·8
	degs.	degs.	degs.	degs.	degs.	degs.	degs.	degs.	degs.	degs.	degs.	degs.	degs.
Mean temperature of the air for each month, and for eight years (Fahr.)	71·4	71·8	69·7	64·8	59·3	55·2	55·2	59·7	65·1	66·6	67·1	70·4	64·7
Mean humidity of the air for each month, and for the year, derived from observations at 9 A.M.	75	72	76	76	72	80	70	66	74	61	66	69	71
Ditto ditto 3 P.M.	73	71	67	64	51	54	46	50	63	58	62	67	60
Ditto ditto 9 P.M.	88	86	87	88	84	83	78	77	86	79	82	83	83
Ditto for eight years at 9 A.M.	71	74	75	75	74	72	68	66	66	71	71	72	71
Ditto ditto 3 P.M.	68	68	64	61	54	49	46	49	56	67	69	71	61
Ditto ditto 9 P.M.	84	85	85	84	81	76	74	75	79	84	84	85	81
	degs.	degs.	degs.	degs.	degs.	degs.	degs.	degs.	degs.	degs.	degs.	degs.	degs.
Mean maximum temperature for eight years,	90·3	91·0	87·5	84·6	79·4	74·7	79·1	84·2	92·1	90·7	91·0	92·3	95·6
Ditto minimum ditto ditto	57·8	58·4	52·2	46·3	39·6	35·8	34·2	38·7	43·5	48·7	50·9	56·4	33·1
Maximum ditto ditto	93·0	97·1	92·8	89·5	85·2	78·2	82·2	89·8	95·4	96·0	97·2	97·6	97·6
Minimum ditto ditto	51·8	55·8	42·0	40·2	35·4	32·0	29·0	34·8	38·0	45·2	45·2	52·2	29·0
Maximum ditto for 1865,	86·8	93·0	90·2	85·0	85·2	71·2	82·2	86·4	94·0	96·0	97·2	97·6	97·6
Minimum ditto for 1865,	56·6	58·0	58·6	53·8	35·8	34·8	37·2	43·0	41·8	45·2	45·2	52·2	34·8
Quantity of rain in inches, 1865,	6·41	4·94	4·57	1·74	0·39	1·28	0·10	1·26	2·44	0·93	3·18	3·84	31·08
Number of days on which rain fell, 1865,	20	13	17	12	2	4	3	6	15	10	17	16	135
Greatest amount of rain in each month for 8 years,	6·63	7·59	5·04	2·02	2·94	1·28	0·74	8·44	3·11	7·21	8·95	6·23	..
Mean fall for eight years,	3·92	4·41	3·29	1·44	0·95	0·26	0·23	0·14	1·32	3·60	4·58	5·04	30·11
Mean number of rainy days for eight years,	16	14	13	9	3	1	2	5	8	17	17	18	123
Thunderstorms, 1865,	8	6	8	6	1	0	1	1	10	4	4	10	59
Mean for eight years,	8	7	5	3	1	0	1	1	4	6	7	8	51
Greatest number in eight years,	12	10	9	7	6	1	2	3	10	14	12	11	..

Mean height of barometer, not corrected for altitude, 27·891 inches; mean maximum, 28·397; mean minimum, 27·406; mean yearly range, 0·991.

CHAPTER VII.

THE TRANSKEIAN TERRITORY, ETC.

THE Transkiean districts comprise the whole of the territory extending from the Great Kei and its tributary the Indwe to Natal, and from Basutoland and the north-eastern extremity of the Colony of the Cape to the sea; the north-western boundary being the Drakensberg and Stormberg Mountains. It is inhabited by various native tribes, under their respective chiefs. "The area of the Transkeian Districts taken together is about 16,000 square miles, being almost equal in shape to Natal. The shape is that of an irregular parallelogram, the length from east to west being from 150 to 170 miles, and the breath, from its inland boundary to to sea, varying from 100 to 110 miles."

"This territory," it is added in the "Handbook to South Africa," which has repeatedly been quoted, "is one of the finest in South Africa with respect to natural features and productive capability. The mountain range at its back rises, at its north-eastern point, to an elevation of 9,657 feet, and is along its whole length of considerable height. From this grand eminence the land slopes down gradually by a hundred miles of grass, forest, and bush to the sands on which the surf of the Indian Ocean perpetually beats. It is watered by a hundred streams, some of which are rivers of magnitude. The Kei, the Bashee, the Umtata, the Umzimvubu, the Umtafuna, and the Umzimkulu are to be classed with the secondary rivers of South Africa. The most considerable of them is the Umzimvubu, which rises at the Giant Kop, receives numerous tributaries from east and west, and, after a course little short in all its wanderings of 250 miles, flows into the sea by a mouth which is navigable for small vessels. The gorges of the Stormberg and Drakensberg are full of fine trees, the uplands are rich in grass, the banks of the rivers bear a thick growth of forest and bush, and the warm lowlands and valleys are favourable to almost any kind of fruit, field, and garden cultivation. The country of the Amapondo is singularly fine

There the grasses grow so luxuriantly as to overtop the head of a tall man, and soil and climate are such as to favour the growth of tropical plants."

In my volume on the "Hydrology of South Africa" I had occasion to state that a glance at Hall's map of south-eastern Africa shows, from the Kei northward to the Delagoa Bay, a river system exactly in accordance to what may be seen in the runnels on the exposed mud bank of a tidal river, from which the ebb has withdrawn the life-like stream; and this is seen, not only in the depicted courses of the rivers on the map, but in the relative depths of the different river beds at the mouths of the rivers, and also at points nearer to their source.

The Rev. Edward Solomon, of Bedford, writing to a friend from the confines of territory occupied by Adam Kok and his tribe of Griquas, while on a clerical mission, having formerly laboured amongst that people as a missionary before their removal from Phillipolis, writes of the adjacent district through which he had passed:—"We have got on our journey pretty well, and are now within a hard day's ride of the Griqua location; but we shall be two days doing the journey. You have no idea of the terrible roads we have had to travel. The old Katberg road that we once passed together is by no means so bad as some parts that we have gone over to-day. You have to walk for miles together, as the roads are such that it is unsafe to sit in the waggon. We have visited three stations of the Church Missionary Society—viz., Mr Patten's 'St. John's,' Mr Waters' 'St. Marks,' and Mr Gordon's 'All Saints,'—two missions of the Moravians and two of the Wesleyans. I am writing this at Mr White's, where we have been very kindly received. I have been much struck with the beautiful rivers we have had to cross in this country. Yesterday evening we crossed the Tsitsi, which is really a magnificent stream. It is one of the main branches of the Umzimvooboo, or St. John's. It is certainly the finest river I have seen, with the exception of the Orange River. Since leaving the Colony we have crossed the Tsomo, the Kromme River, the Umgweli, the Xuka, the Bashee, the Umtata, the Tsitsi, and the Tena. All these are large rivers, and we have passed many smaller ones besides. The country is very grassy, the grass waving like corn, but this is coarse and sour, generally speaking, and it is very bare of wood. But we are constantly passing very fine mountains, with kloofs beautifully wooded. . . . This morning, though anxious to get away, we could not resist the temptation of getting horses and driving down the Tsitsi to see the falls—and I can assure you that we were well repaid for our trouble. The river was running about 100 yards wide, and tumbling down a precipice 375 feet deep, in seven different streams, all looking very beautiful, and sending up a very great vapour."

Crossing the Kei and re-entering the Colony, I find that from East London in 1869 it was reported :—"Agricultural operations during the past year have been carried out in this division most successfully, and the yield of all kinds of grain is abundant and of good quality.

"Cotton is now extensively cultivated, and the growing crops promise well, notwithstanding the unusual quantity of rain which has fallen during the past three months. Coffee is also likely to become, in the course of a few years, a staple article of produce, Many of the farmers have from 100 to 500 trees growing, which promise well. The coast lands from the Keiskama to the Kei are admirably adapted for the growth of both coffee and cotton.

"The cattle and stock of all kinds are in splendid condition, and have greatly increased, as there has been but little sickness among either flocks or herds during the last two years.

"The natives are well off, both in cattle and corn. The results have been but few thefts of stock; but the farmers complain greatly of the difficulty in procuring labour, the natives being too independent."

And in 1875 it was reported :—"The year 1874 having ended in a flood, 1875 was, as might have been expected, an unusually dry time for the coast-lands which form the division of East London, and are generally fertilised by copious showers.

"Agriculture is not carried on in this division to the same extent as in the older divisions of the Eastern Coast of South Africa, but as the population increases with the opening of the harbour and railway, and when markets are established, much larger quantities of produce will no doubt be raised.

"The crops throughout the division have suffered from the unusual scanty rainfall, wheat, barley, and oats were short in the straw, but the former was better and heavier than in the previous year. The mealie and Kaffir-corn crops in some places failed altogether, and generally were very small indeed. The health of stock has upon the whole been good, and with the exception of the appearance of lung sickness here and there among horned cattle, and the loss from various diseases among the flocks of sheep of farmers who moved down to the Crown Lands from the upper districts in the winter, there is but little for the stock farmer to complain of. There are no complaints of the spread of burr-weed, except from the Crown Lands upon the coast, which will be, it is said, soon overrun by that pest if means be not taken for its effectual eradication. Very little scab in sheep has been reported, and the general opinion is, that if the Scab Act were proclaimed in the adjoining districts, the disease would soon be stamped ont. Farmers universally complain of the scarcity of servants, and state that, without a greater certainty of procuring the necessary labour at the right time, they cannot extend their operations. . . .

"I regret having to report that, as the country becomes settled, a change gradually takes place in the character of the grasses upon the pasture lands which appear during nearly the whole year clad with the richest verdure; but there are so many theories started in explanation of the deterioration which occurs, or rather which appears to occur—for where, as in many cases, sheep can no longer be kept, or young stock reared, the larger cattle thrive and are always in excellent condition—that it is difficult to say which is the correct one. This was what took place in the western wards which were first settled by European grantees; and I may mention that in Ward No. 1, which has only 19 farms, some 70,000 sheep were depastured successfully by the first occupants, whereas now, according to the last census returns, not a single head is to be found.

"This gradual unfitness for sheep is creeping eastwards; for there again, upon the Crown Lands, which formerly supported numerous flocks, and were the salvation of thousands of sheep from the upper districts in winter months, the mortality among sheep is increasing year by year, and was during last year very severe indeed.

"According to the census returns, there were in March last 80,600 woolled sheep in the division, but these were almost exclusively confined to the north-eastern and eastern wards—viz., Nos. 8, 9, and 10, which include the highlands of the Kei Randt.

"The population and stock returns are as follows:—

Population, ...	15,425
Horned cattle, ...	37,008
Horses, ...	1,016

Of the former, about one-fifth are of European (viz., 3,735) and the remainder of the native races.

"As to thefts of stock, their diminution, increase, or cessation, I cannot speak with certainty, for the field cornets to the westward report there are none, and those to the eastward complain they are frequent. I can, however, state that those which are complained of at the time, brought to light and punished, are very few in number, considering the quantity of stock in the division, the large body of natives, and the absence of police control.

"Act No. 19 of 1864 has in this division worked well, as far as bringing in a large yearly revenue is concerned, but there have been only two purchases under Act No. 5 of 1870; and I am afraid that, with the late losses of sheep by the lessees, there is but small hope of further purchase under the latter Act; and therefore the settling down of the Crown Land by permanent residents cannot be expected unless those lands are let upon other terms, say those of the Agricultural Lands Act No. 4 of 1870, or some other enactment more favourable to the poorer classes of agriculturists. My own impression is that the whole of the coast lands can only have their resources fully developed by occupiers of small holdings of perhaps 200 or 300 acres each, let upon leases for at least twenty-one years, with the option of re-leasing for further similar terms by the lessee at rentals to be fixed by arbitration, with the proviso that the value of improvements shall be excluded when determining the new rate of rental.

"During the last five years this division has shown marked signs of increasing prosperity. Land which formerly changed hands but slowly at £10, now finds ready purchasers for twenty times that amount, and the revenue from all sources, which was in 1870 but £12,000, has risen to £90,000. This is attributable to the impulse given to all kinds of industry by the trade with the diamond-fields, by a succession of good seasons, and by the prosecution of the large public works in connection with the railway to Queenstown and the harbour at East London.

"The harbour works are also progressing, and it is expected, now that the Hercules has been erected, and the deposit of concrete blocks made possible, that the breakwater will soon show its effect in deepening the water in the mouth of the river.

"In the previous year in the month of December, the river Buffalo was swept clean by an unusually strong freshet, so that every vessel arriving at the port as well as the steamers *Florence*, *Zulu*, and *Kaffir* (the latter two of 1,000 tons each) were able to enter the river, and discharge their cargoes. This continued for some months; and at times quite a fleet of ships enlivened what was then a natural dock, of some three miles in length, 240 yards in width, and with a depth of from 18 to 25 feet."

From King Williamstown in 1874 it was reported :—" In furnishing, for the information of Government, the usual report upon the state of this district, for the Blue-book of 1873, attention was directed to the serious losses sustained by the farming interests, in consequence of the severe drought which had visited this and other parts of the province, and it is to be regretted that the report for 1874 does not in this respect differ from that of the preceding year. The drought has been repeated during the last year, and has been, if possible, more severe, and the losses, consequently, have been greater and more wide-spread.

"In the month of November, the breaking up of the drought was ushered in by a continuous downpour of rain, extending over nearly a week, and producing a flood which rendered all the rivers impassable for weeks, destroyed bridges and drifts, interrupted telegraphic, postal, and all other communication, and, putting a complete stoppage upon the transport of goods, paralyzed the trade and commerce of the country. The destruction of property, the losses in stock and in ripening crops which were swept away by the overflow of rivers and the rush of waters in localities previously regarded as secure against damage from such a source, have been immense. In some parts the soil has been completely removed, laying bare a substratum of ironstone or gravel, and rendering the land all but valueless, whilst in other parts large patches of arable land have been undermined and washed away, so that the month of November, 1874, will be long remembered as associated with the greatest flood that has ever visited this part of the country.

"The roads, both main and divisional, have been very much injured, and a large outlay will be required to restore them to their previous condition; the present taxation for road purposes, levied by the Divisional Council, is quite insufficient to keep these public thoroughfares in even ordinary repair. . . .

" These periodical seasons of drought are invariably followed by extraordinary rainfalls, occasioning heavy losses and great destruction to property.

" The cutting of timber in the forests is carried on year after year, and no provision is made, either for re-producing the quantity of timber annually taken out of our forests, by a systematic course of planting, or by encouraging tree-planting upon a scale of sufficient magnitude to affect the climate, and secure a greater and more uniform and general moisture deposit throughout the Province. The necessity for such a provision is forcing itself upon public attention in a manner that will make the postponement of legislative action in the direction indicated to be attended with ruinous consequences. Such a measure, combined with an irrigation scheme suited to the circumstances of the Colony, are matters affecting so immediately the interests of the agricultural classes, and are of such vital importance to the progress of the Colony by supplying the means of developing the resources of the country, that they deserve and claim the earliest and most earnest attention and consideration of the Parliament. . .

"We have to notice the introduction of two new industries during the year, the one for wood sawing, and the other for pulverising the red claystone, so extensively used by the Red Kaffirs in colouring their bodies, their blankets, and the scant apparel with which they cover their persons. Both these establishments are worked by steam power, and seem to be well supported.

"Wool-washing still continues to be an attractive, because remunerative, investment for capital. The quantity of wool washed or scoured on the banks of the Buffalo is estimated at about five thousand bales, representing two-and-a-quarter millions of grease wool, equivalent, at sevenpence half-penny per pound, to seventy thousand pounds sterling.

"The purchase and leasing of farms by native chiefs is a new feature in frontier affairs, and is regarded by many as ominous of future trouble. The Chief Oba has purchased a farm in the Alice district, where he is located with his followers, and is likely to prove troublesome to the neighbourhood. The Chief Sandilli has made application to be allowed to resume occupancy of his old kraals near Burns' Hill, and expressed his willingness to pay a very large sum for land in that locality; his application, however, was not entertained. The Chief Delima has leased a farm in the Kieskama, at an annual rental of £150. The movements of the chiefs in this direction are looked upon with distrust by many whose knowledge of native character, and acquaintance with their habits, entitle their opinion to considerable weight, but as the chiefs are holders of certificates of free citizenship, interference with them in any other way than by influence of a dissuasive tendency, seems to be excluded. If, in the future, any trouble should arise from the settlement on the farms of these chiefs and their followers, the owners of them will have incurred a grave responsibility by consenting to a transaction, questionable in its character, and productive in its results of most serious consequences. . . .

"Large iron water pipes have been laid during the year in the principal streets of the town, and the water supply is now upon a satisfactory basis, private waterleadings being granted upon application, and the payment of a reasonable rate.

"The bridge over the Buffalo was, it is believed, the only one that was not damaged by the November floods, although the river rose to a level with the deck or flooring.

"The natives are quiet and peaceable. They are becoming comparatively wealthy, from the high price given for wool, and every other product they can bring to the market. Those engaged in transport riding are employed at rates which leave a large margin for profits. The only drawback to their improvement and progressive advancement in civilisation and habits of industry is the facility with which drink is procurable, and the amount of wretchedness and crime which it occasions. This is an evil which should be dealt with by an enactment with more stringent restrictions and heavier penalties than the present Ordinance imposes."

And in 1875:—"The past year has been barren of incidents of sufficient importance to claim notice in an annual report upon the state of the district. Steady progress has been made in all directions.

"The natives, of whom, by the census returns, there are about 100,000 in this division, are gradually turning their attention to agricultural and other industrial pursuits, and are acquiring property either in stock or in land, and, as a natural consequence, are beginning to discover and appreciate the advantages of fixed permanent employment.

"The more general use of the plough for gardening and other purposes, by performing work, which in former years was done solely by the women, is everywhere observable, and may, in my opinion, be regarded as a powerful ally in resisting the extension of polygamy, and obstructing, if not entirely stopping, its progress and advancement. In former times, when gardens and planting of all kinds was imposed upon women, the necessity of having more than one wife, if only for gardening purposes, was at once felt, but the introduction of the plough has tended in some measure to remove this necessity.

"The extensive public works which are now being carried on in this division are providing constant employment to a large number of native labourers at remunerative rates of wages; upwards of 3,000 having passed through the books of the labour agency during the past year, and natives are still coming in from the different tribes to be employed upon the railway and other works, and their employers are satisfied with the work they are able to perform.

"This large and increasing draft upon the labour market for Government works has made it difficult for those engaged in farming pursuits to obtain the requisite native labour supply to meet their requirements, and to carry on farming operations, and much dissatisfaction has, in consequence, been expressed, and much inconvenience felt. But as these native labourers, by their being employed in the harbour works at East London and in the railway works in this division and in the Uitenhage, Port Elizabeth, and Albany divisions, where systematic work is performed under skilled, efficient European overseers, will become expert in the use of the spade, shovel, and pickaxe, and will continually be a large addition to the number of efficient, useful agricultural labourers, and thus supply a want which has been long felt, and in this way there will be, in the course of time, a direct advantage resulting from the employment of these men in the public works, though at present it is attended with considerable inconvenience.

"During the past year the state of affairs beyond the borders required the removal of the F.A.M. Police out of the Colony to the Transkeian Territories. It might reasonably have been expected that the withdrawal of the police for six months would be followed by an increase of crime, especially in stock and horse stealing, but the very opposite has been the result, and it is generally admitted that there has been less stealing in the Colony while the police were in the country beyond the Kei.

"It is possible that the employment of so large a number of natives upon public works in this and in the adjoining divisions, and the presence of the police beyond the borders, preventing or making difficult the exchange of stolen property, may account for this decrease in crime. There is no doubt that constant employment must, to a large extent, counteract the natural tendency to evil, which an indolent and inactive life so greatly excites and encourages.

"There is a general impression prevalent, which finds expression specially in the Divisional Council, that the natives should be compelled, by increased direct taxation, to contribute more largely to the general revenue than they do, and more especially by taxation for road purposes, to contribute towards the expenses incurred by divisional councils for the repair and maintenance of roads.

"The land in the occupation of natives in the Tamacha district, if converted into farms of ordinary dimensions, would probably represent or be equivalent to three or four decent sized farms, which, if in the possession of Europeans, would yield in annual quit-rents from £15 to £20 at the outside, and probably not very much more for road rates. While the hut-tax contributed by the natives for this tract of land, equivalent to about four farms, amounts to between £200 and £300. In the Middle Drift district the hut collection is over £1,600; while in the Tembani district it is over £1,000. It should, moreover, be borne in mind that in addition to the above direct contributions to the general revenue, the native population, which, as previously stated, is about 100,000, must be large consumers of importations, as well as of Colonial products, and in these respects are large contributors to the general revenue. European clothing also is coming into general use, the habits and requirements of civilised society are steadily and noticeably being acquired, and in many ways the revenue of the Colony is largely indebted to the native population for their contributions.

"Many of the natives are lessees of Crown lands, and some of them have become proprietors by purchase under the provisions of the Act, and this may be regarded as a most decided and encouraging indication of their progress and improvement.

"The season has been a remarkable one, the rains having kept off till within a few days of Christmas, and great fears were entertained that there would be no planting and therefore no crops, and a great scarcity of provisions; but these have all passed away, as continuous and abundant rainfalls have characterised the season since the first fall of rain."

CONCLUSION.

In accordance with the design of this volume, there have been brought under consideration all the existing supplies of moisture in South Africa. It has been shown that there are no other supplies upon which we can reckon, nor any means of increasing these as a capital upon which to draw.

From this it may be inferred that it is only by the husbanding of these—the conservation and economic distribution of the water thus supplied—that the aridity which is characteristic of the country can be met to promote the indefinite extension of agricultural operations in the Colony.

It does not come within the scope of this volume to discuss the question—By what works of hydraulic engineering may this be effected?

The author sees nothing to retract or to modify, in statements relative to the effect of vegetation and of trees upon the water supply, in the volumes published by him, entitled respectively :—

"Hydrology of South Africa; or, Details of the former hydrographic condition of the Cape of Good Hope, and of causes of its present aridity, with suggestions of appropriate remedies for this aridity." In which the desiccation of South Africa, from pre-Adamic times to the present day, is traced by indications supplied by geological formations, by the physical geography or general contour of the country, and by arborescent productions in the interior, with results confirmatory of the opinion that the appropriate remedies are irrigation, arboriculture, and an improved forest economy; or the erection of dams to prevent the escape of a portion of the rainfall to the sea; the abandonment or restriction of the burning of the herbage and bush in connection with pastoral and agricultural operations; the conservation and extension of existing forests; and the adoption of measures similar to the *réboisement* and *gazonnement* carried out in France, with a view to prevent the formation of torrents and the destruction of property occasioned by them.*

* London: Henry S. King & Co. 1875.

"*Réboisement* in France; or, Records of the re-planting of the Alps, the Cevennes, and the Pyrenees with trees, herbage, and bush, with a view to arresting and preventing the destructic consequences of torrents." In which are given a *resumé* of Surell's study of Alpine torrents, and of the literature of France relative to Alpine torrents, and remedial measures which have been proposed for adoption to prevent the disastrous consequences following from them,—translations of documents and enactments, showing what legislative and executive measures have been taken by the Government of France in connection with *réboisement* as a remedial application against destructive torrents,—and details in regard to the past, present, and prospective aspects of the work.*

"Forests and Moisture; or, Effects of forests on humidity of climate." In which are given details of phenomena of vegetation on which the meteorological effects of forests affecting the humidity of climate depend,—of the effects of forests on the humidity of the atmosphere, on the humidity of the ground, on marshes, on the moisture of a wide expanse of country, on the local rainfall, and on rivers,—and of the correspondence between the distribution of the rainfall and of forests,—the measure of correspondence between the distribution of the rainfall and that of forests,—the distribution of the rainfall dependent on geographical position, determined by the contour of a country,—the distribution of forests affected by the the distribution of the rainfall,—and the local effects of forests on the distribution of the rainfall within the forest district.†

By sylviculture, carried out upon a scale of which perhaps no one but the author dreams, may devastating floods be prevented, and the distribution of the rainfall over time and space may be to some extent equalised.

Pending the consideration, execution, completion, and progressively advancing effective operation of such measures as are thus referred to, the present generation may pass away without reaping the benefit; but it has here been shown that there is at their command supplies of water which may be utilised at less expenditure and with more speedy returns.

It appears that, however great the aridity of South Africa, however frequent and protracted the droughts which alternate with deluges and floods, and however extensive the regions over which

* London: Henry S. King & Co. 1876.

† Edinburgh: Oliver & Boyd. London: Simpkin, Marshall, & Co. 1877.

extreme aridity almost constantly prevails, there is a supply of water which, if stored and distributed in accordance with the requirements of the colonists, might not only provide abundance for man and beast, but allow of a very great extension of agricultural operations, and of the production of gardens and orchards, orangeries and vineyards, to their delectation and profit. This is the one point which which this volume is designed to establish.

And it has been shown that, with scarcely one exception, the reports in regard to any one division or district of the colonised portion of South Africa may, *mutatis mutandis*, be taken as a report of any other, and as an epitome of the account given of the entire region embraced by the whole—now a deluge, now a drought; now luxuriance, now sterility; now an overflow, now a lack—with everywhere natural facilities for storing up abundance against a day of want.

An esteemed collaborateur wrote to the *Cape Monitor*, in March, 1861:—"It was a favourite saying with the great agricultural philosopher, Sir John Sinclair, that he who makes a blade of corn grow where none grew before is an undoubted benefactor to his country. If Sir John had been bred in South Africa, instead of the 'land of the mountain and the flood,' he might with equal force apply the saying to water. A drop of water in most of our African soil will produce a blade of grass, or nourish an ear of corn, and is far more valuable to the Colonial farmer than manure is to the English; for there is scarcely a year but one or other of our districts loss thousands of cattle from its want. The only remedy that we have as yet adopted to assist us in these periodical calamities, has been a form of fasting and praying—praying, not that we may be guided to make use of what a bounteous Providence never fails to give to the industrious, but that we may have our present necessities supplied with the least possible trouble to ourselves. Famine may not as yet have been actually felt in the Colony, but year after year gives unmistakable evidence of its approach. Hundreds of our achter veldt farmers have often lost the most of their flocks without the knowledge of the good folks of Capetown. And at present one half of them—many the descendants of noble families—do not taste bread during six months of the year, and some of them do not break bread in their families from one end of the year to the other. Their numbers are, however, at present comparatively few, and their wants still less. In severe droughts they can draw out an existence, with the remnant of their flock of goats, in the last place of refuge, the beds of periodical rivers. But when population, say trebles itself, the picture becomes too dismal to dwell upon, if no steps are taken for their relief in seasons of severe drought.

"Could a general scheme of irrigation be wound up and set going like a watch, then we might wait until we see the risk of our railway trucks being drawn by oxen, and our harbour simply used as a harbour of refuge, or, what is more serious still, the utter destitution of some thousands of our farming population. Is it at all likely that we should have in one year a complete failure in our vine, one-half of the Colony prostrated with drought, and the other half swept with horse and lung sickness? Certainly not; and it is only owing to the hitherto sparseness of our rural population that coincidences of this nature have not been noted, or generally experienced.

"Irrigation in South Africa becomes therefore, a matter, not of choice, *but of absolute necessity*. We have yet not only to learn how to get water, but also how to be industrious so as to produce sufficient bread for our families; consequently irrigation with us will be of slow growth, and it is not on the fertile farms on the Hex River, or Worcester, or in the well-watered valley of the Upper Olifant's River, that we can first look to for establishing a system of colonial irrigation. As a general rule, the occupiers of such favoured spots reign in peace and contentment; their forefathers in the good old times did for them the little that nature required to supply their wants.

"Scotland with her comparatively barren soil, and ungenial climate, took the lead of fertile England in agriculture. This apparent anomaly is to a certain extent owing to the stimulus it gives, even with the most phlegmatic, to see unproductive land converted into fruitful fields. The result is not only profitable, but the same pleasure or pride is experienced by the systematic farmer that the native has in his first scientific application of the lever to a mass of rock. We can perceive the pleasure and self satisfaction the like of Mr Hare daily experiences in viewing his but lately barren land clothed with well arranged plantations, and fields of waving corn, compared with the farmer who from a child plucks his requirements from off his stoep, and sits under the oak tree planted by his forefathers.

"I make these remarks as advocating the suggestion contained in my last letter, namely, that portions of Zwartland and Koeberg can be profitably irrigated, and that a successful experiment in a tract of country, so very unlikely for such a purpose, will establish a system of irrigation, which *must sooner or later* be extensively adopted in South Africa. In my next I shall assume such data as are likely to be found in the districts in question, suitable for the construction of artificial lakes, and from this endeavour to show the probable cost of water so collected."

When I was at the Cape of Good Hope I prepared a "Memoir on Irrigation, and its applicability to agricultural operations in South Africa, with notices of Government measures required to facilitate its more extensive introduction throughout the Colony;" and a *resumé* of this Memoir was appended to the Report of the Colonial Botanist for 1866.

In this, after citing evidence given by me before a Parliamentary Committee, appointed "to consider and report whether it is expedient to adopt any, and what, measures in order to increase the construction of works for irrigation and the storage of water, whether river water can be more extensively and equitably distributed, and whether water disputes can be more satisfactorily settled than under the present law," in successive chapters, information was supplied in regard to the supply of water at present available for agricultural operations in the Colony; the forms of irrigation required by different methods of agriculture which are followed in the Colony; the facilities for irrigation existing in different districts of the Colony; the physical difficulties in the way of a more extensive irrigation of the Colony, and the means of overcoming these which are at command; the legal difficulties in the way of more extensive irrigation; and the Government measures required for the extension of irrigation.

In the preceding pages are embodied the information then submitted in regard to the supply of water at present available for agricultural operations in the Colony; and in regard to the facilities for storage of this existing in different districts of the Colony. The information submitted on the other subjects mentioned is forthcoming when required.

PRINTED BY CRAWFORD AND WALKER, HIGH STREET, KIRKCALDY.

Spotted something that isn't quite right?

Let us know and we'll fix it for future reprints.
And as a thank you for your diligence, we'll
send you a free Sothis book of your choice.

publisher@sothis.press

This page intentionally left blank

Made in United States
North Haven, CT
04 November 2026

21764784R00390